STRUCTURE, COHERENCE AND CHAOS IN DYNAMICAL SYSTEMS

Proceedings in nonlinear science

Published in association with the Centre for Nonlinear Studies,
The University, Leeds LS2 9NQ, UK

Series co-ordinator: Arun V. Holden

Also in this series:

Artificial intelligence and cognitive sciences
edited by J. Demongeot, T. Hervé, V. Rialle and C. Roche

Other volumes are in preparation

STRUCTURE, COHERENCE AND CHAOS IN DYNAMICAL SYSTEMS

Edited by
Peter L. Christiansen and Robert D. Parmentier

Manchester University Press
Manchester and New York
Distributed exclusively in the USA and Canada by St. Martin's Press

Published by
Manchester University Press
Oxford Road, Manchester M13 9PL, UK
and Room 400, 175 Fifth Avenue, New York, NY 10010, USA

Distributed exclusively in the USA and Canada
by St. Martin's Press Inc.,
175 Fifth Avenue, New York, NY 10010, USA

British Library cataloguing in publication data
Structure, coherence and chaos in dynamical systems
   1. Nonlinear dynamical systems
   I. Christiansen, Peter L. II. Parmentier, Robert D.
   III. Series
   515.3'5

Library of Congress cataloging in publication data applied for

ISBN 0-7190-2610-5 *hardback*

Printed in Great Britain
by Biddles Ltd, Guildford and King's Lynn

CONTENTS

Preface   x

Contributors   xi

Introduction
                            A.C. Scott   1

1 Optimality, adequacy and the evolution of complexity
                  P.M. Allen & J.M. McGlade   3

2 Pattern selection and competition in near-integrable
   systems
          A.R. Bishop, D.W. McLaughlin & E.A. Overman II   23

3 The latest generation of supercomputers, exemplified
   by the CRAY-2
                           T. Bloch   41

4 Recent work on nonlinear dynamics within MIDIT
        B. Branner, K. Munk Andersen & A. Sandqvist   45

5 Quantum and classical statistical mechanics of
   integrable models in 1+1 dimensions: some recent
   results
             R.K. Bullough, D.J. Pilling & J. Timonen   59

6 Transport and fluctuations in the driven and damped
   sine-Gordon chain
                         M. Büttiker   77

7 Recent work on Josephson tunnel junctions within
   MIDIT
                      P.L. Christiansen   93

8 An interesting non-linear effect in a neural network
                      R.M.J. Cotterill   113

9 Relaxation kinetics in bistable systems
                  G. Dewell & P. Borckmans   123

10 Theory and applications of the discrete self-trapping
   equation
                 J.C. Eilbeck & A.C. Scott   139

Contents

11 The sine-Gordon breather as a model particle: quantized states from classical field theory
U. Enz   161

12 A macroscopic approach to synergetics
H. Haken & A. Wunderlin   169

13 Patterned and irregular activity in excitable media
A.V. Holden & G. Matsumoto   185

14 Dynamics and phase transitions in artificial intelligence systems
B.A. Huberman   199

15 Global stability of the chaotic state near an interior crisis
R.L. Kautz   207

16 Modified Painlevé test for analytic integrability
M. Kruskal   227

17 Complex and chaotic waves in reaction-diffusion systems and on the effects of electric field on them
M. Marek, I. Schreiber & L. Vroblová   233

18 Growth and decay of dendritic microstructures
S.P. Marsh and M.E. Glicksman   245

19 Solitons in optical fibres: an experimental account
L.F. Mollenauer   257

20 Generation of spatially asymmetric, information-rich structures in far from equilibrium systems
G. Nicolis, G. Subba Rao, J. Subba Rao & C. Nicolis   287

21 Prototype long-scale equations of dissipative dynamics
L.M. Pismen   301

22 Time of emergence and dynamics of cooperative gene networks
S. Rasmussen, E. Mosekilde & J. Engelbrecht   315

23 Relationship between classical diffusion in random media and quantum localization
T. Schneider   333

24 Structure and dynamics of replication-mutation systems
P. Schuster   347

25 Nonlinear dynamics in the world economy: the economic long wave
J.D. Sterman   389

26 Computer modelling of thermodynamic properties of very dilute, charged hard spheres in a dielectric continuum
T.S. Sørensen & P. Sloth
415

27 Noise and chaos in driven Josephson junctions
M. Tinkham
427

28 Collective coordinate methods for soliton dynamics
S.E. Trullinger
441

29 Spiral waves in the Belousov-Zhabotinskii reaction
J.J. Tyson
457

30 Model complementarity, visualization and numerical diagnostics
N.J. Zabusky
465

31 An application of Shilnikov's theorem to linear systems with piecewise linear feedback
P. Grabowski
473

32 The self-organization hypothesis for 2D Navier Stokes equations
E. van Groesen
481

33 Chaos and the structure of the basin boundaries of a dissipative oscillator
H.M. Isomäki, J. von Boehm & R. Räty
485

34 Technique to trace bifurcation points of periodic solutions
C. Kaas-Petersen
491

35 Qualitative research on solitons of nonlinear Schrödinger equation with external field
Guan Ke-ying
497

36 Melnikov theory applied to a variable length pendulum
V. Muto, M. Bartuccelli & M.P. Soerensen
503

37 Hopf bifurcation in functional-differential systems - computational aspects
M. Szymkat
511

38 Numerical results on the relationship between classical diffusion and quantum systems
M.P. Soerensen, T. Schneider, A. Politi, E. Tosatti & M. Zanetti
517

39 Construction of integrable Riccati systems by the use of low dimensional Lie algebras
S. Wojciechowski
523

Contents

40 Finite-time thermodynamics
B. Andresen   529

41 Quantum corrections to the specific heat of the easy-plane ferromagnetic chain
H.C. Fogedby, K. Osano & H.J. Jensen   535

42 Blow-up in nonlinear Schroedinger equations
J. Juul Rasmussen & K. Rypdal   541

43 The soliton laser: a computational two-cavity model
P. Berg, F. If & O. Skovgaard   547

44 On the travelling wave solutions of the Maxwell-Bloch equations
J.N. Elgin & J.B. Molina Garza   553

45 Chaos in a Josephson junction mechanical analog
J.A. Blackburn, S. Vik, Yang Zhou-Jing, H.J.T. Smith & M.A.H. Nerenberg   563

46 Soliton dynamics in long inhomogeneous Josephson junction
A.A. Golubov & A.V. Ustinov   569

47 Avalanche breakdown and turbulence in semiconductors
K.M. Mayer, R.P. Huebener, R. Gross, J. Parisi & J. Peinke   575

48 Oscillatory oxidation of CO on Pt(110)
M. Eiswirth, P. Möller, R. Imbihl & G. Ertl   579

49 Unidirectional solidification using the concept of a boundary layer of solute diffusion
M. Hennenberg   585

50 Bifurcations in 3 dimensional Turing systems: prepattern simulation on supercomputers
A. Hunding   591

51 Symmetry-breaking phase transitions and boiling-type turbulence of a simple morphogenetic reaction-diffusion system
B. Röhricht, J. Parisi, J. Peinke & O.E. Rössler   597

52 How to distinguish between chaos and amplification of statistical fluctuations in a chemical oscillator
F.W. Schneider, Th. Kruel & A. Freund   601

53 Parameter space structure of a model chemical oscillator
B. Turcsányi   607

54 Chaos and bifurcations in dynamical systems of nerve membranes

K. Aihara, M. Kotani & G. Matsumoto 613

55 Aspects of quantum mechanical thermalization in the alpha-helix

H. Bolterauer 619

56 Resonant and quasiclassical excitations of solitons in the alpha-helix

H. Bolterauer, R.D. Henkel & M. Opper 625

57 Numerical analysis of the order-chaos-order transition for the N=3 discrete self-trapping equation

S. De Filippo, M. Fusco Girard & M. Salerno 633

58 Coherence, chaos, and consciousness: nonlinear dynamics at "macro" and "micro" extremes of brain hierarchy

S.R. Hameroff, R.C. Watt, A.C. Scott, C.W. Schneiker & P. Jablonka 639

59 Ferroelectric properties of bilayer lipid membranes

K.M.C. Da Silva 645

60 Self-trapping of infrared energy absorbed in acetanilide

A. Tenenbaum 651

Index 659

PREFACE

This volume of the Manchester University Press Series on Nonlinear Science is devoted to Structure, Coherence and Chaos in Dynamical Systems. It contains results from the MIDIT 1986 Workshop on Structure, Coherence and Chaos in Dynamical Systems held at the Technical University of Denmark, DK-2800 Lyngby, Denmark, August 12-16, 1986. The workshop is the first international manifestation of the priority area Modelling, Non-linear Dynamics and Irreversible Thermodynamics established by the Technical University in 1984. In Danish this long name abbreviates to MIDIT. One hundred and fifty-nine scientists from 23 countries participated in the workshop and 31 invited talks and 80 posters were presented. A photograph of the participants is shown on page

Thomas B. Thriges Fond is thanked for its generous support to the MIDIT centre in general and to the workshop in particular. The additional funding provided by the Danish Natural Science Research Council, the Danish Council for Scientific and Industrial Research, NORDITA, Otto Monsteds Fond and IBM Danmark A/S is gratefully acknowledged. The United States Army European Research Office, London is thanked for its sponsorship of the participation of the Laboratory of Applied Mathematical Physics, The Technical University of Denmark in the workshop.

The scientific committee had the following members: H. Haken, University of Stuttgart, West Germany; A.V. Holden, University of Leeds, England; R.D. Parmentier, University of Salerno, Italy; A.C. Scott, University of Arizona, U.S.A.; P.L. Christiansen and T.S. Sorensen, The Technical University of Denmark. The organizing committee consisted of R.D. Parmentier, University of Salerno, Italy; A.C. Scott, University of Arizona, U.S.A.,; P.L. Christiansen, V.L. Hansen, E. Mosekilde, J. Mygind and T.S. Sorensen, The Tecnical University of Denmark.

The administration of the Technical University of Denmark, students at the MIDIT centre and in particular our indefatigable secretaries, Ms. Lise Gudmandsen, Bettina Gyldenfeldt, Lis Kaufmann, Lynne MacNeil, Kirsten Studnitz and Dorthe Thoegersen are thanked for assistance at the workshop and in the preparation of these proceedings.

Robert D. Parmentier
Peter L. Christiansen   Editors

Contributors

Kazuyuki Aihara
Dept. of Electronic Engineering
Tokyo Denki University
2-2 Nishiki-cho, Kanda
Chiyoda-ku, Tokyo 101
Japan

Kurt Munk Andersen
Mathematical Institute
The Technical University of
Denmark, Building 303
DK-2800 Lyngby
Denmark

M. Bartucelli
Mathematics Department
Queen Mary College
University of London
Mile End Road
London E1 4NS
England

A.R. Bishop
Theoretical Division and
Center for Nonlinear Studies
Los Alamos National Laboratory
Los Alamos
NM 87545
U.S.A.

Tor Bloch
Director, Centre de Calcul
Vectoriel pour la Recherche
Ecole Polytechnique
91128 Palaiseau Cedex
France

H. Bolterauer
Justus-Liebig-University
Fachbereich Physik
Institut für Theoretische Physik
6300 Giessen
West Germany

P.M. Allen
Centre of Ecotechnology
Cranfield Institute of
Technology
Cranfield
Bedford MK43 0AL

Bjarne Andresen
Physics Laboratory
University of Copenhagen
Universitetsparken 5
2100 Copenhagen 0
Denmark

P. Berg
Laboratory of Applied
Mathematical Physics
The Technical University
of Denmark, Building 303
DK-2800 Lyngby
Denmark

J. Blackburn
Physics and Computing
Department
Wilfrid Laurier University
Waterloo, Ontario N2L 3C5
Canada

Juhani von Boehm
Department of General
Sciences
Helsinki University of
Technology
SF-02150 Espoo
Finland

P. Borckmans,
Service de Chimie
Physique 2
Campus Plaine U.L.B.
C..P. 231
Boulevard de Triomphe
1050 Bruxelles, Belgium

Contributors

Bodil Branner
Mathematical Institute
The Technical University
of Denmark
Building 303
DK-2800 Lyngby
Denmark

Markus Büttiker
IBM T.J. Watson Research
Center, P.O. Box 218
Yorktown Heights
N.Y. 10598, U.S.A.

Peter Leth Christiansen
Laboratory of Applied Mathematical
Physics
The Technical University of
Denmark
Building 303
DK-2800 Lyngby
Denmark

G. Dewel
Service de Chimie Physique 2
Campus Plaine U.L.B., C.P. 231
Boulevard de Triomphe
1050 Bruxelles
Belgium

Markus Eiswirth
Institut für Physikalische Chemie
Der Universitat München
Sophienstr. 11
8000 München 2
West Germany

Jacob Engelbrecht
Physics Laboratory III
The Technical University of
Denmark, Building 303
DK-2800 Lyngby
Demark

G. Ertl
Fritz-Haber-Institut
1 Berlin (West) 33
West Germany

R.K. Bullough
Institute of Science and
Technology
The University of
Manchester
P.O. Box 88
Manchester M60 1QD
England

D. Camel
Laboratoire d'Etudes de la
Solidification
C.E.N.G., B.P.X., Grenoble
F-38041, France

R.M.J. Cotterill
Department of Structural
Properties of Materials
The Technical University
of Denmark
Building 307
DK-2800 Lyngby
Denmark

J.C. Eilbeck
Heriot-Watt University
Department of Mathematics
Riccarton,
Currie, Edinburgh EH14 4AS
Sctland

John N. Elgin
Imperial College of
Science and Technology
Huxley Building
Queens Gate,
London SW7 2BZ, England

U. Enz
Philips Research
Laboratories
P.O. Box 80.000,
5600 JA Eindhoven
The Netherlands

J.J. Favier
Laboratoire d'Etudes de la
Solidification
C.E.N.G., B.P.X., Grenoble
F-38041, France

S. De Filippo
Dipartimento di Fisica
Teorica e SMSA
Universita' di Salerno
84100 Salerno
Italy

Dr. A. Freund
University of Würzburg
Institute of Physical
Chemistry
Marcusstr. 9/11
D-8700 Würzburg, GFR

M. Fusco Girard
Dipartimento di Fisica
Teorica e SMSA
Universita' di Salerno
84100 Salerno
Italy

A. A. Golubov
Institute of Solid State
Academy of Sciences of
the USSR
Chernogolovka
Moscow District 142432, USSR

E. van Groesen
Department of Applied
Mathematics
Twente University of
Technology
P. O. Box 217
7500 AE Enschede, Holland

H. Haken
Institute for Theoretical
Physics and Synergetics
University of Stuttgart
Pfaffenwaldring 57/IV
D-7000 Stuttgart 80, West Germany
West Germany

Hans Fogedby
Institute of Physics
University of Aarhus
8000 Aarhus C
Denmark

J. B. Molina Garza
Department of Mathematics
Imperial College of
Science and Technology
London SW7 2BZ
England

M. E. Glicksman
School of Engineering
Materials Engineering
Department
Rensselaer Polytechnic
Institute, Troy
New York 12180-3590
U. S. A.

Piotr Grabowski
Institute of Control and
Engineering
Academy of Mining and
Metallurgy
Al. Mickiewicza 30
Krakow 30-059, Poland

R. Gross
Physikalisches Institut II
Universität Tübingen
D-7400 Tübingen, FRG

Stuart Hameroff
Dept. of Anesthesiology
University of Arizona
Medical Center
Tucson AZ 85724, U. S. A.

Contributors

R.D. Henkel
Insitut für Theoretische
Physik
Justus-Liebig-Universität
Giessen
6300 Giessen, FRG

Arun V. Holden
Department of Physiology
University of Leeds
Leeds LS2 9JT

R.P. Huebener
Physikalisches Institut II
Universität Tübingen
D-7400 Tübingen, FRG

Flemming If
Laboratory of Applied
Mathematical Physics
The Technical University
of Denmark, Building 303
DK-2800 Lyngby, Denmark

Heikki Isomäki
Inst. of Mech. Y329
Department of General Sciences
Helsinki University of Technology
SF-02150 Espoo 15, Finland

H.J. Jensen
Institute for Materials Research
McMaster University
1280 Main Street West
Hamilton, Ontario L8S 4M1
Canada

R.L. Kautz
National Bureau of Standards
Electromagnetic Technology Division
Boulder,
325 Broadway, Colorado 80303, U.S.A.

M. Hennenberg
Commissariate à l'Energie
Atomique
Centre d'Etudes Nucléaires
de Grenoble
Lab. d'Etude de la
Solidification
85X, 38041 Grenoble Cedex,
France

B.A. Huberman
Xerox Palo Alto Research
Center
Intelligent Systems Lab.
3333 Coyote Hill Road
Palo Alto, CA 94304
U.S.A.

Axel Hunding
Panum Institute
University of Copenhagen
Blegdamsvej 3B
DK-2200 Copenhagen N
Denmark

R. Imbihl
Fritz-Haber-Institut
Faradayweg 4/6
1 Berlin West 33
West Germany

Paul Jablonka
Computer Aided Engineering
Laboratory
Dept. of Aerospace and
Mechanical Engineering
University of Arizona
Tucson, AZ 85724, U.S.A.

Christian Kaas-Petersen
The School of Mathematics
University of Leeds
Leeds LS2 9JT
England

Guan Ke-ying
Department of Mathematics
Beijing Institute of
Aeronautics & Astronautics
P.O. Box 85, Beijing
P.R. of China

M. Kotani
Tokyo Denki University
Kanda, Chiyoda
Tokyo
Japan

Martin Kruskal
Princeton University
Fine Hall
Princeton NJ 08544, U.S.A.

S.P. Marsh
Materials Engineering Department
Rensselaer Polytechnic Institute
Troy, NY 12180-3590, U.S.A.

K.M. Mayer
Physikalisches Institut II
Universität Tübingen
D-7400 Tübingen, FRG

David McLaughlin
Mathematics Department
University of Arizona
Mathematics Building 89
Tucson, AZ 85721, U.S.A.

P. Möller
Institut für Physikalisches Chemie
8 München 2, FRG

V. Muto
Laboratory of Applied Mathematical
Physics
The Technical University of
Denmark
DK-2800 Lyngby
Denmark

Th. Kruel
University of Würzburg
Institute of Physical
Chemistry
Marcusstr. 9/11
D-8700 Würzburg, FRG

M. Marek
Pragh Institute of
Chemical Technology
Dept. of Chemical
Engineering
166 28 Praha 6
Suchbatarova 5
Czechoslovakia

G. Matsumoto
Electrotechnical Lab.
Analogue Information
Section
Tsukuba Science City
Niihari-gun, Ibaraki 305
Japan

J.M. McGlade
Centre of Ecotechnology
Cranfield Institute of
Technology
Cranfield
Bedford MK43 0AL, England

L.F. Mollenauer
A.T. & T. Bell Laboratries
Holmdel, NJ 07733, U.S.A.

E. Mosekilde
Physics Laboratory III
The Technical University
of Denmark, Building 306
DK-2800 Lyngby, Denmark

M.A.H. Nerenberg
Department of Applied
Mathematics
University of Western
Ontario
Canada

Contributors

C. Nicolis
Institut d'Aéronomie Spatiale
de Belgique
Av. Circulaire, 3
1180 Bruxelles
Belgium

G. Nicolis
Université Libre de
Bruxelles
Faculté des Sciences
Campus Plaine
Boulevard du Triomphe
1050 Bruxelles
Belgium

M. Opper
Institut für Theoretische Physik
Justus-Liebig-Universität Giessen
6300 Giessen, FRG

K. Osano
Matsushita Electric
Industrial Company
Central Research Labs.
Moriguchi, Osaka 570
Japan

E.A. Overman II
Department of Mathematics
Ohio State University
Columbus, OH 43210
U.S.A.

J. Parisi
Physikalisches Institut II
Universität Tübingen
Morgenstelle 14
D-7400 Tübingen 1
West Germany

J. Peinke
Physikalisches Institut II
Universität Tübingen
D-7400 Tübingen, FRG

D.J. Pilling
Department of Mathematics
U.M.I.S.T., P.O. Box 88
Manchester M60 1QD
England

L.M. Pismen
Department of Chemical Engineering
Technion - Israel Institute
of Technology
32000 Haifa
Israel

A. Politi
Istituto Nazionale di
Ottica, Largo E. Fermilo
I-50125 Firenze, Italy

Gita Subba Rao
Department of Biophysics
All India Institute of Medical
Sciences
New Delhi-110029, India

J. Subba Rao
School of Environmental
Sciences
Jawaharlal Nehru
University
New Delhi-110067, India

J. Juul Rasmussen
Physics Department
Association EURATOM-Risø
National Laboratory
DK-4000 Roskilde
Denmark

Steen Rasmussen
Physics Laboratory III
The Technical University
of Denmark, Building 306
DK-2800 Lyngby
Denmark

Dr. Raimo Räty
Department of General Sciences
Helsinki University of Technology
SF-02150 Espoo, Finland

Dr. B. Röhricht
Physikalisches Institut II
Universität Tübingen
D-7400 Tübingen, FRG

Dr. O.E. Rössler
Institut für Physikalisches und
Theoretische Chemie
Universität Tübingen
D-7400 Tübingen, FRG

K. Rypdal
Institute of Mathematical and
Physical Sciences
University of Tromsø
P.O. Box 953, N-9001 Tromsø
Norway

Allan Sandqvist
Mathematical Institute
The Technical Univesity of Denmark
DK-2800 Lyngby, Building 303
Denmark

F.W. Schneider
University of Würzburg
Institute of Physical Chemistry
Marcusstr. 9-11
8700 Würzburg
West Germany

Conrad W. Schneiker
Advanced Biotechnology Laboratory
Optical Sciences Center
University of Arizona
Tucson, AZ 85721
U.S.A.

Peter Schuster
Institut für teoretische Chemie
und Strahlenchemie
der Universität
Wahringerstrasse 17
A-1090 Wien, Austria

O. Skovgaard
Laboratory of Applied Mathematical
Physics
The Technical University of Denmark
DK-2800 Lyngby, Building 303
Denmark

A. Rouzaud
Laboratoire d'Etudes de la
Solidification
C.E.N.G., B.P.X., Grenoble
F-38041, France

Mario Salerno
Dipartimento di Fisica
Teorica
Universita' di Salerno
I-84100 Salerno
Italy

Alwyn C. Scott
Department of Electrical
and Computer Engineering
University of Arizona
Tucson, AZ 85721
U.S.A.

T. Schneider
IBM Research Division
Zürich Research Laboratory
Saumestrasse 4
8803 Rüschlikon
Switzerland

I. Schreiber
Department of Chemical
Engineering
Prague Institute of
Chemical Technology
166 28 Prague 6
Czechoslovakia

K. M. Correia Da Silva
Lab. Biomecanica
Instituto Gulbenkian de
Cienua
Apart 14,
2781 Oeiras Codex,
Portugal

Peter Sloth
Fysisk-Kemisk Institut
(MIDIT)
The Technical University
of Denmark, Building 206
DK-2800 Lyngby, Denmark

H.J.T. Smith

Mads Peter Soerensen

Contributors

Department of Physics
University of Waterloo
Ontario
Canada

Torben Smith Sørensen
Fysisk-Kemisk Institut (MIDIT)
Technical University of Denmark
Building 206
DK-2800 Lyngby
Denmark

Maciej Szymkat
Institute of Automatic Control
Academy of Mining and Metallurgy
Al. Mickiewicza 30
10-059 Krakow
Poland

J. Timonen
Department of Physics
University of Jyväskylä
SF-40100 Jyväskylä
Finland

E. Tosatti
International School for Advanced
Studies
Strada Costiera 11
I-34100 Trieste
Italy

Bela Turcsányi
Central Research Institute for
Chemistry of the Hungarian
Academy of Sciences
P.O. Box 17
H-1525 Budapest
Hungary

A.V. Ustinov
Institute of Solid State Physics
Academy of Sciences of the USSR
Chernogolovka
Moscow district 142432
USSR

L. Vroblová

Laboratory of Applied
Mathematical Physics
The Technical University
of Denmark, Building 303
DK-2800 Lyngby, Denmark

John D. Sterman
System Dynamics Group
Massachusetts Institute
of Technology
Sloan School of
Management
50 Memorial Drive
Cambridge, Mass. 02139
U.S.A.

Alexander Tenenbaum
Department of Physics
University of Rome
Piazzale A. Moro 2
I-00185 Roma, Italy

M. Tinkham
Department of Physics
Harvard University
Lyman Laboratory of
Physics
Cambridge, Mass. 02138
U.S.A.

S.E. Trullinger
Department of Physics
University of Southern
California
Los Angeles, CA 90089/0484
U.S.A.

J.J. Tyson
Department of Biology
Virginia Polytechnic
Institute and State
University
Blacksburg, VA 24061
U.S.A.

S. Vik
Wilfrid Laurier University
Department of Physics &
Computing
Waterloo, Ontario
Canada

Richard C. Watt

Department of Chemical Engineering
Prague Institute of Chemical
Technology
166 28 Prague 6
Czechoslovakia

Stefan Wojciechowski
Department of Mathematics, LITH
Linköping University
S-581 83 Linköping
Sweden

Yang Zhou-Jing
Wilfrid Laurier University
Department of Physics & Computing
Waterloo, Ontario
Canada

M. Zannetti
Dipartimento di Fisica
Universita' degli Studi di Salerno
I-84100 Salerno
Italy

Advanced Biotechnology
Laboratory
Dept. of Anesthesiology
University of Arizona
Tucson, AZ 85724
U.S.A.

Arne Wunderlin
Institute for Theoretical
Physics
University of Stuttgart
Pfaffenwaldring 57/III
D-7000 Stuttgart 80
West Germany

N.J. Zabusky
University of Pittsburgh
MIB
Pittsburgh, PA 15260
U.S.A.

INTRODUCTION

"There is no better, there is no more open door by which you can enter into the study of natural philosophy", Michael Faraday has written, "than by considering the physical phenomena of a candle." Although the wisdom of this remark lay unrecognized for well over a century, it is now rather widely understood that the study of nonlinear diffusion in a candle (that quintessential common object!) is closely related to the dynamics of nerve impulse propagation, the spread of contagious disease, and the beating of the human heart. Thus physicists, physiologists and physical chemists find themselves talking to each other, and the fragmentation of modern science, which has seemed to characterize the present century, tends to diminish.

Indeed, the most striking feature of modern research in nonlinear dynamics is its interdisciplinary nature. The same solitary wave dynamics that John Scott Russell first observed on the Union Canal in August of 1834, governs the evolution of plasma wave solitons and explains the now celebrated computer experiment conducted in the summer of 1953 by Enrico Fermi, Stan Ulam and John Pasta. The same equation that solid state physicists and electrical engineers use to describe the propagation of magnetic flux quanta (called "fluxons") on the Josephson transmission line is also employed by theoretical physicists as a model for elementary particles. Similarly, the soliton on an optical fibre promises to revolutionize the practice of high speed data transmission, and this development is closely related to suggested mechanisms for the transport and storage of biological energy in protein. These are exciting times!

The above remarks should not be supposed to imply that all of nonlinear dynamics has become simplified in the modern perspective. At the same time that applied scientists and mathematicians have recognized that many nonlinear partial differential equations - which were formerly considered almost hopeless as objects for mathematical study - are rather readily analyzed, it has become clear that the phenomena called chaos renders the analytical solution of many nonlinear ordinary differential equations - which were formerly considered almost too trivial to investigate - impossible. Yet the detailed study of chaos in simple dynamical systems has introduced new concepts and has also fostered many interdisciplinary activities. The electrical engineer studying "excess noise" in a Josephson junction mixer (a vital component in the radio telescope and some RADAR receivers) can talk directly with the anaesthesiologist who is investigating the electroencephalogram as a measure of mental state and the cardiologist who is searching for signs of arrhythmia in the electrocardiogram.

The soliton and chaos seem to be emerging as the ying and yang of modern science. Just as nonlinear effects can lead to an unexpected degree of predictability in some situations, they can cause surprising unpredictability in others. These two concepts embody complementary aspects of reality.

In all of these developments, the computer is evolving as an increasingly important research tool. Here it should be emphasized that the numerical power of the computer is not being used in the traditional "number-crunching" mode. It is not a question of calculating π to the next million decimal places: that is not the point. In modern nonlinear science, computer power is used deftly to ask subtle questions of numerical experiments. A well designed research strategy now includes theoretical analysis, physical experiments and numerical experiments on a more or less equal footing.

The workshop on "Structure, Coherence and Chaos in Dynamical Systems" held at the Technical University of Denmark under the aegis of the recently organized initiative area on Modellering, Ikke-linear Dynamik og Irreversibel Termodynamik (MIDIT) from August 12 to 16, provides a fresh and tasty smörgåsbord of current developments in nonlinear science. Physics and physiology, engineering and economics, chaos and quantum theory, fluid dynamics and physical chemistry are all represented. It is hoped that from sampling this fare, the reader will come to appreciate the profound insight of Michael Faraday's advice.

Alwyn C. Scott

OPTIMALITY, ADEQUACY AND THE EVOLUTION OF COMPLEXITY

Abstract

The existing theory and models of evolution are discussed and it
is pointed out that the mechanical models which underlie much of
the neo-Darwinian Synthesis are inadequate to describe a
continuing evolutionary process.   We show that evolution will
select for populations which retain "variability", even though
this variability is, at any given time, loss-making.   Our
simulations suggest that mechanisms for generating and maintaining
variability will be an important part of the evolutionary process,
and that this will lead to much greater natural diversity than is
possible with the usual, mechanical equations of population
dynamics.   Evolution is seen to be "driven" by the noise to which
it leads.

1.1   Introduction

Evolutionary theory is still a subject of bitter controversy one
hundred years after Darwin's death.   Disagreements flourish both
as to the theoretical basis on which evolutionary change should be
understood and also on the evidence in support of one or another
idea [1-19].

However, broadly speaking, the generally accepted view is that
of 'gradualist neo-Darwinism'.   It considers itself to be the
'modern version' of Darwin's theory, including the knowledge
concerning genetics and biology which was not known 100 years ago.
It may be stated as follows:

"Organisms within a population vary with respect to physiology,
morphology and behaviour.   Because resources are limited - food,
space, mates, etc. - selection will favour those organisms whose
behaviour and morphology enhance their access to resources,
relative to other organisms, and hence increase their
reproductive output relative to others in the population.   Thus
the product of natural selection operating on a variable
population is an increased relative abundance of those
individuals with behaviours that enhance access to resources -
in other words, whose behaviour relative to those resources may
be said to be more optimal.   Optimization is therefore an

3

expected outcome of natural selection operating in a world where resources are finite and act as a limiting factor [20].

This offers a rather clear statement of the idea underlying neo-Darwinian theory of evolution, which is often compressed into the more opaque statement 'As a result of evolution, every individual maximizes his inclusive fitness'. However, whichever way it is stated the message is one of **materialistic** optimization resulting from the evolutionary process. Survival is reserved for those who best monopolize and make use of the sources of material existence - energy and the biological building materials.

But, of course, everything hinges on the use of the words 'outcome' and 'optimization'. It would be more correct to replace the last sentence of the definition above by "Improvement in average individual behaviour is therefore expected as a result of natural selection operating in a world where resources are finite and act as a limiting factor".

However, if this change is made, the power of evolutionary theory to explain observed behaviour is greatly weakened. Evolution would only allow us to understand the changes in behaviour if we also assume that evolution has come to an end - all avenues have been explored, and nothing more can happen. And this is indeed the perspective adopted in the usual ideas of 'Neo-Darwinism'. It is implied above, in the word 'outcome'. The evolutionary wisdom of past experiences is supposed to be 'stored' in the genes of each population, the cumulative result of a supposedly infinite number of trials and errors under all possible conditions! A 'perfect information' theory. If that were true, then indeed we would have found some 'extremum' of a pre-existing potential function, and observed bevhaviour could be 'explained' on that basis.

But, what of the 'resources'? If they too are living creatures, then will they not evolve in response to the 'improvements' of their predator? Will not the evolution of the other populations of the system always be such as to counter the adoption of an apparently 'optimal' strategy, thus decreasing its 'pay-off'? And, if this is so, then can evolution ever be said to have 'come to an end'?

If not, then the most that can be said of the behaviour of any particular individual or population is that its continued existence proves only that it is sufficiently effective - not optimal.

We can get a glimpse of the real complexity of these issues from Figure 1 showing a fairly simple ecosystem [2]. Clearly, most species here feed on several prey, and are fed on by several predators. Are the behaviours, strategies and morphologies of any particular species now fixed for all time? Is further improvement and ripost impossible? Or is it simply a question of 'time scale'? Are more 'successful' strategies on their way, but only slowly? If so, what can we say about those present actually? In other words, is evolution limited by a 'lack of suitably imaginative mutations', or is any assumed stability due to selective forces acting on a very complete and thorough 'search' of possible types? Are we stuck going up a 'hill', or really sitting at the top?

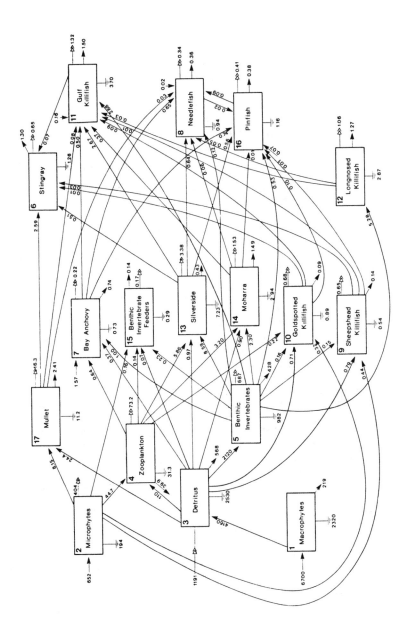

Figure 1. A fairly typical, relatively simple ecosystem. The carbon flow diagram for the Crystal River tidal marsh (Homer and Kemp, 1983).

These are all good questions but, at the moment, we may say that there are no generally accepted answers, only opinions.

Perhaps owing to these difficulties, even among those who accept the idea of evolution leading to 'optimality', opinion differs as to what precisely is being optimized, since there is no agreement as to whether it is biological molecules, genes, individuals, groups, populations, species or even ecosystems which are the dominant unit of selection. The 'fitness' of each 'unit' is in reality, 'hidden' in the interactions of the system, and therefore the persistence of a system will necessarily imply the presence of genes, individuals, groups and populations, all in a hierarchical web of interactions, and clearly it becomes a matter more of opinion than science whch is more 'correct'. The controversy that exists is probably simply due to the fact that the questions "What is the unit of selection?" and "What is being optimized by evolution?" are not the right questions to ask.

Linked to all this is the question of whether evolution is mainly a smooth process of improvement of existing forms, a 'hill-climbing' process resulting from the natural selection of a 'dense cloud' of mutant types, or on the other hand, it is mostly composed of long periods of stasis, when almost no change occurs, separated by relatively rapid 'bursts' of considerable change. This latter view, advanced particularly by Eldridge and Gould [7], has received some backing from the fossil evidence, but the arguement remains unresolved. The underlying problem is that of 'niche filling'. If the gradualists believe that evolution is a smooth progression resulting from the selection of ever more advantageous mutations, then how can the different 'species' that we observe emerge? In other words, if 'hill-climbing' is what evolution is about, then how can it ever jump across 'valleys' in the adaptation space?

There is no such problem with the 'punctuationalist' view, because they suggest that at certain times, large variabilities occur in the phylogeny, and new types are produced which fall quite far from the previous phenotypes. They are thus 'parachuted' over to another area of the 'morphology' space, most probably of course falling on 'stony ground' and corresponding to an inviable form. However, certain types find at least a subsistence in their new character, so that they in their turn occupy part of the ecosystem, and may well adjust somewhat to their niche through the occurrence of some 'hill-climbing' mutations.

In one view, the mutations are viewed as being 'dense and local', while in the other they are viewed as 'sparse and dispersed'. On the whole, opinions differ merely as to the relative importance of the two processes, and the evidence which has been put forward to decide this seems at present to be inconclusive.

However, there is an important difference in the manner in which these different styles of evolutionary processes are modelled, and this is the point of our present paper. The point is that although neo-Darwinism begins by discussing variability and selection, the dynamic models which often are used to represent it, rarely contain both.

The aim of this paper is to show how the interactions between species can be considered, while at the same time considering the problem of the discovery of new 'strategies' or 'niches'. We shall develop a model which can be run with all these features included and which, in addition, can be used to explore the transition from purely 'genetic' evolution, due to selection acting on random variability, to the evolution of culture, which is due to behavioural change due to the adoption of advantageous, or at least satisfying strategies.

This model should throw some light on to another persistent problem – that is natural diversity. If one examines ecosystems which exist in reality as in Figure 1, then we find enormous diversity. As Margalef calls it – the 'baroque' of nature. There are far more species than would be 'required' in order to create a trophic chain capable of pumping energy and matter efficiently through the system. Whatever evolution may be doing, it seems that making streamlined trophic structures with a minimum of morphological diversity is not it.

If a set of mechanical equations is set up which try to represent a real ecosystem, then, when they are run forward in time on a computer, most of the species present are eliminated, and a radical simplification occurs.

Quite generally, as May [9] pointed out, it becomes increasingly difficult to construct a mathematical model which gives rise to a stable coexistence of species as complexity increases. Since the natural systems which surround us do in reality maintain their complexity, we may conclude that the mechanical equations which are used in such models obviously do not capture the real interactions and adaptability of the natural system!

What is missing is the whole wealth of potential adaptations which are hidden in the system – and result from its evolution. For example, we might observe a particular species of bird, and use its average clutch size in our parameters. However, it has been shown that if eggs are removed each day, then the same bird can produce ten times the number of eggs! In other words, there are mechanisms hidden in the system which could radically change the value of a parameter, in the appropriate circumstances. Similarly, an animal which is hungry may decrease the amount of time it basks in the sun, and step up its hunting. In this and many other ways the system has in it hundreds of 'checks and balances' which maintain its complexity, when a machine-like 'model' of it at a particular moment will run down to an unrealistic simplicity. However, 'homeostasis' cannot be enough to 'explain' the system; firstly because there may be many possible such regimes of operation, and secondly because evolution may still be occurring.

In this paper we shall present a simple model which attempts to generate an 'evolutionary population dynamics', where the parameters involved have a plasticity which is the result of evolutionary processes of 'phenotypic discovery' and 'systematic selection'.

1.2 **A simple model of co-evolution**

Let us imagine first a 2-dimensional space of possible morphologies or strategies for a population, where each point, i, corresponds to a particular blend of the 2 dimensions of character. However, below it, we may imagine a surface on which we represent the 'resources' (generally prey species) available to each particular character i. In this way, a population with characteristics of the point i can 'feed on' the resources at and around the point i, which are represented in the plane below. The degree to which this feeding can be spread depends on the degree of specialization of the particular type However, we shall make the assumption that the more 'generalist' a species is, the lower its maximum capture rate at the point i.

Above the 2-dimensional 'character space' of the population considered, we can also imagine a surface which expresses the 'dangers' which will result from a given morphology or strategy – that is from predator populations which consider the population we have chosen as a 'resource'. This is illustrated in Figure 2.

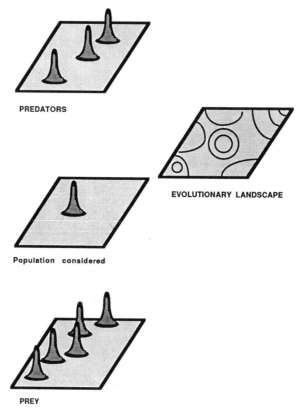

PREDATORS

EVOLUTIONARY LANDSCAPE

Population considered

PREY

Figure 2. The middle plane refers to 2 dimensions of character or morphology. The lower surface shows the 'resources' available to a given character, and the upper one shows the 'dangers involved'.

We may consider that, for any particular 'character' of our population there is a relief map of possible rewards and dangers which could result, given the prey and predator species that exist in the system at that time. This means that the 'character space' is spanned by a kind of 'contour map' of potential 'Malthusian parameters' (birth rate - death rate), which will show the 'selective advantage' of any particular character choice, given the existing circumstances.

The 'evolutionary potential' viewed from any particular point will depend on the size of the prey and predator species in the system and also on the 'degree of flexibility' of the population. The resource map 'seen' by a generalist will be quite different from that of a 'specialist'. A character shift that may correspond to moving 'downhill' in potential for one may be 'uphill' for the other. In this paper we shall simply examine the simplest problem of how evolution leads a population to 'climb' a hill in the evolutionary landscape.

Let us consider the simplest possible case of a single population x. According to its 'situation' in the character space of Figure 2 it will have a certain birth rate and mortality, and the limits of the resource on which it feeds will give rise to a logistic type population dynamics.

The equation will be:

$$dx/dt = b_1 x[1 - x/n)] - m_1 x \qquad (1)$$

where n is the limiting resource related to the prey species in the level below, and b_1 and m_1 the birth and death rates which

correspond to the position in 'character space' - that is on the evolutionary landscape.

Now, what interests us here, is the question of 'evolution'. How does the 'character' of a population change over time in response to the 'rewards and dangers' of particular strategies? In order to examine this, we must consider two important aspects:

a)   Imperfect reproduction leading to 'diffusion' in character space.

b)   The operation of selection in amplifying 'favourable' mutations and suppressing unfavourable ones.

Usually, the models of population genetics or of neo-Darwinian evolution simply consider the second term. That is, they assume that all strategies are present initially, and examine the 'selection' of the 'fittest'. We shall attempt to retain both aspects.

In order to model the evolutionary process, we shall first consider a part of the relief map on the right of Figure 2. We shall take the situation where the local evolutionary landscape consists of a single 'hill' and in this preliminary work we shall consider two aspects of character which affect separately the birth and death rates. In this way we can construct a parameter space of the effectiveness of the two species. For this, we have to consider two indices, i and j, which characterize the effectiveness of b and m respectively. So, we have:

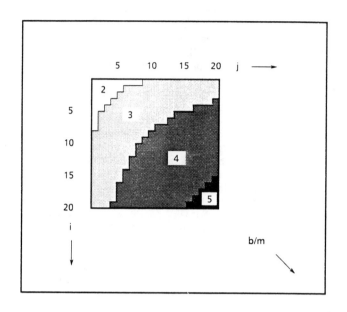

Figure 3. The parameter space in which we can study co-evolution of x and y.

$$b(i) = b^o(1 + \alpha i) \text{ and } m(j) = m^o[1 - (\beta j/(1 + \beta j))] \qquad (2)$$

and for different values of b^o and m^o corresponds to a parameter space in which we can examine evolution of x and y. We have an 'evolutionary landscape' which simply shows increasing effectiveness in a diagonal direction towards the bottom right-hand corner, as in Figure 3.

For species which reproduce perfectly, coexistence is only possible if the two species are of exactly equal effectiveness. If, for example, we place species x at the point i=3, j=5 and y at the point i=5, j=7 and then, for b^o=2 and m^o=1, we find that y eliminates x. This is shown in Figure 4.

This is completely in agreement with the usual theory of competitive exclusion, and of Darwinian selection.

But what if a species does not reproduce perfectly? For example, supppose that only a fraction (f) of the births of x show fidelity to the parents, how will this look in the evolutionary landscape? It seems very reasonable to suppose that, on average, this would give rise to a negative effect. That is, the births of 'mutant' types would be biased towards the less effective.

We shall suppose, as a first assumption, that 3/4 of the mutants are characterized by a **lower** value of i and j, while only 1/4 have **higher values**.

For the population at i and j, good replicate births remain at i and j. For the 'mutants', however, we shall suppose that 3/8 fall into i-1, j and another 3/8 to i, j-1. In agreement with the idea that chance errors more often give rise to negative effects than positive, we suppose that only 1/8 go to i+1, j and 1/8 to i, j+1.

10

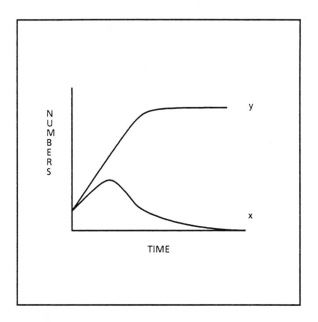

Figure 4. Two species x and y are in total competition, and y is more effective than x. y eliminates x.

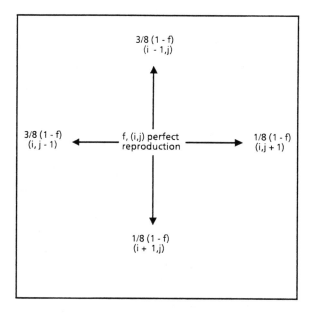

Figure 5. The pattern of reproduction of each population x(i,j) and y (i,j), linking all points of the evolutionary landscape.

The reproductive pattern for **each** population $x(i,j)$ and $y(i,j)$ is given in Figure 5.

Each point in parameter space is connected to its vertical and horizontal neighbours through mutations or stochasticity of behaviour. We have not considered the possibility of diagonal mutations because there seems no reason to suppose that being 'better' or 'worse' at the factors which influence reproduction has any connection with the behaviour which affects longevity.

The equations for the dynamics of our competing species are:

$$dx(i,j)/dt = [\{b(i)(fx(i,j)+3/8(1-f)x(i,j+1)+1/8(1-f)x(i,j-1)\}$$
$$+b(i+1)3/8(1-f)x(i+1,j)+b(i-1)1/8(1-f)x(i-1,j)].C$$
$$- m(j)x(i,j) \qquad (3)$$

and

$$dy(i,j)/dt = [\{b(i)(fy(i,j)+3/8(1-f)y(i,j+1)+1/8(1-f)y(i,j-1)\}$$
$$+b(i+1)3/8(1-f)y(i+1,j)+b(i-1)1/8(1-f)y(i-1,j)].C$$
$$-m(j)y(i,j)$$

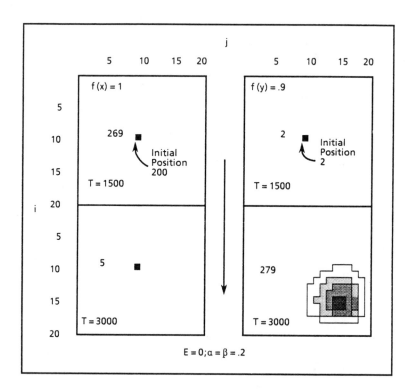

Figure 6. The competition between a perfect reproducer (left) and an imperfect one (right) goes in favour of the imperfect one, given sufficient time. At time 3000 population x=5 and y=269.

where f is the fraction of perfect reproduction (e.g. 90%), and C is the 'crowding' factor: $(1 - [\Sigma \ x(i,j)+\Sigma \ y(i,j)]/n)$.

Now let us examine first the possibility that a perfectly reproducing species $(f(x)=1)$ could be 'invaded' by an imperfectly reproducing one $(f(y)=.9)$. At first sight this would seem to be impossible, because at any given moment (including the initial one) it is less favourable to make 'errors' than not to. Since 3/4 of reproductive errors are 'worse' and only 1/4 are 'better' than the parent type, clearly it is less effective.

In Figure 6 we show the result of initially placing 200 units of x at position 10,10, and then seeing if 2 units of y placed also at 10,10 can invade.

Initially, as can well be imagined, the species x grows, while y remains at a low value. However, the errors in reproduction of y lead to a 'cloud' of small populations around its initial position 10,10. These are mostly at lower values of i and j, but some are higher. As the selection operates, the least effective populations of y are eliminated rapidly, but gradually small amounts of mutant populations 'invade' the more effective regions of the parameter space (3), and after some time they start to multiply at the expense of x, and then rapidly eliminate the perfect reproducer. His 'perfect reproduction' allows no possibility of an adaptive response.

The simulation above showed that even if a population of 'perfect reproducers' had ever evolved, it would have been invaded and replaced by error-makers. However, we can now examine more carefully the effect of 'error-making' on a species' ability to maintain itself or evolve in its evolutionary landscape. If we consider a single species with the mutation scheme as given in Figure 4 above, and successively alter the gradient of the evolutionary landscape from low to high by increasing α and β from .001 to .01 to .03 and finally .05, then we see that with a very low gradient mutations are bad and the net effectiveness of the species drops, whilst with .05 the centroid advances and the overall effectiveness increases (Figure 7).

In other words, if the slope is sufficiently large, then error-making is advantageous, and we find a forward evolution whose rate depends on the slope of the landscape. Obviously the slope will depend on the values of b and m maintained, and it will probably become harder to improve on something that is already effective. From our general view in Figure 2, we can imagine that as we reach the top of a 'hill', selection will operate in favour of lower error rates, because local exploration and discovery have yielded all there is to 'know' about that resource.

But this assumes that the 'resource' itself does not change its character as a result of this increasingly successful exploitation. This is an important point. If we are to adopt a properly 'ecosystemic' view of evolution then we must admit the fact that as our 'error-making' species climbs a 'hill' in his evolutionary landscape, the 'hill' itself will move away as the 'characteristic' of the prey and predators change in view of the success of our 'hill climber'. In this first study, we can represent this complex phenomenon in a rather simple way. We can assume that over time, any population with indices i and j is in

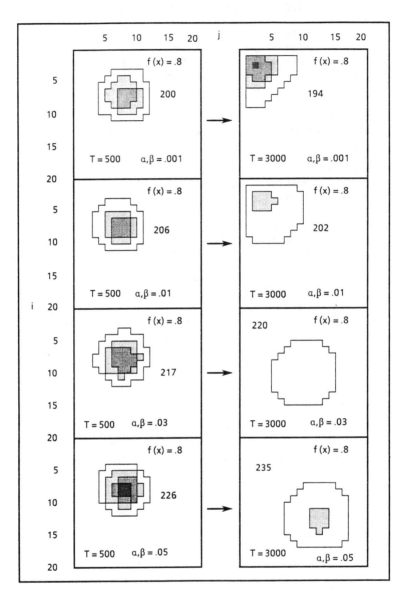

Figure 7. Evolution of an imperfect reproducer, landscapes of varying slope.

fact blown backwards by an 'environmental wind'. There will be a backwards diffusion from i+1 and j+1 to i and j, and from there to i-1 and j-1. In this manner we can say that in order simply to maintain its position on a slope of increasing success, then the combination of error-making and selection must succeed in countering this backward blowing wind.

We can account for this in the following way:

$dx(i,j)/dt = \{b(i)[fx(i,j) + 3/8(1-f)x(i,j+1) +$
$1/8(1-f)x(i,j-1)]$

$\qquad\qquad + b(i+1)3/8(1-f)x(i+1,j)+b(i-1)1/8(1-f)x(i-1,j)\}.C$
$\qquad\qquad - m(j)x(i,j) + E[x(i+1,j)+x(i,j+1) - 2x(i,j)]$

and a similar equation for $y(i,j)$.

Here the term C is the 'crowding' or competition for the limiting factor, n, as before.

The term E is a measure of the instability of the environment and the rate of evolution of the other species in the system. If it is large, then a species which does not evolve in its behaviour will rapidly become less and less effective over time. This term sets the natural limit to evolutionary progress in terms of 'effectiveness'.

With an environmental wind of 0.1, and a gradient across the landscape set by $\alpha=\beta=.1$, a species with a fidelity of .8 becomes less effective through time, i.e. the centroid moves from 15,15 back to 8,8. However, when E is increased to 0.2 the species falls back still further to 3,3 (Figure 8).

Thus, to survive a species must have the necessary error-making ability to explore his environment and also compensate for the evolution of his prey and predators.

In reality, it seems clear that the value of E which a species must counteract, and which keeps it in a situation of sufficient

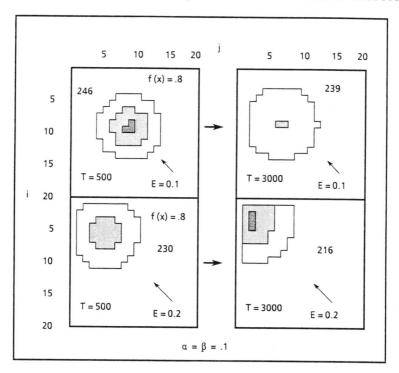

Figure 8. With an environmental wind, E, of .1 the equilibrium position of the population centroid is at 8,8.. With E = .2 it is at 3,3.

turnover, will itself be the result of the other species in the system making similar adjustments. In this way, there is a kind of 'competition' between the different levels which will ensure that E will not go to zero. This is why evolution can probably never be viewed as being 'over'. In reality the mutual linkages will ensure that each population maintains a variability as part of its strategy enabling survival. And this variability will be seen at each instant as being inefficient and sub-optimal!

Using our model above we can also discover the most advantageous rate of error-making in any particular landscape. We can run our model and study the competition between two species which start out with initially identical effectiveness (same i and j) but where the fraction of errors is different. We can see if there is elimination or superiority of one or the other.

For example when we start with two competing species which are initially at 3,5 in a landscape of $\alpha = \beta = .1$, and E = 0, and with $f(x) = .95$ and $f(y) = .85$ the result is that x is at first more efficient than y and multiplies faster. But after some time, the added selection pressure and stochasticity of y enable it to 'push forward' across the landscape faster and leave x behind. Finally, y eliminates x. In this simulation we assumed that 90% of errors were worse, and only 10% better compared to 75% and 25% in the previous cases. Clearly, this does not change the type of behaviour revealed, and we may conclude that whatever the precise value is, the same kind of result will be true.

Competition between a species x with $f(x) = .8$ and y with $f(y) = .6$ results in victory for x. Initially, both populations lose effectiveness, but then x increases and overtakes y. In this case 60% fidelity was insufficient, and the average losses incurred too great. From these simulations we can conclude that the best amount of 'error' would be somewhere between 10% and 20%.

Whatever the precise fraction of superior and inferior 'mutants' is (and this is determined by the natural world not the species or the ecologist), there will exist an 'optimal' rate of mutation which will be selected for by evolution. This evolutionary process will therefore lead quite naturally to a **rate of evolution** which is itself regulated by the evolutionary process, in which the 'progress' being made in the different levels of the system is just counteracted by that in the others!

In order to explore this point a bit further we have to attempt to treat the 'error-making' property of a population a little more carefully. In the above simulations we have assumed, rather unjustifiably, that we could have two separate populations attempting to 'climb a hill' in evolutionary space, and with characteristic degrees of 'variability generation' or 'error-making'. Now this is not quite realistic, and what we can attempt to examine more closely, is the possibility that a population would carry this 'fidelity' of reproduction as an heritable characteristic which could itself vary.

For these simulations we choose the j index to represent the degree of perfect reproduction. When j=0, we have only 5% perfect reproduction and 95% error! But when j=19 we shall suppose that we have 100% fidelity. The i axis will still represent the rate of reproduction of any sub-population, and will still serve to

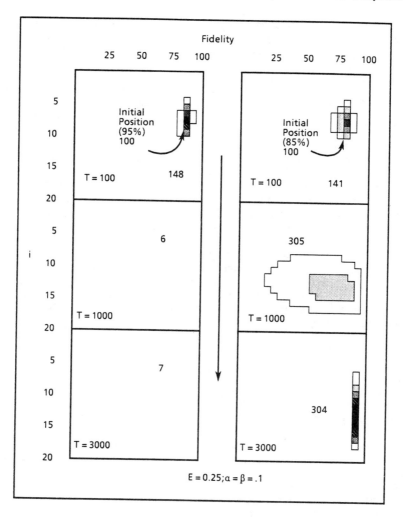

Figure 9. Evolution of fidelity in an evolutionary landscape with a steep slope and strong environmental wind. Here in order to move forward 'variability' is strongly selected for, then high effectiveness is maintained by high fidelity.

represent the 'hill' up which evolution is struggling to climb. Clearly, the higher the value of i attained, the more selection favours it. However, the value of j will dictate how 'true' replication is, and how much 'error-making' is occurring, and how this is helping or hindering in the ability to climb the hill and be 'selected'. Underlying this, then, we assume that 'errors' inreproduction will change the rate of 'error-making'. In this way selection itself will act on the system to decide the degree of 'variability' which is the most effective in climbing the hill.

If we look at two competing species, one initially at 95% perfect replication (j=18) and the other, y, with 85% (j=16), then in a landscape with a very shallow slope given by $\alpha = \beta = .01$, the evolution stays well defined (i.e. does not fall to low fidelities) although the species go through some changes as to which is the most prolific and effective.

When the slope is changed to $\alpha = \beta = .1$, however, variability is rewarded and we see that the species with initially the lower fidelity 'wins'. On its way to 'winning', however, the species goes through an 'identity crisis' and plunges to very low average fidelity, but then gains until it has only a distribution of fidelity that ranges from 85% to 100%.

From these simulations we see that if a certain amount of 'random variability' is a property of a particular population, then selection can operate on it to regulate the variability which is 'necessary' for 'hill climbing' or countering the evolution of other species represented by E.

In the above we show that 'variability' must be part of any 'strategy', and in any given situation, selection would retain the populations with the right degree of variability. Not only that, we show that variability itself could be regulated by evolutionary forces and that the right mechanisms which lead to 'discovery' can also be discovered! In other words what we see as the result of evolution are not populations with optimal behaviour but rather a vast learning system!!

1.3 Discussion

In the preceding section we have developed a simple model of adaptive population dynamics with parameters that have 'plasticity' and can respond to evolutionary processes. An important point of this work is that we do not look at the systems supposing that evolution has come to an end, but instead regard it as an ongoing process. This is in contrast with the neo-Darwinian models, and those of population genetics, where it is supposed that all strategies or genes that can appear in the system are there initially and selection simply operates to reveal the 'fittest'. In our model the space of our Figure 1, which considers the prey and predators of any particular level, we see that this would correspond to supposing that these were of fixed behaviour and abundance and initially all possible 'character types' were present for the population under study. Evolution would simply select for those types which are sitting on a 'b-m' hill - a peak of their Malthusian parameter - their adaptiveness.

We believe, however, that the problem is much subtler than this and could be likened to a problem of 'fishing' [22]. That is, evolution is about discovering hills in the evolutionary landscape, and successfully exploiting them and climbing towards the top. However, obviously in climbing them, the species which constitute resources or predators will also be involved in a similar 'game', and an 'arms race' will result. This means that as a population climbs a hill in the evolutionary landscape, so the hill and the evolutionary landscape move!

In our model we have chosen to examine here the very simple problem of climbing a particular 'hill', and have explored how small mutations or errors in reproduction could achieve this. We find that variability is necessary to success in hill-climbing, even though this necessarily implies a certain level of 'waste' or 'sub-optimality' at any given instant. Furthermore, we show how the mechanism of variability itself could be adjusted by the evolutionary process itself, leading to the idea that evolution is driven by the noise to which it leads.

Given that inaccurate reproduction certainly must precede perfect reproduction, we see that evolution only had to 'tame' the noise already present in the system, in order to get this result. Evolution is then driven from within, each species being part of the environment of others, and strongly connected to a changing physical environment as well. Instead of evolution being simply the selection of 'optimal behaviour', or being in response to a changing environment, we see it as driving itself from the noise within it. Chaos and noise are no longer seen as being only negative, here they are viewed as being perhaps the ultimate source of creativity in the universe. We have called this the 'theory of evolutionary drive'.

Clearly, in reality, each aspect of behaviour should be characterized by a dimension and an index, and evolution will consist in the diffusion outwards into this multi-dimensional space, and the selection of the parts of this that correspond to the same total 'effectiveness'. Co-evolution within an ecosystem will lead to behaviour on the part of individuals and species which is **not optimal** in terms of the simple rationality of maximal exploitation of known resources, or maximal avoidance of death. Variability in behaviour and phenotype, although on average **loss-making**, will lead to the **discovery** of more effective behaviour that ensures the longer term survival of the species.

Observed behaviour is therefore not optimal, but it is at least adequate, and it will be such that the evolutionary process itself is 'optimizing'. Here we make a careful distinction between 'optimizing' and 'optimal', because although there will be a tendency to select for species with the most effective rate of 'mutation', it should be remembered that we are using equations which are deterministic approximations to underlying probabilistic mechanisms with much more uncertainty [23].

Because behaviour is sub-optimal at any given instant, with loss-making parts, selection is relaxed and the species remaining in competition are not operating a 'knife-edge' selection. In other words it is not true that every single activity at every moment is necessary and incompressible - there is some room for basking in the sun, and for inconsequential pleasures. In this way, partly, the rigours of the mechanical models of population dynamics are relaxed, and we can better imagine the emergence and maintenance of the complex and diverse systems of reality. The usual neo-Darwinian approach leads to such a harsh view of the world that the very existence of an individual or species is viewed as proof of an optimized totality of activities, since selection is 'razor sharp', and chooses between pure strategies.

In our view, strategies or characteristics do not 'breed pure'. This means that we cannot assess the 'fitness' of a single population type without considering that of all the types to which it leads through its 'errors'. Since 'fitness' can only be judged over time, then since population types are connected together by their 'future variabilities' we cannot consider them apart. We must consider the 'population cloud' with its average performance and its variability. We have a kind of 'group selection', in that the future of each member of a population is intertwined with that of the others. Similarly, in thinking of the whole ecosystem, we see that the progress made in one sector will set the standard for the others, and again evolution cannot be considered for a population, or a strategy, in isolation. In our view, there is a much greater margin for survival than previously thought, since each species is always producing a mixture of behaviour only some of which is 'optimal', while the rest is stochastic.

Clearly, if we imagine that **each** level of an ecosystem has this type of richness which our two species show above, then there must obviously be an enormous store of 'hidden' adaptability. It is not surprising that the mechanical models which are commonly used to 'understand' ecosystems often fail.

A final point concerns 'risk' and the need to allow mistakes. Our model tells us that evolution leads to behaviour which includes a stochastic search of different behaviours, localities and perhaps values. However, rational anticipation of the result is not possible, or it would have been done and would be part of the system already. Therefore, going beyond the present frontiers of the system, and of knowledge, necessarily involves both 'failures' and 'successes', not just the latter. In view of this, it seems very important that the system should 'allow' such mistakes to be made, without too dramatic a cost, otherwise, exploration and forward evolution will cease. Instead of a view of the world which sees an evolutionary equilibrium of optimal behaviour, based on material efficiency, we see one of flexibility and adaptation, where the margins of survival are not as narrow or demanding as had been previously thought, and where imagination, eccentricity and individuality find a larger place.

Acknowledgements

We wish to thank Guy Engelen for help with the computer program, and the Solvay Institute. This work resulted in part from a project sponsored by the United Nations University, Global Learning Division, and also Fisheries and Oceans, Canada.

References

[1]   Darwin, C. (1859). *The Origin of Species*. John Murray, London.

[2]   Fisher, R.A. (1958). *The Genetical Theory of Natural Selection*. Dover, New York.

[3]   Wright, S. (1968, 1969, 1977, 1978). *Evolution and the Genetics of Populations*. Volumes I, II, III and IV, University of Chicago Press.

[4]   Ford, E.B. (1964). *Ecological Genetics*. Methuen, London.

[5]   Mayr, E. (1982) *The Growth of Biological Thought, Diversity, Evolution and Inheritance*. Belknap Press of Harvard University Press, Cambridge.

[6]   Dawkins, R. (1980)   *The Selfish Gene.*   Oxford University Press, Oxford.

[7]   Eldredge, N. and Gould, S.J. (1972) in *Models of Palaeobiology,* ed.Schopf, T., 82-115, Freeman-Cooper, San Francisco.

[8]   Maynard-Smith, J. (1982). *Evolution and the Theory of Games.* Cambridge University Press, Cambridge.

[9]   May, R.M. (1973). *Stability and Complexity in Model Ecosystems.* Princeton University Press, New Jersey.

[10]   Wynne-Edwards, V.C. (1962). *Animal Dispersion in Relation to Social Behaviour.* Oliver and Boyd, Edinburgh.

[11]   Wynne-Edwards, V.C. (1986). *Evolution through Group Selection.* Blackwell Scientific Publications, Oxford.

[12]   Wilson, D.S. (1980). *The Natural Selection of Populations and Communities.* Benjamin/Cummings, Menlo Park, California.

[13]   Orgel, L.E. and Crick, F.H.C. (1980). *Nature* **286**, 604-607.

[14]   Doolittle, W.F. and Sapienza, C.(1960). *Nature* **284**, 601-603.

[15]   Dover, G. (1982). *Nature* **299**, 111-117, 1982.

[16]   Van Valen, L.(1973). *Evol. Theory* **1**, 1-30.

[17]   Roughgarden, J. (1979). *The Theory of Population Genetics and Evolutionary Ecology: An Introduction.* MacMillan, New York.

[18]   Waddington, G.H. (1967) in *Mathematical Challenges to the Neo-Darwinian Interpretation of Evolution,* ed. Moorehead, P.S. and Kaplan, M.M., 113-115, Wistar Institute Press, Philadelphia.

[19]   Jantsch, E. (1980). *The Self-organizing Universe.* Pergamon Press, New York.

[20]   Pyke, G.H., Pulliam, H.R. and Charnov, E.L. (1977). *Q. Rev. Biol.* 52, 137-154.

[21]   Homer, M. and Kemp. W. (1983) *Mathematical Biosciences* 64, 231, Elsevier Science Publishing, Amsterdam.

[22]   Allen P.M. and McGlade (1986). *J. Can. J. Fish. Aquat.Sci.* **43**, 1187-1200.

[23]   Allen, P.M. and Ebeling, W. (1983). *Biosystems* 16, 113-126.

PATTERN SELECTION AND COMPETITION IN NEAR-INTEGRABLE SYSTEMS

Abstract

Pattern selection and competition in certain near-integrable systems is discussed. These systems provide models for controlled studies of low dimensional attractors in high (infinite) dimensional systems. Four examples from damped, driven pendulum rings are summarized in the order of increasing spatial complexity of their chaotic attractors. These examples illustrate the use of numerical and analytical techniques from soliton mathematics to study properties of chaotic attractors. In particular, the connection of (unperturbed) homoclinic states with instabilities of spatial patterns, with interactions between patterns, and as possible sources of temporal chaos is emphasized.

2.1 Introduction

Complex systems often exhibit rich spatial and temporal patterns, which possess both coherent and chaotic features. This property is found in many physical systems (water waves, plasmas, turbulent fluids, laser light, solid state and electronic devices), as well as in numerical solutions of partial differential equations and cellular automata [1]. Analysis of such complicated systems is extremely limited; thus, usually one is restricted to numerical or physical observations.

Although these complex systems have high (even infinite) dimensional phase spaces, their chaotic attractors are often low dimensional. (At least this low dimensionality seems to apply to the macroscopic transport via collective structure.) Recently theorists, using methods from the mathematical theory of dynamical systems, have made considerable progress in discovering and understanding universal properties of **low dimensional** systems. These theoretical developments have given considerable insight toward the interpretation of observations of the long time behaviour of complex systems. However, a mathematical theory of chaotic attractors for **infinite** dimensional systems is still in its infancy.

Our own recent research has focused on model partial differential equations which are far simpler than realistic physical systems, yet complicated enough to possess low dimensional attractors in an infinite dimensional state space.

Typically we study one dimensional nonlinear wave equations which are nearby "integrable" equations. These model problems permit a detailed (at times even analytical) study of specific properties of chaotic attractors in a high dimensional system. In particular, we use these models to develop intuition about the nature of the chaotic attractors and the mechanisms which cause pattern selection, competition between patterns, and the generation of chaos. Some of these features, which can be understood in detail for the model problems, are rather general and will extend to more realistic systems where such detailed investigations are impossible at present - nor can they be anticipated in the near future.

In this article, we will briefly describe several representative examples, in order of the increasing spatial complexity of their attractors, which illustrate the above philosophy. The reader should consult the references for further details. All of the examples which we discuss here are perturbed sine-Gordon equations, which provide elegant model problems.

2.2 Homogeneously AC Driven, Damped Pendulum Ring

A concrete example is the damped, ac-driven sine-Gordon equation under periodic boundary conditions:

$$\phi_{tt} - \phi_{xx} + \sin \phi = \varepsilon[-\alpha\phi_t + \Gamma \sin(\omega t)], \tag{1}$$

$$\phi(x + L, t) = \phi(x, t), \tag{2}$$

$$\phi(x, t = 0) = \phi_{1n}(x), \tag{3}$$

$$\phi_t(x, t = 0) = v_{1n}(x).$$

Here $0 < \varepsilon \ll 1$, and the control parameters are α (the strength of the dissipation), Γ (the amplitude of the ac driver), ω (the frequency of the ac driver), L (the spatial period), and the initial data (ϕ_{1n}, v_{1n}). This equation models a chain of harmonically coupled nonlinear pendula, with periodic boundary conditions describing a ring configuration. Note that both the dissipation and driving are spatially homogeneous. Historically this system with no spatial structure (i.e. the single pendulum) has provided a useful model to direct and test one's understanding of temporal chaos in deterministic dynamical systems [2,3]. However, these studies of the single pendulum neglect all effects of spatial structure which, when present, can completely alter the nature of the space-time attractor.

In numerical experiments [4,5] with weak homogeneous perturbations and single humped initial data, the fundamental phenomenon is a resonance between a spatially localized excitation (a "breather") and the ac driver. (It should be mentioned that the driving frequency ω is chosen less than unity in order to resonate with a spatially localized breather whose natural frequency ω_{Br} satisfies $0 < \omega_{Br} < 1$. If one drives at an $\omega > 1$, one resonates with extended phonons whose natural frequencies satisfy $1 < \omega_{ph} < \infty$. Above a low threshold driving strength Γ, the initialized excitation adjusts, persists, and locks periodically to the driver. As the system is further stressed by increasing Γ, this locked state loses stability and more complicated spatial structures emerge. As parameter values are varied, the attractors can be temporally periodic, quasi-periodic, subharmonically locked, or chaotic (with intermittency). Depending principally on the frequency ω of the ac driver, the chaotic dynamics is controlled by either (i) "breather $\leftrightarrow$ kink-antikink" or (ii) "breather $\leftrightarrow$ radiation" transitions. These distinct types of nonlinear modes (kinks, breathers, and anharmonic radiation) classify regimes of chaotic evolution.

The reader will notice that the terms "kink", "breather", and "anharmonic radiation" have not been defined. In general such definitions are not precise, involving approximate techniques such as collective co-ordinate methods. It is especially difficult to distinguish between breathers and radiation. A major advantage in working near an integrable soliton equation is that nonlinear spectral transforms exist which provide a precise definition and classification of these distinct nonlinear modes. A detailed description of the application of nonlinear spectral methods to chaotic attractors of systems (2.1) may be found in [6,7]. These nonlinear spectral methods have allowed us to demonstrate that a chaotic attractor can be comprised of a small number of nonlinear modes which undergo collision, annihilation, nucleation, decay, and transition between coherent and extended states. In addition these spectral methods have identified the presence of infinite period homoclinic orbits which are associated with transitions between the three types of nonlinear modes. We believe that these homoclinic states, although less familiar than the pendulum separatrix, play as fundamental a role in this chaotic pendulum chain. On the one hand, these homoclinic states are related to instabilities which generate more complicated spatial patterns; on the other hand, the infinite period states act as sources of extreme sensitivity which can produce chaos. A main result of the nonlinear spectral method is to establish numerically the presence of frequent homoclinic crossings along the chaotic attractor.

We describe two cases. First, we fix the driving frequency ω less than, and not near, unity [4,5,6]. For example, at $\omega = 0.6$ and $\varepsilon\alpha = 0.2$. Then the bifurcation sequence as a function of increasing stress parameter Γ is given symbolically by

$$
\begin{pmatrix} \text{Flat} \\ \text{Periodic} \end{pmatrix} \Rightarrow \begin{pmatrix} \text{1 Excitation} \\ \text{Periodic} \end{pmatrix} \Rightarrow \begin{pmatrix} \text{2 Excitations} \\ \text{Periodic} \end{pmatrix} \Rightarrow \begin{pmatrix} \text{2-4 Excitations} \\ \text{Chaotic} \end{pmatrix}
$$

$$
\Rightarrow \begin{pmatrix} \text{4 Excitations} \\ \text{Periodic} \end{pmatrix} \Rightarrow \quad \ldots
$$

25

In these symbols the top entry of each ordered pair describes the spatial structure of the attractor, while the second entry indicates its temporal behaviour.

Such bifurcation sequences may be understood as follows: As a function of increasing stress, the flat attractor becomes unstable to a $k_1(=2\pi/L)$ Fourier excitation which saturates into an attractor with one localized breather per spatial period. With further stress, this single breather develops a $k_2(= 4\pi/L)$

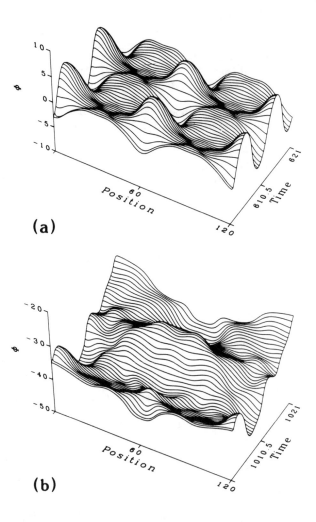

(a)

(b)

Figure 1. *Space-time evolutions of $\phi(x,t)$ for the SG system through two driving periods for $\varepsilon\alpha = 0.2$, $\omega = 0.6$, with periodic boundary conditions, and driving strengths (a) $\varepsilon\Gamma = 0.8$, which results in periodic time evolution; (b) $\varepsilon\Gamma = 1.0$, which results in chaotic kink-antikink motions (nearly repeating every driving period).*

instability which may saturate into two breathers per period or one breather plus a k_2 anharmonic phonon, depending principally on on dissipation strength. (The first situation of period halving into two breathers occurs at larger dissipation.) When the attractor contains a k_2 anharmonic phonon, the temporal behaviour may be quasi-periodic, subharmonically locked, or chaotic depending upon parameter values. Similar distinctions apply as the system is further stressed and four spatial excitations emerge (arising from a $k_4 (= 8\pi/L)$ instability).

In this low frequency ($\omega \sim 0.6$) case, the chaotic regimes are characterized by "breather" to "kink-antikink" transitions since the breathers which resonate with a low ferquency ac driver have large amplitude ($\sim 2\pi$) and thus are near the kink-antikink threshold.

Figure 1 shows how, in the chaotic regime, the basic coherent structures persist, but that their phase-locking relative to each other has been (chaotically) broken so that the structure fails to repeat by a small amount after each driver period. We can decompose the field at each instant of time into radiation and either two "breathers" or two "kinks" and two "antikinks". Furthermore, chaotic evolution of $<\phi(t)>$ (the spatial average of ϕ) through multiples of 2π does not take place via single particle dynamics but rather through the slow diffusion of the kinks (antikinks). Our nonlinear spectral analysis indeed confirms the presence of a small number of localized excitations in the chaotic attractor [6]. This small number of dominating localized modes suggests that the chaos may be governed by a low dimensional strange attractor. We have checked that the correlation dimension [6,8] is indeed low (~ 3). In fact, the chaos observed in this experiment is intermittent between the periodic locked states (Figure 1a and its period-4 analogue) and the irregularly evolving unlocked states (Figure 1b). This intermittency is reflected in the time series for the spatial averages of u and u_t shown in Figure 2.

We close our discussion of this experiment by emphasizing that, at these low driving frequencies, breather to antikink breakup dominates the chaotic attractor. A homoclinic state in the unperturbed system is associated with this transition. This infinite period state admits a familiar physical interpretation and provides a natural source of sensitivity. In the next experiment, we describe a nonlinear Schrödinger (NLS) regime which is dominated by the breather to radiation transition, whose homoclinic states are far less familiar [9,10] but similarly important as potential sources of NLS chaos.

Next, we consider the second case with a higher driving frequency ω **near**, but still less than, unity [7,11]. For example, consider $\omega = 0.87$ and $\varepsilon\alpha \sim 0.04$). (One difficulty with the experiments at smaller ω (e.g., $\omega \sim .6$) is that the interesting bifurcations occur at rather large values of the stress parameter ($\varepsilon\Gamma \sim 1.0$). These perturbations are large enough that analysis based on the integrable sine-Gordon theory is not appropriate. When ω is raised to $\omega = .9$, the interesting bifurcations occur at

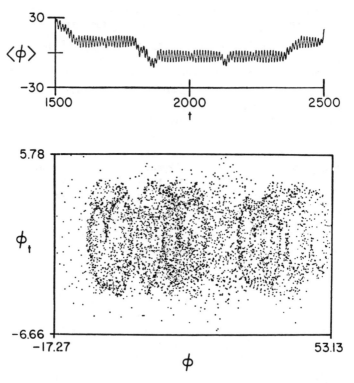

Figure 2. Spatial average ⟨φ⟩(t) and phase-plane for the chaotic regime at εΓ = 1.0 (Fig. 1b), showing intermittency: laminar regimes are time-periodic and spatially locked 2-breather (Fig. 1a).

much smaller values of the stress parameter making analysis based on the integrable theory more appropriate. The choice $0 \ll \omega < 1$ places us in the "nonlinear Schrödinger (NLS) regime". That is, when $\omega < 1$, one can use perturbation methods to approximate a class of equations (which includes the sine-Gordon equation (2.1)) by an NLS equation:

$$u(x,t) = \sqrt{6\varepsilon}[A(\varepsilon t, \sqrt{\varepsilon}x)e^{it} + c.c] .$$

Here the complex amplitude $A(T,X)$ satisfies

$$-2iA_T + A_X + 3AA^*A = i\alpha A + \Gamma^\varepsilon e^{-i(1-\omega)T} ,$$

where $\Gamma^\varepsilon = \Gamma/2\sqrt{6\varepsilon}$. Thus, because of the generic nature of NLS, this particular experiment at $\omega < 1$ actually applies to a wide class of physical problems [12,18].

At these parameter values ($\omega = 0.87$, $\varepsilon\alpha = .04$) numerical experiments on the initial value problem (2.1) reveal a beautiful route to chaos with the following features:

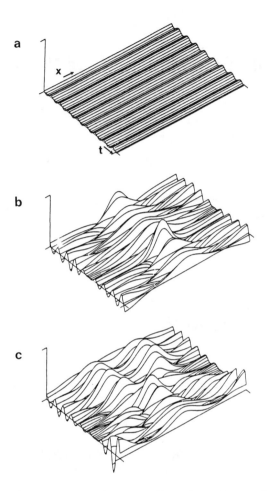

Figure 3. Space-time profiles on the attractor: (a) εΓ = 0.050 (flat in space, periodic in time); (b) εΓ = 0.105 (quasi-periodic); (c) εΓ = 0.110 (chaotic). Note the second hump centred at x = ±12 in the quasi-periodic case. Note also the changes in spatial symmetry in the chaotic case.

(i)   **temporally** – one frequency → two frequencies → chaos;
(ii)   **spatially** – one localized "hump" → two localized "humps";

(iii) symmetry changes and pattern competition;
(iv)   low dimensional, yet chaotic, attractors;
(v)   temporal intermittency, accompanied by spatial symmetry changes.

All of this action occurs at small values of the bifurcation parameter ($εΓ \stackrel{\sim}{<} 0.12$); hence, even the chaotic system is, in some sense, near-integrable. As the stress parameter εΓ increases, the attractor changes according to the following symbolic sequence:

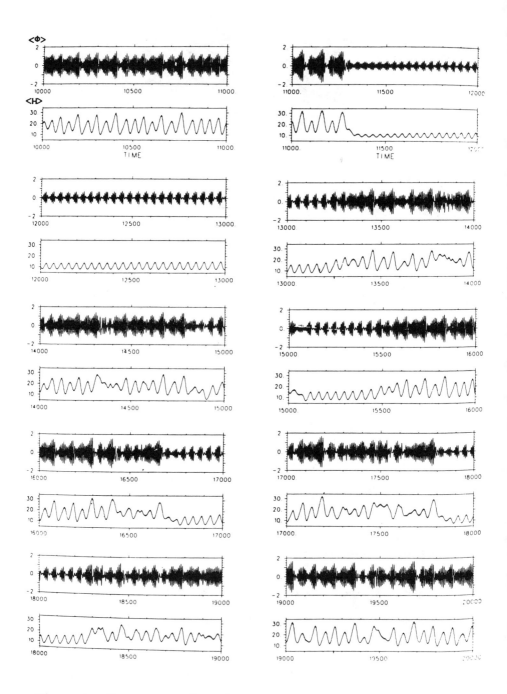

Figure 4. The energy, H, and mean value of the waveform, $\langle\phi\rangle$, as a function of time for $\varepsilon\Gamma = .1055$ (chaotic). Note the long growing laminar region for $11,300 < t < 13,200$ and the chaotic regions which bound it.

$$\begin{pmatrix} \text{Flat} \\ \text{Periodic} \end{pmatrix} \Rightarrow \begin{pmatrix} 1 \text{ Excitation} \\ \text{Periodic} \end{pmatrix} \Rightarrow \begin{pmatrix} 2 \text{ Excitations} \\ \text{Quasi-Periodic} \end{pmatrix} \Rightarrow$$

$$\begin{pmatrix} 1 \text{ Excitation} \\ \text{Periodic} \end{pmatrix} \Rightarrow \begin{pmatrix} 2 \text{ Excitations} \\ \text{Quasi-Periodic} \end{pmatrix} \Rightarrow \begin{pmatrix} 2 \text{ Excitations} \\ \text{Chaotic} \end{pmatrix} .$$

We briefly comment on these attractors. Their space-time behaviour is depicted in Figure 3. For small driving amplitudes ($0.0 < \varepsilon\Gamma < 0.058$), the periodic spatial structure of the initial condition decays as a transient, and the attractor is an x-independent flat state with no spatial structure. This state is periodic in time with the period of the ac driver. As this flat

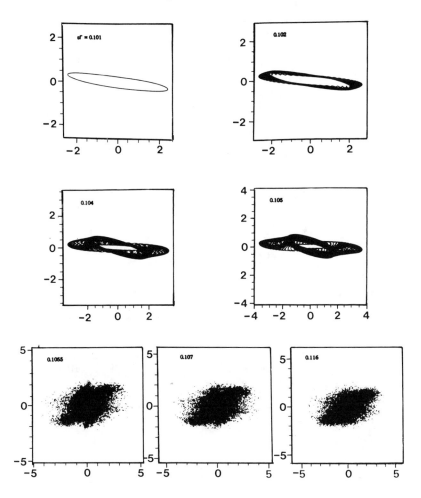

Figure 5. Phase planes for $\varepsilon\Gamma$ = (a) 0.101 (periodic); (b) 0.102 (quasi-periodic); (c) 0.104 (quasi-periodic); (d) 0.105 (quasi-periodic), and (e) 0.1055 (chaotic).

state is further stressed (by increasing $\varepsilon\Gamma$), it loses stability and the new attractor which emerges has the spatial structure of one breather-like hump per spatial period riding over a flat background. Its temporal behaviour remains periodic at the driving frequency. As $\varepsilon\Gamma$ is still further increased, this state of one hump per period loses its stability to a new attractor which has two humps per spatial period (one being localized in space and the second with the character of $k = 2$, extended, anharmonic radiation), and two temporal frequencies (one at the driving frequency $\omega = 0.87$, and the second at much lower frequency). Temporally, the state can be either quasi-periodic or subharmonically locked to the driver. With further increase of $\varepsilon\Gamma$, the attractor returns to the (1Ex,P) state, and then back to (2Ex,QP) (with, again, one localized and one extended hump). We

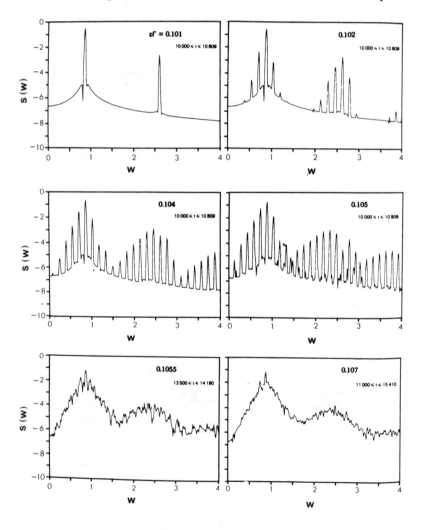

Figure 6(a). The power spectra at selected values of $\varepsilon\Gamma$.

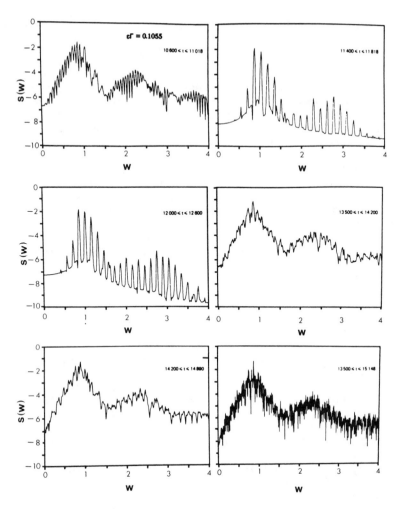

Figure 6(b). Temporal power spectra for selected time intervals in the chaotic regime.

have not measured the structure of such "windows" in detail. As $\varepsilon\Gamma$ increases further, the two humps per period "dance" irregularly, rather than quasi-periodically, and both are localized.

With the stress parameter $\varepsilon\Gamma$ fixed well within the chaotic region, the spatial structure of this chaotic attractor contains two humps per period, just as the milder (2Ex,QP) attractor. The difference between these two spatial structure is as follows: In the quasi-periodic case (2Ex,QP), one of the humps is extended; in the chaotic case (2Ex,D), both humps are localized and the state is almost spatially period halved.   These two localized humps dance, decay into radiation, reform through radiation, focus and grow back into localized states, develop relative centre of mass motion, collide, and generally interact chaotically.

Finally, we describe the attractor at $\varepsilon\Gamma$ values just above the chaotic threshold. Here the attractor is temporally intermittent. The time series for the energy $\langle H \rangle = \int dx [\frac{1}{2}\phi_t^2 + \frac{1}{2}\phi_x^2 + 1-\cos\phi]$ and the average displacement $\langle \phi \rangle = \int dx \phi$ are depicted in Figure 3 for $\epsilon\Gamma = .1055$. Note the long, linearly growing "laminar" regions separated by chaotic bursts. In these laminar regions the attractor is the same as the quasi-periodic attractor at slightly lower values of $\epsilon\Gamma$. It has two humps per period, one localized and the second $k = 2$ anharmonic radiation. Temporally, it acts quasiperiodically. As the $k = 2$ radiation grows and focuses into a more localized hump, the amplitude of the time series (Figure 4) grows linearly; finally, the time series "bursts". In this region of time, the attractor at $\epsilon\Gamma = .1055$ is very similar to the chaotic attractor at higher $\epsilon\Gamma$ values which was described in the preceding paragraph. It contains two localized humps which dance, decay, and interact chaotically. As $\epsilon\Gamma$ is increased from .1055, the percentage of time the attractor resides in the quasi-periodic "laminar" regions decreases, as is to be expected for intermittency.

We have claimed that the temporal bifurcation sequence is P → QP → C. In order to deduce this, we tested the numerical data with standard diagnostics of dynamical systems theory (phase planes, Poincaré sections, temporal power spectra, leading Lyapunov exponent, and correlational dimension). These tests are described in detail elsewhere [7,11]. Here we restrict ourselves to the phase planes of Figure 5, and the power spectra of Figure 6. We note: (i) All diagnostics are consistent with the above scenario. (ii) The correlation dimension changes with increasing $\epsilon\Gamma$ as 1. → 2. → 3.5 → 4.3. The dimension of 3.5 occurs at $\epsilon\Gamma = 1.055$, where the attractor is intermittent, and it can be interpreted as a "weighted average" of 2 and 4.3. The rise from 2 → 3.5 occurs rapidly, between $\epsilon\Gamma = .105$ and $\epsilon\Gamma = .0155$. (iii) The power spectra plots (Figure 6) show that the *laminar region* of the intermittent attractor indeed acts as if it were quasi-periodic in time.

Nonlinear spectral measurements [7,11] show that only a few (~3 or 4) nonlinear modes are appreciably excited - even when the attractor is chaotic. The number of appreciably excited modes increases with $\epsilon\Gamma$, changing when the attractor changes to within the accuracy of our study:

(F,P):	1 Mode - k_0	
(1Ex,P):	2 Modes - k_0; b_1	
(2Ex,QP):	3 Modes - $k_0, k_2 b_1$	
(2Ex,C):	3-4 Modes -	

In this list, k_0 stands for flat, x independent radiation, k_2 represents anharmonic extended radiation similar to $\cos(2.\frac{2\pi}{L}x)$, and b_1 denotes one localized "breather" per period. In the case of the chaotic attractor we do not list the type of nonlinear modes because those which are present change with time; at some times (k_0, k_2, b_1, b_1) are present, while at other times (k_0, k_1, k_2, b_1), etc. (Actually, even in the quasi-periodic attractor, the excitation changes between a localized breather b_1 and extended anharmonic k_1 radiation.) Thus, at a fixed value of the stress $\varepsilon\Gamma$, the chaotic attractor consists in a few nonlinear modes which change their type as a function of time. These type changes constitute interactions and transitions between radiation and localized spatial states.

Homoclinic orbits and homoclinic crossings: The spatially

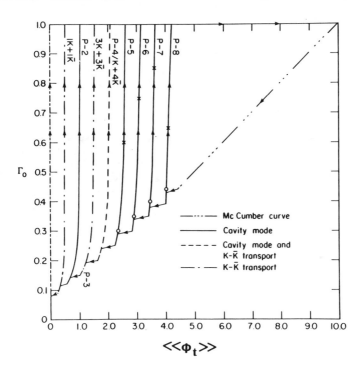

Figure 7. A summary of numerical results. The arrows indicate the direction in which the driving, Γ_{dc}, was adiabatically changed. The x's on the periodic steps (labelled P-N) denote where the periodic symmetry breaks to another P-N state (see Figure 8) when Γ_{dc} is increased, and the o's indicate where the symmetry is regained as Γ_{dc} is decreased. Notice that for $0.15 < \Gamma_{dc} < 0.18$ on the 3K-3K step, the spatial pattern is periodic.

The nomenclature reflects an annular Josephson junction context for definiteness. (See Christiansen's review article and references therein, these proceedings, for the effect of the ϕ_{xxt} dissipation).

periodic, integrable sine-Gordon equation [9,10] has homoclinic orbits separating radiation modes from localized modes. In this NLS regime, the most important homoclinic state is that which separates the breather from an anharmonic k_1 phonon [10,11]. We have used the nonlinear spectral transform to check directly the presence of these homoclinic crossings along this chaotic attractor. As the system is sufficiently stressed to access either large amplitude quasi-periodic or chaotic attractors, these numerical measurements detect frequent homoclinic crossings of k_1 $\leftrightarrow$ b_1 type, which were just discussed.

Thus, homoclinic crossings of k_1 $\leftrightarrow$ b_1 type are present, and these are potential sources of chaos in the NLS regime. While the measurement of such k_1 $\leftrightarrow$ b_1 states is straightforward with nonlinear spectral projections, it is quite inaccessible by standard data analysis such as spatial and temporal profiles, time series diagnostics, or linear spectral transforms.

2.3 DC Driven, Damped Pendulum Ring

One might imagine that a homogeneously dc driven, damped ring of pendula would exhibit little spatial structure. However, this perturbation actually yields a very rich variety of space-time patterns which illustrate a number of important general phenomena including transverse instabilities on moving interfaces and spatial competition among nonlinear modes.

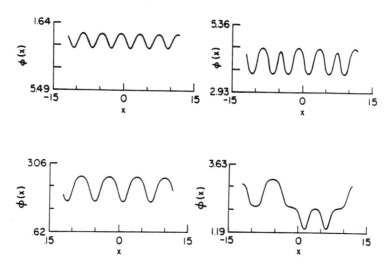

Figure 8. Two examples of coexisting attractors on the same step: (a) periodic 6 (P-6) standing breather train whose lengths compete with higher driving, yielding the amplitude delocking illustrated; (b) the P-4 state may undergo kink-antikink breakup to give the $4K + 4\bar{K}$ state shown, whose kinks and antikinks move transversely. On each of the ϕ axes, ϕ is given modulo 2π.

The model problem is as in equation (2.1), except that we replace "$\Gamma \sin \omega t$" with a dc driver "Γ_{dc}". Again, the most interesting regime is one of small dissipation ($\varepsilon\alpha < 1$) where there is a distinctive hysteresis: Namely as $\varepsilon\Gamma_{dc}$ is increased from 0, there is no mean rotation ($<\phi_t>$) until $\varepsilon\Gamma_{dc} = 1$, where the pendula begin to rotate homogeneously in space and with an oscillatory time dependence about $<\phi_t>$. As $\varepsilon\Gamma_{dc}$ is now decreased, a finite value of $<\phi_t>$ persists for $\varepsilon\Gamma_{dc} < 1$, accompanied by the spontaneous appearance of time dependent spatial patterns. The hysteresis diagram is organized into "steps" (see Figure 7), each of which is associated with specific spatial structures. These may be locked breather wave trains (anharmonic standing waves) or transversely moving kink-antikink pairs.

Linear stability analysis of the x-independent rotating state can be used to predict the periodicity of the emerging structures on the high steps very satisfactorily [19]. These structures

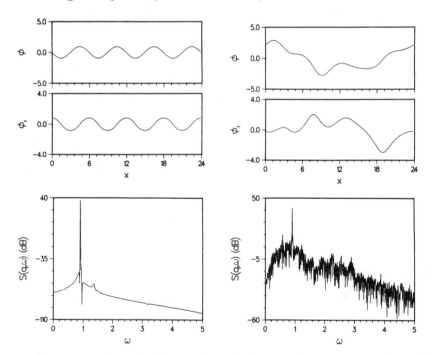

Figure 9. Snapshots of ϕ and ϕ_t, and power spectra, for spatially-dependent driving $\Gamma \sin(\omega t - kx)$, of a sine-Gordon ring with $L = 24$, $\varepsilon\alpha = 0.1$ and $\omega = 0.9$: (a) $2\pi/k = 6$, $\varepsilon\Gamma = 1.1$ and (b) $2\pi/k = 24$, $\varepsilon\Gamma = 0.5$. In case (a), the resonant breather length scale is $\sim 2\pi/k$ and locked, time-periodic response occurs even with large driving strengths. In case (b), the two length scales are very different and a low-dimensional chaotic, spatially-irregular attractor is found even at small driving strengths.

saturate in the nonlinear sine-Gordon potential into breather wave trains, in that their amplitudes and widths have breather characteristics. These nonlinear structures **compete** for transverse space. This competition is resolved in two distinct ways: (1) amplitude delocking of the breather wavetrain producing breathers of different sizes (see Figure 8); or (2) breather break-up into kink-antikink pairs. The latter occurs on the lower steps where the "washboard" frequency is smaller; and hence, the resonant breathers have large amplitudes and thus easily break into kink-antikink pairs.

Clearly there are a large variety of sources of low dimensional chaos in this situation. Mathematically, homoclinic crossings, such as those discussed in section 2, are certainly present. Physically chaos may be induced either through irregular jumping between steps or through transitions between patterns on the same step.

2.4 Inhomogeneously AC Driven, Damped Pendulum Ring

In the previous sections we found: (i) that an ac driver could introduce a characteristic length scale - the width of the saturated (breather) modes which resonate temporally with the driver; and (ii) that competition between spatial patterns results when this length scale is inconsistent with the total length available per hump. In this section we consider an experiment [20] which tunes this competition by controlling two length scales. This control is achieved with a spatially periodic ac driving term of the form $\Gamma \sin (\omega t - qx)$. Again, the frequency ω controls the width of the saturated mode. However, now the driver introduces an additional spatial scale, $2\pi/q$. In effect, this driver resonates with the localized breathers of frequency ω and with phonons of wave number q. If the two length scales are near commensurate, a strong temporally periodic locking of a breather wave train of spatial periodicity $2\pi/q$ occurs, even for very high driving strengths (see Figure 9a). In this manner we can begin to study the temporal evolution of complicated (chaotic?) spatial patterns.

2.5 Conclusions

In this article we have described several examples of coherence and chaos in near integrable systems. Such models are simple enough to allow careful, controlled studies of low dimensional chaos in a high dimensional setting. In particular, the spatial patterns of the attractor are closely related to solutions of the nearby integrable system, and a nonlinear spectral transform can be used to quantify properties of these spatial structures. In some instances such measurements show that the attractor can be represented by a small number of nonlinear modes; however, these modes must continually change their type between radiation and localized states as the phase point evolves along the attractor. Thus, these nonlinear spectral measurements establish the importance of soliton-radiation interactions on the chaotic attractor. In addition, these measurements detect frequent homoclinic crossings as the phase point evolves along the attractor. The homoclinic states are related to instabilities which generate more complicated spatial patterns, and they act as sources of sensitivity which can produce temporal chaos. A main result of the nonlinear spectral measurements of these chaotic attractors is to establish numerically the presence of frequent homoclinic crossings.

Of course, systems near soliton equations must have special properties. Nevertheless, the discovery and verification of the behaviour of their chaotic attractors is much easier than in general situations. Consider, as an example, the detailed information which we described in this article about the homoclinic crossings in the perturbed sine-Gordon equation. Given the importance of homoclinic crossings in the near-integrable situation, we can now begin to ask about the possibility of homoclinic states separating distinct spatial patterns in more general settings.

For the future, it remains to be seen which of the chaotic phenomena identified in near soliton systems will persist in a more general framework. In the near-integrable framework, we are continuing to develop an analytical reduced description of attractors that incorporates the relevant **small** subset of interacting nonlinear modes, as well as the dynamical features of homoclinic orbits.

Acknowledgement

Our viewpoint in this article rests on the theoretical foundation of the geometry of the sine-Gordon equation, work done jointly with N. Ercolani and G. Forest. We gratefuly acknowledge many pleasant working hours and numerous long discusssions with each of these scientists. Without them, this work could not have been done.

References

[1]   See articles in *Spatio-Temporal Coherence and Chaos in Physical Systems,* Proceedings of Los Alamos Workshop, ed. A.R. Bishop, G. Gruner and B. Nicolaenko, *Physica* **23D**, 1986.
[2]   D. D'Humieres, M. Beasley, B. Humberman and A. Libchaber (1982). *Phys. Rev.* **A26**, 3483; J. Crutchfield, J. Farmer and B. Huberman (1983). *Phys. Reports* **92**, 42.
[3]   J. Guckenheimer and P. Holmes (1983). *Nonlinear Oscillations, Dynamical Systems, and Bifurcations,* Springer, New York.
[4]   A.R. Bishop, K. Fesser, P.S. Lomdahl, W.C. Kerr, M.B. Williams and S.E. Trullinger (1983). *Phys. Rev. Lett.* **50**, 1095.
[5]   A.R. Bishop, K. Fesser, P.S. Lomdahl and S.E. Trullinger (1983). *Physica* **7D**, 259.
[6]   E.A. Overman, D.W. McLaughlin and A.R. Bishop (1986). *Physica* **19D**, 1.
[7]   A.R. Bishop, M.G. Forest, D.W. McLaughlin and E.A. Overman (1986). *Physica* **23D**, 293.

[8]   P. Grassberger and I Procaccia (1983).   *Phys. Rev. Lett.*
        50,   346.

[9]   N.   Ercolani,   M.G.   Forest   and   D.W.   McLaughlin   (1986).
        Physica **18D**, 472.

[10]   N.   Ercolani,   M.G.   Forest   and   D.W.   McLaughlin   (1986).
        Homoclinic Orbits for the Periodic Sine-Gordon Equation, in
        preparation.

[11]   A.R.   Bishop,   D.W.   McLaughlin and E.A.   Overman   (1986) *A
        Quasi-Periodic Route to Chaos in a Near Integrable PDE:
        Homoclinic Crossings,* Univ. of Arizona preprint.

[12]   G.D.   Doolen,   D.F.   DuBois and H.   Rose   (1983).   *Phys. Rev.
        Lett.* **51**, 335.

[13]   H.T.   Moon,   P.   Huerre and L.G.   Redekopp   (1983).   *Physica*
        7D, 135.

[14]   D.W. McLaughlin, J.V. Moloney and A.C. Newell (1983).   *Phys.
        Rev. Lett.* **51**, 75.

[15]   J. Wu, R. Keolian and I. Rudnick   (1984).   *Phys. Rev. Lett.*
        521, 1421.

[16]   H.T.   Moon and M.V.   Goldman   (1984).   *Phys. Rev. Lett.* **53**,
        1921.

03 *T. Bloch*

THE LATEST GENERATION OF SUPERCOMPUTERS, EXEMPLIFIED BY THE CRAY-2

3.1 Introduction

Progress in computational science depends to a considerable extent on the availability of bigger and faster computers.

The latest generation of supercomputers: CRAY Research Inc.'s X-MP4 and CRAY-2, the SX-2 from NEC Corporation and, a little further out in the future, the ETA-10 from ETA Corporation adopt a variety of solutions in order to deliver a maximum of usable computational power to the user: one (SX-2) to eight (ETA-10) processors with four in between (X-MP4 and CRAY-2); automatic paging from a huge, shared semiconductor memory into the memories dedicated to each CPU (ETA-10) or "normal" virtual memory management for the scalar calculations coupled with direct memory access for the vector calculations (SX-2) or a main memory with direct access coupled with a big semiconductor "disk" (X-MP4 and SX-2) or just a very, very large main memory (CRAY-2).

We believe that the size of the main memory of the CRAY-2 (268,435,456 words) and the multiprocessor approach taken by all the above machines except the SX-2 represent important trends; the memory size because it brings programming convenience to a field of computation where so much time has been spent by the users in order to organize the input/output for "out-of-core" models and the multiple CPU approach because this is the only possible way in the long run to achieve shorter and shorter execution times for complete programs. (A single CPU can be made to go very fast indeed on vector calculations but for purely scalar works, designs are more limited by physical limits.) The best hope at this stage for significant improvement of total execution speed is to section the codes into several independent pieces which can execute in parallel. Still this remains a difficult problem for most user codes and a heavy user investment in programming and methodology development over the next years is necessary.

3.2 The CRAY-2

The CRAY-2 from CRAY Research Inc. was first delivered to the Magnetic Fusion Energy Laboratory at Livermore in Spring 1985. It uses the same logic circuits as the X-MP4 but its architectural design is significantly different:

(a)   a shorter cycle time (4.1 ns) doing less per cycle.   This increases the vector performance (a speed of 1.6 GFLOPS has been measured on a kernel with all 4 processors going in parallel) but penalizes somewhat scalar performance since more gates are "wasted" in latching;

(b)   a very big memory made out of DRAM: the slowest memory technology for the fastest computer.   A large number of banks (128) and sophisticated access circuits ensure that this memory can deliver up to one word per cycle to each processor - barring bank conflicts, refresh cycles, etc.   The access time is, of course, relatively slow (about 250 ns) and a small local memory of 16384 words with a very fast access has been incorporated into each processor to store intermediate results.   Transfers between this local memory and the registers are under program control.

Ignoring the I/O organization the architecture of the CRAY-2 looks pretty much like four CRAY-1's packed into one box sharing one huge central memory.   Each processor is roughly the same speed as a CRAY-1 on scalar code and up to three times faster on long vector operations.   Chaining (using the output of one pipeline directly as the operand of another operation when it is opportune) has disappeared altogether and memory access patterns have become more sophisticated (scatter/gather, compressed iota) compared to the fixed stride access of the 1976 CRAY-1.

3.3 **User Challenges**

The real potential of the CRAY-2 probably does not lie with its speed but rather with its enormous main memory: 268,435,456 words, the equivalent of about 16 full 6250 bpi tapes.   We believe that this choice is one that will profoundly influence future supercomputing, but it is also one that users will have to adapt to both in the choice of problems, numerical methods and programming methodology.   The optimistic view is that:

(a)   3-dimensional problems can be attacked realistically - "in core" (out-of-core models should probably not be run on a CRAY-2 considering the size and the speeds of disks today);

(b)   it will be more convenient and faster to try out new methods as the I/O part of the out-of-core models disappear;

(c)   codes will be organized to use several processors in long running jobs demanding only limited I/O activity for checkpointing (overall systems performance will thus be greatly improved);

(d)   graphics development will follow suit and interpretation of results will not become a bottleneck;

(e)   the "scalar problem" (the execution time of large codes, becoming dominated by irreducible segments of scalar code) becomes less severe as the size of the problems increases.

But it is also very easy to list the obstacles to overcome and feel pessimistic about how quickly the full potential of a CRAY-2 will be brought to bear on big modelling problems:

(a)   recoding old out-of-core programs implies a considerable effort even when it goes in the direction of simplification of the code and, furthermore, it obliges the scientist to take the risk that they will no longer execute on other, smaller machines (or even future ones, if the big memories have not come to stay);

(b) the size of problems which can be fitted into the huge central memory will tax the computational power available beyond reasonable bounds. The difficulties of presentation and interpretation of the big volume of data available to the user will become a major bottleneck;

(c) the organization of individual codes into independently multitasked pieces and the debugging of such codes is too much work for the user for too little gain in most cases;

(d) the scalar problem is worsened, due to the degradation of the scalar speed relative to the vector speed (compared to the CRAY-1).

Conclusion

The scientist using a CRAY-2 is an integral part of the system — like with any large scientific instrument — and the final answer to the future value of the architectural solutions adopted in the CRAY-2 lies with him. As far as the author is concerned, he believes firmly that big memories and multiprocessors are here to stay, that the users will benefit enormously from the big memories and will, eventually, make the necessary investment to learn how to master the multiprocessing aspects of programming and debugging.

Reference

[1]   T. Bloch (1986). in *Computing in High Energy Physics,* eds. L.O. Hertzberger and W. Hoogland, Elsevier Science Publishers B.B. (North Holland).

RECENT WORK ON NONLINEAR DYNAMICS WITHIN MIDIT

HOLOMORPHIC SURGERY AND SYMMETRIES IN THE MANDELBROT SET

4.1 Introduction

This paper reports on some recent results of joint work with A.
Douady and John H. Hubbard.

Let P_c denote a quadratic complex polynomial and $P_{a,b}$ a cubic
complex polynomial given by

$P_c(z) = z^2 + c$ (critical point 0), $P_{a,b}(z) = z^3 - 3a^2 z + b$ (critical
points $\pm a$).

The Mandelbrot set M and the connectedness locus in degree 3 C_3
is defined as

$$M = \{c \in \mathbb{C} \mid P_c^{\,n}(0) \nrightarrow \infty\} \text{ and } C_3 = \{(a,b) \in \mathbb{C}^2 \mid P_{a,b}^{\,n}(\pm a) \nrightarrow \infty\}.$$

The sets M and C_3 could as well have been defined as the set of
parameters for which the corresponding Julia sets are connected.

4.2 The Mandelbrot Set - as a universal object

The Mandelbrot set is found not only in the parameter space for
quadratic polynomials, but in many other complex one-parameter
families of analytic mappings, for instance in one-parameter
subfamilies of cubic polynomials. The reason is that the mappings
(or some iterates thereof) locally may behave as a polynomial of
degree 2.

The proper definition and theorem is obtained by A. Douady and
J.H. Hubbard in the paper [5]. The main idea is the following:
A one-parameter family $\Lambda \approx D$ of polynomial-like mappings of degree
2

$$f_\lambda : U'_\lambda \to U_\lambda \, , \, \lambda \in \Lambda,$$

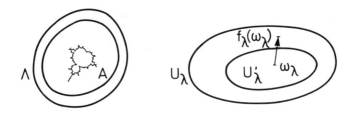

Figure 1.

contains a copy of M if there exists a closed subset $A \subset \Lambda$, $A \approx \bar{D}$, such that $f_\lambda(\omega_\lambda) \in U_\lambda / U'_\lambda$ for all $\lambda \in \Lambda$ int A, where ω_λ is the critical point of f_λ in U'_λ, and such that $f_\lambda(\omega_\lambda) - \omega_\lambda$ turns around 0 once when λ describes the boundary of A, ∂A (see Fig. 1).

4.3 Holomorphic Surgery – from quadratic to cubic polynomials

The limb $M_{1/2}$ is the subset of M attached to the cardioid at the period doubling bifurcation point c = –.75 (see Fig. 2).

When c changes from being inside the cardioid to be in $M_{1/2}$ the fixed point α_c changes from being attractive to repulsive and α_c becomes a pinching point in the Julia set, accessible from outside along the rays $R_c(1/3)$ and $R_c(2/3)$. Figure 3 shows a computer drawing of the Julia set for c = –1 with the important points and rays marked.

Consider the one-parameter subfamily of cubic polynomials given by $b = 2a^3 - 2a$. Each polynomial $Q_a(z) = z^3 - 3a^2z + (2a^3 - 2a)$ maps the critical point +a on to the fixed point –2a. Let F

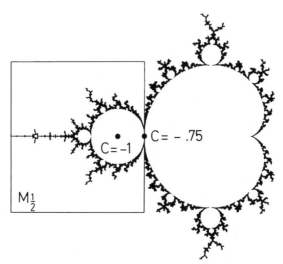

Figure 2.

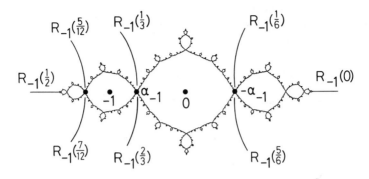

Figure 3.

denote the subset in C_3

$$F = \{a \in \mathbb{C} \mid Q_a^{\ n}(-a) \quad \infty\}$$

and let $F_{1/2}$ be the limb of F attached to the circle $|a| = 1/3$ at
the point $a = -1/3$. When the parameter value a changes at $a = -1/3$ from being inside the circle to be in $F_{1/2}$, the fixed point
$-2a$ changes from being attractive to repulsive and the critical
point $+a$ becomes a pinching point in the Julia set. The set $F_{1/2}$,
which is contained in the boundary of the connectedness locus,
∂C_3, can also be characterized as

$$F_{1/2} = \{a \in F \mid \lim_a R_a(0) = -2a\}.$$

By using holomorphic surgery we construct a map Ψ: $M_{1/2} \quad F_{1/2}$.
The idea in the construction is as follows: For any $c \in M_{1/2}$
remove the part of the plane to the left of $R_c(1/3) \cup \{\alpha_c\} \cup R_c(2/3)$
and glue $R_c(1/3)$ and $R_c(2/3)$ together by equipotential
identification. We define a new map, the first return map, by

$$z \quad \begin{cases} P_c(z) \text{ if } P_c(z) \text{ is in the allowed region} \\ \\ P_c^{\ 2}(z) \text{ if } P_c(z) \text{ is in the removed region.} \end{cases}$$

The first return map is discontinuous along
$R_c(1/6) \cup \{-\alpha_c\} \cup R_c(5/6)$. By smoothing the map, changing the complex
structure and using the theorem of integration by Ahlfors-Bers we
find a polynomial-like mapping of degree 3 and finally an $a \in F_{1/2}$,
where the critical point 0 for P_c becomes the critical point $-a$
for Q_a and $-\alpha_c$ becomes the other critical point $+a$. By surgery we

have created a new critical point.

Theorem: the map Ψ: $M_{1/2} \to F_{1/2}$ is a homeomorphism. The proof can be found in [4].

4.4 Cubic Polynomials - outside the connectedness locus

For every $r > 1$ we have a one-parameter subfamily $L_r^+(0)$, of cubic polynomials characterized by

1) $+a$ escapes to ∞ at the fixed rate, log r,
2) $-a$ escapes to ∞ at a slower rate or not at all,
3) two rays $R_{a,b}(1/3)$ and $R_{a,b}(2/3)$ meet at $+a$.

The set $L_r^+(0)$ is homeomorphic to D and is one of the leaves in a trefoil clover, see [1] and [2]. The subset

$$E_r^+(0) = \{(a,b) \in L_r^+(0) \mid P_{a,b}^n(-a) \not\to \infty\}$$

has infinitely many connected components, of which infinitely many are copies of M, due to the fact described in 2. The proof will occur in [3].

The amazing fact is that all the components in $E_r^+(0)$ in the limit, following the stretching rays as $r \downarrow 1$, connect to form $F_{1/2}$ and therefore to a pattern we already knew, namely the limb $M_{1/2}$ of the Mandelbrot set. The reason for the existence of part of the Mandelbrot set, $M_{1/2} \approx F_{1/2}$, on the boundary ∂C_3 is completely different from the reason for all the copies of M in $L_r^+(0)$, or in M. It is therefore possible to extract some new knowledge of $M_{1/2}$ from our knowledge of $\mathbb{C}^2 - C_3$.

4.5 Symmetries in the Mandelbrot Set

To each $\theta \in \mathbb{R}/\mathbb{Z}$ we have a one-complex parameter subfamily, $L_r^+(0)$, of cubic polynomials. The turning of the complex structure introduced in [2] induces a homeomorphism between any two of the sets $L_r^+(\theta)$ In particular we get the turning operator of a full turn

$$\tau: \quad L_r^+(0) \to L_r^+(0).$$

In general a copy of M in $E_r^+(0)$ is turned by τ to another copy of M in $E_r^+(0)$. The period of $-a$ associated with the centre of such a copy of M is an invariant under τ. Details will appear in [3].

Let $\mathcal{M}_{1/2}$ denote the set of Mandelbrot sets contained in $M_{1/2}$, primitive with respect to tuning (defined in [5]). The turning operator τ induces, by means of the map ψ described in 4.3, an operator $\tilde{\tau}: \mathcal{M}_{1/2} \to \mathcal{M}_{1/2}$. It can be shown that the period of the critical point 0 associated with the centre of an M in $\mathcal{M}_{1/2}$ is an invariant under $\tilde{\tau}$.

Similarly the limb $M_{1/3}$ of the Mandelbrot set is homeomorphic to a subset of fourth degree polynomials with only 2 critical points.

4.6 Conclusion

The homeomorphism $\Psi: M_{1/2} \to F_{1/2}$ is one of the first results about the boundary ∂C_3. The main interest in understanding the turning operator is to get a description of different parts of ∂C_3.

Another result is that J. Milnor has observed that ∂C_3 is non-locally-connected.

REFERENCES

[1]   B. Branner (1986). The Parameter Space for Complex Cubic Polynomials, *Proceedings of the Conference on Chaotic Dynamics and Fractals*, 169-179, Academic Press.
[2]   B. Branner and J.H. Hubbard, *The Iteration of Cubic Polynomials, Part I: The Global Topology of Parameter Space*, to appear in Acta Mathematica.
[3]   B. Branner and J.H. Hubbard, *The Iteration of Cubic Polynomicals, Part II: Patterns and Parapatterns*, in preparation.
[4]   B. Branner and A. Douady (1986). *Surgery on Complex Polynomials*, in preparation for the Proceedings of the Symposium on Dynamical Systems, Mexico.
[5]   A. Douady and J.H. Hubbard (1985). On the Dynamics of Polynomial-like Mappings, *Ann. Sc. de l'E.N.S*, **18**, 287-343.
[6]   J. Milnor (1986). *The Cubic Mandelbrot set*, preprint.

ON A SPECIAL TYPE OF TWO-STROKE OSCILLATIONS

4.7 Introduction

A **two-stroke oscillator** is a one-dimensional model governed by an equation

$$\ddot{x} + f(x,\dot{x})\dot{x} + g(x) = 0, \qquad (1)$$

having one nontrivial periodic solution (up to translations in time), along which the energy $1/2\ \dot{x}^2 + \int_0^x g(\xi)d\xi$ has exactly one maximum and one minimum per period. This notion goes back to P. le Corbeiller [2,5]. Here we consider the simple case

$$\ddot{x} + f(x)\dot{x} + x = 0, \qquad (2)$$

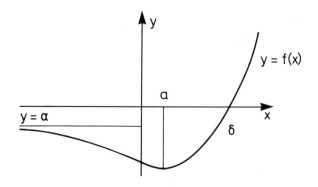

Figure 4.

where $f(x)$ is a C^1-function, defined for all real x, and satisfying the conditions
(i) $f(x)(x-\delta) > 0$ if $x \neq \delta$
(ii) $f'(x)(x-a) > 0$ if $x \neq a$,
where $0 \leq a < \delta$, a and δ being real constants (see Fig. 4).
From (i) the existence of $\alpha = \lim_{x \to -\infty} f(x)$ and $\beta = \lim_{x \to \infty} f(x)$ follow, and from (ii) it further follows that $-\infty < \alpha \leq 0$, $0 < \beta \leq +\infty$. By a combination of classical methods and methods of the authors it will be shown that there exists at most one nontrivial periodic solution (up to translations in time). Moreover there exists (exactly!) one if (α,β) belongs to the domain A of the parameter plane indicated on Fig. 5, and none if $(\alpha,\beta) \notin A$.

Furthermore the corresponding closed trajectory in the $(x,\dot{x})$ plane is positively asymptotically orbitally stable from both sides, globally from the outside. Finally we show that the energy $1/2\dot{x}^2$ + $1/2x^2$ has one maximum and one minimum per period along the closed trajectory. In other words Eq. (2) is a two-stroke oscillator if and only if $(\alpha,\beta) \in A$.

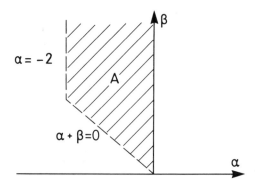

Figure 5.

4.8 Discussion in the Phase Plane

Eq. (2) is equivalent with the system

$$\dot{x} = y, \quad \dot{y} = -f(x)y-x \; ; \; (x,y)\in R^2. \tag{3}$$

In [3] (see Theorem 2) we have shown that if for a general autonomous C^1-system

$$\dot{x} = P(x,y), \quad \dot{y} = Q(x,y) \; ; \; (x,y)\in R^2 \tag{4}$$

it is possible to find a C^1-function $V(x,y)$, $(x,y)\in R^2$, such that the function, defined by

$$\tilde{V} = (P,Q).\,grad \; V - v.\,div(P,Q)$$

fulfils the condition $\tilde{V}(x,y) \geq 0$ for all $(x,y)\in R^2$, then the components of the set $\{(x,y)\in R^2 \mid V(x,y)>0\}$, the so-called p-cells, are all positively invariant. Moreover, if a p-cell Ω does not contain critical points of (4), and if V does not vanish identically in any open annular subdomain (i.e. a domain, whose boundary is two Jordan curves surrounding a critical point) of Ω, then Ω contains at most one closed trajectory for system (4). Now for system (3) take the C^1-function

$$V(x,y) = y^2 + f(x)(x-a)y + x^2 - a\delta, \quad (x,y)\in R^2 \; .$$

By a straightforward calculation one gets

$$\tilde{V}(x,y) = f'(x)(x-a)y^2 + af(x)(x-\delta), \quad (x,y)\in R^2.$$

From (i) and (ii) the propositions $\tilde{V}(x,y) \geq 0$ for all $(x,y)\in R^2$ and $\tilde{V}(x,y) = 0 \iff [a > 0 \land (x - \delta \land y = 0)]_\lor [a = 0 \land (x=0_\lor y=0)]$ immediately follow. The p-cell Ω containing the half line $y = 0$, $x\in]\sqrt{a\delta}, + \infty[$ does not contain the only critical point $(0,0)$ of (3), since $V(0,0) = -a\delta \leq 0$. Hence Ω is positively invariant and contains at most one closed trajectory. Now

$$\frac{d}{dt} (1/2\dot{x}^2 + 1/2x^2) = y\dot{y} + x\dot{x} = -f(x)y^2 \; , \tag{5}$$

which shows that $\dot{x}^2 + x^2 = y^2 + x^2$ and hence the distance $r = (x^2+y^2)^{-1/2}$ from $(0,0)$ increases strictly along the trajectories in the half plane $x < \delta$. Hence any closed trajectory Γ must surround $(0,0)$ and meet the vertical line $x = \delta$. Since $\sqrt{a\delta} < \delta$ we conclude that Γ must meet $(\Omega$ which in this case also contains the half line $y = 0$, $x\in] -\infty,, - \sqrt{a\delta} [$ and surround $(0,0))$ and consequently be unique. We have shown that there exists at most one closed trajectory to system (3). In case of existence, the closed trajectory Γ meets the line $x = \delta$ once for $y > 0$ and once for $y < 0$. Hence Eq. (5) shows that the energy exhibits the behaviour of a two-stroke oscillator. Moreover $\Gamma = L(\Gamma_1^+)$ for any nonsingular trajectory Γ_1 inside Γ, i.e. Γ is positively asymptotically stable from inside (this follows from Eq. (5) combined with the uniqueness: $L(\Gamma_1^+)$ must be a closed trajectory and hence be Γ).

In some cases it is quite easy to show that there actually is no closed trajectory at all. Consider namely the field $(\dot{x},\dot{y})$ on the halfline $y=x$, $x<0$. Then

$$(\dot{x},\dot{y}) \cdot (-1,1) = -y-f(x)y-x = -x(f(x) + 2) .$$

If $\alpha \leq -2$ (see Fig. 1) we get $(\dot{x},\dot{y}) \cdot (-1,1) < 0$ for all $x < 0$. Then the sector $y \leq x \leq 0$ is positively invariant, and no closed trajectory exists.

If $-2 < \alpha < 0$, $\alpha + \beta < 0$ nonexistence can be shown by considering the phase portraits of the two linear systems

$$\dot{x} = y , \; \dot{y} = -\alpha y - x \; ; \; x < 0$$
$$\dot{x} = \dot{y} , \; y = -\beta y - x \; ; \; x > 0 \qquad (6)$$

The phase portraits are shown on Fig. 6 (we just refer to the well known feature; recall that $-2 < \alpha \leq 0$ and $0 \leq \beta \leq -\alpha < 2$).

By a standard calculation involving the explicit solutions of (6) it can be shown that

$$y_1 = -y_o \exp(-\alpha \pi [4 - \alpha^2]^{-1/2})$$
$$y^1 = -y^o \exp(-\beta \pi [4 - \beta^2]^{-1/2}) ,$$

where y_1, y_o, y^1, y^o are defined on Fig. 6.

Starting at the point $(0,y^o)$ and following the trajectories of

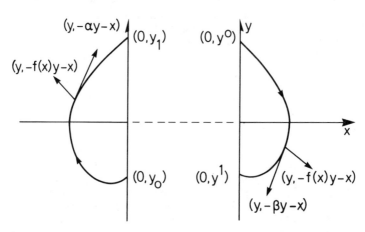

Figure 6.

(6) around (0,0) (one turn), the next point of the positive y-axis will be $(0, y_1)$ where

$$y_1 = y^o \exp(-\alpha\pi[4 - \alpha^2]^{-1/2} - \beta\pi[4\beta^2]^{-1/2}) .$$

From $-2 < \alpha \leq 0$, $\alpha + \beta \leq 0$ we get $y_1 \geq y^o$, i.e. the "combined" phase-portraits of (6) consist of spirals going outwards or a system of closed curves. The field $(y, -f(x)y-x)$ of system (3) is directed outwards on these "trajectories" as seen from

$$(y, -f(x)y-x) . (\alpha y + x, y) = (\alpha - f(x))y^2 \geq 0 \text{ if } x < 0.$$

$$(y, -f(x)y-x) . (\beta y + x, y) = (\beta - f(x))y^2 \geq 0 \text{ if } x > 0.$$

By a standard argument we conclude the nonexistence of closed trajectories. The remaining case $-2 < \alpha \leq 0$, $\alpha + \beta > 0$ (including $\beta = +\infty$) is treated in the next section.

4.9 Discussion in the Lienard Plane

Introduce the function

$$F(x) = \int_0^x f(\xi)d\xi, \quad x \in R .$$

Clearly $F(x) \to +\infty$ if $x \to +\infty$. It is well known that Eq. (2) also is equivalent with the system

$$\dot{x} = y - F(x) , \quad \dot{y} = -x ; \quad (x,y) \in R^2 . \tag{7}$$

Let the even and odd part of $F(x)$ be respectively

$$\phi(x) = 1/2[F(x) + F(-x)], \quad x \in R$$

and

$$\psi(x) = 1/2[F(x) - F(-x)], \quad x \in R.$$

In [1] (see Theorem 3 and Criteria 1 and 3) we have shown that if for some $x_o > 0$

(a) $\limsup\limits_{x \to -\infty} \psi(x) < \liminf\limits_{x \to +\infty} \psi(x)$

(b) $\limsup\limits_{x \to +\infty} [\log(1/2\, x^2) - \int_{x_o}^x 1/\xi\, \phi'(\xi)d\xi] = +\infty ,$

then there exists a simple Jordan curve K surrounding (0,0) such that all the trajectories of (7) coming from a point outside K will move inwards, reach the curve K and eventually stay in its interior.

Suppose now that $-2 < \alpha \leq 0$ and $\alpha + \beta > 0$, $\beta < +\infty$ holds.

Choose $\varepsilon > 0$ such that $\alpha + \beta - \varepsilon > 0$. Then $x_\varepsilon > 0$ can be chosen such that

$$\alpha - 1/2\varepsilon < f(x) < \alpha \quad \text{if} \quad x \leq -x_\varepsilon$$
$$\beta - 1/2\varepsilon < f(x) < \beta \quad \text{if} \quad x \geq x_\varepsilon.$$

By integration this yields

$$F(x) < - \int_x^{-x_\varepsilon} (\alpha - 1/2\varepsilon)d\xi + \int_0^{-x_\varepsilon} f(\xi)d\xi$$

$$= (\alpha - 1/2\varepsilon)(x + x_\varepsilon) + a_\varepsilon \quad \text{if} \quad x \leq -x_\varepsilon$$

$$F(x) \geq \int_{x_\varepsilon}^x (\beta - 1/2\varepsilon)d\xi + \int_0^{x_\varepsilon} f(\xi)d\xi$$

$$= (\beta - 1/2\varepsilon)(x - x_\varepsilon) + b_\varepsilon \quad \text{if} \quad x \geq x_\varepsilon$$

where a_ε and b_ε are constants. Then

$$\psi(x) \geq 1/2[(\beta - 1/2\varepsilon)(x - x_\varepsilon) + b_\varepsilon$$

$$- (\alpha - 1/2\varepsilon)(-x + x_\varepsilon) - a_\varepsilon]$$

$$= 1/2(\alpha + \beta - \varepsilon)x + c_\varepsilon \quad \text{if} \quad x \geq x_\varepsilon \ ,$$

where c_ε is a constant. Hence $\Psi(x) \to +\infty$ if $x \to +\infty$, and since $\psi(x)$ is odd proposition (a) follows. To investigate (b), choose $x_0 = x_\varepsilon$:

$$\log(1/2x^2) - \int_x^x 1/\xi \ \phi'(\xi)d\xi$$

$$= \log(1/2x^2) + 1/2 \int_{x_\varepsilon}^x 1/\xi[-f(\xi) + f(-\xi)]d\xi$$

$$> \log(1/2x^2) + 1/2 \int_{x_\varepsilon}^x 1/\xi[-\beta + \alpha - 1/2\varepsilon]d\xi$$

$$= 1/2(4 + \alpha - \beta - 1/2\xi) \log x + d_\varepsilon \quad \text{if} \quad x \geq x_\varepsilon \ ,$$

where d_ε is a constant. If $\beta - a < 4$ we may choose ε such that also $4 + \alpha - \beta - 1/2\varepsilon > 0$, and then proposition (b) follows. In

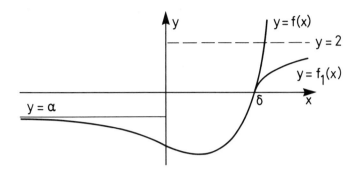

Figure 7.

the domain $-2 < \alpha \le 0$, $-\alpha < \beta < \alpha + 4$ propositions (a) and (b) are
both valid. We can conclude the existence of the above mentioned
curve K. Since the only critical point $(0,0)$ of system (7) is
negatively asymptotically stable (this is shown in section 2;
recall the equivalence between systems (3) and (7)) a closed
trajectory must exist. Hence a **unique** closed trajectory exists,
and it will be positively asymptotically stable from both sides
(as a consequence of uniqueness and a remark in section 2) - and
globally from outside, since K is "globally attracting". This
result holds in particular if $-2 < \alpha \le 0$ and $\beta = 2$. If $-2 < \alpha \le 0$
and $2 < \beta \le +\infty$ the existence of a closed trajectory to system (7)
follows by comparison. Choose namely some C^1-function $f_1(x), x\in R$
as shown on Fig. 7: $f_1(x) \le f(x)$ for all $x\in R$, $\lim_{x\to-\infty} f_1(x) = \alpha$ and
$\lim_{x\to+\infty} f_1(x) = 2$. Then the system

$$\dot{x} = y - F_1(x) , \quad \dot{y} = -x ; \quad (x,y)\in R^2 , \qquad (8)$$

where $F_1(x) = \int_0^x f_1(\xi)d\xi$, $x\in R$, has a (unique) closed trajectory
Γ_0, which in particular is positively asymptotically stable from
outside.
  Consider the field $(\dot{x},\dot{y})$ of system (7) on the trajectories of
system (8) (see Fig. 8):

$$(\dot{x},\dot{y})\cdot(x,y-F_1(x))=x(F(x)-F_1(x))=x \int_0^x(f(\xi)-f_1(\xi))d\xi \ge 0 \text{ for all } x\in R.$$

Hence the curve Γ_0 may play the role of the curve K in the
argument above. We get the same conclusion.
  Summarizing we have shown that in case $-2 < \alpha \le 0$, $\alpha + \beta \ge 0$, $0 < \beta \le + \infty$ there exists a unique closed trajectory Γ_0 to system
(7). Moreover Γ_0 is positively asymptotically orbitally stable
from both sides, and globally from outside. By the equivalence
between systems (3) and (7) the same is valid for system (3) in

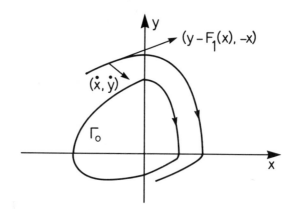

Figure 8.

the phase plane.   Hence the corresponding periodic solution(s) to Eq. (2) is a two-stroke oscillation (see section 4.7).

4.9   Conclusion

We have proved the statement of the introduction under conditions (i) and (ii).   This result can be proved along the same lines as above if the following slightly weaker conditions hold

$$(i')\quad f(x)\ (x-\delta) \geq 0 \text{ if } x \neq \delta$$

$$(ii')\quad f'(x)\ (x-a) \geq 0 \text{ if } x \neq a$$

and if $f(x)$ is not identically zero in any neighbourhood of $x=0$ (in which case $(0,0)$ is a centre for system (3)).

As a specific example we mention the case $f(x) = -\rho(1-x)\exp x$, where ρ is a positive constant.   Here $a=0$, $\delta=1$, $\alpha=0$ and $\beta=+\infty$.   We infer that Eq. (2) then is a two-stroke oscillator.   This example was studied in [2].

It finally should be noticed that if $a=0$ in (ii') - i.e. $f'(x)x\geq0$ if $x \neq 0$ - then the $\tilde{V}$-technique used in section 2 applies also if (i') does not hold.   Hence if $f'(x)x\geq0$ if $x \neq0$ and if $f(x)$ is not identically zero in any neighbourhood of $x=0$, then Eq. (2) has at most one nontrivial periodic solution (up to translations in time).   This classical result goes back to Sansone and Massera (see e.g. [4] p.307), and was proved by a different approach.

References

[1]   K. Munk Andersen and A. Sandqvist (1985).   MAT-REPORT NO. 1985-09 Mathematical Institute, The Technical University of Denmark.

[2]   P. le Corbeiller (1960).   *IRE Trans. Circuit Theory CT-17,* 387.

[3]   A. Sandqvist and K. Munk Andersen (1980).   *IEEE Trans. Circuits and Systems CAS-27,* 1251.

[4]   G. Sansone and R. Conti (1964). *Non-linear Differential Equations,* Macmillan, New York.

[5]   Zeng Xian-wu (1983). *Acta Math. App. Sinica,* 6(1), 13.

05   *R.K. Bullough, D.J. Pilling and J. Timonen*

QUANTUM AND CLASSICAL STATISTICAL MECHANICS OF INTEGRABLE MODEL IN 1+1 DIMENSIONS: SOME RECENT RESULTS

5.1 Introduction

The title of this meeting is "Structure, Coherence and Chaos in Dynamical Systems", and the programme shows the wide range of disciplines to which these concepts are now being applied.   This particular paper is concerned with the **statistical mechanics** of the so-called "integrable models".   Typically these models are model dynamical systems whose equations of motion admit as "structure" the very coherent soliton solutions [1]. We use this presentation to summarize a number of results on the statistical mechanics (SM) of integrable models we have reported recently [2,3].

The example we shall be concerned with most of all is the "sine-Gordon model": this has as classical equation of motion the sine-Gordon equation (s-G)

$$\phi_{xx} - \phi_{tt} = m^2 \sin \phi \tag{1}$$

and $\phi = \phi(x,t)$ is a real field in 1+1 dimensions (one space x and one time dimension t);   we use ϕ_x for $\partial\phi/\partial x$, etc. and m is a "mass".   The s-G is a natural nonlinear generalization of the linear Klein-Gordon equation (K-G)

$$\phi_{xx} - \phi_{tt} = m^2\phi. \tag{2}$$

This has a history in relativistic field theories where $mc\hbar^{-1}$ is a Compton wavelength.   We choose units such that $\hbar = c = 1$.   The K-G has harmonic solutions A exp-i$(\omega t-kx)$ with $\omega(k) = (m^2+k^2)^{1/2}$.   The s-G has the so called "kink", "antikink" and "breather" solutions; but it also has solutions of harmonic type with the same "linearized dispersion relation" $\omega(k)$ as K-G [4].   These solutions are called "phonons" in this paper.   The kink (antikink) solutions are [4]

$$\phi = 4 \tan^{-1} \exp \{\pm m(x - Vt)(1 - V^2)^{-1/2}\} \tag{3}$$

They are coherent and cohere even in collisions for they are solitons [1,4]. In this sense the "breathers" (bound kink-antikink pairs) are also solitons. These coherent motions are very different from the "chaotic" solutions which a damped driven s-G can exhibit [5].

Closely related to s-G is the sinh-Gordon (sinh-G) equation

$$\phi_{xx} - \phi_{tt} = m^2 \sinh \phi. \tag{4}$$

This does not have soliton solutions but like s-G can be solved, as a classical field theory, by the inverse scattering (or spectral transform, ST) method [1,4]. (For sinh-G the ST is self-adjoint so there are no solitons [1,4]). Evidently sinh-G → s-G by $\phi \to i\phi$.

A third equation of interest is the "Nonlinear Schrödinger equation" (NLS)

$$-i\phi_t = \phi_{xx} - 2c \ \phi|\phi|^2 \tag{5}$$

in which ϕ is complex. It has two forms. In (5) as written a real coupling constant c appears (nothing to do with the velocity of light c): when c > 0 the model is the "repulsive" NLS; when c < 0 it is the "attractive" NLS (c=0 is the trivial linear wave equation). The "attractive" NLS has soliton solutions [1,4]: the "repulsive" NLS does not. The s-G and sinh-G relate in exactly the same way as the two NLS equations; moreover (set $\phi \to \sqrt{\gamma_0}\phi$ and let $\gamma_0 \to 0$) both s-G and sinh-G are linear K-G when the coupling constants $\gamma_0 \to 0$.

All four models are **completely integrable Hamiltonian** dynamical systems in the sense of Liouville-Arnold [1,4,6]. Kruskal [7] raises the question of integrability *per se*, but complete Hamiltonian integrability, at least for a finite number of degrees of freedom, is well understood [6]; and problems can arise in this connection only because $\phi(x,t)$ has a continuously infinite number of degrees of freedom. Otherwise a problem is that these integrable systems are so "coherent" one can treat their statistical mechanics only by overtly or covertly admitting perturbations - and even this may not be enough because of KAM theory [8]. The conventional view is, nevertheless, that for the system in the steady state (but coupled to a heat bath) it is sufficient to compute the partition function Z. This paper is therefore concerned with the calculation of both the classical and quantum partition functions Z for s-G, sinh-G, and the two NLS equations. Several results in this connection have been briefly reported recently [2,3,9,10].

Details of the calculations form a highly technical exposition in mathematical physics [2,3,9,10,11]. This is of interest in itself, but the main point of the presentation now is to show that the quantum and classical SM of the integrable models has a certain **universal** character. Nevertheless we also illustrate aspects of the purely technical character of the work for their wider application. Particularly, the methods of functional integration we have developed may be applicable outside the domain of equilibrium thermodynamics [12]. Haken [13] describes an approach to the fundamental, and very general, problem of "far from equilibrium thermodynamics". We think functional integral methods will be helpful here as they are already in the linear regime [14] (and see [15]). In a different direction we have suggested the results reported here may have some bearing on the thermalization of solitons in biological molecules [16].

5.2 **The Background and its Interconnections**

Our recent results on the SM of integrable models [2,3] depend on two new methods. These relate to two other methods already available for the quantum integrable models namely the method of Bethe ansatz (BA) [17] and the quantum inverse method (QIM) [19]. We therefore quote a needed result of the QIM and BA later in this section and use it to establish a number of inter-connections. We also use this section to bring together (i) a number of technical ideas, (ii) a number of apparently different integrable models. Specifics (iii) follow after that. The Section 5.3 summarizes the recent results on s-G, sinh-G and NLS.

(i) Although we restrict to 1+1 dimensions only, the field is already "rich" in that it involves and appeals to a wide range of technical ideas and properties. As mentioned we are concerned with classical and quantum integrability, for the former with "complete Hamiltonian integrability", therefore with action-angle variables, and so with the spectral or scattering data of the ST [1,2,3,4,17,18]: we then need quantum spectral data for the quantum theories [17,19]. We are concerned with solitons but not with their familiar boundary conditions "decaying at infinity" [1,4]. We actually need a thermodynamic limit which is well defined under periodic boundary conditions [17]. We shall show that though decaying b.c.s. at infinity are wrong for this purpose (and lead to wrong results) the action-angle variables calculated under these b.c.s. can still be used. To prove this, however, we need action-angle variables for integrable lattices under periodic b.c.s of period L, the lattice giving a count of degrees of freedom as $L \to \infty$. The role of coupling constants (γ_o for s-G and

sinh-G, c for the two NLS) is important. The problem of functional integration under periodic b.c.s arises and is solved [2,3,11]. Finally we are concerned with the important physical [20] (and mathematical [21]) concept of "bose-fermi equivalence". This relates to "supersymmetry" and string theory [22]. So we see still another wide range of concepts and ideas albeit now of rather technical character. Bose-fermi equivalence in particular seems to be an intrinsic property of solitons even at a classical level [21].

(ii) The reasons for studying the classical and quantum SM of s-G and sinh-G are: the empirical model of the ferromagnetic $CsNiF_3$, an S = 1 spin chain, in the presence of an applied transverse magnetic field B^x, reduces approximately in classical and continuum limit to the classical s-G [23]: its excitations in

61

the temperature range $3^{\circ}K \leq T \leq 11^{\circ}K$ have been studied by neutron scattering [24]. For $B^X = 0$ only, the empirical model relates more exactly to the classical Landau-Lifshitz model which is integrable [4]. The quantum case of the model when $S = 1$ and $B^X = 0$ is solved for its quantum SM and neutron scattering cross-section [25] and involves two spin-1/2 XYZ models [4,25]. The quantum spin-1/2 XYZ model [4] plays a fundamental role in the theory of the classical SM of the 8-vertex model and thus of the 6-vertex and Ising models in 2+0 dimensions (two Euclidean dimensions, no time) [26]. Because the Hamiltonian of spin-1/2 XYZ commutes with the row-by-row transfer matrix of 8-vertex [26], solution of either model solves the other! Moreover the quantum spin-1/2 XYZ model (a spin model) maps by Jordan-Wigner transformation [20] to the quantum massive Thirring model (a massive fermion model) and is then equivalent, by Coleman-Luther correspondence [20], to the **quantum** s-G (a massive boson model). Thus this is an outstanding example of bose-fermi equivalence. It is also an astonishing example of equivalent coherent (=integrable) models in different contexts.

(iii) We come to specifics. The Hamiltonian of classical s-G is [1,4]

$$H[\phi] = \gamma_o^{-1} \int [\tfrac{1}{2}\gamma_o^2 \Pi^2 + \tfrac{1}{2} \phi_x^2 + m^2(1 - \cos \phi)] \ dx, \ \gamma_o > 0 \tag{6}$$

with Poisson bracket $\{\Pi, \phi\} = \delta(x-x')$. Hamilton's equations $\phi_t = \{\phi, H\}$, $\Pi_t = \{\Pi, H\}$ reduce to (1). The coupling constant γ_o "controls the nonlinearity": canonical transformation by $\phi \rightarrow \gamma_o^{1/2}\phi$, $\Pi \rightarrow \gamma_o^{-1/2}\Pi$ shows that, by developing in powers of γ_o, $H[\phi]$ is that for the linear KG, (2), when $\gamma_o \rightarrow 0$. Further under $\phi \rightarrow i\phi$, $\gamma_o \rightarrow -\gamma_o$, $H[\phi]$ becomes that of sinh-G (3) and s-G and sinh-G are related by analytical continuation in γ_o (or $\sqrt{\gamma_o}$) while both equations linearize to K-G as $y_o \rightarrow 0$ as noted.

The classical integrable models in 1+1 dimensions are Hamiltonian integrable and generalize the simpler exemplifications of the Liouville-Arnold theorem for a finite number of degrees of freedom [6]. Their Hamiltonians can be expressed in terms of a continuous infinity of commuting independent action-variables, and motion is on tori and generalized tori [4]. For s-G the new Hamiltonian is [1,2,4,18]

$$H[p] = \sum_{i=1}^{N_k} (M^2+p_i^2)^{1/2} + \sum_{j=1}^{N_{\bar{k}}} (M^2+\bar{p}_j^2)^{1/2} + \sum_{\ell=1}^{N_b} (4M^2\sin^2\Theta_\ell+\hat{p}_\ell^2)^{1/2}$$

$$+ \int_{-\infty}^{\infty} \omega(k)P(k) \ dk \tag{7}$$

where $M = 8m\gamma_o^{-1}$ and $\omega(k) = (m^2+k^2)^{1/2}$. It is important to notice that the part

$$H[p] = \int_{-\infty}^{\infty} \omega(k)P(k)dk \qquad (8)$$

from (7), for which $0 \le P(k) < \infty$, $0 \le Q(k) < 2\pi$, and $\{P(k),Q(k')\}$ = $\delta(k-k')$ [4,18], is also the Hamiltonian of a bunch of oscillators namely of the linear K-G, (2) [4]. This is the free field, and any quantum theory derived from it (by the quantization $[\Pi,\phi] = -i\delta(x-x')$) would be trivial. It is therefore remarkable that (8) is exactly the Hamiltonian of sinh-G in action variables even though sinh-G is nonlinear and involves γ_o ((8) does not involve γ_o). A similar contrast arises in the two cases of the NLS for which $H[p]$ is given by (8) alone with $\omega(k) = k^2$ when $c >$ 0. This is also $H[p]$ for $-i\phi_t = \phi_{xx}$. The problem is resolved [2,3,4,11] by imposing periodic boundary conditions. The action-angle variables just described are all found under "decaying boundary conditions at infinity" [1,2,17,18]. These boundary conditions give a zero density, and therefore the wrong, thermodynamic limit [17]. It is generally assumed that nature **requires** a finite density limit.

The existence of action-angle variables provides a trivial quantization. For example, the breather phase space proves to have [4,18] action-angle variables $4\gamma_o^{-1}\theta_\ell$, Φ_ℓ with $0 \le 4\gamma_o^{-1}\theta_\ell <$ $2\pi\gamma_o^{-1}$, $0 \le \Phi_\ell < 8\pi$: thus the quantisation ($\hbar = 1$) $\oint 4\gamma_o^{-1}\theta_\ell d\Phi_\ell =$ $2\pi n_\ell \Rightarrow \theta_\ell = n_\ell\gamma_o/16$ where n_ℓ is an integer ($0 \le n_\ell \le [8\pi\gamma_o^{-1}] =$ integral part) which agrees with the exact result [19,27] when renormalization is taken into account (whereby $\gamma_o \rightarrow \gamma_o'' = \gamma_o[1 - \gamma_o/8\pi]^{-1}$). For sinh-G, however, only the "oscillators" are available for quantization and $P(k) = n_{ph}(k)$, where $n_{ph}(k) =$ $0,1,2, \ldots$ for each k for bosons and would be $0,1$ only for fermions ("ph" means "phonons", and we use "phonons" to mean these oscillations henceforth). The same picture emerges for the two NLS equations.

The problem to be addressed is the computation of the partition function Z for s-G, sinh-G and the two NLS equations. To define a natural Z we consider the Feynman propagator G which, for a field $\phi(x,t)$, is a functional integral [4]. We take G in Hamiltonian form [4]

$$G(\phi,\phi_o;T) = \int \mathcal{D}\Pi\mathcal{D}\phi \exp i\, S[\phi] \qquad (9a)$$

$$S[\phi] = \int_0^T dt \left[\int dx\, \Pi(x,t)\phi_t(x,t) - H[\phi]\right]$$

$$\phi = \phi(x,T), \quad \phi_0 = \phi(x,0) . \tag{9b}$$

Feynman's usual, Lagrangian, form follows by performing the $\mathcal{D}\Pi$ in (9) when $H[\phi]$ is quadratic in Π [4]. But (9) itself runs over a phase space and has a simpler measure [4]. Then to evaluate the eigenspectrum, for example, one needs the trace of G. This means trajectories are periodic in time, period T. The free energy F is computed from the Wick rotated propagator $\bar{G}$ obtained by $t \to -i\tau$, $0 \le \tau < \beta$, β^{-1} = temperature [4]. Then $Z = \text{Tr } \bar{G}$ and $F = -\beta^{-1}\ell nZ$ and "trajectories" (in τ) are periodic of period β. For the finite density and thermodynamic limit we compute the free energy per unit length as $\lim\limits_{L \infty} FL^{-1}$: L is chosen as a (spatial) period for periodic boundary conditions. Thus we finally work on a space-time torus with periods β and L.

We note that in classical limit ($\hbar \to 0$) [4]

$$Z = \int \mathcal{D}\Pi\mathcal{D}\phi \exp - \beta H[\phi]. \tag{10}$$

A possible calculational procedure is therefore canonically to transform to (simpler) action-angle variables so that

$$Z = \int \mathcal{D}\mu \exp - \beta H[p] \tag{11}$$

in which the proper measure $\mathcal{D}\mu$ is to be determined [2,3,4,11]. The thermodynamic limit now seems to mean we need action-angle variables under periodic b.c.s. These are available for the integrable Toda lattice [28] but not for s-G, sinh-G or the NLS: we have therefore found these in forms sufficient for our purpose [2,3,11] though we then show [2,3] how we can replace them by the corresponding ones under decaying b.c.s. (In passing we note that Opper [29] in effect solves the SM of the Toda lattice by using the coordinates of [28] directly). The quantum case offers the same possibility and the same problems as the classical case. It also shows that, for a real free energy FL^{-1}, we **must** quantize as sketched above, that is for sinh-G for example $\oint PdQ = 2\pi n_{ph}(k)$ (or $2\pi(n_{ph}(k)+1/2)$) for each k. This is made explicit at (33) in 5.3 below.

It is well known that the classical partition function (10) can be evaluated by the Transfer Integral Method (TIM) [4]. However, it has proved both important and instructive to be able to evaluate (11) instead: the action-angle variables under decaying b.c.s are determined by the classical spectral data of the ST. But spectral data are intrinsic to the QIM also and underlie the BA method. We sketch some results in this connection.

The QIM for the repulsive NLS (c > 0) follows the following line: in the classical case first of all one works [19] with the "monodromy" (= "transition" = "transfer") matrix

$$T(\lambda) = \begin{bmatrix} A(\lambda) & B(\lambda) \\ C(\lambda) & D(\lambda) \end{bmatrix}, \quad \lambda \in \mathbb{C} \tag{12}$$

for a **lattice** of small spacing Δ under periodic b.c.s.:

$$T(\lambda) = \prod_{n=(N-1)}^{-N} L(n|\lambda); \quad L(n|\lambda) = \begin{bmatrix} 1-i\lambda\Delta/2, & -i\sqrt{c}\Delta\phi_n^* \\ i\sqrt{c}\phi_n\Delta, & 1+i\lambda\Delta/2 \end{bmatrix} + O(\Delta^2) \tag{13}$$

and $\phi_n \equiv \Delta^{-1} \int_{x_{n-1}}^{x_n} \phi(x)dx$, $\{\phi_n, \phi_m^*\} = i\delta_{nm}\Delta^{-1}$. Then $\tau(\lambda) \equiv \text{Tr}T \equiv$ $A(\lambda)+D(\lambda)$ is a constant of the motion and thus a generator of Hamiltonians. T contains all of the spectral data and $A(\lambda)$, etc. go over to the usual spectral data $a(\lambda)$, $b(\lambda)$ etc. in a careful limit when $L \to \infty$ [4,19]. The quantum case under periodic b.c.s. is then [19]

$$[\phi(x), \phi^\dagger(x)] = \delta(x-x') \iff R(\lambda,\mu)T(\lambda) \otimes T(\mu) = T(\mu) \otimes T(\lambda)R(\lambda,\mu) \tag{14}$$

where $\otimes$ is Kronecker product and a 4×4 matrix, the R-matrix [19], is introduced:

$$R \equiv \begin{bmatrix} f(\mu,\lambda) & 0 & 0 & 0 \\ 0 & g(\mu,\lambda) & 1 & 0 \\ 0 & 1 & g(\mu,\lambda) & 0 \\ 0 & 0 & 0 & f(\mu,\lambda) \end{bmatrix}; \quad \begin{aligned} f(\mu,\lambda) &= 1 - \frac{ic}{\mu-\lambda}, \\ g(\mu,\lambda) &= \frac{ic}{\mu-\lambda}. \end{aligned} \tag{15}$$

Note that the usual quantization $[\phi, \phi^\dagger] = \delta(x-x')$ induces corresponding commutators for the elements of the monodromy matrix T through (14): these elements A, B, etc. are now operators. The result follows by defining T as the ordered operator $T = \overset{\frown}{\prod} L(n|\lambda)$: $L(n|\lambda))$ is an operator because ϕ and $\phi^\dagger (\phi^* \to \phi^\dagger)$ are operators. This operator T acts on a pseudo-vacuum $|0\rangle$: $|0\rangle = \prod_n |0_n\rangle$ in terms of lattice site vacua so that $T(\lambda)|0\rangle = \prod_n L(n|\lambda)|0_n\rangle$ with

$$L(n|\lambda)|0\rangle_n = \begin{bmatrix} 1 - i\lambda\Delta/2 & -i\sqrt{c}\ \phi_n^\dagger\Delta \\ 0 & 1+i\lambda\Delta/2 \end{bmatrix} |0\rangle_n. \tag{16}$$

Then $C(\lambda)|0> = 0$, $A(\lambda)|0> = \alpha(\lambda)|0>$, $D(\lambda)|0> = \delta(\lambda)|0>$ and (for $(2N+1)\Delta = L$ and $\Delta \to 0$)

$$\alpha(\lambda) \sim \exp -i\lambda L/2 , \quad \delta(\lambda) \sim \exp i \lambda L/2 . \qquad (17)$$

One then proves that $\prod_{j=(N-1)}^{-N} B(\lambda_j)|0>$ is an eigenstate of $A(\mu)+D(\mu)$

and a necessary and sufficient condition for this is that the values of λ are constrained to values λ_n given by (j,n integers)

$$\lambda_n L = 2\pi n - \sum_{j \neq n} \Delta(\lambda_n, \lambda_j) ,$$

$$\Delta(\lambda, \lambda') = -2 \tan^{-1} [c/(\lambda - \lambda')] . \qquad (18)$$

One shows too that $\hat{Q} \equiv \int \phi^{\dagger}(x)\phi(x)dx$, $\hat{P} \equiv i\int \phi^{\dagger}(x)\phi_x(x)dx$ and $\hat{H} \equiv \int (\phi^{\dagger}_x \phi_x + c\phi^{\dagger}\phi^{\dagger}\phi\phi)dx$, the number, momentum and Hamiltonian operators, have eigenvalues $Q_N = 2N$, $P_N = \sum_{j=-N}^{(N-1)} \lambda_j$, and $H_N = \sum_{j=-N}^{(N-1)} \lambda_j^2$ respectively, and all of these results coincide with those found by the method of Bethe ansatz (BA) for the repulsive NLS [17].

The crucial result is (18): it is a consequence of periodic b.c.s. Indeed, for a bounded sum $\sum_{j \neq n}$, (18) is $\lambda_n = 2\pi n L^{-1}$ for $L \to \infty$, so the λ_n are the free modes $k_n = 2\pi n L^{-1}$ such that $k_n \to k$, the modes of k-space, as $L \to \infty$. We find, nevertheless, that coupling of these modes through the phase shifts Δ in the $\sum_{j \neq n}$ in (18) changes this picture in thermodynamic limit (where the sum is actually O(L)).

It is important to notice that we interpret $\Delta(\lambda, \lambda')$ as the **smooth** branch $\Delta_f(\lambda, \lambda')$ of the $\tan^{-1}$ function such that $\Delta_f \to 0$, $\lambda - \lambda' \to +\infty$, and $- 2\pi$, $\lambda - \lambda' \to -\infty$. This induces a description in terms of fermions [17]. We find [2,3,4] there is a boson description also which uses $\Delta_b = \Delta_f + 2\pi\theta(\lambda'-\lambda)$ (where θ is the step function) for Δ: Δ_b is the true 2-body S-matrix phase shift [18].

In a short but remarkable paper Yang and Yang [30] calculated FL^{-1} for the repulsive quantum NLS: their model consisted of N **bosons** on a line with repulsive δ-function interactions of strength c. In first quantization the quantum mechanical problem is

$$\left\{ -\sum_{j=1}^{N} \frac{\partial^2}{\delta x_i^2} + 2c \sum_{i>j} \delta(x_i - x_j) \right\} \psi = E\psi \tag{19}$$

but in second quantization the problem is exactly the quantum NLS for $c > 0$ and commutation $[\phi, \phi^\dagger] = \delta(x-x')$ as at (14). Using (18) with $\Delta \equiv \Delta_f$ Yang and Yang found the elegant result

$$\lim_{L \to \infty} FL^{-1} = \mu\bar{n} - (2\pi\beta)^{-1} \int_{\infty}^{\infty} \ell n[1+e^{-\beta\bar{\epsilon}(k)}]dk \tag{20a}$$

$$\bar{\epsilon}(k) = \omega(k) - \mu + (2\pi\beta)^{-1} \int_{-\infty}^{\infty} \{d\Delta_f(k,k')/dk'\} \ell n[1+e^{-\beta\bar{\epsilon}(k')}]dk' \tag{20b}$$

where $\bar{n} = \lim_{N \to \infty} NL^{-1}$, μ is a chemical potential, and $\omega(k) \equiv k^2$. If we put the fermi oscillator contributions in (20a) in bose form (see below), (20a) is the result for $c = 0$, i.e. for the free field $-i\phi_t = \phi_{xx}$: but (20b) shows the modes are not free. The main result of our recent work [2,3,11] is to show that expressions like (20) govern the free energies of all of the quantum integrable models: moreover, though (20) is in fermi form, there is an equivalent bose form; and this bose form has the natural classical limit which is the result too for the Maxwell-Boltzmann statistics of the classical integrable models. The results apply whether the classical model has soliton solutions or not: thus quantum s-G has a form like (20) for FL^{-1} and classical s-G does so too as 5.3 makes plain.

A result like (20) for quantum s-G was found some five years ago by the BA method: such expressions were given for spin-1/2 XYZ [31], and in 1982 the mapping from spin-1/2 XYZ to quantum s-G was used [32,33] to find FL^{-1} for quantum s-G. Another route is from quantum MTM [34,35]. One of our recent results [3] is to derive the form (20) for sinh-G with $\omega(k) = (m^2+k^2)^{1/2}$ and a different expression for Δ_f: it is quoted in bose form at (31) in 5.3. The BA result for quantum s-G is considerably more complicated than (20) despite the connection of classical s-G to sinh-G. This is because of the many more coherent quantum excitations in the s-G case. We summarize the simplest form of the result for s-G next.

For quantum s-G it proves convenient [31,32,33,34,35] to work with coupling constant $\mu = \pi(1 - p_o^{-1})$: $p_o = 8\pi\gamma_o^{-1}$ in terms of γ_o. It then emerges that the analogue of (20b) in the BA description involves an actual infinity of coupled integral equations when μ (γ_o) $\times$ π^{-1} is irrational. However, when p_o = n = integer these reduce to n−1 coupled integral equations [32,34,35,36] involving n−1 energies $\tilde{E}_j(x)$, j = 1, ..., n−2 and j = s instead of the single $\bar{\epsilon}$ in (20b). Precisely for p_o = n, these are [34,35,36]

$$\tilde{E}_j(x) = E_j(x) + (2\pi\beta)^{-1} \sum_{k=1}^{n-2} \int_{-\infty}^{\infty} dx' \frac{d}{dx'} B_{jk}(x'-x) \ln[1+e^{-\beta\tilde{E}_k}]$$

$$+ 2(2\pi\beta)^{-1} \int_{-\infty}^{\infty} dx' \frac{d}{dx'} B_{js}(x'-x) \ln[1+e^{-\beta\tilde{E}_s}]$$

$$\tilde{E}_s(x) = E_s(x) + (2\pi\beta)^{-1} \sum_{k=1}^{n-2} \int_{-\infty}^{\infty} dx' \frac{d}{dx'} B_{ks}(x'-x) \ln[1+e^{-\beta\tilde{E}_k}]$$

$$+ 2(2\pi\beta)^{-1} \int_{-\infty}^{\infty} dx' \frac{d}{dx} B_{ss}(x'-x) \ln[1+e^{-\beta\tilde{E}_s}] \quad (21)$$

in which $E_j \equiv M_j \cosh x$, $M_j \equiv 2M \sin [\pi j/2(n-1)]$, $E_s(x) \equiv M \cosh x$ (relativistic covariance is exploited by using rapidities x). The free energy is

$$FL^{-1} = -(2\pi\beta)^{-1} \sum_{j=1}^{n-2} \int_{-\infty}^{\infty} dx \, E_j(x) \ln[1+e^{-\beta\tilde{E}_s}]$$

$$-2(2\pi\beta)^{-1} \int_{-\infty}^{\infty} dx E_s(x) \ln[1+e^{-\beta\tilde{E}_s}] \quad . \quad (22)$$

There are now n−2 distinct quantum excitations, tentatively identified as quantum "breathers", and two, each with energy $\tilde{E}_s$, identified as quantum solitons, namely quantum kink and antikink. The phase shifts are complicated but are shown [35] to coincide with the computed S-matrix shifts [37]. For example, the "breather-breather" shifts are

$$B_{jk}(x) = \theta(x, j+k) + \theta(x, |j-k|) + \sum_{\ell=1}^{\min(j,k)-1} \theta(x, j+k-2\ell)$$

68

$$\theta(x,j) = -i \ \ell n \left\{ \frac{\sinh x - i \ \sin(\pi j/2(n-1))}{\sinh x + i \ \sin(\pi j/2(n-1))} \right\} \qquad (23)$$

and the B_{js} and B_{ss} are "breather-soliton" and "soliton-soliton" shifts [36]. Since it is proved [38] that F is analytic in μ (in γ_o) it is remarkable that the system (21) becomes an infinite number of equations when p_o is irrational. This case which involves further excitations called "long strings" [10,35] has still to be treated but a simplification arises when $p_o = n+\epsilon$ ($\epsilon =$ infinitesimal > 0). In this case the infinite system reduces to n coupled equations in $\tilde{E}_j$, $1 \le j \le n$ and the definition of $\tilde{E}_s$ through

$$[1+e^{-\beta \tilde{E}_s}] = [1+e^{-\beta \tilde{E}_{n-1}}][1+e^{-\beta \tilde{E}_n}]^{1/2} \qquad (24)$$

reduces this system to the n-1 equations (21) precisely [10,35].

Analyticity in μ can also be checked by taking the classical limit of the system (21) with (22). We summarize two recent calculations [9,10] in 5.3 next and remark here that if, instead, we evaluate Z given by (10) (not (11)) by the Transfer Integral Method [4] we now know [2,3] that the complete low temperature expansion is the asymptotic expansion

$$FL^{-1} = (F^{(1)} + F^{(2)} + F^{(3)} + \ldots) + F_{KG} \qquad (25)$$

$$F^{(1)} = -\beta^{-1}m \left[\frac{8}{t\pi}\right]^{1/2} e^{-1/t} \left[1 - \frac{7}{8}t - \frac{59}{128}t^2 - \frac{897}{1024}t^3 \ldots\right]$$

$$-\beta^{-1}m \left[\frac{1}{4}t + \frac{1}{8}t^2 + \frac{3}{16}t^3 + \frac{53}{128}t^4 + \ldots\ldots\right] \qquad (26a)$$

$$F^{(2)} = \frac{8m}{\pi} Me^{-2t}\left\{\ell n \frac{4C}{t} - \frac{5}{4}t[\ell n \frac{4C}{t} +1] - \frac{t^2}{32}[13 \ \ell n \frac{4C}{t} +2]+\ldots\right\}$$

$$\qquad (26b)$$

$$F^{(q)} = O[e^{-q/t}] \ , \ q = 3,4,\ldots \qquad (26c)$$

where $M \equiv 8m\gamma_o^{-1}$, and $t \equiv (M\beta)^{-1}$; and C is Euler's constant. Also $F_{KG} = \beta^{-1}a^{-1}(\ell n\beta a^{-1} + \frac{1}{2}ma)$ and uses the **lattice** dispersion relation $\omega_n^2 = m^2 + (\sin \frac{1}{2}k_n a/\frac{1}{2}a)^2$; $k_n = 2\pi n \ L^{-1}$, $-N \le n \le N$ and the lattice spacing is a. On the other hand we get for sinh-G by the same method [3,39,41]

$$FL^{-1} = + \beta^{-1} m [\frac{1}{4} t - \frac{1}{8} t^2 + \frac{3}{16} t^3 - \frac{53}{128} t^4 + \ldots] + F_{KG}$$

$$(27)$$

which is the analytic continuation of (25) with (26) in γ_o: note that soliton (kink and antikink) contributions no longer arise. A nice question is now whether the series in (27) and (26a) are really due to phonon contributions. We show (5.3) that the series in (26a) arises from (21) with (22) as breather-breather contributions. But we know from (8) that the series can only arise in (27) for sinh-G from "phonon" contributions – which it does (5.3).

5.3 **The New Results**

(i)   Perhaps the main result of the new work [2,3] is that the form of (20) for FL^{-1} exemplifies the whole class of quantum integrable models whose classical integrable forms do not have soliton solutions. All of these classical integrable models have the expression (8) for H[p] with different $\omega(k)$.

(ii)   There are, however, a fermion description like (20), an equivalent boson description, and a classical Maxwell-Boltzmann particle description all of the same general form as (20). Thus in the classical case of sinh-G we use classical Floquet theory [4,11,28] and a sinh-Gordon lattice, spacing a, [3] under periodic b.c.s. to show that, for period $L < \infty$, allowed modes $\tilde{k}_n$ are

$$\tilde{k}_n = k_n - \frac{1}{L} \sum_{m \neq n} \Delta_c(\tilde{k}_n, \tilde{k}_m) P_m \qquad (28)$$

with $\Delta_c = -\frac{1}{4}\gamma_o m^2 / (k\omega(k') - k'\omega(k))$, while

$$H[p] = \sum_n \omega(k_n) P_n + O(L^{-1}) . \qquad (29)$$

Since $\omega(k) = (m^2 + k^2)^{1/2} + O(a)$, $k_n = 2\pi n L^{-1}$ and $P_n = 2\pi L^{-1} P(\tilde{k}_n)$ $\leftrightarrow dkP(k)$ as $L \to \infty$, (29) describes oscillators coupled by (28) for $L < \infty$. Evidently (28) generalizes (18) to the classical problem, while (29) goes to (8) for $L \to \infty$. Note how the action variables for decaying b.c.s. at infinity are contained in (28) and (29) even though the b.c.s. are periodic b.c.s. In the quantum case

$$\Delta_c \to \Delta_b = -2 \tan^{-1} \{\sin (\frac{1}{8}\gamma_o'')/(k\omega(k') - k'\omega(k))\} \qquad (30)$$

and $\oint P_m dQ_m = 2\pi n_m \Rightarrow P_m = 0,1,2, \ldots$. The $\tan^{-1}$ in (30) has a jump at $k = k'$ so Δ_b is the bose shift. In fermion description $P_m = 0,1$ and $\Delta_c \to \Delta_f = \Delta_b + 2\pi\theta(k'-k)$ so (28) is identically (18) (except applied to sinh-G not NLS).

(iii) From the results (ii) two methods of further calculation are now developed [2,3]. The method of "generalised BA" [2,3] is to start from (28), quantum or classical, then as in [30] define an entropy S, and a free energy $FL^{-1} = (E - \beta^{-1}S)L^{-1}$ and then minimize FL^{-1}. The entropy S is written down for fermions, bosons or M-B particles depending on the choice for (28). The energy E is given by (29) for $L \to \infty$ by the steps [12] $L^{-1}(2\pi L^{-1}P(\tilde{k}_n))$ $\leftrightarrow L^{-1}dkP(k) \leftrightarrow \rho(k)dk$ and $EL^{-1} = \int_{-\infty}^{\infty}\omega(k)\rho(k)dk$: thus $\rho(k)$ is the number of quantum particles per unit length in the quantum cases and is the number of M-B particles per unit length in the classical case even though P(k) is an action variable. The extra factor L^{-1} involved guarantees a proper finite density thermodynamic limit.

The result (in boson description) is [3]

$$\lim_{L\to\infty} FL^{-1} = (2\pi\beta)^{-1} \int_{-\infty}^{\infty} \ln[1-e^{-\beta\epsilon(k)}]dk$$

$$\epsilon(k) = \omega(k) + (2\pi\beta)^{-1} \int_{-\infty}^{\infty} \frac{d}{dk} \Delta_b(k,k') \ln[1-e^{-\beta\epsilon(k')}]dk'$$

$$(31)$$

and the equivalent fermion description follows directly from this by $(1 + e^{-\beta\epsilon(k)}) = (1 - e^{-\beta\epsilon(k)})^{-1}$ and $\Delta_b \to \Delta_f$. Alternatively fermions can be used to start with at (28). The classical result uses M-B particles and Δ_c as above and the result is

$$\lim_{L\to\infty} FL^{-1} = (2\pi\beta)^{-1} \int_{-\infty}^{\infty} \ln\{\beta\epsilon(k)\} dk$$

$$\epsilon(k) = \omega(k) + (2\pi\beta)^{-1} \int_{-\infty}^{\infty} \frac{d}{dk} \Delta_c(k,k') \ln\{\beta\epsilon(k')\} dk' \quad (32)$$

which is also the classical limit $\Delta_b \to \Delta_c$, $\ln(1-e^{-\beta\epsilon(k)}) \to$ $\ln\{\beta\epsilon(k)\}$ of (31). It is then noteworthy that iteration of (32) yields the TIM result (27) exactly.

(iv)   The second method, the "method of functional integration", again starts from (28) and (29) which are iterated to evaluate the functional integral. It then turns out that the constraint (28) is just what is needed to make the development of $-L^{-1}\beta^{-1} \ln Z$ the iteration of the system (32). This is the classical SM analysis and the quantum cases simply work from the quantum forms of (28), bose or fermi, and regain (31) (bose case) or its fermi equivalent. Because a condition like (28) is the crucial feature of both the "generalized BA" (iii) and the "method of functional integration" it is plain that the role of the functional integral itself is no more than to minimize FL^{-1}. It is therefore remarkable that in the functional integral method one averages out the P_n and Q_n while in the generalized BA $\rho(k_n) = P_n L^{-1}$ is the particle density. Seen this way, that both methods reach the same integral equations is then rather striking.

It is also remarkable that the quantum form of (11) to be used is ($h = 2\pi$)

$$Z = \int \mathcal{D}\mu \, \exp \int_0^\beta d\tau \, \{i\int P(k)Q_\tau(k)dk - H[p]\} \qquad (33)$$

with $\mathcal{D}\mu = \lim_{N \to \infty} \prod_{m=-N}^{(N-1)} dP_m \, dQ_m \, (2\pi)^{-1}$. Then, for real $\ln Z$, we **must** have $\oint P_m \, dQ_m = 2\pi n_m$ ($n_m = 0,1,2, \ldots$ for bosons, or perhaps $2\pi(n_m + 1/2)$), while the real quantum mechanics resides at the quantum form of (28) where Δ_c is Δ_b (or Δ_f), the 2-body S-matrix phase shifts. This structure is rather different from the simple form of Feynman's prescription for quantum Z which is $Z = \int \mathcal{D}\phi \, \exp$ i $S[\phi]$ and $S[\phi]$ is simply the Lagrangian form of the classical action.

(v) Both methods, generalized BA and method of functional integration, extend naturally to the case where the classical model has solitons. Thus in the case of s-G Floquet theory shows [2,11] that (28) is replaced by

$$L\tilde{k}_n = Lk_n - \sum_m' \Delta(\tilde{k}_n, \tilde{k}_m)P_m + \sum_i \Delta_k(\tilde{k}_n, \tilde{p}_i) + \sum_j \Delta_{\bar{k}}(\tilde{k}_n, \tilde{\tilde{p}}_j)$$

$$L\tilde{p}_k = Lp_k - \sum_m \Delta_k(\tilde{k}_m, \tilde{p}_k)P_m - \sum_i' \Delta_{kk}(\tilde{p}_k, \tilde{p}_i) - \sum_j \Delta_{k\bar{k}}(\tilde{p}_k, \tilde{\tilde{p}}_j)$$

$$\qquad (34)$$

and a similar expression for $L\tilde{\tilde{p}}_\ell$: the Σ' omits the index on the left sides. All of the classical phase shifts Δ_{kk} (kink-kink) etc. are calculated from ST theory [4,11] just as Δ_c in (28) is [3,4,39,41].

Again it is action-variables under decaying b.c.s. which are written into (34) but an interesting point is that we have replaced (7) for H[p] by $\tilde{H}[p] = \Sigma_i(M^2 + \tilde{p}_i^2)^{1/2} + \Sigma_j(M^2 + \tilde{\tilde{p}}_j^2)^{1/2} + \Sigma_n \omega(\tilde{k}_n)P_n$. This feature is due to the duplication of roles played by 'phonons' and 'breathers' in the s-G theory. Note that the quantum expressions (20) involved (so-called) "breathers" but not "phonons"; $\tilde{H}[p]$ involves phonons but not breathers.

By using the classical shifts in (34) and bose statistics with quantization conditions $\oint P_m dQ_m = 2\pi n_m$, but fermi statistics for the kinks and antikinks one reaches the **semi**-classical system [2, 10, 35]

$$FL^{-1} = (2\pi\beta)^{-1} \int_{-\infty}^{\infty} dx\, \omega(x)\, \ell n[1-e^{-\beta\tilde{\epsilon}(x)}] -$$

$$- 2(2\pi\beta)^{-1} \int_{-\infty}^{\infty} dx E_s(x)\, \ell n[1+e^{-\beta\tilde{\epsilon}_s(x)}]$$

$$\tilde{\epsilon}(x) = \omega(x) + \gamma_o(8\pi\beta)^{-1} \int_{-\infty}^{\infty} dx'\, \frac{d}{dx}\, \Delta(x-x')\, \ell n\, [1-e^{-\beta\tilde{\epsilon}(x')}]$$

$$+ 2(\pi\beta)^{-1} \int_{-\infty}^{\infty} dx'\, \{\cosh(x'-x)\}^{-1}\, \ell n[1+e^{-\beta\tilde{\epsilon}_s(x')}]$$

$$\tilde{E}_s(x) = E_s(x) - (\pi\beta)^{-1} \int_{-\infty}^{\infty} dx'\, \{\cosh(x'-x)\}^{-1}\, \ell n[1-e^{-\beta\tilde{\epsilon}(x')}]$$

$$- 8(\pi\gamma_o\beta)^{-1} \int_{-\infty}^{\infty} dx'\, \frac{d}{dx}\, \Delta_{kk}(x'-x)\, \ell n[1+e^{-\beta\tilde{\epsilon}_s(x')}]$$

$$(35)$$

where $\omega(x) = m\cosh x$, $E_s(x) = M\cosh x$, $\Delta(x) = (\sinh x)^{-1}$; also $\frac{d}{dx}\Delta_{kk}(x) = \ell n\{\cosh x + 1\} - \ell n\{\cosh x - 1\}$. The classical expressions [2] can now be derived for M-B particles ("kinks", "anti-kinks" and "phonons") or by $\ell n(1-e^{-\beta\tilde{\epsilon}}) \to \ell n\beta\tilde{\epsilon}$, $\ell n(1+e^{-\beta\tilde{E}_s}) \to e^{-\beta\tilde{E}_s}$ in (35). Iteration of this classical system then yields (25) with (26) exactly, in agreement with the TIM. Again both the generalized BA method and the functional integral method yield the same results essentially as in the case of sinh-G. But it is also possible to show from the classical functional integral analysis [2] why the so-called "ideal gas phenomenology" [40] works as well as it does and why it failed to give some of the correct "dressing" terms.

(vi) The system of integral equations (21) found by the usual BA method indicates the scale of complexity involved in the case of quantum s-G. We note the relatively simple form of expressions (35) in semiclassical limit where (apparently) only one bose field (the phonons) and two fermion fields (kink plus antikink) are involved. It is therefore striking that, in taking the classical limit of (21) with (22) we first define a semiclassical limit of the phase shifts and then (with [9]) find [10,35] a transformation of the n-2 fermions with energies $\tilde{E}_j(1 \leq j \leq n-2)$ to the single bose field. Then (35) follows exactly. After that the reduction of (35) to classical form and the iteration of that classical form yields (25) with (26) exactly as described already. This way we show [10] that (21) with (22) reduces to (25) with (26) exactly in classical limit in agreement with the TIM for s-G with Z evaluated from (10). This calculation finally agrees completely with that by Chen et al [9,36].

(vii) Everything we have described above for sinh-G has its equivalent in the case of the repulsive NLS model: the changes are only that $\omega(k) = k^2$ and $\Delta = -2\tan^{-1}\{c/(k-k')\}$, bose or fermi, or $\Delta_c = -2c/(k-k')$ [3,12,39].

(viii) There are wholly equivalent forms of the s-G theory for "attractive NLS". However, the quantum attractive NLS has the special difficulty that there is no stable ground state. This seems to mean that a relativistically covariant model (namely s-G) is the proper physical model to use rather than the attractive NLS.

(ix) Our functional integral method provides the first rigorous justification of the *ad hoc* but pioneering methods of Dashen et al (DHN) [27] for quantum s-G.

All of the considerable technical detail of these various results is being reported elsewhere [2,3,10,11,35,39]. The results as given in (i) ... (ix) above indicate the 'universal' forms of the free energy FL^{-1} for "all" of the integrable models. Note that at zero temperature ($\beta^{-1} = 0$) these results for the quantum SM become the quantum theories, simply, of these models. Thus, e.g. those for quantum s-G and sinh-G fit into the scheme connecting s-G to the 8-vertex model described in 5.2 (though precisely where sinh-G fits here is open). It is open too whether for $\beta^{-1} > 0$ there is a still grander scheme of connected models (note however that, renormalized against K-G, only the renormalized coupling constant γ_o/β appears in results like (25) and (27)) while still other open questions are:- Extension of the theory to the 2+1 dimensional integrable models: the classical integrable models like KP, DS and 3-wave interaction equations [42] are completely integrable but without any phase shifts [42] - so would seem to yield only free field quantum theories. On the other hand classical KP has a fundamental bose-fermi equivalence exhibited in representations of an underlying Kac-Moody-Lie algebra $g\ell(\infty)$ [21,22]. It is interesting that even in 1+1 dimensions the bose-fermi equivalence of the descriptions of FL^{-1} listed in 5.3 is not obviously quite the same as the bose-fermi equivalence of MTM and quantum s-G described in 5.2 The significance of this is also open and needs further exploration. The problem of the correlation functions is still to be explored within the present framework.

References

[1] For example: Solitons, eds. R.K. Bullough and P.J. Caudrey (1980). *Topics in Current Physics 17*, Springer-Verlag,

Heidelberg; M.J. Ablowitz and H. Segur (1981). *Solitons and the Inverse Scattering Transform,* SIAM Studies in Appl. Maths., SIAM, Philadelphia.

[2]   J. Timonen, M. Stirland, D.J. Pilling, Yi Cheng and R.K. Bullough (1986). *Phys. Rev. Lett.* **56**, 2233.

[3]   R.K. Bullough, D.J. Pilling and J. Timonen (1986). *J. Phys. A: Math. Gen.* **19**, 955.

[4]   R.K. Bullough, D.J. Pilling and J. Timonen (1985) in *Nonlinear Phenomena in Physics,* 70-102, 103-128, and references therein, ed. F. Claro, Springer-Verlag, Heidelberg.

[5]   Refer to McLaughlin's paper this Meeting; also in general terms see P. Cvitanović's paper this Meeting; and P. Cvitanović: Universality in Chaos (Adam Hilger Ltd., Bristol, 1984); and R.K. Bullough (1984) in *Nonlinear Electrodynamics in Biological Systems,* eds. W. Ross Adey and Albert F. Lawrence, Plenum, New York.

[6]   V.I. Arnold (1978). *Mathematical Methods of Classical Mechanics,* Springer-Verlag, Heidelberg.

[7]   M.D. Kruskal, this Meeting.

[8]   V.I. Arnold and A. Avez (1968). *Ergodic problems of classical mechanics,* W.A. Benjamin, New York; J. Moser (1973). *Stable and random motions in dynamical systems,* Ann. Math. Studies 77, Princeton UP; J. Ford (1975) in *Fundamental Problems in Statistical Mechanics III,* ed. E.G.D. Cohen, North Holland, Amsterdam; M.J. Berry (1978). *Am Inst. Phys. Conf. Proc.* **46**, 16.

[9]   Niu Niu Chen, M.D. Johnson and M. Fowler (1986). *Phys. Rev. Lett.* **56**, 907.

[10] J. Timonen, R.K. Bullough and D.J. Pilling (1986). *Phys. Rev. B* **34**, 6525.

[11] R.K. Bullough, Yi Cheng, D.J. Pilling and J. Timonen, to be published.

[12] Equilibrium theory is extended to a dynamical theory in Quantum Kinetic Heisenberg models, in *Coherence, Cooperation and Fluctuations* (1986), eds. F. Haake, L.M. Narducci and D.F. Walls, CUP, Cambridge, pp. 18-34.

[13] H. Haken, this Meeting.

[14] L. Onsager and S. Machlup (1953). *Phys. Rev.* **91**, 1505; S. Machlup and L. Onsager (1953). *Phys. Rev.* 91, 1512.

[15] H. Haken (1976). *Z. Physik* **B24**, 327; B.H. Lavenda (1977). *Riv. Nuovo Cim.* 7, 229; R. Graham (1980) in *Functional Integration,* eds. J.P. Antoine and E. Trapegui, 263-280 and references, Plenum, New York.

[16] R.K. Bullough (1985) in *The Living State-II,* 458-466, ed. R.K. Mishra, World Scientific, Singapore.

[17] H.B. Thacker (1981). *Rev. Mod. Phys.* **53**, 253 and references.

[18] L.A. Takhtadzhyan and L.D. Faddeev (1974). *Teor. Mat. Fiz.* 21, 160; R.K. Dodd and R.K. Bullough (1979). *Phys. Scr.* 20, 514.

[19] E.K. Sklyanin, L.A. Takhtadzhyan and L.D. Faddeev (1979). *Teor. Mat. Fiz.* 40, 194; L.D. Faddeev (1983) in *Recent Advances in Field Theory and Statistical Mechanics,* eds. R.

Stora and J.B. Zuber, 561-608, North Holland, Amsterdam.

[20] A.H. Luther, in *Solitons* Ref. [7] pp. 355-372.

[21] V.G. Kac and D.H. Peterson, *Lectures on the Infinite Wedge Representation and the MKP hierarchy*, given at the Montreal Summer School, August 1985; also *Proc. Nat. Acad. Sci. USA* 78, 3308, 1981. Paper by M. Jimbo and T. Miwa in Ref. [22], p. 275.

[22] J. Lepowsky, S. Mandelstam and I.M. Singer (eds.) (1985). *Vertex Operators in Mathematics and Physics,* Springer-Verlag, Heidelberg.

[23] H.J. Mikeska (1978). *J. Phys.* C11, **L29**; J. Timonen and R.K. Bullough (1981). *Phys. Lett.* **82A**, 183.

[24] J.K. Kjems and M. Steiner (1978). *Phys. Rev. Lett.* **41**, 1137.

[25] J. Timonen and A. Luther (1985). *J. Phys.* **C18**, 1439.

[26] R.J. Baxter (1982). *Exactly Solved Models in Statistical Mechanics*, Academic Press, New York.

[27] R.F. Dashen, B. Hasslacher and A. Neveu (1975). *Phys. Rev.* **D11**, 3424.

[28] H. Flaschka and D.W. McLaughlin (1976). *Prog. Theor. Phys.* **55**, 438.

[29] M. Opper, *Solution of a random chain problem - an approach using canonical variables of an integrable system*, see this Meeting and submitted to *J. Phys. A*; also see *Phys. Lett.* **112A**, 201, 1985.

[30] C.N. Yang and C.P. Yang (1969). *J. Math. Phys.* **10**, 1115.

[31] M. Takahashi and M. Suzuki (1972). *Prog. Theor. Phys.* **48**, 2187.

[32] M. Fowler and X. Zotos (1982). *Phys. Rev.* **B24**, 2634; **B25**, 5806.

[33] M. Imada, K. Hida and M. Ishikawa (1982). *Phys. Lett.* **90A**, 79; (1983). *J. Phys.* **C16**, 35.

[34] S.G. Chung and Y.-C. Chang (1983). *Phys. Lett* **93A**, 230.

[35] J. Timonen, D.J. Pilling and R.K. Bullough, to be published.

[36] M.D. Johnson, N.N. Chen and M. Fowler, *Phys. Rev.* B, to be published.

[37] V.E. Korepin (1979). *Teor. Mat. Fiz.* **41**, 169.

[38] H. Araki (1969). *Commun. Math. Phys.* **14**, 120.

[39] R.K. Bullough, Yi Cheng, D.J. Pilling, M. Stirland and J. Timonen, to be published.

[40] For example, A.R. Bishop (1983) in *Physics in One Dimension*, eds. J. Bernasconi and T. Schneider, Springer-Verlag, Heidelberg; N. Theodorakopoulos (1982). *Z. Phys.* **B46**, 367 and (1984). *Phys. Rev.* **B30**, 4071.

[41] R.K. Bullough, D.J. Pilling, M. Stirland and J. Timonen (1986). *Physica* **18D**, 368.

[42] Z. Jiang, R.K. Bullough and S.V. Manakov (1986). *Physica* **18D**, 305, and to be published.

TRANSPORT AND FLUCTUATIONS IN THE DRIVEN AND DAMPED SINE-GORDON CHAIN

6.1 Introduction

In this paper, we review some key features of transport in the driven and damped sine-Gordon chain.   We consider a chain of particles separated from each other in the x direction, and subject to displacement in an orthogonal direction θ.   The particles are in a potential

$$V = V_o(1 - \cos(\theta)) - F\theta \qquad (1)$$

consisting of a periodic sinusoidal term and a driving force potential − Fθ.   We assume that the coupling between adjacent particles is strong enough such that neighbouring particles remain close to each other in the θ direction.   The equation governing the motion of the chain is

$$I\partial^2\theta/\partial t^2 + \Gamma\gamma\theta/\partial t = -\partial V/\partial\theta + \kappa\partial^2\theta/\partial x^2 + \zeta. \qquad (2)$$

Here I is the inertia of the particles and κ is the coupling constant of adjacent particles.   Particles experience a frictional force proportional to their velocity with a damping constant γ and experience a thermal fluctuation force ζ, with $\langle\zeta\rangle$ = 0 and

$$\langle\zeta(x,t)\zeta(x',t')\rangle = 2\gamma kT\delta(t - t')\delta(x - x'). \qquad (3)$$

We are interested in two questions: first we discuss the average speed $\langle\partial\theta/\partial t\rangle$ of a particle in the chain as a function of the applied field and temperature.   This question has been of interest in the theory of dislocations for more than three decades [1-3].   Stimulated by more recent work [4], we put forward an answer [5-7] closely related to concepts already developed in the dislocation literature.   The second problem we shall address is fluctuations Δθ(x,t) = θ(x,t) − $\langle\theta\rangle$ away from the ensemble average motion of the carriers.   This problem we believe is of more recent interest [8,9] and our contribution [7,10-13] to it has been to apply the insights gained in the theory of dislocations.   We will mostly, but not exclusively, focus on the limit of high damping,

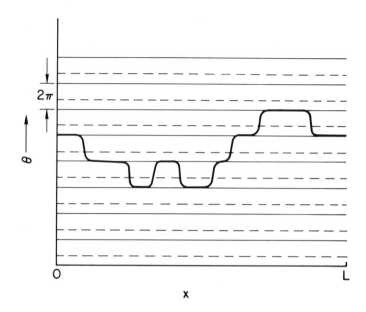

Figure 1.   Typical low temperature configuration of the displacement field θ(x,t) at a given instant of time. The solid lines correspond to local minima of the potential, and thin broken lines correspond to local maxima of the potential given by Eq. (1).

such that the inertial term in Eq. (2) can be neglected compared to the frictional term.

To pose these questions in a more explicit form, we consider Fig. 1.   Figure 1 shows an instantaneous configuration of the chain at low temperatures and for a driving force $|F| < V_o$.   For driving forces in this range, the potential V, Eq. (1), exhibits a sequence of local minima separated by local maxima.   Following the dislocation literature [1-3], we call a local maximum a Peierls hill and a local minimum a Peierls valley.   At low temperatures, most of the chain lies in the Peierls valleys of the potential. Segments of the chain lying in different valleys are connected by transition regions called kinks and antikinks.   The first spatial derivative of the displacement field is positive for a kink and negative for an antikink.   In the limit of large friction, in the cae that the inertial term of Eq. (2) can be neglected, motion of the chain is governed by three processes: since the potential energy of consecutive Peierls valleys decreases in the direction of the force, the total potential energy of the system decreases if the segments of the chain lying in valleys with lower potential energy are expanded and the segments of the chain lying in potential valleys with higher potential energy are contracted. This has the consequence that kinks move on the average to the left and antikinks to the right.   The magnitude of their average velocity u(F) is determined by the balance of the energy gained by the motion along x and the energy dissipated in this process.   A kink passing a particle located at x, advances the displacement of this particle by 2π. Similarly, an antikink passing this particle also advances the particle by 2π. Hence, the time-development of the displacement of the particle is determined by the number of kinks and antikinks which pass this particle per unit time. The current of kinks is j_k = -um where u is the kink velocity and m is the density of kinks in the chain. The current of antikinks is j_a = un where n is the density of antikinks. Hence, the average speed with which a particle advances is determined by

$$<\partial\theta/\partial t> = -2\pi<j_k - j_a> = 2\pi u <m + n>. \qquad (4)$$

To find the average speed of a particle of the chain, we have thus to find the kink velocity as function of the driving force and we have to determine the density of kinks and antikinks as a a function of the field and of temperature. The velocity of kinks is determined solely by the deterministic equation of motion. Fluctuations in the velocity away from the average velocity are unimportant. In contrast, the determination of the kinks and antikink densities is a statistical problem. In the chain with damping, kinks and antikinks have a finite lifetime. In the limit of heavy damping, a kink and an antikink colliding will annihilate each other. All the kinetic and potential energy of a colliding kink antikink pair is dissipated. There is a complementary process to the annihilation of kinks which is the nucleation of kink-antikink pairs. Due to the fluctuation force ζ, in Eq. (2), a segment of the chain is occasionally thrown over a Peierls hill into the next lower Peierls valley. If the transferred piece of chain is large enough to overcome the energy needed to form a kink and antikink pair, the fluctuation keeps growing and has thus given rise to the birth of a new kink-antikink pair. In the steady state, the nucleation rate of kink-antikink pairs j_{nuc} must be balanced by their annihilation rate. The recombination rate of kink-antikink pairs can be found by the following argument: the probability that a kink encounters an antikink in the time interval dt is determined by the probability that there is an antikink in the range 2udt swept out by the motion within that time interval. Since the density of antikinks is n, the probability is 2undt. The rate of recombination per unit time and length is thus 2unm if there are m kinks per unit length. Since kinks and antikinks are created in pairs and recombined in pairs, the difference N-M of the number of kinks M and antikinks N is conserved. Since we have periodic boundary conditions, the number of kinks and antikinks present must be equal. Hence, the average kink and antikink densities are equal, $n_o \equiv <m> = <n>$. Thus the balance between nucleation events and recombination events is

$$j_{nuc} = 2un_o^2. \qquad (5)$$

This can be used to express the steady state kink density in terms of u and j_{nuc}. Using Eq. (5) we find that the average velocity of a particle is determined by

$$\langle \partial\theta/\partial t\rangle = 2\pi(2uj_{nuc})^{1/2}. \tag{6}$$

The solution sketched here differs in two essential points from that of [2]. Seeger and Schiller did not accept Eq. (5), but instead considered a set of coupled equations treating the generation and recombination of kink-antikink pairs on each Peierls hill separately [14]. Reference [7] showed that this is not necessary and that the argument given above, which has also been invoked in the theory of crystal growth [15], is correct. Another essential point, on which we depart from the early work on dislocations [1,2], is the calculation of the nucleation rate. This will be discussed subsequently in greater detail. First, however, we discuss the propagation of kinks in the presence of damping and the driving force. This is followed by a discussion of the nucleation rate for kink-antikink pairs. Finally, we address fluctuations away from the average behaviour.

6.2 Velocity of Driven Kinks

We start by considering the purely viscous chain (1=0 in Eq. (2)) in the absence of fluctuations and examine solutions of the form $\theta(x,t) = \theta(z)$, where $z=x+ut$. Here, u is the propagation velocity in the minus x direction. The waves $\theta(z)$ are solutions of the ordinary differential equation

$$\kappa d^2\theta/dz^2 - u\gamma d\theta/dz = -dU/d\theta, \tag{7}$$

where U is the original potential V turned upside down, $U = -V$. This equation describes the motion of a single particle [16] with mass κ in a tilted sinusoidal potential U subject to damping $\eta = -u\gamma$. A kink, describing a transition from one Peierls valley to an adjacent one, corresponds to a particle which starts at a maximum of U with zero velocity and reaches an adjacent maximum with zero velocity. To find such trajectories, we have to select the appropriate friction constant $\eta(F)$. We have determined the critical friction for the occurrence of such solutions. The result, $u(F) = -\eta(F)/\gamma$, is of the form [6]

$$u(F) = u_0\phi(F/V_0). \tag{8}$$

Here $u_0 = (\kappa V_0)^{1/2}/\gamma$ is a unit of velocity and the function ϕ, defined by the curve labelled $\chi = 0$ in Fig. 2, is independent of the parameters of Eq. (2). For small fields the propagation velocity is linear in the field giving rise to the kink mobility,

$$\mu = u(F)/F = (\pi/4\gamma)(\kappa/V_0)^{1/2}. \tag{9}$$

The propagation velocity increases monotonically with increasing field to a value [6,17,18] $u^* = u_0\phi^*$ at $F = V_0$, where $\phi^* \sim 1.19$. Interestingly, there is a kink structure for every propagation

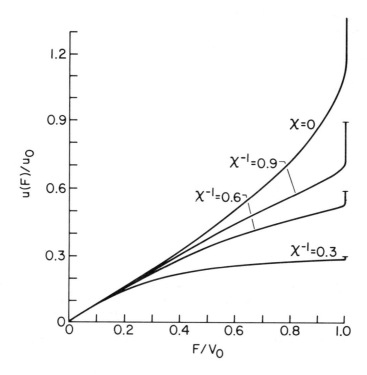

Figure 2. Propagation velocity u of a kink in the sine-Gordon chain. The u(F)-curve for the purely viscous chain (χ=0) defines the function φ in Eq. (8). Additional curves obtained with Eq. (10) give the propagation velocity of kinks in the chain with inertia.

velocity $u > u^*$.

Consider next the equation of motion of the chain with inertia in the absence of noise. The waves $\theta(x + ut)$ are also obtained from an ordinary differential equation which is second order in derivatives with respect to z, like Eq. (7). In fact, these equations can be mapped on to one another. Thus to each solution of the purely viscous chain there exists a similar solution of the equation with inertia. In particular, the velocity of a kink of the chain with inertia can also be expressed with the help of the function ϕ and an additional parameter which is a measure of the inertia, $\chi = (IV_o)^{1/2}/\gamma$. We find a kink velocity [19]

$$u(F) = u_o\phi(F/V_o)/[1 + \chi^2\phi^2(F/V_o]^{1/2}. \qquad (10)$$

The small field mobility Eq. (9) is unaffected by the inertial term. The maximum velocity of the kink in the chain with inertia is given by the velocity of sound of the chain in the absence of the periodic potential, $u_c = (\kappa/I)^{1/2} = u_s/\chi$. Figure 2 shows the kink velocity as a function of field for a number of chains with

81

differing degrees of inertia [20].

The similarity between the viscous chain and the chain with inertia extends to all the properties of the solutions. As an additional example, we mention here the width of the kink defined as $\delta = 2\pi/|d\theta/dz|_{max}$. For a kink of the purely viscous chain, the kink width as a function of the propagation velocity can be expressed in the form [20]

$$\delta(u) = \delta_0(|u|/u_0)\Delta(|u|/u_0). \qquad (11)$$

Here $\delta_0 = (\kappa/V_0)^{1/2}$ is a unit of length and Δ is a parameter independent function with the properties $\Delta(0) = \infty$ such that $\Delta\phi = \pi$ as u tends to zero and $\Delta(\infty) = \pi$. For a chain with inertia we obtain [20]

$$\delta(u) = \delta_0(|u|/u_0)\Delta(|u|/[u_0^2 - \chi^2u^2]^{1/2}). \qquad (12)$$

Figure 3 shows the width of kinks in chains with differing inertia. For the purely viscous chain, $\chi = 0$, the width of the kink increases monotonically. For small inertia the kink width first decreases and, after reaching a minimum, increases again with increasing velocity. For the kink travelling at the velocity of sound, the kink width is $\delta(u_s) = \pi\delta_0/\chi$. In chains with large inertia, the kink width decreases monotonically with increasing propagation velocity. Minima in the kink width as a function of the kink velocity are also discussed in [21] but the scaling

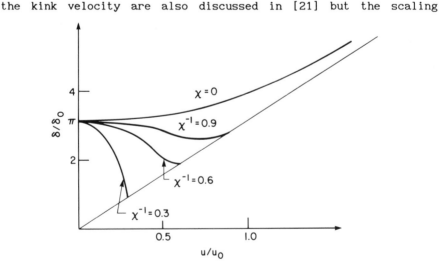

Figure 3. Width of the driven kinks in the sine-Gordon chain. The curve labelled $\chi = 0$ determines the scaling function $\Delta(\phi)$ of Eq. (11) and the other curves are generated by Eq. (12). The straight line gives the width $\delta(u_s) = \pi\delta_0/\chi$ of the kink propagating with the velocity of sound u_s.

behaviour is not addressed. The similarity laws discussed here have recently been applied by Magyari [22] to a chain of particles in a double well potential.

Many discussions in the literature of kink velocities invoke an energy balance equation [16,23-25] which is evaluated using a solution of the Hamiltonian equation of motion. This gives the correct zero field mobility [16] but leads to $\phi = F/V_o$ which, in contrast to the correct result shown in Fig. 2, is linear in the field also for large driving forces. We will subsequently invoke Eq. (8) to describe the average velocity of kinks in the heavily damped chain subject to thermal noise, Eq. (2). Effects of noise on the average motion have been considered in [26] and found to be small for temperatures small compared with the rest energy of a kink in agreement with our expectation [5,16].

6.3 Nucleation of Kink-antikink Pairs

Consider a long segment of the sine-Gordon chain lying in a Peierls valley. There is a lower lying state where each particle is displaced by 2π in the direction of the applied field. In the absence of noise, the chain must stay in the original valley. In the presence of thermal fluctuations, however, the chain can make a noise activated transition over a sequence of intervening states of higher energy, to move toward the state with lower energy [1-3, 5-6]. The section of the chain in the lower energy state is connected through a kink and antikink to the segment in the valley with higher energy as shown in Fig. 4. If the transferred section is small, then the attraction between the newly formed kink and

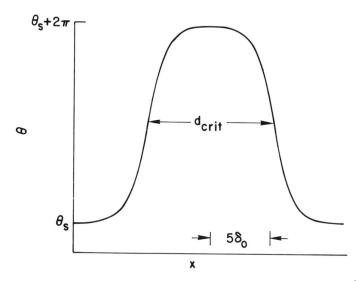

Figure 4. *Critical nucleus for a low field* $F/V_o = 10^{-4}$. *The displacement field of the initial valley is* θ_s. *The width of the nucleus is* $d_{crit} \sim -\delta_o \log F/V_o$, *where* $\delta_o = (\kappa/V_o)^{1/2}$.

antikink outweighs the energy which can be gained by moving the pair farther apart. The fluctuation collapses, and the initial state is restored. If, however, the transferred segment is long enough, such that the energy gained by moving the newly formed kink and antikink away from each other outweighs the attractive energy between the pair, the transferred region of chain grows.

Thus to create a segment of transferred chain, fluctuations have to pass through a saddle point configuration (Fig. 5). There is an energy barrier $\Delta E_N(F)$ which has to be surmounted. Figure 5 shows a qualitative sketch of the energy of a fluctuation as a function of the distance between the kink-antikink pair. Initially, the chain lies in the Peierls valley of higher energy (A). Thermal fluctuations must supply energy to form the kink and antikink and to separate them, initially. The critical separation d_{crit} characterizes the critical nucleus. In region c, the applied force outweighs the attractive force between the pair, the kink and antikink move apart with a velocity $2u(F)$. At D, a recombination event occurs. The location of the recombination event of course depends not only on the separation d of the original pair, but also on the location of the third kink (antikink) involved in the recombination.

The excess energy of the critical nucleus over that of the

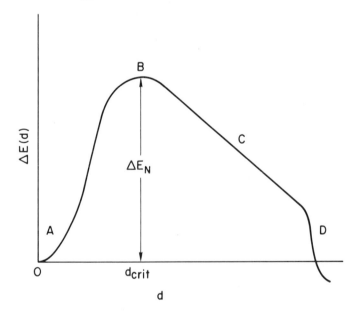

Figure 5. *Qualitative sketch of the energy of a kink-antikink pair separated by a distance d. In the Peierls valley (A) the chain is uniform. Energy must be supplied to pull the kink and antikink apart until a critical separation d_{crit} is reached given by the width of the critical nucleus (B). In the range (C) the interaction is dominated by the applied force. At D recombination with another kink terminates the motion of one of the original partners.*

initial state is shown in Fig. 6 as a function of the applied field.

The picture outlined here has been clearly understood by the dislocation physicists [1-3]. Calculations of the nucleation rate were presented invoking either one of the following two approaches: in the simple approach, a one-dimensional problem is solved by considering a Fokker-Planck equation for the separation of the kink-antikink pair; that is, Fig. 5 is taken at face value. A second approach is a real many dimensional solution of the nucleation problem invoking the transition state approach put forward by Vineyard [27]. This approach does not take into account the damping and the diffusive motion of the fluctuations through configuration space. Thus, it overestimates the nucleation rate [28]. To achieve a consistent approach, both nucleation and recombination events must be evaluated for the same degree of damping. We are interested in the case of heavy damping. The formalism to treat a many dimensional nucleation problem in the limit of strong damping has been put forth by Brinkman [29], Landauer and Swanson [30] and Langer [31]. We refer to this formalism as BLSL approach and this is what Petukhov and Pokrovskii [3], and References [5] and [6] invoke. The nucleation rate per unit time and length is given by

$$j_{nuc} = \Omega e^{-\Delta E_N(F)/kT}. \qquad (13)$$

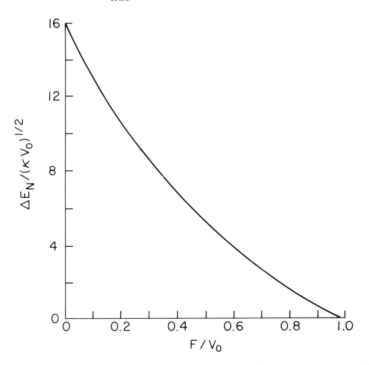

Figure 6. *Activation energy barrier for the nucleation of kink-antikink pairs as a function of the field. The curve shown is independent of the parameters in Eq. (2).*

As is typical in thermal activation problems, all the subtleties [28] of such a calculation are concerned with the prefactor Ω. Our results [5-6] for the heavily damped and driven sine-Gordon chain are valid for $\Delta E_N \gg kT$ and $F \gg n_o kT$, where n_o is the kink density. The first condition is required by the BLSL approach.

The second condition, which yields a lower limit, for the driving force ensures that a fluctuation, once it has reached the region C, actually moves away from the saddle point configuration. For fields $F < n_o kT$, region C becomes very flat, and at the same time, the critical nucleus has a separation d_{crit} which exceeds the average distance between kinks. Thus, for very low fields, the interaction with the other kinks present in the chain must be taken into account. These many kink problems have not been addressed by authors presenting results for the nucleation rate at very low fields, except for the recent discussion in [32].

The average velocity of a particle in the driven and damped sine-Gordon chain is shown in Fig. 7 for various temperatures. The low field asymptotes are calculated in the following way: for low fields, where the velocity is $u = -\mu F$, with μ given by Eq.

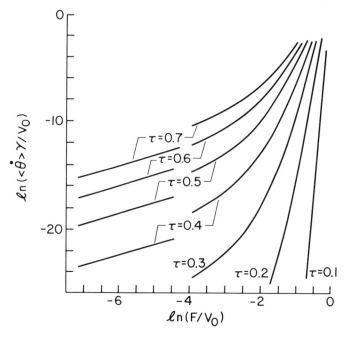

Figure 7. Velocity of particles in the sine-Gordon chain as function of the driving force for different temperatures $\tau = kT/(\kappa V_o)^{1/2}$. The straight lines on the left show the extrapolation of $\langle \partial\theta/\partial t \rangle = 4\pi n_{eq} \mu F$ to higher fields (see Eq. (14)). The other curves at larger values of F/V_o, are based on a computational evaluation of Eqs. (6) and (13).

(9), we can use Eq. (4) directly to obtain $<\partial\theta/\partial t> = 4\pi n_{eq}\mu F$, where

$$n_{eq} = (2/\pi)^{1/2}(V_o/\kappa)^{1/2}(E_o/kT)^{1/2}\exp(-E_o/kT) \qquad (14)$$

is the equilibrium kink density as determined by Seeger and Schiller [33] and later by others [34-36]. In Eq. (14) $E_o = 8(\kappa V_o)^{1/2}$ is the energy of a static kink.

6.4 Fluctuations in the Sine-Gordon Chain

In this section we treat the long time and long range fluctuations of the sine-Gordon chain away from the ensemble averaged behaviour. We have presented two methodically different discussions. Here we focus on the **hydrodynamic** approach and describe our efforts to substantiate a set of simple equations written down by Brailsford [37]. We have treated fluctuations in chains of finite length and also fluctuations in the infinitely long chain. Here we focus on the latter.

First, we discuss the fluctuations of the chain as a function of x at a given instant of time. Consider again Fig. 1. As we follow the chain, moving away from a given position x_o in the positive x direction, we see that each kink increases the difference in displacement, $\theta(x) - \theta(x_o)$, by 2π, and each antikink decreases the difference by 2π. In [7] it was shown that given N kinks and N antikinks, each geometrical sequence of kinks and antikinks is equally likely [38]. Thus, as we follow the chain, we see an uncorrelated sequence of jumps, each of magnitude 2π. Hence each configuration of the chain at a given instant of time is a realization of a diffusion process with x playing the role of time and a diffusion constant [7,13]

$$D_\theta = (1/2)(2n_o)(2\pi)^2. \qquad (15)$$

Here the total kink and antikink density, $2n_o$, gives the jump rate per unit length. Thus the mean square displacement per unit length grows as

$$<(\theta(x) - \theta(x_o))^2> = 2D_\theta(x - x_o). \qquad (16)$$

Since each configuration of the chain at a given instant of time is a realization of a diffusion process, the probability for a particular realization [12] must be proportional to

$$P(\theta(x)) = \exp[-\frac{1}{4D_\theta}\int_0^L (\partial\theta/\partial x)^2 dx]. \qquad (17)$$

This can be used to derive the static structure factor of the displacement field [12] and other static correlation functions

[13].

To discuss time-dependent correlation functions we consider the local kink and antikink densities $m(x,t)$ and $n(x,t)$. These local kink densities change due to the three basic processes discussed in the introduction: the propagation of kinks, and the nucleation and the recombination of kink-antikink pairs. Thus the equations for these densities are according to Brailsford [37]

$$\partial m/\partial t + \partial j_k/\partial x = j_{nuc} - j_{rec} + \xi, \qquad (18)$$

$$\partial n/\partial t + \partial j_a/\partial x = j_{nuc} - j_{rec} + \xi. \qquad (19)$$

Here j_k and j_a are the kink and antikink currents, j_{nuc} and j_{rec} are the nucleation rate and the recombination rate and ξ is a generation recombination shot noise. Since kinks and antikinks are generated in pairs and recombine in pairs, the right hand side of the two Eqs. (18,19) is identical. The imbalance between the kink and antikink densities $\rho = m - n$ obeys a continuity equation

$$\partial \rho/\partial t + \partial j/\partial x = 0 \qquad (20)$$

with a current $j = j_k - j_a$. Eq. (20) reflects the fact that $N - M = \int \rho dx$ is a constant of the motion.

Let us first consider the driven case for fields $F > nkT$. In this case, the motion of the kinks gives rise to currents $j_k = -um$ and $j_a = -un$. Furthermore, the nucleation rate is determined by Eq. (13) and the recombination rate is, as discussed in the introduction, given by $j_{rec} = 2umn$. Reference [7] showed that, in the steady state, the generation and recombination events occur randomly in space and time. Since in the steady state $j_{nuc} = j_{rec}$, the generation and recombination events contribute equally and independently to the shot noise. On the time and length scale of interest here the stochastic force ξ has a very short correlation time and correlation length. Reference [7] found $\langle \xi(x,t)\xi(x',t')\rangle = 2j_{nuc}\delta(x-x')\delta(t-t')$. Thus we have completely specified Eqs. (18,19). The long term motion of the displacement field can now be obtained in the following way. As discussed in the introduction, the temporal change of the displacement field at a given location is determined by the kink and antikink currents,

$$\partial \theta/\partial t = -2\pi j = -2\pi(j_k - j_a). \qquad (21)$$

The spatial variation of the coarse grained displacement field is related to the kink-antikink imbalance

$$\partial \theta/\partial x = 2\pi\rho. \qquad (22)$$

For the Fourier transform of the displacement field, we find using Eqs. (18,19) and (21,22) a dynamic structure factor [7,13]

$$S^{\theta\theta}(\omega,q) = \frac{4D_\theta D_\rho}{\omega^2 + D_\rho^2 q^4}, \qquad (23)$$

where $D_\rho = u/4n_0$ and D_θ is given by Eq. (15). Using Eq. (23) we find for the mean square displacement at a given instant of time Eq. (16). For the mean square displacement of a given particle as function of time, Eq. (23) yields [7]

$$<(\theta(x,t) - \theta(x,0))^2> = 16\pi^{3/2} n_0 (D_\rho t)^{1/2}. \qquad (24)$$

where $\delta\theta(x,t) = \theta(x,t) - <\theta(x,t)>$ is the deviation from the ensemble averaged motion. More general correlation functions can also be calculated [13].

Consider next the sine-Gordon chain at equilibrium [10-11]. As pointed out in section 6.3, our results for the recombination rate and the nulceation rate are not applicable in this case. But at equilibrium it is sufficient to consider Eq. (20). At equilibrium the kinks and antikinks move purely diffusively with a diffusion constant [10] which is determined by the Einstein relation $D = \mu kT/2\pi$. The factor 2π arises because μF is the velocity along the x direction but the field conjugate to F is θ and not x. There are of course many ways to derive this diffusion constant without appealing directly to the Einstein relation [39]. The kink and antikink currents at equilibrium are $j_k = D\partial m/\partial x + \zeta$ and $j_a = -D\partial n/\partial x + \zeta$, where ζ_k and ζ_a are fluctuations in the densities n and m due to the Brownian motion of kinks and antikinks in the x direction. The average of these forces is zero and the correlation for the kinks is $<\zeta_k(x,t)\zeta_k(x',t')> = n_{eq}\partial(x-x')\partial(t-t')$ and similar for the antikinks. Furthermore, there are no correlations between ζ_k and ζ_a. This completely specifies the current j in Eq. (20) and, since at equilibrium $\partial\theta/\partial t$ depends only on ρ, this completely specifies the long term fluctuations of the chain at equilibrium. For the Fourier transform of the displacement field with respect to q and ω, we obtain again Eq. (23) but with $D_\rho = D = \mu kT/2\pi$ and $D_\theta = 4\pi^2 n_{eq}$. This yields for the mean square displacement of a particle at equilibrium [10,11,13]

$$<(\theta(x,t) - \theta(x,0))^2> = 16\pi^{3/2} n_{eq} (Dt)^{1/2}. \qquad (25)$$

The $t^{1/2}$ behaviour given by Eq. (25) differs from the predictions of both [8] and [9]. Schneider and Stoll [8] found growth proportional to $t^{4/3}$ based on computer simulations of Eq. (2). Gunther and Imry [9] found $t^{1/2}$ on a time scale short compared to the life time of a kink and argued that for long times the mean square displacement should grow as t. Later, Schneider and Stoll

[40] reanalyzed their data and predicted that in a Hamiltonian chain growth should be proportional to $t^{2/3}$ and in a chain with damping proportional to $t^{1/2}$. More recent work [41,42] also agrees with us on the $t^{1/2}$ behaviour. Recently there have even been experiments [43] which seem to agree best with a $t^{1/2}$ behaviour [44]. In real systems Eq. (2) is unlikely to be realized but has to be modified [45] to account for possible nonuniformities (impurities) in the potential Eq. (1) and in the coupling constant along x.

In [13] we have pointed out that there exists a whole class of systems which exhibit the same long term behaviour as the sine-Gordon equation. All these systems exhibit a conserved quantity and thus an associated density whose time evolution is governed by a continuity equation (see Eq. (20)) and which is like the kink-antikink imbalance ρ related to the physical field of interest (see Eq. (22)). In this class of systems, ordinary diffusion is inhibited through some form of interaction between the diffusing entities.

References

[1]   A. Seeger (1956). *Phil. Mag.* **1**, 651;   J. Lothe and J.P. Hirth (1959). *Phys. Rev.* **115**, 543.

[2]   A review of the statistical mechanics of solitons as it emerged from the theory of dislocations is given by A. Seeger and P. Schiller (1966) in *Physical Acoustics, Vol III,* ed. W.P. Mason, p. 361, Academic Press, New York.

[3]   B.V. Petukhov and V.L. Pokrovskii, *Soviet Phys. JETP* **36**, 336;   A.P. Kazantsev and V.L. Pokrovskii, *Soviet Phys. JETP* (1970) **31**, 362.   This work has come to our attention only recently.

[4]   S.E. Trullinger, S.E. Miller, M.D. Guyer, R.A. Bishop, F. Palmer and J.A. Krumhansl (1978).   *Phys. Rev. Lett* **40**, 206; **40**, 1603.

[5]   M. Buttiker and R. Landauer (1979).   *Phys. Rev. Lett.* **43**, 1453.

[6]   M. Buttiker and R. Landauer (1981).   *Phys. Rev.* **A23**, 1397.

[7]   C.H. Bennett, M. Buttiker, R. Landauer and H. Thomas (1981). *J. Stat. Phys.* **24**, 421.

[8]   T. Schneider and E. Stoll (1978).   *Phys. Rev. Lett.* 41, 1429.

[9]   L. Gunther and Y. Imry (1980).   *Phys. Rev. Lett.* **44**, 1225.

[10]   M. Buttiker and R. Landauer (1980).   *J. Phys.* **C13**, L325.

[11]   M. Buttiker and R. Landauer (1981).   *Phys. Rev. Lett.* **46**, 75.

[12]   M. Buttiker (1981).   *Phys. Lett.* **81A**, 391.

[13]   M. Buttiker and R. Landauer (1982) in *Nonlinear Phenomena at Phase Transitions and Instabilities,* 361, ed. T. Riste, Plenum Publishing Corporation.

[14]   A Seeger and P. Schiller, in Ref. [2], p. 481.

[15]   F.C. Frank (1974).   *J. Cryst. Growth* **22**, 233.

[16]   R. Landauer (1977).   *Phys. Rev.* **A15**, 2117.

[17]   M. Urabe (1955).   *J. Sci. Hiroshima Univ. Ser.* **A18**, 379.

[18]   P. Marcus and Y. Imry (1980).   *Solid State Commun.* **33**, 345.

[19] M. Buttiker and H. Thomas (1980). *Phys. Lett.* **77A**, 372.
[20] M. Buttiker and H. Thomas (1988). *Phys. Rev.* **A37**, 235.
[21] S. Ferrigno and S. Pace (1985). *Phys. Lett.* **12A**, 77.
[22] E. Magyari (1984). *Z. Phys.* **B55**, 137; (1984). *Phys. Rev. Lett.* **52**, 767; (1985) *Z. Phys.* **B62**, 113.
[23] D.W. McLaughlin and A.C. Scott (1978). *Phys. Rev.* **A18**, 1652.
[24] D.J. Bergman, E. Ben-Jacob, Y. Imry and K. Maki (1983). *Phys. Rev.* **A27**, 3345.
[25] A. Davidson and N.F. Pederson (1984). *Appl. Phys. Lett.* **44**, 465; F. If, M.P. Soerenson and P.L. Christiansen (1984). *Phys. Lett.* **100A**, 68; A. Davidson, B. Dueholm, B. Kryger and N.F. Pederson (1985). *Phys. Rev. Lett.* **55**, 2059.
[26] D.J. Kaup (1983). *Phys. Rev.* B27, 6787.
[27] G.H. Vineyard (1957). *Phys. Chem. Solids* 3, 121.
[28] M. Buttiker, E.P. Harris and R. Landauer (1983). *Phys. Rev.* **B28**, 1268.
[29] H.C. Brinkman (1956). *Physica* **22**, 149.
[30] R. Landauer and J.A. Swanson (1961). *Phys. Rev.* **121**, 1668.
[31] J.S. Langer (1969). *Ann. Phys. (N.Y.)* **54**, 259.
[32] F. Marchesoni (1986). *Phys. Rev.* **B34**, 6536.
[33] See Ref. [2], p. 446.
[34] See Ref. [6], Appendix A.
[35] A.R. Bishop, J.K. Krumhansl and S.E. Trullilnger (1980). *Physica* **D1**, 1.
[36] J. Timonen, M. Stirland, D.J. Pilling, Yi Cheng and R.K. Bullough (1986). *Phys. Rev. Lett.* **21**, 2233.
[37] A.D. Brailsford (1961). *Phys. Rev.* **122**, 778.
[38] Methods similar to those of Ref. [7] have been applied to arrays of bistable systems by K. Kawasaki and T. Nagai (1983). *Physica* **121A**, 175, and K. Seikomoto (1984). *Physica* **A125**, 261.
[39] M. Reimoissenet (1978). *Solid State Commun.* 27, 681; M. Salerno, E. Joergensen and M.R. Samuelsen (1984). *Phys. Rev.* **B30**, 2635.
[40] T. Schneider and E. Stoll (1980). *Phys. Rev.* **B22**, 395.
[41] K. Maki (1981). *Phys. Rev.* **B24**, 335.
[42] M.J. Gillan (1985). *J. Phys.* **C18**, 4885.
[43] J.P. Boucher, F. Mezei, L. Regnault and J.P. Renard (1985). *Phys. Rev. Lett.* **55**, 1778.
[44] K. Sasaki and K. Maki (1987). *Phys. Rev.* **B35**, 257.
[45] V.M. Vinokur (1986). *J. Phys. (Paris)* **47**, 1425.

RECENT WORK ON JOSEPHSON TUNNEL JUNCTIONS WITHIN MIDIT

7.1 Introduction

One of the research activities within MIDIT for many years has been and continues to be the study of the dynamics of Josephson tunnel junctions.   The scope of these studies comprises experimental, computational and analytical work.   In this chapter we shall review briefly some of the most recent investigations. The colour plates 1 and 2 show a Josephson tunnel junction and a computer solution of the perturbed sine-Gordon equation.   The author acknowledges the contributions to this report from J. Bindslev Hansen, R.D. Parmentier and N.F. Pedersen.

7.2 Autonomous, but distributed systems: Coherence, Solitons, Intrinsic Dynamics

7.2.1 *Intermediate length, linear overlap junctions*

In this section we review some of the results obtained in [1] concerning instabilities of dynamic states on the intermediate length junction.   The mathematical model of the overlap Josephson junction is, in normalized form, the perturbed sine-Gordon equation

$$\phi_{xx} - \phi_{tt} - \sin\phi = \alpha\phi_t - \beta\phi_{xxt} - \Gamma \qquad (1a)$$

with boundary conditions

$$\phi_x(0,t) = \phi_x(\ell,t) = \eta. \qquad (1b)$$

Here, $\phi(x,t)$ is the usual Josephson phase variable, x is distance along the junction, normalized to the Josephson penetration length λ_J, and t is time, normalized to the inverse of the Josephson plasma angular frequency ω_o.   The model contains five parameters: α, β, Γ, ℓ, and η.   The term in α represents shunt loss due to quasiparticles crossing the junction, the term in β represents dissipation due to the surface resistance of the superconducting thin-films, Γ is the uniform bias current normalized to the maximum zero-voltage Josephson current I_o, η represents the normalized external magnetic field, and the normalized length of the junction is denoted by ℓ. It is assumed that the width of the junction is much smaller than λ_J.

We first consider the case of homogeneous boundary conditions, i.e., $\eta = 0$ in Eq. (1b). If $\alpha = \beta = \Gamma = 0$ the McCumber solution of Eqs. (1) is exactly [2]

$$\phi = \phi_0(t) = 2\text{am}[t/k; k], \tag{2}$$

where am is the Jacobian elliptic amplitude function of modulus k. For nonzero α, β, and Γ, we assume that Eq. (2) solves Eqs. (1) in the power-balance approximation. This yields the following expressions for the McCumber branch of the I-V characteristic of the junction:

$$\Gamma = \frac{4\alpha E(k)}{\pi k}, \tag{3a}$$

$$V = \langle \phi_t \rangle \equiv \omega = \frac{\pi}{kK(k)}, \tag{3b}$$

where $K(k)$ and $E(k)$ are, respectively, the complete elliptic integrals of first and second kinds.

Following Burkov and Lifsic [3], we now express solutions of Eqs. (1) in the vicinity of the McCumber solution as

$$\phi(x,t) = \phi_0(t) + \tilde{\phi}(x,t), \tag{4}$$

where ϕ_0 is given by Eq. (2) together with the conditions of Eqs. (3), and $\tilde{\phi}$ is a small perturbation of the form

$$\tilde{\phi}(x,t) = y(t)\exp(ibx) \tag{5}$$

with b constant. Inserting Eqs. (5) and (4) into Eqs. (1), we obtain an ordinary differential equation for $y(t)$:

$$\ddot{y} + (\alpha + \beta b^2)\dot{y} + \{b^2 + \cos[\phi_0(t)]\}\, y = 0, \tag{6a}$$

where

$$b = n\pi/\ell, \quad n = 0, 1, 2, \ldots \tag{6b}$$

and overdots denote derivatives with respect to t.

Equation (6a) is a damped Hill's equation; the stability boundaries are given by [1]

$$b^2 = \left[\frac{\omega}{2}\right]^2 + \frac{1}{2\omega^2}\left[1 - \frac{1}{8\omega^4}\right]$$

$$\pm \frac{1}{2}\left[\left[1 - \frac{1}{8\omega^4}\right]^2 - \omega^2(\alpha + \beta b^2)\right]^{1/2} - \frac{1}{8\omega^2}\left[1 - \frac{1}{8\omega^4}\right]^2 , \tag{7}$$

with b given by Eq. (6b). For given values of α, β, ℓ, and n, Eq. (7) gives two values for ω, say ω_+ and ω_-, which are the stability boundaries, provided that the argument of the square root in Eq. (7) is positive. Using Eqs. (3) the voltage stability boundaries ω_+ and ω_- can be translated into the corresponding current values, say Γ_+ and Γ_-.

For relatively short junctions, $1 \lesssim \ell \lesssim 5$, the multimode theory developed by Enpuku et al. [4] is used to treat the magnetic case $\eta \neq 0$ in Eq. (1b). The stability analysis under the assumption of small amplitudes of the spatial modes then leads to an expression similar to Eq. (7) but containing η as a parameter [1].

Figure 1 shows a portion of an I-V characteristic calculated numerically from Eqs. (1) using a multimode expansion. Parameter values are $\alpha = 0.05$, $\beta = 0.02$, $\ell = 2$, and $\eta = 0$. Both the

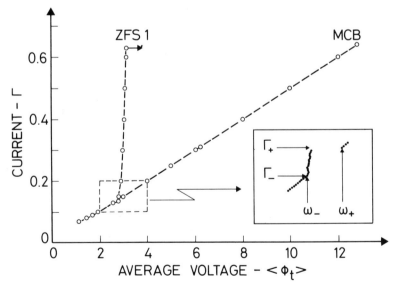

Figure 1. I-V characteristic calculated from Eqs. (1) using $\alpha = 0.05$, $\beta = 0.02$, $\ell = 2$, and $\eta = 0$, showing the McCumber background curve (MCB) and the first zero-field step (ZFS1). Inset shows detail of region where ZFS1 joins the MCB.

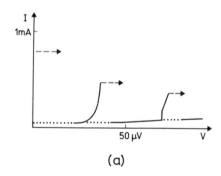

(a)

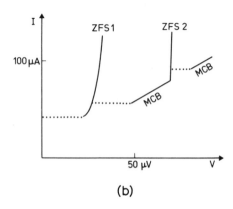

(b)

Figure 2. (a) Detail of experimental I-V characteristic of Nb-NbO$_x$-Pb tunnel junction measured at a temperature slightly below the transition temperature of the lead electrode and in zero magnetic field. Arrows indicate switching to the gap voltage. (b) Same characteristic with 10 x magnified current scale. Dotted lines indicate switching from higher-voltage to lower-voltage states.

McCumber background curve (MCB) and the first Zero Field Step (ZFS1) are evident. The inset shows in more detail the region where ZFS1 joins the McCumber curve.

Using this numerical procedure, we find that the stability boundaries associated with ZFS1 for the parameter values used are, expressed in terms of bias current, $\Gamma_+ = 0.1712\pm0.0005$ and $\Gamma_- = 0.1401\pm0.0001$. Inserting the same parameter values, together with n = 1 in Eq. (6b), in Eq. (7) and (3), we get $\Gamma_+ = 0.1711$ and $\Gamma_- = 0.1404$. Considering that this instability region occurs at the very lower end of the asymptotic linear region in the McCumber curve, for which the analysis was developed, the agreement is more than satisfactory.

Figure 2a shows a portion of the experimental I-V characteristic of a Nb/Pb Josephson junction measured at a temperature somewhat below the transition temperature of the lead electrode and in zero magnetic field. The dashed arrows indicate switching from the zero-voltage state and from the first ZFS's to the gap voltage. Figure 2b is the same characteristic with a 10 x magnification of the current scale. The dotted lines in Fig. 2 indicate switching from higher-voltage to lower-voltage states.

The characteristics shown in Fig. 2 correspond to a normalized junction length of about $\ell = 3.2$ and an α-loss term, estimated from the slope of the McCumber curve, of about $\alpha = 0.03$. The experimental determination of the β-loss term is subject to rather large uncertainties, and so we have treated β as an adjustable parameter in what follows.

Figure 3 shows a comparison between the experimentally determined stability boundaries (circles) in a magnetic field associated with ZFS1 and those obtained from the stability theory, shown as solid lines. The experimental values of voltage and magnetic field were normalized using the expressions $\langle \phi_t \rangle = V/\Phi_o f_o$ and $\eta = 2\pi\Phi_{ext}/\Phi_o\ell$, where V is the voltage, Φ_o is the magnetic flux quantum, Φ_{ext} is the applied magnetic flux threading the junction, and f_o is the plasma frequency. The experimental data were taken at a temperature for which $\alpha = 0.026$ and $\ell = 3.16$, and these same parameter values were used in the theoretical

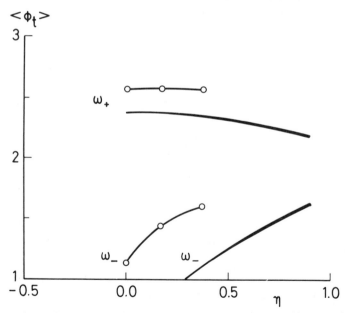

Figure 3. *Stability boundaries in average voltage $\langle \phi_t \rangle$ as a function of magnetic field η measured experimentally (circles) and calculated from stability theory (solid curves) for (a) ZFS1. Fixed parameter values: $\alpha = 0.026$ and $\ell = 3.16$. $0 \leq \beta \leq 0.07$ giving rise to the widening of the solid curves for large η.*

expression. The β value used in the stability theory calculation was varied between 0 and 0.07; the effect of this variation is indicated by the slight thickening of the curves in Fig. 3. The agreement between the experimental and theoretical values for ω_{+} is reasonable.

Thus a linear-stability analysis of the perturbed sine-Gordon equation which describes the dynamics of Josephson tunnel junctions indicates that the mechanism that determines the experimental observation of ZFS's may be described in terms of the growth of parametrically excited instabilities of the McCumber curve. This analysis gives good agreement with both numerical and experimental results in the asymptotic linear region of the McCumber curve and for sufficiently small values of the applied magnetic field.

Numerical integration of the multimode equations verifies that the parametrically excited instabilities evolve into fluxon oscillations. The multimode approach is a useful alternative to the direct numerical simulation of Eqs. (1) inasmuch as it gives reasonably reliable results at a considerably reduced computing cost.

7.2.2 *Analytic results for a triangular current phase relation*

The Josephson transmission line is modelled by the perturbed sine-Gordon equation (1) with the periodic boundary conditions

$$\phi_x(0,t) = \phi_x(\ell,t)$$

$$\phi_t(0,t) = \phi_t(\ell,t). \tag{8}$$

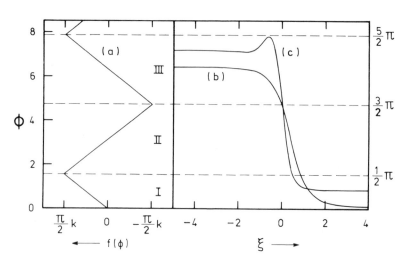

Figure 4. (a) Triangular current-phase relation f(ϕ). (b,c) Fluxon line shapes with α = 0.02, β = 0.01, and k = 2/π. (b) Fluxon velocity u = 0.9 giving Γ = 0.079. (c) u = 1.0 giving Γ = 0.569.

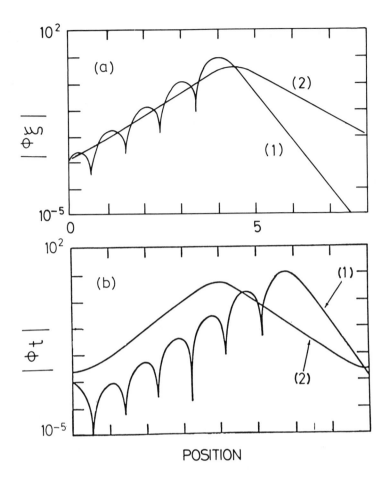

Figure 5. Fluxon line shapes. (a) Analytical results. (b) Numerical integration of Eq. (1). The parameters are: α = 0.02, β = 0.01, k = 2/π, (a-1) u = 0.9, Γ = 0.079; (a-2) u = 1, Γ = 0.569, (b-1) Γ = 0.1, (b-2) Γ = 0.675.

In what follows we review the contents of [5] and will consider the case of a junction which is so long that we may neglect the effects of the boundaries. We shall therefore study the stationary fluxon motion.

In general when α, β and Γ are different from zero Eq. (1a) cannot be solved analytically. To extract some analytical information one can approximate the sine term in Eq. (1a) by using a triangular current-phase relation having 2π periodicity. As shown in Fig. 4a $f(\phi)$ is defined by

$$f(\phi) = \begin{cases} k(\phi - 2n\pi), & -\pi/2 + 2n\pi < \phi < \pi/2+2n\pi \\ -k(\phi - \pi-2n\pi), & \pi/2 + 2n\pi < \phi < 3\pi/2+2n\pi. \end{cases} \qquad (9)$$

The main idea of this procedure is that now Eq. (1a) is a

piece-wise linear equation, i.e. the ϕ-axis is divided in regions (regions I, II and III in Fig. 4) in which Eq. (1a) is linear and the corresponding solution can be found by elementary analysis. Caution has to be taken to join properly the solutions at the boundaries of the different ϕ-regions. The procedure, requiring rather lengthy calculations, has been carried out in [6], using a travelling-wave assumption of the solution to Eq. (1a), which is then reduced to a third order ordinary differential equation:

$$-\beta u\phi_{\xi\xi\xi} + (1 - u^2)\ \phi_{\xi\xi} + \alpha u\phi_\xi - f(\phi) + \eta = 0 \qquad (10)$$

where $\xi = x-ut$, u being the velocity of the travelling wave. The solution in the three regions of Fig. 4 is given by

$$\phi = \begin{cases} A_1\exp(q_1\xi) + \eta/k & \text{(I)} \\ B_1\exp(r_1\xi) + B_2\exp(r_2\xi) + B_3\exp(r_3\xi) - \eta/k + \pi & \text{(II)} \\ C_2\exp(q_2\xi) + C_3\exp(q_3\xi) + \eta/k + 2\pi & \text{(III)} \end{cases} \qquad (11)$$

where A_1, C_2, C_3, B_1, B_2, B_3 are constants and q_1, q_2, q_3 and r_1, r_2, r_3 are the roots of the characteristic polynomial $P^\pm$ of Eq. (10) (the q's are related to P^+ and the r's to P^-)

$$P^\pm = \beta ux^3 - (1 - u^2)x^2 - \alpha ux \pm k = 0. \qquad (12)$$

Due to the form of $P^\pm$ we have, for P^+, always a negative real root q_1 and for q_2 and q_3 either two positive real roots or a pair of complex conjugate roots with positive real part. For P^- we have always a positive root r_3 and for r_1 and r_2 either two negative roots or a pair of complex conjugate ones with negative real part.

This information, together with the asymptotic conditions on the solution ϕ at $\xi\to\pm\infty$ gives Eq. (11). The values of the constants A_1, B_1, B_2, B_3, C_2, C_3 and η are determined by the matching conditions for ϕ, ϕ_ξ and $\phi_{\xi\xi}$ at the borders of the regions in Fig. 4 [6]. The condition for the observation of the overshoot in the fluxon shape (two complex conjugate roots in P^+) is given by

$$\beta > 4\alpha^3/27k^2. \qquad (13)$$

In the presence of overshoot, see Fig. 4c, the period of the oscillations and their decay rate can be computed from the imaginary and the real parts of the two complex conjugate roots q_2 and q_3 of P^+, respectively. In Fig. 5 the absolute value of ϕ_ξ is shown on a logarithmic scale; for both the period and the decay rate of the oscillations a surprising similarity with the numerical results of the integration of Eq. (1a) with periodic boundary conditions, Fig. 5 [7], is found. The current-velocity relationship for the fluxon can be derived analytically [6] and in Fig. 6 a comparison of the computed Γ - u curve with the results of a numerical simulation of Eq. (1a) is shown. In this calculation the value of k (Eq. (9)) is chosen to be $8/\pi^2$ to make the junction coupling energy per unit length equal to that of Eq. (1a) with the original $\sin\phi$ term. As shown in Fig. 6 the analytical results obtained are very close to the numerical ones for the original system.

7.3 Non-autonomous, Distributed Systems: Long Driven Junctions

In this section we consider several effects on fluxon dynamics of externally applied driving signals. Both sinusoidal and stochastic drivers are studied, the first representing microwave irradiation and the second representing internal thermal noise in the junction. A number of distinctly different effects are observed, depending on the values of the various system parameters: (a) for a relatively weak driver, of either sinusoidal or stochastic type, introduced through the bias current, numerical solutions of the model equation with periodicboundary conditions show the major effect to be a slight destabilization of the fluxon propagation velocity, leading to a broadening of the linewidth of the associated radiation; (b) for a stronger microwave drive, experimental measurements show a phase

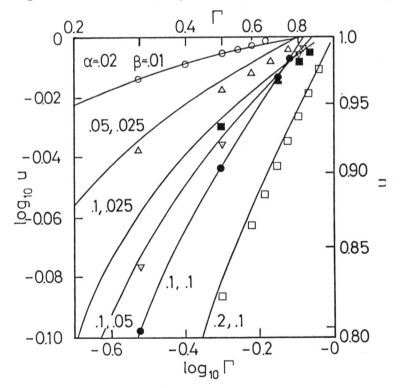

Figure 6. Comparison of the Γ - u curves computed analytically (solid curves) and the results [7] from numerical integration of Eq. 1 (marks) for various values of the parameters α and β.

locking to the external radiation of the internal fluxon
oscillations corresponding to both Fiske steps and zero-field
steps; (c) for a moderate sinusoidal drive, introduced through
the boundary conditions, numerical solutions of the model equation
for an overlap geometry junction show a rich variety of both
intermittency-type and period-doubling approaches to chaos.

7.3.1 *Annular junctions with sinusoidal signal and noise*

In this section we review some of the results obtained in [8] and
[9]. The narrow linewidth of the circular junction oscillator is
demonstrated by computer simulations, in agreement with
experimental results.

The perturbed sine-Gordon equation (1) is considered with a
driving term of the form

$$\Gamma = \Gamma_{DC} + \Gamma_{AC} \sin(\Omega t), \tag{14a}$$

representing a constant bias current, Γ_{DC}, and a sinusoidal
microwave irradiation, $\Gamma_{AC}\sin(\Omega t)$, or

$$\Gamma = \Gamma_{DC} + \Gamma_n(x,t) \tag{14b}$$

where $\Gamma_n(x,t)$ is Gaussian white noise with zero mean, $\langle \Gamma_n(x,t) \rangle =$
0, and autocorrelation function $R_\Gamma(\zeta,\tau) = \langle \eta(x,t)\ \eta(x+\zeta,t+\tau) \rangle =$
$\sigma_n^2 \delta(\zeta)\zeta(\tau)$ where σ_Γ^2 is the variance of the noise. Periodic
boundary conditions (8) are used.

The very narrow linewidth of the radiation emitted from a
Josephson-junction oscillator (less than 1 kHz at 10 GHz) suggests
that a relative numerical accuracy of at least 10^{-7} is essential.
We solve Eqs. (1a), (8), and (14) numerically by using
a pseudospectral method on a CRAY-1-S. This method, a Fourier
transform treatment in space together with a leapfrog scheme in
time, has the advantage of simplicity and high-order accuracy in
the approximations to the space derivatives.

As a measure of the frequency fluctuations, which is essentially
the linewidth of the oscillator, we have calculated the standard
deviation of the fluxon revolution frequency $\sigma_f = [\langle (f_n - \langle f_n \rangle)^2 \rangle]^{1/2}$ for values of the cyclic driving frequency Ω between
0.4 and 2.0 in the microwave case Eq. (14a). The solid curve in
Fig. 7 shows the results from the numerical solution.

A perturbation theory using

$$\phi(x,t) = \phi^S(x,t) + \tilde{\phi}(t), \tag{15}$$

where $\phi^S(x,t)$ is the travelling soliton and $\tilde{\phi}(t)$ is a small
background, leads to the ordinary differential equation for $\tilde{\phi}(t)$

$$-\overset{..}{\widetilde{\phi}} - \sin\,\widetilde{\phi} = \alpha\overset{.}{\widetilde{\phi}} + \Gamma_{DC} + \Gamma_{AC}\,\sin(\Omega t) \qquad (16)$$

as well as a determination of the soliton velocity $u(t)$. The soliton revolution time, T_n (and the corresponding frequency, $f_n = 1/T_n$) has first been determined from the equation

$$\frac{2\pi}{\ell}\,\int_{t}^{t+T_n}\,u(t')dt'\,=\,2\pi, \qquad (17)$$

which yields the dashed-dotted curve in Fig. 7 in poor agreement with the solid curve. Realizing that the main contribution to the linewidth stems from the background radiation, $\widetilde{\phi}(t)$, and **not** from perturbations of the soliton velocity, $u(t)$, we replace Eq. (17) by

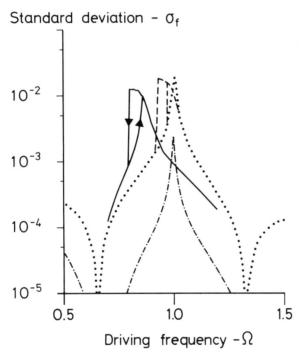

Standard deviation - σ_f

Driving frequency - Ω

Figure 7. *Standard deviation of electromagnetic radiation frequencies versus driving frequency Ω. Parameters $\alpha = 0.01$, $\beta = 0$, $\Gamma_{DC} = 0.02$, $\Gamma_{AC} = 0.01$, and $\ell = 8$. Solid curve: Computational solution of Eqs. (1a), (8), and (14a). Dashed-dotted curve: Kink model result. Dotted curve: New perturbative result using linearized version of Eq. (16). Dashed curve: New perturbative result using the full Eq. (16). Dotted and dashed curves overlap away from resonance region.*

$$\frac{2\pi}{\ell} \int_{t}^{t+T_n} u(t')dt' + \tilde{\phi}(t+T_n) - \tilde{\phi}(t) = 2\pi \qquad (18)$$

and find the dotted curve, when a linearized version of Eq. (16) is used. The hysteresis phenomenon is recovered (dashed curve in Fig. 7), when the full Eq. (16) is used. The level of the linewidth is now predicted correctly by the perturbation theory while the location of the maximum is still predicted at a slightly too high frequency.

In the case of Gaussian white noise, Eq. (14b), an estimate of the standard deviation, σ_f, based on Hamiltonian perturbation theory leads to the curves shown in Fig. 8b. Figure 8a shows the corresponding results from the numerical simulation. As can be

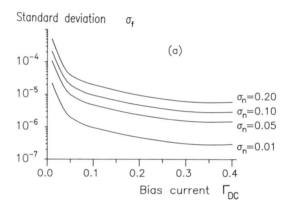

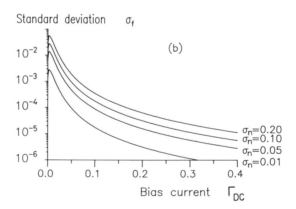

Figure 8. Standard deviation of revolution frequency σ_f for white Gaussian noise drive as a function of bias current Γ_{DC}, for $\alpha = 0.01$, $\beta = 0$, $\ell = 8$, and $\sigma_\Gamma = 0.01$, 0.05, 0.10, and 0.20. (a) Numerical simulation. (b) Hamiltonian perturbation theory.

seen, the perturbation theory is able to predict the right qualitative dependence on the length, the noise amplitude, and the bias, but it predicts an overall standard deviation that is about a factor of 10 too large. The reason for this discrepancy is at present not known.

7.3.2 *Phase locking to an external microwave signal*

Here we review the phase locking experiments reported in [10]. Phase locking phenomena in nonlinear oscillators are of interest both intrinsically and for applications.   In particular, for Josephson junction oscillators, phase locking offers the possibility of raising output power to practically useful levels. To this end, we consider the dynamic behaviour of an intermediate-length, overlap geometry Josephson junction exposed to microwave radiation in the 26-40 GHz band.   Experimental data are reported for a Nb-NbOx-Pb junction. The mathematical model is that of Eqs. (1) and (14a) with $\ell \sim 1$. For such junctions, there are three fundamental regimes of behaviour, each of which gives rise to specific structures in the dc current-voltage (I-V) characteristic of the junction:

7.3.2.a *Oscillations associated with Zero Field Steps (ZFS).*
Here we have $\Gamma_{AC} = 0$ and $\eta = 0$ or $\eta > 0$ but small.   This oscillation may be interpreted as a parametrically excited resonance in which the exciting term is due to the spatially homogeneous part of ϕ, and where the frequency of the excited, spatially inhomogeneous term is one-half of that of the exciting term.

7.3.2.b *Oscillations associated with Fiske Steps (FS).* Here we have $\Gamma_{AC} = 0$ and $\eta > 0$ and not necessarily small.   This oscillation may be interpreted as a nonlinear cavity resonance in which the exciting term is again due to the spatially homogeneous part of ϕ, but the frequency of the excited, spatially inhomogeneous term is now equal to that of the exciting term. This means that the voltage location of FS1 is one-half that of the ZFS1, even though they have the same fundamental oscillation frequency.

7.3.2.c *Oscillations associated with RF-induced steps.* Here we have $\Gamma_{AC} \neq 0$ and $\eta > 0$. This oscillation is simply a direct phase locking of the spatially homogeneous part of ϕ to the applied microwave signal. In the I-V characteristic it gives rise to a constant-voltage step at a voltage corresponding to the frequency of the microwave signal (and harmonics).
   We now consider what happens if we apply a microwave signal with a frequency comparable to that of the ZFS and FS resonances. Experimentally, one observes the formation of vertical (constant voltage) regions on both FS1 and ZFS1, depending on the value of the applied magnetic field.   The amplitude of these regions increases with increasing microwave power.   Although only a

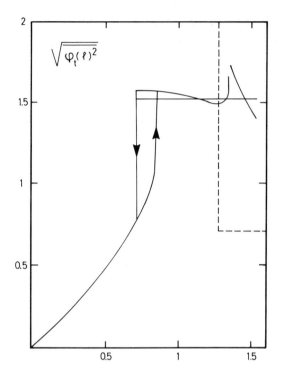

Figure 9.   (See facing page).

qualitative theoretical understanding of these phenomena is presently available, it is clear that phase locking effects play a major role.

7.3.3   *Pattern competition and chaotic phenomena in large, rf-driven junctions*

Here we review the numerical simulation studies reported in [11].

The transition to chaos in a driven, damped sine-Gordon model of a long, microwave-driven Josephson junction is shown to exhibit an infinite sequence of period-doubling bifurcations leading to chaos [12] - the first bifurcation has been found experimentally [13]. This first bifurcation occurs, however, after a narrow regime characterized by an intermittency type chaos and quasiperiodic oscillations resulting from the drastic change in pattern from the stationary period-one solution to the stationary period-two solution.   The intermittency arises in a process of pattern competition where quasi-stable patterns described by the presence of one and two breather-like modes are frequency locked to the driving signal.   The patterns describing the stationary solutions in the following period doublings are very similar to the pattern describing the period-one solution.

We consider the perturbed sine-Gordon Eq. (1a) with $\beta = \Gamma = 0$ and boundary conditions

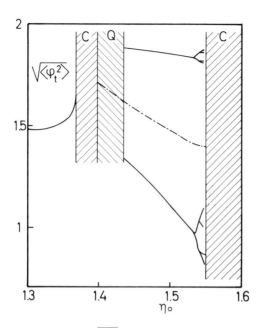

Figure 9.   The rms value $\sqrt{<\phi_t^2>}$ vs. the amplitude of the applied field η_o.   Full lines are the rms values, the dot dashed lines are the average in the region with quasi-periodic and period doubled solutions.   In (b) C indicates chaotic regimes, Q indicates quasi-periodic regime.   The first period doubling occurs through a chaotic regime.

$$\phi_x(0,t) = 0 \qquad\qquad (19a)$$

and

$$\phi_x(\ell,t) = \eta_o \sin \Omega t. \qquad\qquad (19b)$$

In Fig. 9a (see facing page) we show a plot of the rms value of ϕ taken at $x = \ell$ as a function of the amplitude η_o.   The sampling interval follows the period of the driving signal Eq. (19b). Figure 9b shows the region from $\eta_o = 1.3$ to $\eta_o = 1.6$ in enlarged form.   The full lines are the rms values, the broken line is the average rms value.   The cross hatched regions marked C and Q are characterized by chaotic and quasiperiodic motion, respectively. In the first C region (to the left) the motion is chaotic due to pattern competition while in the C region to the right the chaotic motion has been preceded by an infinite sequence of bifurcations. The first bifurcation is seen to occur through the C and Q regions to the left.   Furthermore, the average rms value increases as the first C region is approached.   Summarizing, it has been shown that in the present system competition between patterns gives rise to intermittency and chaos.

7.4   Non-autonomous, But Spatially Homogeneous Systems: Subharmonic Generation in Small Josephson Junctions

The small Josephson tunnel junction has attractive properties as a model system for the study of nonlinear dynamics in (0 + 1)-dimensions. In this section we review results presented in [15] and [16] regarding subharmonic generation, in particular period-doubling and period-tripling, in small microwave-driven junctions obtained both experimentally and through digital computer simulations, including effects of noise and near-resonant perturbations on a period-doubling bifurcation. One of the predictions for small microwave-driven junctions is the existence of narrow bands of even- and odd-subharmonic solutions in the microwave frequency-amplitude plane, often separated by regions of chaos [17]. Measurements were made on underdamped Nb-NbOx-Pb overlap tunnel junctions. The linear dimension was small compared with the Josephson penetration depth. For a given pump frequency the one-third harmonic signal appeared in a narrow range of pump power. An example at about 4 GHz is given in Fig. 10, with the pump signal at about 12 GHz superimposed for comparison. The signal was as much as 50 dB avbove the noise level of the detector with a linewidth of less than 100 Hz. In the amplitude-frequency plane the one-third harmonic generation occurred for ranges of microwave current and frequency that were in reasonable agreement with the results of digital computer simulations [15].

The phenomenon of small signal amplification in bifurcating

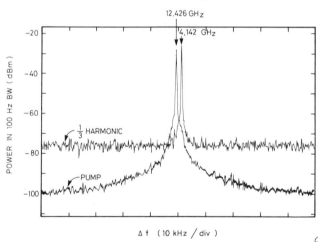

Figure 10. Experimental observation of the one-third harmonic generation in small Josephson junction. Typical spectrum analyzer display showing one-third harmonic generation in a Josephson tunnel junction dc biased in the zero-voltage state. For comparisons of the linewidths the stored display of the microwave pump signal at 12.426 GHz and the one-third harmonic signal emitted by the junction are shown on the same frequency span. As seen the one-third harmonic generation does not lead to noise broadening of the line (within the 100 Hz resolution bandwidth of the spectrum analyzer).

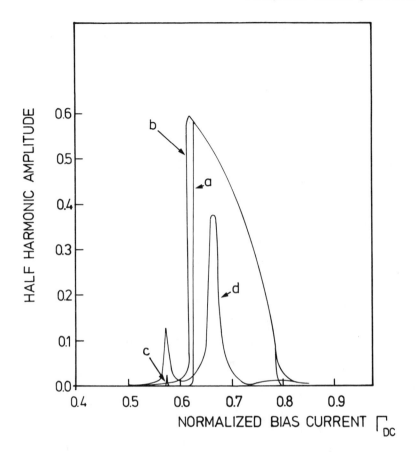

Figure 11. Example of a calculation of the dynamic stabilization. Numerical results for the amplitude of the period-doubled signal versus the dc bias, Γ (the control parameter).

a) no noise $\quad (\Gamma_n=0)$, no perturbation $\quad (\Gamma_p=0)$.

b) with noise $\quad (\Gamma_n>0)$, no perturbation $\quad (\Gamma_p=0)$.

c) no noise $\quad (\Gamma_n=0)$, with perturbation $(\Gamma_p>0)$.

d) with noise $\quad (\Gamma_n>0)$, with perturbation $(\Gamma_p>0)$.

Parameters: $\alpha = 0.2$, $\Gamma_d = 1.0$, $\Omega_d = 1.6$, $\Gamma_p = 0.01$. $\Omega_p = (127/256)1.6$ and $2ek_BT_{noise}/\hbar I_o = 10^{-4}$.

dynamical systems has recently received renewed attention [18]. A related problem is that of dynamic stabilization of a nonlinear system against period-doubling bifurcations by a periodic perturbation. We consider a dynamical system driven at a frequency Ω_d. At a particular value, μ_o, of a control parameter μ the system undergoes a period-doubling bifurcation to a limit cycle of frequency $\Omega_d/2$. A periodic perturbation with a frequency

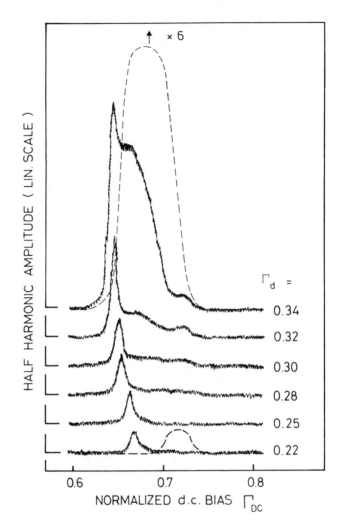

Figure 12. Experimental observation of the dynamic stabilization. Experimental data for the amplitude of the periodic-doubled signal versus the dc bias for a number of values of the drive amplitude, Γ_d. Dashed curves: $\Gamma_p = 0$ (no perturbation). Full curves $\Gamma_p = 6.4 \times 10^{-3}$. (Parameters: $\alpha = 0.05$, $\Omega_d = 1.66$, $\Omega_d/2 - \Omega_p = 10^{-5}$ and $2ek_BT_{noise}/\hbar I_o = 5 \times 10^{-3}$).

Ω_j in the vicinity of $\Omega_d/2$ is, however, shown to suppress the period-doubled limit cycle, giving rise to stable motion beyond μ_o. Furthermore the perturbation itself may be amplified when the control parameter is just below μ_o [18].

With the Josephson tunnel junction as a model system we have carried out both numerical simulation (including noise) and measurements on junctions in order to study this dynamical stabilization [16]. In natural normalized units the equation of motion for the phase difference across the junction is:

$$\phi_{tt} + \alpha\phi_t + \sin\phi = \Gamma_{DC} + \Gamma_d \sin\Omega_d t + \Gamma_j \sin\Omega_j t + \Gamma_n(t). \quad (20)$$

Some results of numerical simulations of this equation shown in Fig. 11 illustrate the influence of noise and a stabilizing periodic signal on the period-doubling bifurcation. Here the calculated amplitude of the periodic-doubled signal is plotted versus the control parameter, which is the normalized dc bias, Γ_{DC}. We note that the apparent importance of the noise in trace (d) is remarkable since in the non-stabilized case of trace (b) the effect of the noise is only minor and quantitative. The lesson to learn from such simulations is that for a meaningful comparison with experiments the theory of the dynamic stabilization must include the effects of noise. For comparison we have carried out an experimental investigation of the effect of a small near-resonant perturbing microwave signal on the threshold for half-harmonic generation in Josephson tunnel junctions. A typical set of experimental traces is shown in Fig. 12. The suppression of the periodic-doubled signal does not occur abruptly as predicted in [18] but rather more continuously. The most interesting result of our experimental investigation is, however, the appearance of a new instability region close to the original bifurcation point (showing up as new satellite peaks in Fig. 12). We assume that this instability is of the same type as the one seen in the digital simulations (Fig. 11, trace (d)).

Acknowledgements

The financial support of the Danish Natural Sciences Research Council and the European Research Office of the United States Army (through contract No. DAJA-45-85-C-0042) to part of the work in the present paper is acknowledged.

References

[1] S. Pagano, M.P. Soerensen, R.D. Parmentier, P.L. Christiansen, O. Skovgaard, J. Mygind, N.F. Pedersen and M.R. Samuelsen (1986). *Phys. Rev.* **B33**, 174.

[2] R.D. Parmentier (1978) in *Solitons in Action,* ed. K. Lonngren and A.C. Scott, 173-199, Academic Press, New York.

[3] S.E. Burkov and A.E. Lifsic (1983). *Wave Motion* **5**, 197.

[4] K. Enpuku, K. Yoshida and F. Irie (1981). *J. Appl. Phys.* **52**, 344.

[5] S. Pagano, N.F. Pedersen, S. Sakai and A. Davidson (1987). *IEEE Trans. on Magn.* **MAG-23**, 1114.

[6] S. Sakai and N.F. Pedersen, *Phys. Rev. B*, in print.

[7] A. Davidson, N.F. Pedersen and S. Pagano (1986). *Appl. Phys. Lett.* **48**, 1306.

[8] F. If, P.L. Christiansen, R.D. Parmentier, O. Skovgaard and

M. P. Soerensen (1985). *Phys. Rev.* **B32**, 1512.

[9]   M. Fordsmand, P. L. Christiansen and F. If (1986). *Phys. Lett.* **A71**.

[10]   G. Costabile, G. Eriksen, S. Pagano and R. D. Parmentier (1986). *MIDIT 1986 Workshop*, Lyngby, Denmark.

[11]   O. H. Olsen and M. R. Samuelsen (1986). *MIDIT 1986 Workshop*, Lyngby, Denmark.

[12]   O. H. Olsen and M. R. Samuelsen (1985). *Appl. Phys. Lett.* **47**, 1007.

[13]   Yu. Ya. Divin, J. B. Bindslev Hansen, J. Mygind, O. H. Olsen, N. F. Pedersen and M. R. Samuelsen (1985). *SQUID-85 Proceedings of the Third International Conference on Superconducting Quantum Devices*, 507 pp, Walter de Gruyter, Berlin.

[14]   O. H. Olsen and M. R. Samuelsen (1986). *Phys. Rev.* **B34**, 3510.

[15]   J. Bindslev Hansen, J. Clarke, J. Mygind, G. A. Ovsyannikov and H. Svensmark (1986). *Appl. Phys. Lett.* **48**, 1744.

[16]   H. Svensmark, J. Bindslev Hansen and N. F. Pedersen (1987). *Phys. Rev.* **A35**, 1457.

[17]   N. F. Pedersen and A. Davidson (1981). *Appl. Phys. Lett.* **39**, 830.

[18]   K. Wiesenfeld and B. McNamara (1985). *Phys. Rev. Lett.* **55**, 13; (1986). *Phys. Rev.* A33, 629.

AN INTERESTING NON-LINEAR EFFECT IN A NEURAL NETWORK

Abstract

The response to a variety of input patterns has been studied in a vector model assembly of interconnected neurons. The time evolution of the injected signal was followed, attention being paid to both its subsequent topology and phase. The model was realistic in that it included action potential impulses in the axon regions, statistically distributed synaptic delays, and electrotonic waves in the dendrites. It was observed that the fate of an injected signal was sensitive to the relative phases of the signal's constituent parts, and that the neural network was thus acting as a non-linear discriminator.

8.1 Introduction

There has recently been shown much interest in the properties of neural network models [18,9,1,14,12,2]. This article addresses a well-defined issue: does coherent firing of individual neurons play a role in the function of the cerebral cortex? Its main purpose is to present the results of a computer simulation of a neural network in which the exact timing of impulses is indeed of paramount importance. And it is demonstrated that such a network would have potentially useful powers of discrimination and recall.

It is reasonably clear that the timing of the arrival of nerve impulses at a given neuron cannot be a matter of total indifference. The voltage across a neural membrane relaxes back towards its resting value, once a stimulus has been removed, so it is easy to envisage situations in which incoming impulses will fail to provoke a response unless they can act in unison by arriving simultaneously, or nearly so, at the somatic region.

The wherewithal to probe the extent of correlated neural firings should soon be at hand. A number of centres have recently developed facilities which permit simultaneous recording from several neurons [6,16,11], and there has already been some discussion of the interpretation of the voluminous data which these set-ups will no doubt produce [10,8,7]. We should soon know whether correlated trains of impulses do occur in the cortices of alert experimental animals, under the influence of appropriate stimuli.

Meanwhile, one must resort to simulations of the type described

in this article, which is divided up in the following manner. The
necessary neural background is presented in the next section, and
there then follows a discussion of the concept of coherence and
correlation, in the neural context. The computer model is then
described, and separate sections are devoted to the discussion of
possible applications to memory and autism. The article closes
with a general discussion of the issue of coherence.

8.2 Coherence and Correlation in the Neural Context

The central idea in this article is that the exact timing of the
arrival of nerve impulses at the extremities of a given neuron is
critically important. The docking of neurotransmitter molecules
with receptors on the post-synaptic membrane touches off the
electrotonic wave which flows down the dendrite towards the soma.
But this wave is attenuated in both space and time and if it is to
make a contribution to the collective influence of all incoming
waves, at the soma, it must arrive at approximately the same time
as waves which have travelled along other dendrites. If waves
arrive out of phase with one another, their potential for raising
the somatic voltage above its threshold value will not be
realized.

There are actually three time constants to be considered, in
connection with this issue, and we may take them in the order in
which they are encountered as information flows from one soma, out
along the axon and axonal branch, across the synapse, and finally
along the dendrite towards the next soma. It is a common fallacy
to imagine the first of these steps as occupying an appreciable
amount of time; in one's mind's eye, the impulse is seen as a
well defined blip travelling out along the axon, not unlike a
soliton wave. This is quite misleading, as the following little
calculation will demonstrate. The duration of an action potential
pulse is about 1 millisecond, and the pulse travels at very
roughly 20 metres per second (though actually faster or slower
than this depending upon the degree of myelination). The spatial
extent of the pulse, measured between its leading and trailing
edges is thus about 20 millimetres. But a typical axon, in the
cortex, might be a mere 0.5 millimetres, so the trailing edge of
the impulse will not even have left the soma by the time the
leading edge has reached the extremities of the axon branches.

The other two time constants are appreciably longer, and it is
through them that pulses initially coherent with each other can
get out of phase. This is true of the synaptic delay, with its
approximately one millisecond duration, and it is true to an even
greater extent of the electrotonic delay, which is typically ten
times longer.

It is possibly quite significant, in this context, that some
synapses connect axon branches directly with somatic regions [19].
These axonal-somatic synapses would enable one soma to pass
information on to another soma without the dendritic delay
intervening, and this could permit the presynaptic neuron to act
as a sort of pacemaker, as discussed later in this article.

8.3 **The New Computer Model**

We will now describe a recently constructed computer model which aims at testing the idea of coherence, and at elucidating possible consequences of this mode of action. The model comprises a series of layers each consisting of the same number of cells, and with all possible combinations of the cells in two adjacent layers having unidirectional synaptic contacts. A single axonal input is assumed to feed into each synapse, and the latter is assumed to be followed by a single dendritic pathway to the subsequent somatic region. Because of the unidirectionality, an input pattern to the first layer, consisting of action potential pulses or a lack of these, will give rise to further patterns of firings and failures to fire, travelling down through the model layer by layer. Whether a particular synapse is excitatory or inhibitory is chosen by a random number generator, and this type of choice is also applied to the initial synaptic strengths, to their maximum values, and also to the time constants and maximum amplitudes of the electrotonic responses in the associated dendritic regions. Finally, the random number generator is also used to select a distributed set of values for the synaptic delays.

At any time t, the input voltage at the soma of the jth cell in the ith layer will be given by

$$V_{j,i}(t) = \sum_k \Phi_{k,i-1}(t)\, S_{j,i,k,i-1}\, D_{j,i,k,i-1}(t)$$

where the k summation is over the n neurons in layer i-1, and $\Phi_{k,i-1}(t)$ is the output voltage of the kth neuron in that layer. This output voltage can take only the (normalized) values of 1 or 0, depending on whether or not the input voltage to that cell reached the threshold value θ. The factors S are measures of the current synaptic strengths, while the terms D(t) take account of the temporal variation of the dendritic voltages. At each time step the value of $V_{j,i}(t)$ is compared with θ, and the corresponding output voltage $\Phi_{j,i}(t)$ is set at 1 or 0 depending upon whether that threshold has or has not been exceeded.

A study of the properties of this system, by computer simulation, has revealed several interesting modes of behaviour, one of which was certainly quite unexpected. The natural time-constant of such a system is determined by the minimum possible time lapse between successive pulses generated in a given cell, that is to say by the refractory period. A periodic input is given to the first layer of the system, and one then studies the successive generation of impulses in cells in the lower layers. If, for instance, some of the cells in the first layer are given impulses which are coincident with one another, it is found that there is a periodic generation of impulses in the lower layers, at the same frequency as the input frequency. But if cells in the first layer are given impulses which are temporarily offset from one another, a new phenomenon is observed, namely that after an initial transient period, it appears to be impossible for the cyclic state to maintain itself beyond a certain level in the system. This has been given the tentative name "pinch-out", and the phenomenon is illustrated in Fig. 1. It is observed for all phase offsets lying in the aproximate range $150°$–$210°$, for the particular conditions of this investigation.

It is interesting to speculate on the possible advantage, to the brain, of such a phenomenon. It could indicate that signals are unable to penetrate to higher regions of the cortex, unless there is the requisite degree of synchronization between the various inputs to the first layer of cells. This, in turn, could mean that the appropriate regions act as a sort of coherence discriminator. Indeed, there is the suggestion that one could have a piece of cerebral hardware designed to respond to correlations between various inputs.

```
    1            8           14           20           21           61
*000000*     .000000.     .000000.     .000000.     00000000     *000000*
00000000     000000*0     000000.0     000000.0     000000.0     00000000
00000000     00000000     0*000000     0.000000     0.000000     00000000
00000000     00000030     00000000     00000000     00000000     00000000
00000000     00000000     00000000     00000q00     00000000     00000000
00000000     00000000     00000000     00000000     00000000     0.0000.0
00000000     00000000     00000000     00000000     00000000     .00000..
00000000     00000000     00000000     00000000     00000000     000000..
00000000     00000000     00000000     00000000     00000000     .0..00..
00000000     00000000     00000000     00000000     00000000     *00..000
00000000     00000000     00000000     00000000     00000000     000.0000
00000000     00000000     00000000     00000000     00000000     00000000
00000000     00000000     00000000     00000000     00000000     00000000
00000000     00000000     00000000     00000000     00000000     00000000
00000000     00000000     00000000     00000000     00000000     00000000
00000000     00000000     00000000     00000000     00000000     00000000

   82          200          421          436          781          961
00000000     .000000.     *000000*     .000000.     *000000*     *000000*
000000.0     000000.0     00000000     000000.0     00000000     00000000
0.000000     0.000000     00000000     0.000000     00000000     00000000
00000000     00000000     00000000     00000000     00000000     00000000
00000000     00000000     00000000     00000000     00000000     00000000
0.0000.0     0.0000.0     0.0000.0     0.0000.0     0.0000.0     0.0000.0
.00000..     .00000..     .00000..     .00000..     .00000..     .00000..
.0000...     .0000...     .0000...     .0000...     .0000...     .0000...
.0...0*0     00..000.     00..000.     00..00..     00...000.    00...000.
.00..000     000.0000     000*0000     000.0000     000*0000     000*0000
00.*00.0     000.00.0     000.00.0     000.00.0     000.00.0     000.00.0
nnnnn000     000000..     000000..     000000..     000000..     000000..
00000000     000.*.00     000...00     000*..00     000...00     000...00
00000000     00.0..0.     00.0..0.     00.0..0.     00.0..0.     00.0..0.
00000000     .00.00.0     *00.00.0     .00.00.0     *00.00.0     *00.00.0
```

Figure 1(a).   This computer model consists of fifteen topological layers, each comprising eight cells. Synaptic contacts are made between each cell in a given layer and all eight cells in the following layer. (There are thus 64 synapses between each pair of adjacent layers.) The transmission of information is unidirectional, from top to bottom, and the state of each cell is indicated by an 0, for the quiescent state, * for the moment of firing off an action potential, and a dot (.) is used if the cell is in its refractory period, which is of a standard length of 20 computational time steps. The electrotonic time constants were randomly selected, and uniformly distributed in the interval 1–75 time steps, while the synaptic delays were in this case all a standard single time step. The synapses were either excitatory or inhibitory, this being chosen at random. In real time, one time step is about 0.5 ms. The periodic input pattern consisted of simultaneous firings of the first and eighth cells in the first layer, with a period of 60 time steps. As can be seen from the situations at these twelve different instants, the network achieves a cyclic state, the period of which matches that of the input.

```
        1              17             20             21             30             37
   *0000000       .0000000       .0000000      00000000      0000000*      0000000.
   00000000       000000*0       000000.0      000000.0      00000.0       0.0000*0
   00000000       00000000       00000000      00000000      0.000000      0.000000
   00000000       00000000       00000000      00000000      00000000      00000000
   00000000       00000000       00000000      00000000      00000000      00000000
   00000000       00000000       00000000      00000000      00000000      00000000
   00000000       00000000       00000000      00000000      00000000      00000000
   00000000       00000000       00000000      00000000      00000000      00000000
   00000000       00000000       00000000      00000000      00000000      00000000
   00000000       00000000       00000000      00000000      00000000      00000000
   00000000       00000000       00000000      00000000      00000000      00000000
   00000000       00000000       00000000      00000000      00000000      00000000
   00000000       00000000       00000000      00000000      00000000      00000000
   00000000       00000000       00000000      00000000      00000000      00000000
   00000000       00000000       00000000      00000000      00000000      00000000

        61             200            421            450            781            990
   *0000000       .0000000       *0000000      0000000*      *0000000      0000000*
   00000000       000000.0       00000000      000000.0      00000000      000000.0
   ..00.000       .0000000       ..00.000      0.00000       ..00.000      0.00000
   .0*0.000       .0.00000       .0*0.000      .0.00000      .0*0.000      .0.00000
   000.0000       0.000000       0.0.0000      0.00000       0.0.0000      0.000000
   0.0000..       000.0..0       000.0...      000.0..0      000.0...      000.0..0
   .0000000       00000.00       00000.00      00000.00      00000.00      00000.00
   000000.0       0000..00       0000..00      0000..00      0000..00      0000..00
   .0000000       0000.00.       0000*.0.      0000..0.      0000*.0.      0000...0.
   00000000       00..0000       .0.00000      .0.00000      .0.00000      .0.00000
   00000000       000.00.0       00.00000      00.00000      00.00000      00.00000
   00000000       000000.*       00000000      00000000      00000000      00000000
   00000000       000...00       00000000      00000000      00000000      00000000
   00000000       00*0*.0.       00000000      00000000      00000000      00000000
   00000000       .00.00.0       00000000      00000000      00000000      00000000
```

Figure 1(b).   An antiphase input to the model shown in 1(a) gave a dramatically different response.   The first cell in the first layer is made to fire at times 1, 61, 121 ....., while the eighth cell in that layer is made to fire at times 31, 91, 151 .....   As can be seen from these selected situations, information is able to reach the fifteenth layer only during an initial transient period.   Thereafter, despite continuation of the input, nothing is able to penetrate beyond the eleventh layer.   This phenomenon has been dubbed "the pinch-out effect".

8.4 Possible Application to a Pathological Condition

Any model which claims to reflect reality must win its spurs by explaining reality's exceptions and idiosyncracies.   The question thus arises as to whether a model of the type described in this article has anything to say about pathological conditions.   It has been suggested that autism, which usually manifests itself during a patient's first four years of life, might be caused by unduly long synaptic delays [3].   Synaptic delays are a parameter in the present model, so this issue could indeed be investigated.

The autistic child is most often quite free from physical abnormality, and the chief symptom is a gross reticence or inability to interact with the environment.   The patient appears apathetic to both people and objects, and in the early stages this can be mistaken for contentment.   The condition apparently has an organic aetiology [4,15] with hereditary origins [5].   Possibly the strongest recent endorsement of the organic view comes from the widely observed, but inadequately documented, fever effect [20].   When autistics have a moderate fever, they invariably display dramatically more normal behaviour patterns, including a greater desire or ability to communicate.   The effect appears to reach a maximum for fevers of around 2°C.   It seems unlikely that such a modest rise could apreciably influence the rates of either the metabolic processes or the molecular diffusion involved in neural function. But temperature changes of as little as $1^{O}C$ can markedly alter the fluidity of membranes [21], such as those which form the synapses and the neurotransmitter-charged presynaptic vesicles.

The time-limiting factor in the synaptic delay is usually the kinetics of the calcium ion entry process which precedes vesicle-synapse fusion, and the necessity for multiple Ca^{2+} binding to trigger release [13]. But in the case of an incorrect lipid composition, the vesicle-synapse fusion might be unusually sluggish, and an increase in membrane fluidity would lower the vesicle-synapse fusion time, and thereby decrease the synaptic delay.

It is clear that these issues would be amenable to investigation with the aid of computer models of the type described in this paper, and such studies have recently produced some most interesting results. The model had thirty-two cells in each of its fifteen topological layers, in this case, and the parameter of interest was, of course, the synaptic delay. Figure 2 shows the dramatic result of changing the mean value of the latter from 0.5 ms to 2.0 ms. For the longer synaptic delay, the pinch-out effect is again observed.

This is a most intriguing result because it offers a particularly direct explanation of what might lie at the heart of the autistic syndrome. At normal body temperature, the patient's faulty lipid profile gives the synaptic delays that are too long and information is not able to traverse some critical part of the brain because of the pinch-out effect. During a sufficiently high fever, the increased synaptic membrane fluidity gives lower synaptic delays: the pinch-out effect disappears, the information gets through, and the patient appears to recover almost dramatically, only to go back into his or her invisible shell once the fever subsides.

8.5 Discussion

The main result of the present study is, of course, the observation of what has been provisionally called the pinch-out effect. Its discovery was serendipitous, but in retrospect its existence might have been anticipated, particularly if the key role of non-linearity in life's processes had been properly appreciated. One has only to contemplate the intricacies of tissue differentiation, for example, to see how abrupt changes of regime are so central to an organism's structure and function. In the case of the neuron the regime in question is its firing state, and the action potential threshold sets the watershed for the two totally different modes of behaviour. And although the changes in the firing patterns of those other neurons which feed into a given neuron might appear minor, their net effect could be to shift the latter neuron's somatic voltage across the threshold and thus either quench its activity or produce activity where there was previously none. This is non-linearity par excellence, and the pinch-out effect is its stark manifestation.

An important question immediately arises, however: does Nature try to compensate for such an eventuality? Pinch-out is an

```
0000000000000000.0.0.0.0.0.0.0.      0000000000000000.0.0.0.0.0.0.0.
0000.000000000000000000000*00.0      000000000000000000000000000.0..0
0000000000000000000000000000.00      0000000000000000.000000000000000
000000000.00000000000000000000       000000000.000000000000.000000
000000000000000000000000000000       000000000000000000000000000000
000000000000000000000000000000       000000000000000000000000000000
000000000000000000000000000000       000000000000000000000000000000
000000000000000000000000000000       000000000000000000000000000000
000000000000000000000000000000       000000000000000000000000000000
000000.,0000000000000.000000000      000000000000000000000000000000
.00.0000...00.00000.000..000.000     000000000000000000000000000000
00..0*0.0000.0.00.00..000.0...00     000000000000000000000000000000
000..0..000000...0000.....*..00.     000000000000000000000000000000
00000000..0000......0...0.*0..000    000000000000000000000000000000
0.0.0.00000.000.00.0.0.000000.0...   000000000000000000000000000000
```

Figure 2. The pinch-out effect can also be induced by increasing the spread in the synaptic delays. The mean delay in the model was a single time unit for the situation shown at the left, in which information is clearly able to penetrate to the lower layers. But an increase of the mean synaptic delay to four time units gives pinch-out, as seen at the right. The model comprised fifteen layers, each with 32 neurons, and both pictures correspond to the situation at 800 time steps.

attractive proposition, as we shall discuss shortly, but it behoves us to play the Devil's Advocate and ask what might oppose its occurrence. We have indeed already invoked the buffering function of the interneurons, encouraging more activity where little is in evidence and dampening things down when the activity level becomes excessive. With the interneurons acting in the way described in section 8.4 of this paper, pinch-out is nevertheless observed. But this does not rule out the possibility that the brain's control mechanisms are so exquisite as to preclude pinch-out in practice. More comprehensive computer investigations will be required, to elucidate this important issue.

If pinch-out really does play a role in the brain, its utility would seem to be reasonably transparent. One of the central questions in neuroscience concerns the nature of higher processing in the cortex. The mechanisms underlying the early sensory processing seem well on the way to being understood, but the manner in which the resulting information is subsequently handled remains a virtually total mystery. Because such target structures as the muscles and the glands have to be informed whether or not they are to respond to a given set of sensory inputs, the task of the brain's higher processing will, in the final analysis, be one of passing on or not passing on information, and this suggests an important mandate for the pinch-out effect: to provide a mechanism whereby brain components discriminte between information to be transmitted to target structures and information to be blocked or ignored.

If the brain wants to exploit the regulatory possibilities inherent in coherence discrimination, it has the necessary circuitry at its disposal. We have already noted that the dendritic delay is avoided when an axonal branch synapses directly on to the somatic region of the following cell, and the point can now be made that the presynaptic cell in such a case acts like a

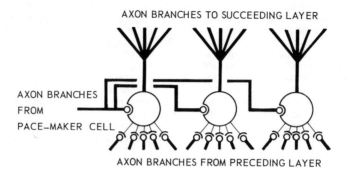

AXON BRANCHES TO SUCCEEDING LAYER

AXON BRANCHES FROM PACE—MAKER CELL

AXON BRANCHES FROM PRECEDING LAYER

Figure 3. A hypothetical (and highly schematic) layout for forced synchronous firing of a number of cells in a topological layer. The thin straight lines are dendrites, and the thick straight lines are axon branches. The larger circles are somatic regions, while the small circles and their associated crescents indicate synapses. The pace-maker cell could be a spontaneous emitter of action potential bursts, but it could also be collectively driven by the cells from the layer preceding those shown in the diagram.

pace-maker. Consider, for instance, the situation in which neither the depolarization of the target cell's somatic membrane by the incoming dendritic waves nor that due to the incoming signal across the conjectured axonal-somatic synapse is sufficent, by itself, to exceed the threshold. If, however, their combined effect is such as to provoke a super-critical depolarization, the result will be not merely an output pulse but one whose timing is dictated by the signal arriving via the axonal-somatic synapse. Indeed, if a cell makes axonal-somatic synapses with a number of target cells it will, like the sergeant on a parade ground, keep all the latter in step. A highly schematic layout of such a situation is depicted in Fig. 3.

But pace-maker neurons, even if they exist, can only be part of the story; we must ask about their location with respect to the sensory input. Two possibilities suggest themselves. One is that the pace-makers are independent and spontaneous emitters of bursts of action potential impulses. It is worth emphasizing, in fact, that recent evidence points to an intracellular origin of such bursts, rather than cyclic patterns due to feed-back effects in an assembly of neurons [17]. The most obvious alternative would place the pace-maker in a topological position analogous to that of an interneuron. This would be particularly intriguing; it implies that the approaching rabble of haphazard signals, in the various input cells, collectively waken the sleeping sergeant, who responds to the reveille by getting the troops into step.

It might appear that the intercellular coherence generated by the hypothetical pace-maker cells would preclude phase differences occurring, but it must be borne in mind that phase differences will also arise because of dendritic path differences, and that these, in turn, will be linked to different input patterns. The induced coherence would thus serve to provide an invariant base-line, against which the various phases woud be revealed and distinguished.

Acknowledgements

It is a pleasure to acknowledge stimulating discussions with John Clark, and with the members of the Lundtofte Brain Study Group: Oyvind Aabling, Stig Benthin, Ulla Binau, Steen Sloth Christensen, Eva Gamwell Davids, Uffe Hansen, Kenneth Hebel, Michael Hebel Malene Dal Jensen, Allan Krebs, Carlo Lund, Stephan Mannestaedt, Kaare Olsen, Lars Petersen, Tonny Haldorf Petersen, Henrik Kaare Poulsen, Carsten Rogaard, Nils Svendsen and Jorgen Villadsen.

References

[1]   S. I. Amari and M. A. Arbib (eds.) (1982). *Competition and Cooperation in Neural Nets,* Springer, Berlin.

[2]   E. Bienstock, F. Fogelman-Soulie and G. Weisbuch (eds.) (1986). *Disordered systems and biological organisation,* Springer, Berlin.

[3]   R. M. J. Cotterill (1985). *Nature* **313,** 426.

[4]   A. R. Da Masio (1978). *Arch. Neurol.* **35,** 777-788.

[5]   S. Folstein and M. Rutter (1977). *Nature* **265,** 726-728.

[6]   G. L. Gerstein, M. J. Bloom, I. E. Espinosa, S. Evanczuk and M. R. Turner (1983). *IEEE Trans. on Systems, Man and Cybernetics* **13,** 668-676.

[7]   G. L. Gerstein, D. H. Perkel and J. E. Dayhoff (1985). *J. Neuroscience* **5,** 881-889.

[8]   A. S. Gevins, J. C. Doyle, B. A. Cutillo, R. E. Schaffer, R. S. Tannehill, J. H Ghannam, V. A. Gilcrease and C. L. Yeager (1981). *Science* **213,** 918, 922.

[9]   U. an der Heiden (1980). *Analysis of neural networks,* Springer, Berlin.

[10]   C. K. Knox and R. E. Poppele (1977). *J. Neurophysiology* **40,** 616, 625.

[11]   M. Kuperstein and H. Eichenbaum, Unit activity, evoked potentials and slow waves in the rat hippocampus and olfactory bulb recorded with a 24-channel microelectrode. *Neuroscience,* in press.

[12]   W. B. Levy, J. A. Anderson and Lehmkuhle (1985). *Synaptic modification, neuron selectivity and nervous system organisation,* Erlbaum, Hillsdale.

[13]   A. R. Martin (1985). in *Handbook of Physiology: Section I - The nervous system, Volume 1, Cellular biology of neurons, Part 1, Chapter 10,* American Physiological Society, Bethesda.

[14]   G. Palm (1982). *Neural Assemblies,* Springer, Berlin.

[15]   L. R. Piggott (1979). *J. Autism Dev. Disorders* 9, 199-218.

[16]   H. J.P Reitboeck (1983). *IEEE Trans. on Systems, Man and Cybernetics* **13,** 676-683.

[17]   D. F. Russell and D. K. Hartline (1978). *Science* 200, 453-456.

[18]   A. C. Scott (1977). *Neurophysics,* John Wiley, New York.

[19]   G. M. Shepherd (1979). *The Synaptic Organization of the Brain,* Oxford University Press.

A neural network

[20] R.C. Sullivan (1980).   *J. Autism Dev. Disorders* 10, 231-241.
[21] H. Trauble, M. Tuebner, P. Wooley and H. Eibl (1976).
Biophys. Chem. **4**, 319-337.

RELAXATION KINETICS IN BISTABLE SYSTEMS

9.1 Introduction

The occurrence of different steady states for the same operating
conditions is a characteristic feature of many driven systems.
This property first appeared in relation to the thermal explosion
problem [1,2].   Thereafter multistability has been presented as
the simplest example of a nonequilibrium transition: i.e. an
instability taking place in systems with characteristic feedback
mechanisms   and   maintained   sufficiently   far   from   thermal
equilibrium [3].   The analogy with equilibrium phase transitions
has therefore been emphasized by various authors [3,4].   However
it is only recently that experiments have been performed which are
accurate enough to allow a quantitative test of this analogy.
The relaxation of a perturbed system to its steady state
provides a useful method for characterizing its essential dynamic
features [5].   In particular, this method has been applied to open
systems exhibiting bistability or critical points.
In this note, we give an interpretation of these relaxation
experiments which stress their general significance.

9.2 Bistability and Critical Points in the C.S.T.R.

The continuous flow stirred tank reactor (C.S.T.R.) has proved a
particularly suitable experimental device for the study of the
dynamics of open chemical systems.   The reactants are fed into the
reactor at a constant flow rate Q, they enter the reservoir with
the input concentration X_o and leave together with their products
at their actual concentrations X(t).   The time evolution of some
species can be followed either by the electrochemical potential of
a selective electrode or by the optical density measured by
spectrophotometric methods.   When the system is well stirred
(ideal reactor) it can be modellized by a set of ordinary
differential equations.   In the isothermal case they take the
general form:

$$\frac{\partial X}{\partial} = F(x) + k_o(X_o - X) \tag{1}$$

where $F(X)$ describes the chemical reaction velocity and $k_o = Q/V_r$

$= t_{res}^{-1}$ is the reciprocal residence time (V_r is the volume of the reactor).

In the presence of exothermic reactions, the energy balance has to be done, it yields an equation for the temperature:

$$C\frac{\partial T}{\partial t} = -h(T-T_w) - k_0 C(T-T_0)$$
$$+ \sum_i (-\Delta H_i)r_i + S(I_0, X) \qquad (2)$$

where C is the heat capacity, T_w and T_0 are respectively the wall and the feed temperature, h decribes the heat transfer, ΔH_i is the actual enthalpy change of the ith reaction characterized by the rate r_i, and S accounts for the source terms due for instance to a laser irradiation. In the non-isothermal case the transcendental subtlety introduced by the Arrhenius form for the rate constants: $k_i \propto T^{\alpha_i} \exp{-(E_i/RT)}$ leads to a great richness in behaviour even in the case of a single first order reaction (FONI model) [6].

Complicated periodic and chaotic behaviours have been experimentally discovered in these C.S.T.R. In this paper, we consider only transient relaxations to steady states in situations like that illustrated in Figs. 1 and 2.

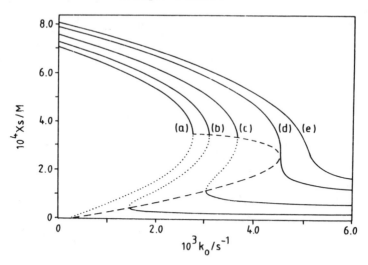

Figure 1. Steady state iodide concentration X_s as a function of the reciprocal residence time k_0 in the C.S.T.R. for a one variable model of the bistable iodate-arsenous acid reaction for various iodide inflow concentration. Solid and dotted curves respectively represent stable and unstable steady states. The dashed curve shows the locus of transition points and the critical point. After Ref. [12].

As the flow rate k_o is increased from zero, the system moves along the thermodynamic branch. At low k_o the stability of this steady state is guaranteed by the minimum entropy production theorem [3]. On the other hand at high flows the system is in a steady state corresponding to a low conversion (flow branch). According to the feed concentrations, the transition between these states may be either smooth (curves (e) and (f) in Figs. 1 and 2) or the two states may coexist over a range of k_o (curves (a), (b) and (c) in Figs. 1 and 2). In general the plot of conversion as a function of flow rate can exhibit Z-shaped (as in Figs. 1 and 2) or S-shaped curves. The crossover between these regimes is approached through a "cusp catastrophe" at (k_{oc}, X_c) (curve (d) in Figs. 1 and 2):

$$\frac{\partial k_o}{\partial X_s}\bigg|_{X_c} = \frac{\partial^2 k_o}{\partial X_s^2}\bigg|_{X_c} = 0.$$

Other patterns of stationary states including mushrooms and isolas (isolated branches) are exhibited in the non-isothermal cases [6] or in the isothermal C.S.T.R. with an additional flow of solvent [7].

The analogy with equilibrium phase transitions is striking. The curves of Figs. 1 and 2 are clearly analogous to the isotherms of a liquid-vapour transition or the isotherms in the magnetization-magnetic field plane in the case of a ferromagnetic

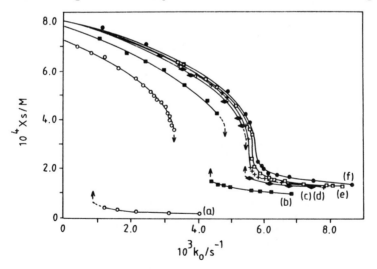

Figure 2. Measured steady state iodide concentration X_s as a function of k_o for various concentrations of iodide in the reactant stream. After Ref. [12].

transition [8].   Similarly close to the point (k_{oc}, X_c) the concentration of some intermediate species can be shown to obey:

$$X = X_c + (k_{oc} - k_o)^{1/3}.$$

The exponent $\delta = 3$ corresponds to a magnetic-field like approach to the critical point (k_{oc}, X_c).   The locus of the hysteresis limits forms a curve (cf. Fig. 1) which is reminiscent of the spinodal curve introduced in the mean field description of first order phase transitions.   However contrary to the equilibrium case, it has been shown that in well stirred reactors, the transitions between the various branches take place at the hysteresis limits.   Nucleation is thus absent in these systems.

9.3   Relaxation Behaviour Near the Hysteresis Limits

9.3.1 *Modelling of the relaxation experiments in the C.S.T.R.*

In the absence of autocatalytic reactions, Heinrichs and Schneider [9] have shown that the largest relaxation time is equal to the residence time t_{res}.   In the presence of positive feedbacks, however, the relaxation time can greatly exceed t_{res} in particular near the marginal stability points.   The hysteresis limits correspond to a saddle-node bifurcation where one stable and one unstable steady state annihilate.   In the vicinity of such an instability point, the dynamics of physically dissimilar systems can be described by amplitude equations the form of which depends only on a few fundamental factors such as the number of marginal modes and the symmetry [4].

In the case of an isolated zero eigenvalue of the corresponding linearized matrix, the amplitude equation assumes the simple form

$$\frac{\partial x}{\partial \tau} = \varepsilon - gx^2 \ , \ \tau = t_o^{-1}t \tag{3}$$

where

$$x(t) = X(t) - X_{ms}$$

$$\varepsilon = k_o - k_{oms}.$$

X_{ms} and k_{oms} are respectively the values of the concentration and the flow rate at marginal stability.   This equation can be obtained directly by projecting the constitutive equations on the zero eigenvector of the transposed Jacobian matrix.

9.3.1.a   *Bistable region ($\varepsilon > 0$).*   The relaxation of a perturbation $u(\tau)$ from the stable steady state

$$x_s = (\varepsilon/g)^{1/2} \tag{4}$$

can be obtained by integrating Eq. (3) to yield the following relaxation curve

$$u(\tau) = \frac{u(0)\, e^{-2x_s g\tau}}{1 + \dfrac{u(0)}{2x_s}(1-e^{-2x_s g\tau})} \tag{5}$$

For small perturbations and/or sufficiently far from the marginal stability point ($u(0)/x_s \ll 1$) the decay becomes exponential with a diverging relaxation time t_{rl}

$$t_{rl} = \frac{\tau_o}{2(ge)^{1/2}} \propto e^{z_1}, \quad z_1 = -1/2. \tag{6}$$

Such a dramatic increase in relaxation times near hysteresis limits ("spinodal slowing down") has been experimentally observed for the first time in the bistable iodate-arsenous acid reaction [10].

In analogy with critical phenomena, a dynamic exponent z_1 characterizing the singularity of the linear response can be introduced [11]. Because of the generic character of the amplitude equation (3), this exponent $z_1 = -1/2$ would be independent of the details of the chemical mechanism responsible for the instability ("UNIVERSALITY"). The experimental determination of this exponent is however difficult. Indeed on approaching the hysteresis limit ($u(0)/x_s$) gets larger and deviations from the exponential behaviour becomes increasingly important as shown for instance in Fig. 3.

In the nonlinear regime, the definition of the relaxation time is no longer unique. An important question is then whether the various definitions have the same singular behaviour for $|\varepsilon| \to 0$.

Experimentally, one generally measures the relaxation half-time $t_{1/2}$ (i.e. the time taken for the perturbation to decay to one-half of its initial value). From Eq. (3) one readily gets

$$t_{1/2} = \frac{\tau_o}{2gx_s} \ln\left[\frac{4x_s + u(0)}{2x_s + u(0)}\right]. \tag{7}$$

In particular at the marginal stability point $\varepsilon = 0$ ($x_s=0$) one gets

$$t_{1/2} = \frac{\tau_o}{gu(0)}. \tag{8}$$

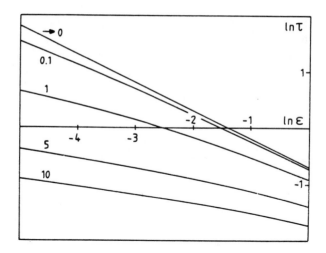

Figure 3. Behaviour of the relaxation time defined by Eq. (11) as one approaches the hysteresis limits for various amplitudes of perturbation u(0). The straight line corresponding to u(0) → 0 has slope -1/2. The nonlinear contributions, when u(0) becomes of the order of x_s lead to a crossover in behaviour.

The curves are indexed by the value of u(0).

This behaviour corresponds to an algebraic decay for $\varepsilon = 0$ $(u(0)>0)$ to the state $X_s = X_{ms}$

$$u(\tau) \propto \frac{1}{\tau}. \tag{9}$$

Because the nonlinear terms act to speed up the relaxation, $t_{1/2}$ has no singularity at $\varepsilon = 0$ contrary to the linear response case. Since its asymptotic value (8) is dependent on the initial value of the perturbation u(0), $t_{1/2}$ is not a convenient quantity to probe the universal form of the relaxation near $\varepsilon = 0$. These difficulties are well illustrated in the experiments in the iodate-arsenous acid system [12].

Another useful definition of the nonlinear relaxation time introduced in the theory of phase transitions [13] is

$$t_s = \int_o^\infty \frac{u(t)}{u(o)} \, dt. \tag{10}$$

From Eq. (3) one gets immediately

$$t_s = \frac{\tau_o}{u(o)g} \ln \left(1 + \frac{u(o)}{2x_s}\right). \tag{11}$$

The various definitions (Eqs. (6), (7) and (11)) yield of course the same value $z_1 = -1/2$ in the linear regime. On the other hand when $u(0)/x_s \gg 1$, t_s presents a logarithmic singularity $t_s \propto \ln|\varepsilon|$, for $\varepsilon \to 0$, corresponding now to an exponent $z=0$. The increasing curvature, as $u(0)$ grows, of the curves in Fig. 3 characterizes the crossover between these regimes. As a result, great care must be exercised in the determination of the dynamical exponent z which depends on the definition of the relaxation time sufficiently close to the hysteresis limit.

By using an extrapolation technique to get the relaxation times of perturbations of very low magnitude Laplante et al. [14] have obtained the value $z_1 = -0.48$ for the dynamic exponent near marginal stability on the flow branch of the chlorite-iodide system. Similarly a value approaching $-1/2$ has been obtained in the limit of small perturbations in the iodate-arsenous acid reaction [12].

9.3.1.b *Plateau behaviour:* *($\varepsilon < 0$).* When $\varepsilon < 0$ Eq. (3) admits two complex conjugated roots with a small imaginary part

$$X = X_{ms} \pm iy \;,\; y = (|\varepsilon|/g)^{1/2}. \tag{12}$$

When perturbed the system relaxes to the stable lower branch X_1 (in our example $X_1 < X_{ms}$) which has not been included in the amplitude equation (3). For $-1 < \varepsilon < 0$ and $X(0) > X_{ms}$ this relaxation is dominated by the transit of the trajectory near the hysteresis limit ($\varepsilon = 0$, $X = X_{ms}$) where the dynamics can still be described by Eq. (3), the integration of which now gives

$$-\tau = \frac{1}{(|\varepsilon|g)^{1/2}} [\tan^{-1}(\sqrt{\frac{g}{|\varepsilon|}} x) - \tan^{-1}(\sqrt{\frac{g}{|\varepsilon|}} x(0))]. \tag{13}$$

In the course of this evolution, when the variable comes close to X_{ms}, so that $x(\tau) \simeq (|\varepsilon|/g)^{1/2}$, the relaxation is indeed slowed down as

$$X(t) \simeq X_{ms} + \frac{\pi}{2}\left[\frac{|\varepsilon|}{g}\right]^2 - |\varepsilon|\tau_o^{-1}t. \tag{14}$$

In this regime, the concentration presents a slow linear time variation with a coefficient proportional to $y^2 \propto |\varepsilon|$. The lifetime of this "plateau" t_{pl} is also very sensitive to the proximity of the hysteresis limit. According to Eqs. (13) and (14) t_{pl} is given by

$$t_{pl} \propto \frac{\tau_o}{(|\varepsilon|g)^{1/2}} \propto \frac{1}{y} \qquad (15)$$

which is independent of X(0).

These long transients (several times t_{res}) could be classified as metastable states according to the dynamical definition of metastability based on the flatness property of the relaxation function [13]. Such a definition is indeed particularly constructive in far from equilibrium systems where the conditions for the existence of a generalized potential (Lyapunov functional) are rarely satisfied.

The plateaus have been observed in many different bistable systems: isothermal and non-isothermal reactions in C.S.T.R., illuminated chemical systems (see section 9.3.b and Fig. 6) and in another field, that of optical bistability.

In a C.S.T.R. the macroscopic nature of the lifetime of these states can lead to an inaccurate experimental determination of the bistability limits. In the problem of thermal explosions [15], they are associated to the time for ignition and their formula

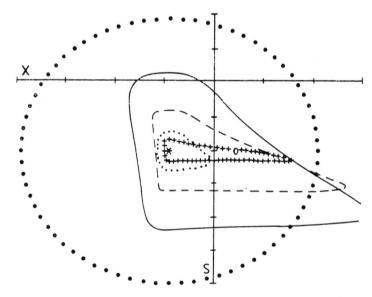

Figure 4. Relaxation diagram in a two variable (X=[I$^-$], S=[I$^-$] + [IO$_3^-$]) model (see [7]) for the iodate-arsenous acid system exhibiting the temporary attraction by the unphysical roots of the stationarity condition above the hysteresis limits. The system relaxes to the stable steady state (*) from a set of initial conditions (•) equidistant from it. The shape is deformed as the points which transit in the neighbourhood of the real part of the complex conjugate roots (o) are slowed down (——— t=5, - - - - - t=10, ++++ t=15 in reduced units) near the hysteresis limit. When the system is further from the marginal point relaxation is much faster and less deforming (..... t=5).

(15) corresponds to Frank-Kamenetskii's law. In optically bistable devices [16], systems which present two transmitted steady states for the same input intensity P in the range $P_1 < P < P_2$, the long transients [17] have definite practical implications by limiting the time response of the device. By increasing the input intensity slightly above the upper turning point P_2, one observes the appearance of significant delay times before the device switches to the upper branch of the bistability curve. In some experiments [18], the power law (15) relating the switching time t_{pl} to the intensity increment $\Delta P = P - P_2$ has been verified:

$$t_{pl} \propto \Delta P^{-1/2}.$$

These long transients can be interpreted as resulting from a temporary attraction by the unphysical roots (12) of the stationarity condition. This effect is even more dramatically visualized by the examination of the phase portrait of a two variable system (such as that proposed for the iodate-arsenous acid system [7]). One may then follow the deformation through the flow of the dynamical evolution of a set of initial conditions to the stable steady state. In Fig. 4 this set is taken equidistant from this steady state. Near the hysteresis limit, the points transiting in the neighbourhood of the real part of the complex non-physical roots are slowed down and the shape is considerably deformed. This effect tends to vanish as one moves away from the stability limit.

Such attraction is also important to apprehend the relaxation behaviour of a system where steady states fall outside the reaction polyhedron defined by balance limitations (such as competitive surface coverage in heterogenous reactions) and the condition of non-negative concentration [19]. Thus the position of **all** roots of a kinetic system is informative to understand the dynamics of relaxation (in particular, the reasons of induction periods).

9.3.2 *Spinodal slowing down in a thermochemical bistable system*

In this section we discuss briefly relaxation experiments which have been performed in a chemical system closed to mass flow. The acid-base reaction of ortho-cresol phthalein, $OCP \rightleftharpoons OCP^- + H^+$, has been shown to exhibit bistability when the solution is illuminated with light absorbed only by the OCP^- molecules. The plot of the absorption A versus the laser power P has an S-shaped form. In order to investigate the phenomenon of spinodal slowing down, Kramer and Ross [20] measured the instantaneous velocity v_{ms} = $\left. \dfrac{\partial A}{\partial t} \right|_{A_{ms}}$ at the value A_{ms} corresponding to the left marginal stability point $P = P_{ms}$. All the measurements give speeds which decrease with the proximity of P_{ms} according to (see Fig. 5):

$$\frac{1}{v_{ms}} \quad |\varepsilon|^{n} \ , \ n = -1. \tag{16}$$

Here $\varepsilon = (P-P_{ms})/\Delta P$ and ΔP is the laser power difference between the marginal points. They obtain the same exponent on both sides of the marginal point. The dynamics of this bistable system near P_{ms} can again be described by the amplitude equation (3) where now $x(\tau)=A(\tau)-A_{ms}$. In the bistable region ($\varepsilon > 0$) the speed v_{ms} of the relaxation of negative perturbations can be obtained directly from Eq. (3)

$$v_{ms} = \frac{\partial x}{d\tau}\bigg|_{x=0} = \tau_{o}^{-1}\varepsilon. \tag{17}$$

On the other hand in the region $\varepsilon < 0$, we have shown that the relaxation presents a plateau for initial conditions $A(0) > A_{ms}$ (see Eq. (14))

$$A(t) = A_{ms} - \tau_{o}^{-1}|\varepsilon|t. \tag{18}$$

From Eq. (18) it is thus obvious that the velocity obeys the same

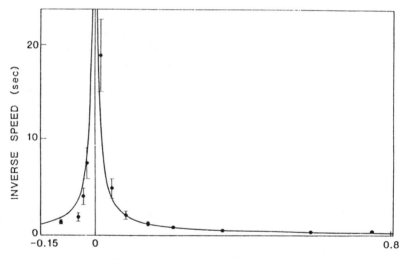

FRACTIONAL DISTANCE FROM M. S.

Figure 5.   Theoretical and experimental speeds of the trajectories of transition between branches of stationary states evaluated at the absorption value of the left marginal stability point plotted against fractional distance from the left marginal stability point for the ortho-cresolophthalein system. Critical slowing down is evident both above and below the left marginal stability point.   The uncertainty in each measurement is displayed, and where no error bar appears, the size of the dot is greater than the estimated uncertainty.   After Ref. [20].

power law as for $\varepsilon > 0$.

As a result, Eq. (3) predicts that the inverse velocity $1/v_{ms}$ presents a symmetric divergence when P_{ms} is approached from either side, $1/v_{ms} \propto \varepsilon^{-1}$ in agreement with the experimental results displayed in Fig. 5.

Because it is directly related to Eq. (3) the exponent n=-1 which characterizes the slowing down of the velocity in the nonlinear regime has also a universal character. This technique provides an elegant and rapid means to demonstrate the phenomenon of spinodal slowing down. The plateaus have also been observed in the OCP system where, as shown in Fig. 6 their lifetime strongly depends on the proximity of P_{ms}.

9.4 RELAXATION KINETICS NEAR A CRITICAL POINT [21]

As already pointed out, when the hysteresis limits merge, the system presents a "nonequilibrium critical point". Showalter et al. [12] have made careful measurements of the relaxation half-time $t_{1/2}$ of the iodate-arsenous acid system in this regime.

Near such a point, the dynamics of systems of dissimilar chemical nature can be described by a generic amplitude equation which takes the following form (Ginzburg-Landau equation)

$$\frac{\partial x}{\partial t} = C_1 + C_2 \varepsilon x - x^3 \tag{19}$$

where $\varepsilon = k_{oc} - k$, $x = X - X_c$. X_c and k_{oc} are the values of the concentration and flow rate at the critical point. The corresponding stationarity equation admits one real and two complex conjugate roots, $x_{1,2} = -x_r/2 \pm iy$. Near $\varepsilon = 0$ one has

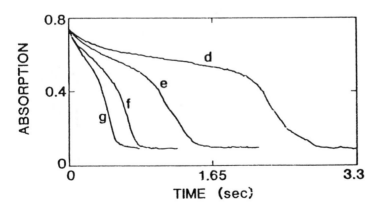

Figure 6. Plot of absorption vs. time for measured relaxation trajectories in the ortho-cresolphthalein system at powers below the left marginal stability point. The distance from this point increases from (d) to (g). After Ref. [20].

133

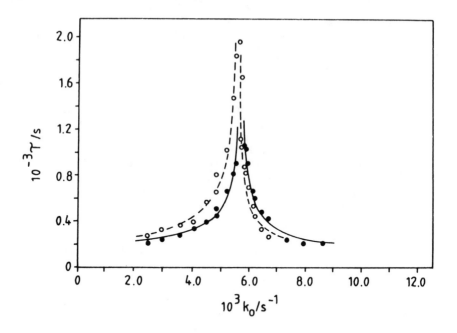

Figure 7. *Relaxation half-time as a function of k_o around k_{oc}. Experimental conditions and reactant concentrations are given in [12]. Circles show half-times for positive perturbations, dots for negative perturbations ($u(o) = \pm 2. \quad 10^{-5} M$). Curves are least-square fits to $\tau = A|k_o - k_{oc}|^z$. After Ref. [12].*

$$x_r \propto C_1^{1/3} \varepsilon^{1/3} \quad , \quad y \propto \frac{3^{1/2}}{2} x_r. \tag{20}$$

The relaxation half-time $t_{1/2}$ can be calculated by integrating Eq. (19) to give

$$t_{1/2} = -\frac{1}{3x_r^2} \ln \left\{ \frac{1}{2} \left[\frac{(3x_r + 2u(o))^2 + 3x_2^r}{(3x_r + u(o))^2 + 3x_2^r} \right]^{1/2} \right\}$$

$$+ \frac{1}{\sqrt{3}x_r^2} \left\{ \tan^{-1} \left[\frac{3x_r + u(o)}{\sqrt{3}\, x_r} \right] - \tan^{-1} \left[\frac{3x_r + 2u(o)}{\sqrt{3}\, x_r} \right] \right\} \tag{21}$$

For small perturbations such that $x_r/u(0) \ll 1$, Eq. (21) becomes

$$t_{1/2} \propto \frac{\ln 2}{3C_1^{1/2}\varepsilon^{2/3}} - \gamma \frac{u(o)}{\varepsilon} \qquad \gamma > 0. \qquad (22)$$

The linear response diverges as $\varepsilon \to 0$ (critical slowing down). Such a lengthening of $t_{1/2}$ has been observed experimentally [12] (see Fig. 7). A dynamic critical exponent $z_c = -2/3$ can also be introduced. This particular value corresponds to the mean-field dynamic exponent of a magnetic field-like approach to the critical point. A similar value is obtained experimentally when the extent of relaxation increases [12]. Here also the experimental determination of z_c is difficult. In the nonlinear regime $t_{1/2}$,

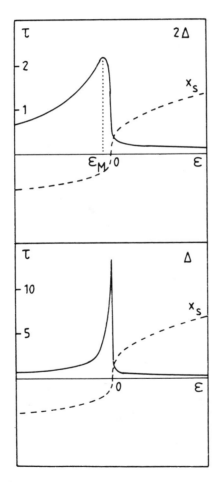

Figure 8.   Relaxation half-time evaluated from Eq. (21) as a function of the distance to the critical point, for two values of the perturbation $u(o) = \Delta$ and 2Δ in reduced units.   The dashed line represents the corresponding steady state.

predicted by Eqs. (21) and (22), not only depends on the magnitude but also on the direction of perturbation. The nonlinear contributions lead for $u(0) > 0$ to a decrease for $\varepsilon > 0$ (to an increase for $\varepsilon < 0$) of the lengthening of $t_{1/2}$ as the critical point is approached. The opposite is predicted for negative perturbations ($u(0) < 0$). This asymmetry is well illustrated in the experiments displayed in Fig. 7.

On the other hand at the critical point, $t_{1/2}$ is equal to a constant $t_{1/2} = 3/2u(0)^2$ corresponding again to a power law decay $x(\tau) \quad 1/\tau^{-1/2}$. Moreover as shown in Fig. 8, $t_{1/2}$ presents for $u(0) > 0$ a maximum for $\varepsilon_{M1} > 0$ whereas for $u(0)$, 0 this maximum occurs for $\varepsilon_{M2} > 0$. These maxima and the asymmetry in the plot of $t_{1/2}$ versus ε are due to the existence of a critical plateau of relaxation. When $x(\tau) \simeq -x_r/2$, the evolution of $x(\tau)$ becomes very slow, proportional to $|\varepsilon|\tau$ and correspondingly the relaxation can become slower than at $\varepsilon = 0$.

9.5 Conclusion

In summary, in ideal chemical reactors metastable states with infinite lifetime exist up to the marginal stability point (spinodal point) where the transition between states takes place via the decay of unstable modes. This behaviour is reminiscent of equilibrium systems with long range forces [13] where the surface energy cost for the formation of a droplet prevents any true nucleation process. It is interesting that the ideal C.S.T.R. provides one of the few experimental examples where the mean field theory of first order transition can be applied. When the stirring rate is lowered, however, it has been discovered experimentally that the width of the bistable region can sometimes be reduced in a dramatic way [22]. In this case the inhomogeneities introduced by the injection process can induce a transition prior to the hysteresis limit.

On the other hand, the influence of fluctuations in the vicinity of marginal stability points has been analyzed in a large number of theoretical papers. In particular, the phenomenon of "transient bistability" has been predicted by Nicolis et al. [23]. It has been observed experimentally in the case of absorptive optical bistability [24]. It was shown that there is an observable time interval during which the probability distribution of the transmitted field is doubly peaked. The most probable and mean plateau lifetime are also shortened in the presence of strong external fluctuations.

References

[1]   D.A. Frank-Kamenetskii (1939). *Acta Phys.-Chim. URSS* **10**, 365.

[2]   Y.B. Zeldovitch and U.A. Zysin (1941). *J. of Tech. Phys.* **11**, 501.

[3]   P. Glansdorff and I. Prigogine (1971). *Structure, Stability and Fluctuations,* Wiley, New York.

[4]   H. Haken, *Synergetics* (1978). Springer, Berlin.

[5]   M. Eigen and L. De Maeyer (1963) in *Technique of Organic Chemistry,* Eds. S.L. Friess, E.S. Lewis and A. Weissburger, Vol. VIII, Part 2, Wiley Interscience, New York.

[6]   R. Aris (1986). *Lect. Appl. Math.* **24**, 75.

[7]   N. Ganapathisubramanian and K. Showalter (1984). *J. Chem. Phys.* **80**, 4177.

[8]   H.E. Stanley (1971). *Introduction to Critical Phenomena,* Oxford University Press.

[9]   M. Heinrichs and F.W. Schneider (1981). *J. Phys. Chem.* **85**, 2112.

[10]   N. Ganapathisubramanian and K. Showalter (1983). *J. Phys. Chem.* **87**, 1098.

[11]   G. Dewel, P. Borckmans and D. Walgraef (1984). *J. Phys. Chem.* **88**, 5442.

[12]   N. Ganapathisubramanian and K. Showalter (1986). *J. Chem. Phys.* **84**.

[13]   K. Binder (1973). *Phys. Rev.* **B8**, 3423.

[14]   J.-P. Laplante, P. Borckmans, G. Dewel, M. Gimenez, J.-C. Micheau (1987). *J. Phys. Chem.* **91**, 3401.

[15]   P. Gray and W. Kodeylewski (1985). *Chem. Eng. Sci.* **40**, 1577, 1703.

[16]   E. Abraham and S.D. Smith (1982). *Rep. Prog. Phys.* **45**, 815.

[17]   D.E. Grant and H.J. Kimble (1983). *Optics Comm.* **44**, 415.

[18]   S. Barbarino, A. Gozzini and F. Maccarrone (1982). *Nuevo Cimento* **71B**, 183.

[19]   V.I. Elokhin, G.S. Yablonskii, A.N. Gorban and V.M. Cheresiz (1980). *React. Kinet. Catal. Lett.* **15**, 245.

[20]   J. Kramer and J. Ross (1985). *J. Chem. Phys.* **83**, 6234.

[21]   G. Dewel, P. Borckmans and D. Walgraef (1985). *J. Phys. Chem.* **89**, 4670.

[22]   J.-C. Roux, P. De Kepper and J. Boissonade (1983). *Phys. Lett.* **97A**, 168.

[23]   F. Baras, G. Nicolis, M. Malek-Mansour and J.W. Turner, *J.* (1983). *Stat. Phys.* **32**, 1.

[24]   W. Lange, F. Mitschke, R. Deserno and J. Mlynek (1985). *Phys. Rev.* **A32**, 1271.

THEORY AND APPLICATIONS OF THE DISCRETE SELF-TRAPPING EQUATION

10.1   Introduction

In this chapter we discuss the equation

$$(i \frac{d}{dt} - \omega_o)\bar{A} + \gamma \, D(|A|^2)\bar{A} + \varepsilon M\bar{A} = 0 \tag{1}$$

which we have called the **discrete self-trapping** (DST) equation
[1].   In (1), A is the n-component vector

$$\bar{A} = col(A_1, \, A_2, \ldots, A_n) \tag{2}$$

where the A_i are complex.   If the real parameters γ and ε are both
equal to zero, then each component of A rotates at a frequency of
ω_o radians/sec.   If ε is not equal to zero, **dispersion** of these
modes is introduced by the nxn, real, symmetric dispersion matrix
M.   If γ is not equal to zero, **anharmonicity** is introduced by the
diagonal matrix

$$D(|A|^2) \equiv \begin{bmatrix} |A_1|^2 & & & 0 \\ & |A_2|^2 & & \\ & & \ddots & \\ 0 & & & |A_n|^2 \end{bmatrix} \tag{3}$$

The DST equation is very much in the spirit of modern nonlinear
dynamics.   It provides a simple yet physically relevant context to
explore the interplay between dispersion and anharmonicity.
Applications of the DST equation include the following:
   (a)   **The Polaron.**   In 1933 Landau suggested that an effect of a
localized electron in a crystal would be to polarize the crystal
which would then lower its energy [2].   This effect has been
studied by Pekar [3], Fröhlich [4], Holstein [5] and many others
since 1970 and can be modelled by the DST equation.
   (b)   **Davydov's soliton.**   In 1973 Davydov suggested that the
amide-I (CO stretching) of peptide groups located along the

alpha-helix of protein might become self-trapped through interaction with low frequency phonons [6]. Extensive numerical models of this effect are based upon the DST equation [7].

(c)   **Globular Protein.** Leaving the organized structure of alpha-helix, the DST can be applied to study the self-trapping effect proposed by Davydov in the context of a real globular protein [8].

(d)   **Self-Trapping in Crystalline Acetanilide** (ACN). Closely related to Davydov's soliton is the discovery that an anomalous component of the amide-I (CO stretching) spectrum of crystalline ACN can be assigned to a self-trapping mechanism [9]. This mechanism differs from that proposed by Davydov because optical modes contribute to the self-trapping [9,10]. Careful numerical studies of this effect are based upon the DST equation [11]. The same effect has been observed in N-methylacetamide [12]. (A photograph of a piece of crystalline ACN being irradiated by an argon laser beam during a Raman scattering experiment is shown in colour Plate 3.)

(e)   **Small Molecules.** The DST equation can be used to study the local mode structure of vibrations in small molecules such as water, ammonia, methane, etc. [13]. Here the anharmonicity is related to the nonlinear restoring forces for hydrogen stretching vibrations.

(f)   **Natural Engines.** To explore the properties of thermodynamic heat engines on the molecular scale, Wheatley and coworkers have constructed a thermoacoustic heat engine consisting of a chain of twelve coupled nonlinear acoustic vibrators [14]. This structure demonstrates self-trapping of acoustical energy as would be predicted from the DST equation.

(g)   **The Nonlinear Schrodinger (NLS) Equation.** Taking Eq. (1) to the limit $\gamma \ll 1$, $n \to \infty$, with εM of the form

$$\varepsilon M \equiv \begin{bmatrix} 0 & 1 & & & & & \\ 1 & 0 & 1 & & & \mathbf{0} & \\ & 1 & 0 & 1 & & & \\ & & 1 & \cdot & & & \\ & & & & \cdot & & \\ & \mathbf{0} & & & & 0 & 1 \\ & & & & & 1 & 0 \end{bmatrix} \tag{4}$$

reduces it to the NLS equation which arises in the theory of solitons [15]. Thus the DST equation is closely related to those physical problems that are modelled by the NLS equation such as: self-trapping of electromagnetic energy in a plasma [16], light energy in an optical fibre [17], and hydrodynamic energy in a wave tank [18].

(h)   **Anderson Localization.** With anharmonic effects eliminated ($\gamma=0$), Eq. (1) has been studied in detail by Anderson to describe the diffusion of charge carriers in amorphous semiconductors and the diffusion of spin states in spin glasses [19]. The main result of these studies is that localized states occur **without** anharmonicity if the probabilistic spread of the diagonal components of M is greater than the largest off diagonal components. The DST equation provides a means for studying how this "Anderson localization" might be influenced by anharmonic effects [1,5].

We hope that this list of applications leaves the reader with the impression that Eq. (1) is an important object for mathematical study. In the remainder of this chapter we will review some of the salient properties of the DST equation, discuss the quantum theory of the DST equation, and describe some details of its application to the study of CH stretching vibrations in benzene.

10.2 **General Properties**

Writing the dispersion matrix in the form [1]

$$\varepsilon M = [m_{ij}] \tag{5}$$

solutions of DST conserve the **energy**

$$H = \sum_{i=1}^{n} (\omega_0 |A_i|^2 - \frac{1}{2} \gamma |A_i|^4) - \sum_{i \neq j} m_{ij} A_i^* A_j \tag{6}$$

and the **number**

$$N = \sum_{i=1}^{n} |A_i|^2. \tag{7}$$

Equation (6) is a **Hamiltonian** since

$$i \, A_j = \partial H / \partial A_j^* \tag{8}$$

and its complex conjugate are equivalent to DST [1].

In general DST is not integrable and solution trajectories are typically **chaotic** with broad power spectra [1], noninteger or fractal dimension [20], and positive Liapunov characteristic exponents [21]. There are four special cases, however, in which DST is completely integrable: (i) in the limit $\varepsilon=0$, (ii) in the limit $\gamma=0$, (iii) in the nonlinear Schrodinger limit which was discussed in the Introduction, and (iv) for $n \leq 2$.

Complete integrability for $n=2$ is demonstrated in detail in reference [1]. From another perspective, Elgin has defined the "density" matrix [20]

$$\rho = [\rho_{jk}] \tag{9}$$

with elements

$$\rho_{jk} = A_j A_k^* . \tag{10}$$

Then DST is equivalent to the matrix equation .

$$\dot{\rho} = i[\rho, B] \tag{11}$$

where

$$B = \gamma \ D(|A|^2) - \varepsilon M. \tag{12}$$

In the case n=2, this formalism has been reduced to a pendulum equation. This approach has recently been described in greater detail by Kenkre and Campbell [22].

  Stationary solutions of DST are time-periodic solutions of the form

$$\bar{A}(t) = \bar{\phi} \ \exp \ (i\omega t) \tag{13}$$

where ω is a real constant and $\bar{\phi}$ is a real or complex constant n-vector. Clearly ω and $\bar{\phi}$ satisfy the nonlinear eigenvalue problem (assuming that $\omega_o = 0$ and $\varepsilon = 1$)

$$\omega\bar{\phi} = \gamma D(|\bar{\phi}|^2)\bar{\phi} + M\bar{\phi} = 0 \tag{14}$$

  In certain cases it is possible to find analytical solutions to (14) [1,13,23,24] while in other cases numerical methods may be required [1,11,25]. To study the stability of stationary solutions one can introduce a perturbation u(t) of the form [23] $\bar{A}(t) = [\bar{\phi} + \bar{u}(t)]\exp(i\omega t)$. The linearized equation satisfied by u is then

$$i\bar{u} - \omega\bar{u} + Mu + \gamma \ \mathrm{diag}(2|\phi_k|^2 u_k + \phi_k^2 u_k^*) = 0 \tag{15}$$

from which either necessary conditions for stability or sufficient conditions for instability can be calculated. Techniques for establishing stability conditions for $\bar{\phi}$ real and for $\bar{\phi}$ complex are discussed in detail in [23].

  As an example, consider DST with n = 6 and

$$M = \begin{bmatrix} 0 & 1 & 0 & 0 & 0 & 1 \\ 1 & 1 & 1 & 0 & 0 & 0 \\ 0 & 1 & 0 & 1 & 0 & 0 \\ 0 & 0 & 1 & 0 & 1 & 0 \\ 0 & 0 & 0 & 1 & 0 & 1 \\ 1 & 0 & 0 & 0 & 1 & 0 \end{bmatrix}. \tag{16}$$

This could be considered as a simple model for the dynamics of the CH stretch vibrations in benzene which includes only nearest neighbour interactions. Known analytic solutions are discussed in detail in [26]. For simplicity we set N = 1 and $\varepsilon = 1$; solutions for different values of $\varepsilon > 0$ can easily be calculated by scaling. (For $\varepsilon<0$, which corresponds to the physical case of benzene, it is necessary to scale all the solutions by ϕ_2, ϕ_4, $\phi_6 \rightarrow -\phi_2$, $-\phi_4$, $-\phi_6$.) Thus the locus of stationary solutions can be indicated on an "$\omega-\gamma$ plane" and this is done for real solutions in Fig. 1. Although some of these loci were calculated analytically, most were obtained numerically using the path finding tehniques

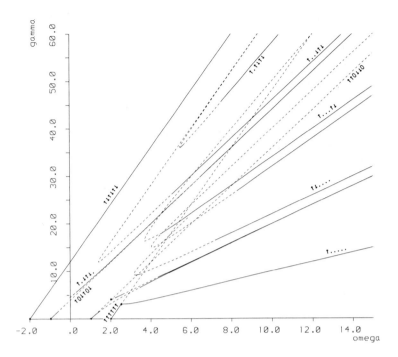

Figure 1.   Bifurcation diagram for stationary solutions of DST with n=6, M as in Eq. (16), ε = 1 and N = 1.   (See text for details of branch notation.)

described in [1].   Dashed loci fail the linearized stability test posed by Eq. (15).   Near $\gamma = 0$, the slopes by these loci can be calculated from perturbation theory as [1]

$$\omega(\gamma) = \omega^{(o)} + \gamma \frac{\sum\limits_{i=1}^{n} |\phi_i^{(o)}|^4}{\sum\limits_{i=1}^{n} |\phi_i^{(o)}|^2} + O(\gamma^2) \tag{17}$$

where $\omega^{(o)}$ and $\phi^{(o)}$ are the corresponding solution to Eq. (14) with $\gamma=0$.

Finally, a word about the notation in Fig. 1.   The symbol ($\uparrow\uparrow\uparrow\uparrow\uparrow\uparrow$) indicates that all six oscillations are in phase which corresponds to the symmetry species A_{1g}.   The symbol ($\uparrow_\downarrow\uparrow_\downarrow\uparrow_\downarrow$) indicates that adjacent oscillations are 180° out of phase which corresponds to the symmetry species B_{1u}.   The symbol ($\uparrow0_\downarrow\uparrow0_\downarrow$) indicates that $\phi_2 = \phi_5 = 0$ and corresponds to the symmetry species E_{2g};   there are two modes of this symmetry.   The symbol ($\uparrow\uparrow0_\downarrow\downarrow0$) indicates that $\phi_3 = \phi_6 = 0$ and corresponds to the symmetry species E_{1u}; there are two modes of this symmetry. Other modes display symmetry that is "broken" by the anharmonicity of DST. For example, the symbol ($\uparrow$.....) indicates that the energy tends to be concentrated on oscillator number one with ϕ_2, ϕ_3, ϕ_4, ϕ_5 and ϕ_6 or as $\gamma \to \infty$. This can be viewed as a "soliton" state or a "local mode" state depending upon one's point of view.

Once we drop the requirement that solutions DST have the stationary property indicated in Eq. (13), we must turn to numerical calculations. The solutions we obtain will depend upon γ and the initial conditions A(0). We are guided by a suggestion of Collins [27] that benzene may support soliton modes circulating around the ring. Thus we use the single set of initial conditions:

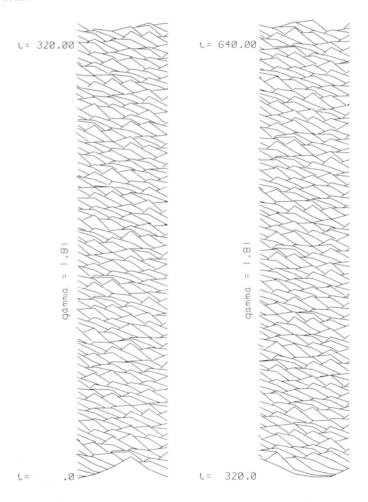

Figure 2.   Time dependent calculation of solutions of the DST equations, with n = 6, M as in Eq. (16), initial conditions as in Eq. (18), and γ = 1.81.

$$A_1(0) = .1370 + .0058i$$
$$A_2(0) = .2643 - .0572i$$
$$A_3(0) = .4043 - .0730i \tag{18}$$
$$A_4(0) = .5520 + .0769i$$
$$A_5(0) = .3260 + .1615i$$
$$A_6(0) = .1277 + .1377i$$

This apparently arbitrary set of initial data was derived as follows. A long chain (n=50) with nearest neighbour interactions and $\gamma = 3.4$ was set up as described in [28]. Down this chain was launched an exact soliton of the NLS aproximation to DST. Once this had settled down to a steadily progressing soliton-like

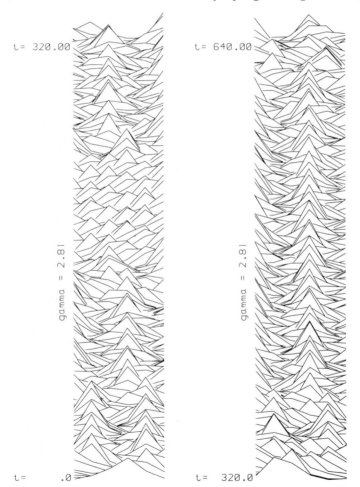

Figure 3. Time dependent calculations of solutions of the DST equations, with n = 6, M as in Eq. (16), initial conditions as in Eq. (18), and $\gamma = 2.81$.

solution on the **discrete** chain, the size values closest to the maximum were recorded as in Eqs. (18). Hence these values should, in some sense, aproximate a moving soliton-like pulse on the smaller ring of six lattice sites.

With these initial conditions, and for various values of γ, the DST equation was integrated numerically using an Adams-Bashforth-Moulton 4th order predictor-corrector method. The value of the invariant N was monitored to check accuracy, and if N deviated by more than a fraction of a percent from unity the calculation was repeated with a smaller time step.

The results of some of these calculations are displayed in Figs. 2 - 5 for γ = 1.81, 2.81, 3.81 and 4.31 respectively. In all these cases a value of h = 1/32 was taken. These values of γ are close to but not identical to those discussed in [26]. Comparison

Figure 4. Time dependent calculations of solutions of the DST equations, with n = 6, M as in Eq. (16), initial conditions as in Eq. (18), and γ = 3.81.

with the figures in [26] shows well that although the detailed dependence depends sensitively on the value of γ, the overall qualitative behaviour does not.

In each figure time is advancing from bottom to top: the left hand plot shows a plot of $|A_i|^2$ for $0 < t < 320$ and the right hand plot shows the continuation for $320 < t < 640$. The horizontal axis is the 6 lattice sites.

For low values of γ, as in Fig. 2, the solution evolves immediately into a shallow progressive wave which travels round the ring with a small chaotic component. For $\gamma = 2.81$ (Fig. 3), a more interesting structure develops. Initially the pulse evolves into a solution which first "hops" between sites 2 and 5, then between sites 1 and 4. At about $t \approx 225$ the pulse reverts to a

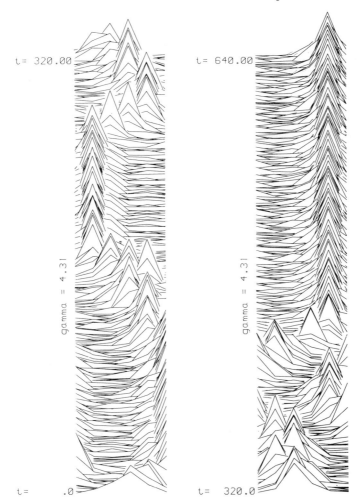

Figure 5. Time dependent calculations of solutions of the DST equations, with n = 6, M as in Eq. (16), initial conditions as in Eq. (18), and γ = 4.31.

hopping mode, then a running mode at t $\approx$ 320, then again a hopping mode for most of the remaining period of calculation.

For γ = 3.81 (Fig. 4) the initial pulse quickly forms a sharp soliton-like mode which exhibits a random-walk type of motion around the lattice. In this mode the pulse switches in an irregular manner between running modes in both clockwise and anticlockwise directions, hopping modes, or stationary modes where the pulse is pinned to one site for a short interval.

For larger values of γ as shown in Fig. 5 (γ = 4.81), the pulse spends much more time in the stationary mode, but occasionally breaks out of this mode at randomly spaced intervals and switches between travelling modes and hopping modes for short periods of time.

10.3 Quantum Theory

We turn now to the problem of quantizing the DST equation. Following Louisell [29] we change the complex mode amplitudes, A_i and A_i^*, to boson annihilation and creation operators $\hat{B}_i$ and $\hat{B}_i^+$. If $|k_i\rangle$ is an eigenfunction of a particular mode, $\hat{B}_i^+|k_i\rangle = \sqrt{k_i + 1}\ |k_i + 1\rangle$, $\hat{B}_i|k_i\rangle = \sqrt{k_i}|k_i - 1\rangle$, and $\hat{B}_i|0\rangle = 0$. Since the ordering is not specified in Eqs. (6) and (7) we take

$$|A_i|^2 \rightarrow \frac{1}{2}\ (\hat{B}_i^+\hat{B}_i + \hat{B}_i\hat{B}_i^+) \tag{19}$$

$$|A_i|^4 \rightarrow \frac{1}{6}\ (\hat{B}_i^+\hat{B}_i^+\hat{B}_i\hat{B}_i + \hat{B}_i^+\hat{B}_i\hat{B}_i^+\hat{B}_i + \hat{B}_i^+\hat{B}_i\hat{B}_i\hat{B}_i^+$$

$$+ \hat{B}_i\hat{B}_i^+\hat{B}_i\hat{B}_i^+ + \hat{B}_i\hat{B}_i\hat{B}_i^+\hat{B}_i^+ + \hat{B}_i\hat{B}_i^+\hat{B}_i^+\hat{B}_i). \tag{20}$$

Using the commutation rule $\hat{B}_i\hat{B}_i^+ = \hat{B}_i^+\hat{B}_i + 1$, Eq. (7) becomes the **number operator** [30]

$$\hat{N} = \sum_{i=1}^{n} (\hat{B}_i^+\hat{B}_i + \frac{1}{2}) \tag{21}$$

and Eq. (6) becomes the **energy operator**

$$\hat{H} = \hbar(\omega_0 - \frac{1}{2}\gamma)\hat{N} - \frac{1}{2}\hbar\gamma \sum_{i=1} \hat{B}_i^+\hat{B}_i\hat{B}_i^+\hat{B}_i \cdot$$

$$- \hbar\varepsilon \sum_{i\neq j} m_{ij}\ \hat{B}_i^+\hat{B}_i. \tag{22}$$

The **ground state** eigenfunction of $\hat{N}$ and $\hat{H}$ is

$$|\psi_0\rangle = |0\rangle|0\rangle\ \ldots\ |0\rangle$$

$$\text{n times} \tag{23}$$

$$\equiv [0\ 0\ \ldots\ 0]$$

with number and energy eigenvalues equal to $n/2$ and $\hbar(\omega_0 - \frac{1}{2})n/2$ respectively. A number eigenfunction of the **first excited state** can be written in the form

$$|\psi_1\rangle = C_1[1\ 0\ 0\ 0\ \ldots\ 0] + C_2[0\ 1\ 0\ 0\ \ldots\ 0]$$

$$+ \ldots + C_n[0\ 0\ 0\ 0\ \ldots\ 1] \tag{24}$$

with number eigenvalue $(1 + n/2)$. The constants (the C_i's) in Eq. (24) can be chosen to make $|\psi_1\rangle$ an eigenfunction of the energy operator. This is effected by requiring that the vector

$$\bar{C} = \mathrm{col}(C_1,\ C_2,\ \ldots,\ C_n) \tag{25}$$

satisfy the matrix equation

$$[\hbar(\omega_0 - \gamma)(1 + n/2) - \hbar\varepsilon M - E_i I]\bar{C} = \bar{0}. \tag{26}$$

Furthermore the condition $\langle\psi_1|\psi_1\rangle = 1$ requires that

$$\sum_{i=1}^{n} |C_i|^2 = 1. \tag{27}$$

For the **kth excited state** the number eigenfunction is constructed as

$$|\psi_k\rangle = C_1[k,0,0,\ldots,0] + C_2[k-1,\ 1,\ 0,\ \ldots,\ 0]$$

$$+ \ldots + C_p[0,0,0,\ldots,k] \tag{28}$$

where p is an integer equal to the number of ways that k balls can be placed in n urns. Thus

$$p = \frac{(k + n - 1)!}{(n-1)!\ k!}. \tag{29}$$

Eq. (28) becomes a normalized energy eigenfunction if the vector $\bar{C} = \mathrm{col}\ (C_1,\ C_2,\ \ldots,\ C_p)$ is an eigenvector of the $p \times p$ matrix obtained from the requirement that

$$A|\psi_j\rangle = E_k|\psi_k\rangle \tag{30}$$

and

$$\sum_{i=1}^{p} |C_i|^2 = 1. \tag{31}$$

If there are two interacting modes $(n=2)$, $p = k+1$, and the

149

matrix equation equivalent to Eq. (30) can be written explicitly as

$$[((k+1)\ \hbar\omega_0 - E_k)I - \hbar Q_k]\bar{C} = \bar{0} \tag{32}$$

where Q_k is a tridiagonal matrix. With k an odd number

$$Q_k = \begin{bmatrix} A(1) & B(1) & 0 & 0 & 0 & 0 & 0 & 0 \\ B(1) & A(2) & B(2) & 0 & 0 & 0 & 0 & 0 \\ 0 & B(2) & . & . & 0 & 0 & 0 & 0 \\ 0 & 0 & . & A(\frac{k+1}{2}) & B(\frac{k+1}{2}) & 0 & 0 & 0 \\ 0 & 0 & 0 & B(\frac{k+1}{2}) & A(\frac{k+1}{2}) & 0 & 0 & 0 \\ 0 & 0 & 0 & 0 & . & . & B(2) & 0 \\ 0 & 0 & 0 & 0 & 0 & B(2) & A(2) & B(1) \\ 0 & 0 & 0 & 0 & 0 & 0 & B(1) & A(10 \end{bmatrix} \tag{33}$$

$$(k+1)\times(k+1)$$

where

$$A(j) = \frac{\gamma}{2}\ [k+1 + (k+1 - j)^2 + (j - 1)^2] \tag{34}$$

and

$$B(j) = \varepsilon\ \sqrt{j(k+1-j)}. \tag{35}$$

In particular

$$A(\frac{k+1}{2}) = \frac{\gamma}{2}\ [k+1 + (\frac{k+1}{2})^2 + (\frac{k-1}{2})^2] \tag{36}$$

and

$$B(\frac{k+1}{2}) = \varepsilon(\frac{k+1}{2}). \tag{37}$$

With k an even number

$$Q_{k=} \begin{bmatrix} A(1) & B(1) & 0 & 0 & 0 & 0 & 0 & 0 & 0 \\ B(1) & A(2) & B(2) & 0 & 0 & 0 & 0 & 0 & 0 \\ 0^{\bullet} & B(2) & . & . & 0 & 0 & 0 & 0 & 0 \\ 0 & 0 & . & . & B(\frac{k}{2}) & 0 & 0 & 0 & 0 \\ 0 & 0 & 0 & B(\frac{k}{2}) & A(\frac{k}{2}+1) & B(\frac{k}{2}) & 0 & 0 & 0 \\ 0 & 0 & 0 & 0 & B(\frac{k}{2}) & . & . & 0 & 0 \\ 0 & 0 & 0 & 0 & 0 & . & . & B(2) & 0 \\ 0 & 0 & 0 & 0 & 0 & 0 & B(2) & A(2) & B(1) \\ 0 & 0 & 0 & 0 & 0 & 0 & 0 & B(1) & A(1) \end{bmatrix} \tag{38}$$

$$(k+1)(k+1)$$

where

$$A(\frac{k}{2} + 1) = \frac{\gamma}{2} [k+1 + \frac{k^2}{2}] \tag{39}$$

and

$$B(\frac{k}{2}) = \sqrt{\frac{k}{2}(\frac{k}{2} + 1)}. \tag{40}$$

In the nonlinear Schrodinger (NLS) limit, which was discussed in the introduction, A_i and A_i^* are replaced by annihilation and creation operators for boson fields, $\hat{\phi}$ and $\hat{\phi}^+$. At equal times these obey the commutation relations $[\hat{\phi}(x), \hat{\phi}(y)] = [\hat{\phi}^+(x), \hat{\phi}^+(y)] = 0$ and $[\hat{\phi}(x), \hat{\phi}^+(y)] = \delta(x-y)$. Neglecting the ground state energy, normal ordering, using units for which $\hbar = 1$, and scaling out ω_o [31], the number operator is

$$\hat{N} = \int dx \; \hat{\phi}^+\hat{\phi} \tag{41}$$

and the energy operator is

$$\hat{H} = \int dx [\hat{\phi}_2^+\hat{\phi}_2 - \frac{1}{2}\gamma \; \hat{\phi}^+\hat{\phi}^+\hat{\phi}\hat{\phi}] \tag{42}$$

Since translational symmetry is introduced in the NLS limit, momentum is conserved. The corresponding momentum operator is

$$\hat{P} = \int dx \; \hat{\phi}^+\hat{\phi}_x. \tag{43}$$

Construction of exact eigenfunctions of $\hat{N}$, $\hat{H}$, and $\hat{P}$ is discussed in detail in [32] and [33]. A particularly lucid exposition of the relation between this problem and the "inverse scattering" transformation of soliton theory has recently been presented by Wadati [34]. Briefly, the number eigenvalue is a positive integer (N), the momentum eigenvalue is Np where p is a real number, and the energy eigenvalue is $[Np^2 + \gamma^2(N-N^3)/48]$.

10.4 Application to Benzene

Although normal mode (NM) analysis of vibrations in polyatomic molecules has proven to be of great value in understanding single quantum transitions [35], it has not been so useful in describing multiple quantum (overtone) transitions where vibrational energy tends to become localized on a single bond. To describe this situation a technique called local mode (LM) analysis has been developed in which eigenfunctions are constructed on a basis of individual bond vibrations [36]. The DST equation provides a model for the vibrational analysis of polyatomic molecules that reduces to either the NM or LM description in appropriate limits [13]. In this section we use DST to analyze the CH or CD stretch overtones of benzene (C_6H_6 or C_6D_6). We choose this example for two reasons; (i) much experimental data is available on both the fundamental [37,38] and the overtone spectra [39-45], and (ii) there have been several LM studies of CH stretching in benzene [46-50].

Of the thirty vibrational modes of benzene we consider only the six CH stretching modes. In the notation of Herzberg [35] these are $v_1(A_{1g})$, $v_5(B_{1u})$, two $v_{12}(E_{1u})$ and two $v_{15}(E_{2g})$. Our dispersion matrix takes the form

$$\varepsilon M = \begin{bmatrix} 0 & \varepsilon_1 & \varepsilon_2 & \varepsilon_3 & \varepsilon_2 & \varepsilon_1 \\ \varepsilon_1 & 0 & \varepsilon_1 & \varepsilon_2 & \varepsilon_3 & \varepsilon_2 \\ \varepsilon_2 & \varepsilon_1 & 0 & \varepsilon_1 & \varepsilon_2 & \varepsilon_3 \\ \varepsilon_3 & \varepsilon_2 & \varepsilon_1 & 0 & \varepsilon_1 & \varepsilon_2 \\ \varepsilon_2 & \varepsilon_3 & \varepsilon_2 & \varepsilon_1 & 0 & \varepsilon_1 \\ \varepsilon_1 & \varepsilon_2 & \varepsilon_3 & \varepsilon_2 & \varepsilon_1 & 0 \end{bmatrix}. \tag{44}$$

From the quantum considerations of the previous section, the ground state eigenfunction is $|\psi_0\rangle = [000000]$ with an energy eigenvalue of $3(\omega_0 - \frac{1}{2}\gamma)$ and a number eigenvalue equal to 3. Eigenfunctions of the first excited state can be written in the form

$$|\psi_1^{\,1}\rangle = C_{11}[100000] + C_{12}[010000] + C_{13}[001000]$$
$$+ C_{14}[000100] + C_{15}[000010] + C_{16}[000001]. \tag{45}$$

Then $\hat{H}|\psi_1^{\,i}\rangle = E_i|\psi_1^{\,i}\rangle$ requires

$$\{[\omega_0 - \gamma - (E_i - E_0)]I - M\}\bar{C}_i = 0 \tag{46}$$

where

$$\bar{C}_i \equiv \text{col}(C_{11}, C_{12}, C_{13}, C_{14}, C_{15}, C_{16}) \tag{47}$$

and I is the identity matrix. Solutions of (46) correspond to the six normal modes of CH stretch vibration. These are listed in Table 1.

From direct substitution of the $\bar{C}_i$ in Table 1 into (46) one finds that

$$\begin{bmatrix} -2 & -2 & -1 & +1 \\ +2 & -2 & +1 & +1 \\ -1 & +1 & +1 & +1 \\ +1 & +1 & -1 & +1 \end{bmatrix} \begin{bmatrix} \varepsilon_1 \\ \varepsilon_2 \\ \varepsilon_3 \\ \omega_0 - \gamma \end{bmatrix} = \begin{bmatrix} v_1 \\ v_5 \\ v_{12} \\ v_{15} \end{bmatrix}. \tag{48}$$

Table 1. <u>Coefficient Vectors for the First Excited Levels</u>

$\bar{C}_i$	Species	$E_i - E_o$
$(1,1,1,1,1,1)$	A_{1g}	ν_1
$(1,-1,1,-1,1,-1)$	B_{1u}	ν_5
$[e^{\pi i/3}, e^{2\pi i/3}, -1, e^{4\pi i/3}, e^{5\pi i/3}, 1]$	E_{1u}	ν_{12}
(complex conjugate)	E_{1u}	ν_{12}
$[e^{2\pi i/3}, e^{4\pi i/3}, 1, e^{8\pi i/3}, e^{10\pi i/3}, 1]$	E_{2g}	ν_{15}
(complex conjugate)	E_{2g}	ν_{15}

Thus the components of the dispersion matrix can be calculated from the fundamental stretch frequencies as

$$\varepsilon_1 = \frac{1}{6}(-\nu_1 + \nu_5 - \nu_{12} + \nu_{15})$$

$$\varepsilon_2 = \frac{1}{6}(-\nu_1 - \nu_5 + \nu_{12} + \nu_{15}) \qquad (49a,b,c)$$

$$\varepsilon_3 = \frac{1}{6}(-\nu_1 + \nu_5 + 2\nu_{12} - 2\nu_{15}).$$

Table 2. <u>Fundamental CH Stretch Frequencies</u> [37]

	Vapour		Liquid	
	C_6H_6	C_6D_6	C_6H_6	C_6D_6
ν_1	3073.5	2302.5	3062	2294
ν_6	3057	2285	3048	2275
ν_{12}	3068	2288	3057	2276
ν_{15}	3055	2275	3047	2267

The fundamental stretch frequencies have been calculated by Brodersen and Langseth from experimental data [37] and are listed for convenience in Table 2.

From (49) and the data of Table 2, the components of εM are calculated and listed in Table 3.

Table 3.   Components of the Dispersion Matrix $\underline{M}$

	Vapour		Liquid	
	C_6H_6	C_6D_6	C_6H_6	C_6D_6
ε_1	-4.9	-5.1	-4.0	-4.7
ε_2	-1.3	-4.1	-1.0	-4.3
ε_3	+1.6	+1.4	+1.0	-0.2

Assume that the elements of the dispersion matrix, M, are equal to zero. Then an exact solution of the classical system (1) is the pure local mode.

$$\bar{A} = \mathrm{col}(\sqrt{N},0,0,0,0,0).\tag{50}$$

From (6) the classical energy of this local mode is

$$H = \omega_o N - \frac{1}{2}\gamma N^2.\tag{51}$$

An exact eigenfunction of the Hamiltonian operator (22) is

$$|\psi_k\rangle = [k00000].\tag{52}$$

This corresponds to the kth excited state of a pure local mode. The energy above the ground state is

$$\begin{aligned}E_k - E_o &= \langle\psi_k|\hat{H}|\psi_k\rangle - \langle\psi_0|H|\psi_0\rangle\\ &= [\omega_o -\frac{1}{2}\gamma]\,k - \frac{1}{2}\gamma k^2.\end{aligned}\tag{53}$$

Thus the effect of quantization on the classical energy (51) is to replace ω_0 by $\omega_0 - \gamma/2$. With this in mind we modify the classical expression for energy (6) to

$$\tilde{H} = \sum_{i=1}^{6} [[\omega_0 - \tfrac{1}{2}\gamma] |A_i|^2 - \tfrac{1}{2}\gamma|A_i|^4] - \sum_{i \neq j} m_{ij} A_i^* A_j \qquad (54)$$

which gives the correct energy for all $m_{ij} = 0$ and close to the correct energy for the $m_{ij} \ll \gamma$.

Using numerical methods that have been described in detail above and in [1], we assume $A_i = \phi_i \exp(+i\omega t)$ and solve the nonlinear eigenvalue problem

$$-(\omega + \omega_0)\bar{\phi} + \varepsilon M \bar{\phi} + \gamma \mathrm{diag}(|\phi_1|^2, \ldots, |\phi_6|^2)\bar{\phi} = \bar{0} \qquad (55)$$

where $\bar{\phi} \equiv \mathrm{col}(\phi_1, \phi_2, \phi_3, \phi_4, \phi_5, \phi_6)$. Substitution of these solutions into (7) and (54) permits us to calculate the quasiclassical energy as a function of N with parametric dependence upon ω_0 and γ. Thus

$$\tilde{H} = \tilde{H}(N; \omega_0, \gamma). \qquad (56)$$

Since experimental data on overtone frequencies is of the form $\nu(N) \pm \Delta(N)$, where $\Delta(N)$ is the estimated experimental error, our procedure is to fit the calculations of $\tilde{H}$ to the data by minimizing the weighted squared difference

$$G(\omega_0, \gamma) = \sum_{N=2}^{N_{max}} \frac{1}{\Delta(N)} [\nu(N) - \tilde{H}(N; \omega, \gamma)]^2 \qquad (57)$$

through variation of ω_0 and γ.

In Table 4 we list all the experimental data used together with our results. These include: (i) optimal values of ω_0 and γ, (ii) corresponding calculations of $\tilde{H}(N; \omega_0, \gamma)$ for $N \geq 2$, and (iii) calculated components of $\bar{\phi}$ for each value of N (note that $\phi_6 = \phi_2$ and $\phi_5 = \phi_3$).

In general the agreement between measured overtone frequencies and our semiclassical calculations is within the estimated experimental error except possibly for the first overtone. This agreement is especially good at the higher overtones which suggests that there is no need for additional anharmonicities (i.e. N^3 dependence [44]) at these energy levels.

Using the notations: (HV) for (C_6H_6 vapour), (DV) for (C_6D_6 vapour), (HL) for (C_6H_6 liquid) and (DL) for (C_6D_6 liquid), we find that the frequency ratios: $\omega_0(HV)/\omega_0(DV) = 1.345$ and $\omega_0(HL)/\omega_0(DL) = 1.352$. These are within a percent of the theoretical ratio: $\sqrt{13/7} = 1.363$. Corresponding ratios of the anharmonicity constants are: $\gamma(HV)/\gamma(DV) = 1.931$ and $\gamma(HL)/\gamma(DL)$

Table 4. <u>Overtone Calculations for Benzene</u>

<u>C_6H_6 (vapour)</u> $\omega_0 = 3159.11$ $\gamma = 114.59$

N	$\nu(N)$	$\Delta(N)$	ref.	$\tilde{H}(N)$	Φ_1	ϕ_2	ϕ_3	ϕ_4
2	5972	10	45	5974.0	1.4135	-.0300	-.0078	.0106
3	8786	10	"	8789.3	1.7317	-.0246	-.0064	.0084
4	11498	5	"	11490.1	1.9997	-.0213	-.0056	.0072
5	14072	5	"	14076.3	2.2359	-.0191	-.0050	.0064
6	16550	5	"	16547.9	2.4494	-.0174	-.0046	.0058
7	18904	5	"	18904.9	2.6456	-.0161	-.0042	.0054
8	21146	5	"	21147.3	2.8283	-.0151	-.0040	.0050
9	23276	5	"	23275.1	2.9999	-.0142	-.0037	.0047

<u>C_6D_6 (vapour)</u> $\omega_0 = 2332.65$ $\gamma = 59.33$

N	$\nu(N)$	$\Delta(N)$	ref.	$\tilde{H}(N)$	ϕ_1	ϕ_2	ϕ_3	ϕ_4
2	4497	5	45	4485.9	1.4101	-.0581	-.0464	.0247
3	6644	5	"	6640.5	1.7298	-.0481	-.0385	.0181
4	8734	5	"	8735.9	1.9985	-.0420	-.0336	.0147
5	10763	5	"	10771.8	2.2350	-.0377	-.0302	.0126
6	12743	5	"	12748.5	2.4487	-.0345	-.0277	.0112
7	14672	5	"	14665.9	2.6451	-.0320	-.0257	.0102
8	16525	5	"	16523.9	2.8297	-.0300	-.0241	.0094

<u>C_6H_6 (liquid)</u> $\omega_0 = 3156.73$ $\gamma = 117.45$

N	$\nu(N)$	$\Delta(N)$	ref.	$\tilde{H}(N)$	ϕ_1	ϕ_2	ϕ_3	ϕ_4
2	5893	2	43	5960.8	1.4138	-.0239	-.0058	.0064
3	8760	3	"	8765.2	1.7318	-.0196	-.0048	.0051
4	11442	3	"	11452.1	1.9998	-.0170	-.0042	.0044
5	14015	3	"	14021.5	2.2360	-.0152	-.0038	.0039
6	16467	3	"	16473.6	2.4494	-.0139	-.0034	.0036
7	18810	3	"	18808.1	2.6457	-.0128	-.0032	.0033
8	21040	7	"	21025.2	2.8284	-.0120	-.0030	.0031
9	23120	50	44	23124.9	3.0000	-.0113	-.0028	.0029

<u>C_6D_6 (liquid)</u> $\omega_0 = 2335.56$ $\gamma = 63.11$

N	$\nu(N)$	$\Delta(N)$	ref.	$\tilde{H}(N)$	ϕ_1	ϕ_2	ϕ_3	ϕ_4
2	4484	20	40	4480.6	1.4111	-.0493	-.0448	.0044
3	6623	2	"	6626.8	1.7303	-.0411	-.0374	.0019
4	8710	2	"	8709.9	1.9988	-.0360	-.0328	.0008
5	10730	2	"	10730.0	2.2352	-.0324	-.0296	.0003

= 1.861.

Our semiclassical calculations indicate that both normal and deuterated benzene are close to the local mode limit even for the first overtone. This is in agreement with the local mode calculations of Halonen [50]. For example at N = 2 (the first overtone) the fraction of total energy in a single bond is: 99.90% for HV, 99.52% for DV, 99.94% for HL and 99.56% for DL.

Considering the level of sophistication that has currently been attained by LM analysis [51], one might ask what we are doing that is new. The answer, we feel, is twofold. First, we make no assumptions concerning the nature of linear interactions between modes (e.g. potential energy coupling, kinetic energy coupling, electromagnetic coupling, etc.). For benzene, the three parameters describing this coupling (ε_1, ε_2 and ε_3) are exactly computed from the fundamental CH stretch frequencies in Eq. (49). Second, our description can be exactly quantized at any level of excitation. All that is required is the inversion of a real, symmetric matrix of order p given by Eq. (29). Thus we expect this picture to be useful in comparing classical and quantum calculations of chaotic molecular dynamics.

References

[1]   J.C. Eilbeck, P.S. Lomdahl and A.C. Scott (1985). *Physica* **16D**, 318.

[2]   L.D. Landau (1933). *Phys. Zeit. Sowjetunion* **3**, 664.

[3]   S. Pekar (1946). *J. Phys. U.S.S.R.* **10**, 341, 347.

[4]   H. Fröhlich (1954). *Adv. in Phys.* **3**, 325.

[5]   T. Holstein (1959). *Ann. Phys.* **8**, 325, 343.

[6]   A.S. Davydov (1973). *J. Theor. Biol* **38**, 559; *Physica Scripta* **20**, 387; (1982). *Biology and Quantum Mechanics,* Pergamon, Oxford; (1982). *Sov. Phys. Usp.* **25**, 899 and references therein.

[7]   J.M. Hyman, D.W. McLaughlin and A.C. Scott (1981). *Physica* **3D**, 23. A.C. Scott (1982). *Phys. Rev.* **A26**, 578; (1983). **A27**, 2767; (1982). *Physica Scripta* **25**, 651. L. MacNeil and A.C. Scott (1984). *Physica Scripta* **29**, 284. V.A. Kuprievich (1985). *Physica* **14D**, 395. H. Bolteraur and R.D. Henkel (this volume).

[8]   P.S. Lomdahl (1984) in *Nonlinear Dynamics in Biological Systems,* eds. W.R. Adey and A.F. Lawrence, 143, Plenum, New York.

[9]   G. Careri, U. Buontempo, F. Galluzzi, A.C. Scott, E. Gratton and E. Shyamsunder (1984). *Phys. Rev.* **B30**, 4689. A.C. Scott, E. Gratton, E. Shyamsunder and G. Careri (1985). *Phys. Rev.* **B32**, 5551.

[10]   D.M. Alexander and J.A. Krumhansl (1986). *Phys. Rev.* **B33**, 7172.

[11]   J.C. Eilbeck. P.S. Lomdahl and A.C. Scott (1984). *Phys. Rev.* **B30**, 4703.

[12]   J.H. Jensen, P.L. Christiansen, O. Skovgaard, O.F. Nielsen and I.J. Bigio, *Phys. Lett.* (in press).

[13]   A.C. Scott, P.S. Lomdahl and J.C. Eilbeck (1985). *Chem. Phys. Lett.* **113**, 29. A.C. Scott and J.C. Eilbeck (1986).

Chem. Phys. Lett. **131**, 23.

[14] J. Wheatley, T. Hofler, G.W. Swift and A. Migliori (1983). *Phys. Rev. Lett.* **50**, 499; (1983) *J. Acoust. Soc. Am.* **74**, 153. J. Wheatley, D.S. Buchanan, G.W. Swift, A. Migliori and T. Hofler (1985). *Proc. Natl. Acad. Sci. USA* **82**, 7805.

[15] V.E. Zakharov and A.B. Shabat (1971). *Zh. Eksp. Teor. Fiz.* **61**, 118 [*Sov. Phys. JETP* 34, 62, 1972].

[16] V.E. Zakharov (1972). *Zh. Eksp. Teor. Fiz.* **62**, 1745. [*Sov. Phys. JETP* 35, 908, 1972]. J. Gibbons, S.G. Thomhill, M.J. Wardrop and D. ter Haar (1977). *J. Plasma Phys.* **17**, 153.

[17] A. Hasegawa and F. Tappert (1973). *Appl. Phys. Lett.* **23**, 142. C.F. Mollenauer, R.H. Stolen and J.P. Gorden (1973). *Phys. Rev. Lett.* 23, 142. A. Hasegawa and Y. Kodama (1981). *Proc. IEEE* **69**, 1145. C.F. Mollenauer (this volume).

[18] J. Wu, R. Koolian and I. Rudnick (1984). *Phys. Rev. Lett.* **52**, 1421. A. Larraza and S. Putterman (1984). *Phys. Lett.* **103A**, 15.

[19] P.W. Anderson (1958). *Phys Rev.* **109**, 1492; (1978). *Rev. Mod. Phys.* **50**, 191. S. Yoshino and M. Okazaki (1977). *J. Phys. Soc. Japan* **43**, 415.

[20] J.H. Jensen, P.L. Christiansen, J.N. Elgin, J.D. Gibbon and O. Skovgaard (1985). *Phys. Lett.* **110A**, 429.

[21] S. De Filippo, M. Fusco Girard and M. Salerno (1987). *Physica* **26D**, 411.

[22] V.M. Kenkre and D.K. Campbell (1986). *Phys. Rev.* **34**, 4959.

[23] J. Carr and J.C. Eilbeck (1985). *Phys. Lett.* **109A**, 201.

[24] I. Nussbaum and S.F. Fischer (1986). *Phys. Lett.* **115A**, 268.

[25] A.C. Scott and L. MacNeil (1983). *Phys. Lett.* **98A**, 87.

[26] J.C. Eilbeck (1987) in *Physics of Many-Particle Systems* 12, 41, ed. A.S. Davydov.

[27] M.A. Collins (1983). *Adv. Chem. Phys.* **53**, 225.

[28] J.C. Eilbeck (1986) in *Computer Analysis for Life Science - Progress and Challenges in Biological and Synthetic Polymer Research,* eds. Chikao Kawabata and A.R. Bishop, 12, Ohmsha, Tokyo.

[29] W.H. Louisell (1962). *J. Appl. Phys.* **33**, 2435.

[30] A.C. Scott and J.C. Eilbeck (1986). *Phys. Lett.* **A119**, 60.

[31] A.C. Scott (1985). *Phil. Trans. R. Soc. Lond.* **A315**, 423.

[32] E.K. Sklyanin and L.D. Faddeev (1978). *Soviet Phys. Dokl.* **23**, 902.

[33] H.B. Thacker and D. Wilkinson (1979). *Phys. Rev.* **D19**, 3660.

[34] M. Wadati (1985) in *Dynamical Problems in Soliton Theory,* ed. S. Takeno, 68, Springer-Verlag.

[35] G. Herzberg (1945). *Molecular Spectra and Molecular Structure, Vol. 2. Infrared and Raman Spectra of Polyatomic Molecules,* Van Nostrand, Princeton.

[36] W. Siebrand and D.F. Williams (1968). *J. Chem. Phys.* **49**, 1960. R.J. Hayward and B.R. Henry (1974). *J. Molec. Spectrosc.* **50**, 58. B.R. Henry (1976). *J. Phys. Chem.* **80**, 2160. D.F. Heller (1979). *Chem. Phys. Letters* **61**, 583. H.S. Muller and O.S. Mortensen (1979). *Chem. Phys. Letters* **66**, 539. B.R. Henry (1981) in *Vibrational Spectra and Structure,* ed. J.R. Durig, p. 269, Elsevier, Amsterdam.

M. L. Sage and J. Jortner (1981). *Advan. Chem. Phys.* **47**, 293. M. S. Child and R. T. Lawton (1982). *Chem. Phys. Lett.* **87**, 217.

[37] S. Brodersen and A. Langseth (1959). *Mat. Fys. Danske. Skr. Vid. Selsk.* **1**, No. 1; ibid. No. 7, 1959.

[38] S. J. Daunt and H. F. Shurvell (1976). *Spectrochimica Act* **32A**, 1545. F. Seifert et al. (1984). *Chem. Phys. Letters* **105**, 635. K. L. Oehme et al. (1985). *Chem. Phys.* **92**, 168.

[39] J. W. Ellis (1929). *Trans. Faraday Soc.* **25**, 888.

[40] T. E. Martin and A. H. Kalantar (1968). *J. Chem. Phys.* **49**, 235.

[41] R. L. Swofford, M. E. Long and A. C. Albrecht (1976). *J. Chem. Phys.* **65**, 179.

[42] R. G. Bray and M. J. Berry (1979). *J. Chem. Phys.* **71**, 4909.

[43] C. K. N. Patel, A. C. Tam and R. J. Kerl (1979). *J. Chem. Phys.* **71**, 1470.

[44] N. Yamamoto, N. Matsuo and H. Tsubomura (1980). *Chem. Phys. Letters* **71**, 463.

[45] K. V. Reddy, D. F. Heller and M. J. Berry (1982). *J. Chem. Phys.* **76**, 2814.

[46] B. R. Henry and W. Siebrand (1968). *J. Chem. Phys.* **49**, 5369.

[47] R. J. Hayward, B. R. Henry and W. Siebrand (1973). *J. Molec. Spec.* **46**, 207.

[48] M. S. Burberry and A. C. Albrecht (1979). *J. Chem. Phys.* **70**, 147.

[49] I. Schek, J. Jortner and M. L. Sage (1979). *Chem. Phys. Letters* **64**, 209.

[50] L. Halonen (1982). *Chem. Phys. Letters* **87**, 221.

[51] M. S. Child and L. Halonen (1984). *Adv. Chem. Phys.* **57**, 1.

11   *U. Enz*

THE SINE-GORDON BREATHER AS A MODEL PARTICLE: QUANTIZED STATES FROM CLASSICAL FIELD THEORY

Abstract

The breather solution of the sine-Gordon equation can be viewed to be an extended oscillator moving as a whole with constant velocity. The fundamental insights of L. de Broglie (1924) leading to quantum mechanics were based on the study of extended moving oscilators. In this context we investigate the properties of the breather solution and show that a "de Broglie wavelength" can be attributed to the breather. In this simple, entirely classical case the momentum of the breather is proportional to its wave vector, and the total energy is proportional to the oscillator frequency. We distinguish between the periodic, but locally anharmonic oscillations of the breather and a harmonic plane wave of equal wave vector but constant amplitude, a distinction similar to that made by de Broglie between the classical, localized u-wave and the probabilistic ψ-wave of quantum mechanics. We conclude that the breather solution represents an example of a classical, non-linear field theory of an extended particle in rectilinear motion. This model theory is more general than the description of a particle by the ψ-wave because it covers the internal constitution of the particle as well as its wave aspects. The model is applied to the treatment of a breather confined in a square well potential. We find discrete energy levels identical with those found in quantum mechanics for a particle in such a potential. The present particle model opens the way to identify classical (three dimensional) solitons performing internal oscillations or rotations with elementary particles.

The aim of this paper is to demonstrate the existence of discrete energy levels of confined solitons of the "breather" type. In particular, the sine-Gordon breather moving in a square-well potential will be treated. The energy levels obtained in this case turn out to be identical with those found in quantum mechanics for a mass point in such a potential well. It is of great interest that our result is obtained in the framework of a classical, nonlinear field theory. The implications of this finding concerning the possibility of a description of particles

161

as classical objects will be discussed below.   In the following we
first recall the results of analogy considerations [1] of moving
breathers and moving particles.   The potential well is then
defined in terms of intersoliton forces.   With the aid of Backlund
transformations perfectly reflecting walls can be constructed.
For a breather confined in the potential well thus obtained, the
discrete energy levels are easily found.

Breathers and particles share properties like rest mass, kinetic
energy and interaction on the one hand, and a wave character on
the other hand.   Clearly, a close analogy exists [1]. (The dual
nature of the breather solution has also been emphasized by J.J.
Klein [7], which came to my knowledge only recently.)   We recall
the breather properties by departing from the sine-Gordon equation

$$u_{xx} - c^{-2} u_{tt} = d^{-2} \sin u. \tag{1}$$

The function $u(x,t)$ is interpreted as a classical scalar field.
Eq. (1) governs the evolution of this field in space and time.   If
we identify c with the velocity of light and consider d as a
constant elementary length, Eq. (1) has the form of a field theory
describing particles.   We focus on the breather solution, which is
considered to represent a physical object or particle moving with
velocity v (β = v/c).   The breather solution of Eq. (1) reads [2]

$$u = 4 \tan^{-1} \left(\frac{s \sin[(r/d)(c(t-t_o)-\beta x)(1-\beta^2)^{-1/2}}{r \cosh[(s/d)(x-x_o-vt)(1-\beta^2)^{-1/2}]} \right). \tag{2}$$

The parameter q (s = sin q, r = cos q) appearing in Eq. (2) is
considered to be a constant.   Both the oscillation frequency of
the breather and its spatial extension are determined by q.   The
constants x_o and t_o define the position of the breather and its
phase at time t = 0.   It is important to note that the field u
given by Eq. (2) is infinitely extended in space and thus
represents an extended object.   As seen from Eq. (2) the
oscillation frequency ω_b is

$$\omega_b = \frac{rc}{d} (1-\beta^2)^{-1/2} \tag{3}$$

The oscillation is highly anharmonic near the centre of the
breather, but harmonic far from it.   Eq. (2) also yields a wave
vector k_b equal to

$$k_b = \frac{rv}{dc} (1-\beta^2)^{-1/2}. \tag{4}$$

In the non-relativistic limit, this expression reduces to a
wavelength

$$\lambda_b = \frac{2\pi cd}{r} \frac{1}{v}, \tag{5}$$

representing, at a fixed time, the distance between positions of equal phase on the x-axis. If we define the "size" of the breather as the maximum separation b of the soliton-antisoliton pair, i.e. the maximum separation of the two points on the x-axis where $u = \pi$, we find from Eq. (2)

$$r \approx 2 \exp \left(-\frac{b}{2d} (1-\beta^2)^{-1/2}\right). \tag{6}$$

Eq. (6) defines b in terms of r and d. The size b is small compared with the wavelength (5) ($d < b < \lambda_b$). The position of the breather, its centre, is defined as the coordinate $x_o + vt$.

The total energy of the breather is found by integration over the energy density [1,3]. The resulting rest energy is

$$E_o = 16 \text{ s Gd}. \tag{7}$$

The constant G appearing in (7) has the dimension of a force. The energy (7), which for the larger part is localized in a space region of the order of the size b, is equivalent to a rest mass $m_b = E_o/c^2$, the mass of the breather. The energy of the moving breather is

$$E_b = E_o(1-\beta^2)^{-1/2}, \tag{8}$$

corresponding to a momentum $p_b = E_b v/c^2$. Summarizing we see that the moving breather is governed by a frequency ω_b and a wave vector k_b on the one hand, and an energy E_b and a momentum p_b on the other hand. By analogy with Planck's constant, a constant $\hbar_b$ has been introduced [1] reading

$$\hbar_b = \frac{16 \text{ s Gd}^2}{rc} = E_o d/rc, \tag{9}$$

so that the following relations hold:

$$E_b = \hbar_b \omega_b \tag{10}$$

and

$$p_b = \hbar_b k_b. \tag{11}$$

The relations (10) and (11) demonstrate clearly the wave aspect of the analogy.

Up to this point we have considered a breather or particle in rectilinear uniform motion. We now turn to the study of the breather as a particle confined in a square well potential. This type of potential is characterized by perfectly reflecting walls, in our one-dimensional case by two walls at positions x = 0 and x = L. Therefore we assume that the function u vanishes at x = 0 and x = L, resulting in the condition

$$\pi n/k_b = L, \qquad (12)$$

where n is an integer. A possible way to deal with this problem is to postulate the existence of such reflecting walls and to solve Eq. (1) for the boundary conditions u(o) = 0 and u(L) = 0. An equivalent but simpler way is to simulate the reflecting walls by the superposition of two trains of breathers of proper phase moving in opposite directions (Fig. 1). Both trains consist of equidistant breathers (distance 2L), one train moving with velocity +v, the other with velocity -v. The breathers of each one of the trains oscillate in phase, the phase of the two trains being opposite. The breather positions are such that the points x = 0 and x = L, the positions of the walls, are centres of antisymmetry. The above construct not only defines the reflecting walls within the framework of soliton theory but also leads directly to the solution.

In the first instance we assume that the width of the potential well is large compared with the size b of the breather: L >> b. The breather-breather interaction inside each of the trains can then be neglected. We now focus on the two breathers B and B' (Fig. 1) which are going to meet at the position x = 0. The evolution of the field u (x,t) can be found by applying the Backlund transformations [3], which read in our notation

$$\frac{1}{2}\left(\frac{\partial \sigma}{\partial x} + \frac{\partial \sigma'}{\partial x} + \frac{1}{c}\frac{\partial \sigma}{\partial t} + \frac{1}{c}\frac{\partial \sigma'}{\partial t}\right) = \frac{a}{c}\sin\left(\frac{\sigma-\sigma'}{2}\right) \qquad (13)$$

and

$$\frac{1}{2}\left(\frac{\partial \sigma}{\partial x} - \frac{\partial \sigma'}{\partial x} - \frac{1}{C}\frac{\partial \sigma}{\partial t} + \frac{1}{C}\frac{\partial \sigma'}{\partial t}\right) = \frac{1}{ad}\sin\left(\frac{\sigma+\sigma'}{2}\right), \qquad (14)$$

where σ and σ' are two breather solutions of Eq. (1) describing the meeting free breathers B and B' (a=1). The same argument applies for the breathers B and C' meeting at x = L etc.

From Eq. (13) and (14) we see that the function u describing the interacting breathers remains zero at x = 0 and x = L. We interpret this result as a reflection of the breathers at these points, whereby their phase is reversed. The breather is completely deformed during reflection, but emerges undisturbed after having travelled a distance of the order of several times b. We have thus constructed perfectly reflecting walls. In the space region 0 < x < L, the region of interest, there is one and only one breather at any time, moving to and fro, so that our procedure indeed describes a confined "particle", in fact a one-particle state. Here too it is important to realize that the breather is an extended object, but in this case the extension is limited to

164

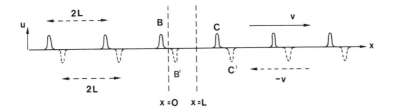

Figure 1. Simulation of breather confinement in the space region $0 < x < L$ by interaction of two trains of equidistant breathers having opposite phase and moving in opposite directions.

the region $0 < x < L$. A schematic representation of the solution is given in Fig. 2. Without proof we note that the above procedure has a more general validity and applies for small breather-breather distances as well, thus including interaction. (A.C. Scott has recently pointed out that a complete solution has been given by Zakharov et al. [8] confirming the existence of a strictly periodic state.)

The energy levels of the confined breather, being in a stationary state, are now easily found. Similar to the familiar linear problem, a stationary state is defined as a state of strictly periodic motion. The condition (12) has been shown to be fulfilled. We therefore have discrete values of the wave vector according to

$$k_{b,n} = \frac{n\pi}{L}. \tag{15}$$

From Eqs. (4), (8) and (15) we find the corresponding energy levels, reading, in the non-relativistic approximation

$$E_{b,n} - E_o = E_o \frac{d^2}{r^2} \frac{\pi^2 n^2}{2 L^2} = \frac{\hbar_b^2}{m_b} \frac{\pi^2 n^2}{2 L^2}. \tag{16}$$

The energy levels (16) are identical with those found in quantum mechanics for a particle of mass m_b moving in a square well potential. (If we relax the assumption $b \ll L$ and include small breather-breather distances in each of the trains, the breather-breather interaction can no longer be neglected. Additional energy terms then occur. The corrections will be of the order b/L or higher orders. These corrections may be compared to the corrections of quantum electrodynamics.) However, there is a great difference: whereas the ψ-wave of quantum mechanics is a plane wave of constant amplitude governed by a linear Klein-Gordon equation (the Compton wave length of which is the length d/r and not the fundamental length d of Eq. (1)), the u-wave is a a field with high but finite amplitude in a small region of space, governed by the nonlinear equation (1). The only

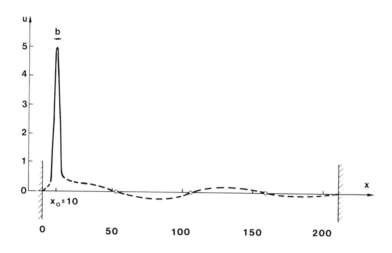

Figure 2.   Schematical representation of the solution describing the confined breather with $c=d=1$, $t=t_o=0$, $x_o=10$, $v=0.3$ and $r=sin$ 0.2.   The figure is not on scale as the amplitude u decreases exponentially and therefore is very small far from the breather site.   The walls of the potential well are situated at $x=0$ and $x=210,8$ $(=2\lambda)$.   The breather or "particle" is identical with the region of space where the values of the field u and its energy density are high.   The larger part of the energy and thus mass is concentrated in a region of the size b $(b \ll L)$.   This "particle" moves with constant velocity $|v|$.   In the stationary state the field u has equidistant modes determined by Eq. (12) $(n=4)$.

common property is the phase. The interpretation differs completely as well: whereas the ψ-wave is subject to the statistical interpretation, there are no arguments that would prevent a completely classical interpretation of the u-wave. It is interesting to note [1] that the solution of the moving breather given above represents an explicit example of a particle model, the moving oscillator, proposed by de Broglie [4] in the early days of quantum mechanics. Our example exhibits the properties postulated by de Broglie, i.e. a phase equal to the phase of the ψ-wave, localization of field energy (and thus mass) in a small space region, and a classical interpretation.

It is worthwhile to take a closer look at the momentum and the position of the breather. In the stationary state the motion of the confined breather is periodic, its velocity is either +v or -v. As the mass is mainly localized in a region of the order of b, the momentum is localized in this region too. The "particle" represented by the breather reaches, in the course of time, all positions between x = 0 and x = L. A "measurement" of position would lead to the following result: a single experiment performed at an arbitrary moment yields an arbitrary value x_b $(0 < x_b < L)$,

but a large number of observations produces all x-values in this range with equal probability. (In fact the peak value of u at the

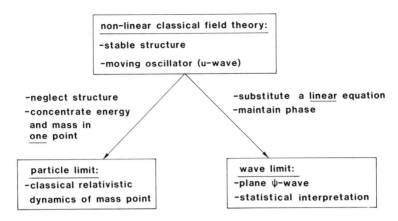

Figure 3. Hierarchy of particle descriptions: the familiar picture of the relativistic mass point on the one hand and the plane wave description of quantum mechanics on the other hand follow from the more general nonlinear description by making the simplifying assumptions indicated in the figure.

"centre" of the breather is, as a function of x, sinusoidally modulated. Nevertheless the breather performs a uniform translational motion reaching all positions x_b. This implies moreover that the breather centre passes the nodes. It does so while u = 0 at the nodes for all times t. D. Bohm and B. J. Hiley [5] have recently discussed the corresponding quantum mechanical problem of a particle in a box. As ψ (being real in that case) is sinusoidally modulated, the probability is zero at the nodes, which means that the particle cannot pass the nodes. This complication is not present in our model.) We conclude that the "particle" has a momentum ± $m_b v$ and at the same time a position defined with a precision of ≈b. The interference effects arise from the fact that u is an extended field interfering simultaneously with both boundaries. However, if the "particle" is to be localized, i.e. confined by experiment to a region smaller than L, the interference is lost, and the momentum becomes indetermined, similar to the quantum mechanical result. (In contrast to quantum mechanics, however, the breather has always a well-defined position. We therefore also conclude that the Heisenberg relation is violated in our model. The product $\Delta p \Delta x$ is smaller than $\frac{1}{2} h_b$ by a factor of the order of L/b.)

The essential difference between our phenomenon of confined breather and the corresponding quantum mechanical case is that we deal with a classical, extended, non-linear field u as the single basic notion, whereas the quantum mechanical description presupposes the (linear) wave ψ *and* the additional notion of "particle" with empirically given properties. Both descriptions are internally consistent, but we think that the former is to be preferred, not because it reproduces the correct energy levels, but because it combines this result with an answer to the question of stability, internal structure and discrete mass of the particle. Hence we conclude that there is a **hierarchy** (Fig. 3) of descriptions of the phenomenon "particle": the description proposed is the most fundamental one, special cases of limited validity follow by simplifying assumptions, i.e. the classical relativistic particle picture by neglecting the internal structure of the u-field and by concentrating the total particle mass in one point; and the plane wave description of quantum mechanics by substituting the plane ψ-wave while maintaining the proper phase of the u-wave. These special cases are not limits in a mathematical sense, but simpler concepts. **Both of them are incomplete, however.**

Finally, we want to stress the general validity of the argument presented. While basing all conclusions presented on the specific equation (1) we claim that the same reasoning holds for a much wider class of classical non-linear equations, provided they are relativistically invariant and admit soliton-like solutions with internal oscillations. A three-dimensional example of a realistic particle model of this type has been proposed recently [6].

This paper is a slightly augmented version of a paper which appeared in *Physica* 21D, 1, 1986.

References

[1]   U. Enz (1985). *Physica* **17D**, 116; (1963). *Phys. Rev.* 131, 1392.

[2]   A.C. Scott (1979). *Phys. Scr.* **20**, 509.

[3]   G.L. Lamb (1980). *Elements of Soliton Theory*, Wiley, New York.

[4]   L. de Broglie (1926). *Comp. rend.* **183**, 447, 1926.

[5]   D. Bohm and B.J. Hiley (1985). *Phys. Rev. Lettr.* **55**, 2511.

[6]   U. Enz (1980). *Physica* **1D**, 329; (1978). *J. Math. Phys.* **19**, 1304.

[7]   J.J. Klein (1976). *Can. J. Phys.* **54**, 1383.

[8]   V.E. Zakharov, La. Takhtadzhyan and L.D. Faddeev (1975). *Sov. Phys. Dokl.* **19**, 824.

A MACROSCOPIC APROACH TO SYNERGETICS

Abstract

We discuss the relation of the slaving principle to Jaynes' principle of maximum calibre and show that in the vicinity of critical points the latter can be obtained as a consequence of the slaving principle. Jaynes' principle then is used in order to determine probability distributions for complex systems when knowledge on the system is possible on macroscopic scales only. As an example we treat the single mode laser. A systematic method is presented how to identify order parameters from a purely macroscopic point of view. Finally a path integral formulation and its relation to the corresponding Fokker-Planck equation is given for the time development of such systems on the basis of Jaynes' principle.

12.1 Introduction

According to its definition, synergetics deals with systems which are composed of many subsystems and which can produce macroscopic spatial, temporal, or functional structures [1,2]. The main goal of synergetics is the search for universal principles underlying the processes of self-organization. The occurrence of self-organization in quite different systems and the search for general principles has become by now rather commonplace. But when we gave this definition of synergetics more than 15 years ago, the situation had been quite different. Nevertheless we feel that a good deal of research is still to be done to elaborate on these general principles or even to find new ones.

12.1.1 *Systems of Synergetics*

In synergetics we consider open systems which are composed of many subsystems and which are driven away from thermal equilibrium by a flux of matter or energy. We are then interested in the spontaneous self-organization on macroscopic scales. A fundamental principle by which self-organization can be understood is the slaving principle [1-3]. According to it, the behaviour of systems in the neighbourhood of critical points is dominated by a few collective degrees of freedom which are also called order parameters. These order parameters slave the subsystems and just give the total system its specific structure or order. In particular, it turns out that the detailed nature of subsystems becomes unessential which in turn means that the slaving principle makes a universal statement about the behaviour of systems. Traditional synergetics is based on a microscopic approach which starts from equations of motion at the microscopic or mesoscopic level. For given control parameters first the stationary state is determined. For instance, for systems in or close to thermal equilibrium the corresponding state vector is homogeneous in space and time and can thus easily be determined. But in a number of cases all such state vectors are available. In the next step the stability of stationary states is studied by linear stability analysis. In this way one obtains critical regions of the control parameters as well as the formation of collective modes. They can be divided into two kinds, namely unstable modes which then serve as order parameters, or the stable modes which will become the slaved modes. According to the slaving principle which applies to the fully nonlinear theory including fluctuations, the slaved modes can explicitly be expressed by the amplitudes of the unstable modes, i.e. the order parameters including fluctuations. By the slaving principle we may eliminate all the slaved modes so that we are eventually left with the order parameter equations alone. They are capable of determining the macroscopically evolving state.

12.2 Need For Macroscopic Synergetics

Quite often the subsystems become by themselves very complicated so that it is difficult or even impossible to formulate the mesoscopic equations explicitly. When we go to the extreme, namely the human brain, the subsystems are, for instance, the nerve cells (neurons) which are by themselves complicated systems. A nerve cell contains its soma, an axon and up to 80 thousand dendrites by which the nerve cells are connected with other cells. In the human brain there are about 10 billions of nerve cells. It is suggested by synergetics that even in spite of this enormous complexity a number of behavioural patterns can be treated by means of the order parameter concept, where now the order parameter equations are established in a phenomenological manner.

Recently we could find a paradigm, namely the coordination of hand movements and especially involuntary changes between hand movements [4]. Though the subsystems are quite numerous and consist of nerve cells, muscle cells and other tissue, the behaviour can be described by a single order parameter. In these experiments performed by Kelso a test person is asked to move his or her fingers in an antiparallel fashion at a given oscillation frequency. When the oscillation frequency is increased suddenly beyond a critical value of ω a new mode sets in, namely a parallel or in other words symmetric hand or finger movement. This transition could be modelled quantitatively in all details in the following way.

Let us call the control parameter ω. The elongation of the fingertips may be then written in the form

$$x_i = r_i \cos(\omega t + \phi_i), \quad i = 1, 2. \tag{1}$$

Because of the abrupt transition between different relative phases it suggests the use of the relative phase

$$\phi = \phi_1 - \phi_2 \tag{2}$$

as order parameter. For a single order parameter the typical order parameter equation of synergetics reads

$$\dot{\phi} = -\frac{\partial V(\phi)}{\partial \phi}. \tag{3}$$

The form of V can be fixed by the requirement of symmetry between the two hands

$$V(\phi) = V(-\phi). \tag{4}$$

By the periodicity of V as a function of ϕ which follows from (1) we obtain

$$V(\phi) = v(\phi + 2\pi) \tag{5}$$

and finally we make the postulate of simplicity, i.e. we look for the simplest function obeying (4) and (5) which gives a nontrivial result. This is then made possible by the hypothesis

$$-V = a \cos \phi + b \cos 2\phi. \tag{6}$$

The potential V is exhibited for various control parameter values in Figure 1. This model is capable of explaining all experimental

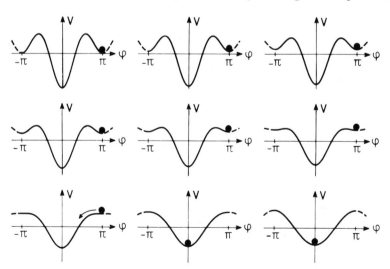

Figure 1. The potential V of (3) for various values of the control parameter.

results, namely hysteresis, critical behaviour and even critical fluctuations, provided a noise source is added to (3). This result encourages one to look for a macroscopic formulation of typical phenomena in synergetics. To this end we shall consider in the next section the maximum calibre principle.

12.3 The Maximum Calibre Principle

To formulate a macroscopic approach in a general form, we start from Jaynes' principle which reads: "If any macrophenomenon is found to be reproducible then it follows that all microscopic details that were not under the experimenter's control must be irrelevant for understanding and predicting it" [5]. As was outlined by Jaynes and other authors, this principle means that we are dealing with an experiment from the theoretical point of view by inference rather than by deduction from a microscopic level.

In order to understand how this principle can be cast into a mathematical form let us recall Gibbs' principle or rather Jaynes' generalization of that principle. Let macroscopic information be given by the vector f where

$$f = (f_1, f_2, \ldots, f_N) \tag{7}$$

are measured macroscopic quantities. Note that the index may distinguish different variables but also the same variable at different times. The calibre is then defined by

$$i_f = \ln W_f \tag{8}$$

where W_f is the number of microstates which belong to the macroscopic state with the given values of f. Knowing f we now wish to make a prediction on a different variable or on the same variable but for different times.

$$g = (g_1, g_2, \ldots, g_M). \tag{9}$$

When now the calibre is determined under the constraints (7) and (9) in general we have to expect the inequality

$$i_{fg} \le i_f. \tag{10}$$

It is possible to find (9) provided the corresponding calibre is positive

$$i_{f,g} > 0. \tag{11}$$

The principle now states that state $\underset{\sim}{g}$ is realized or can be found experimentally for which

$$i_{f\underset{\sim\sim}{g}} = \max. \tag{12}$$

is valid. We now remind the reader of the general mathematical formulation. We denote the probability for state j of the system by p_j. The information or calibre is then defined by the following original formula of Shannon by means of

$$i = - C \sum_j p_j \ln p_j \tag{13}$$

where C is a constant. The macroscopic knowledge is represented by constraints in the form

$$f_k = \sum_i p_i f_i^{(k)} \tag{14}$$

where

$$(f_k) = \underset{\sim}{f}. \tag{15}$$

We introduce the set of Lagrange multipliers

$$(\lambda_k) = \underset{\sim}{\lambda} \tag{16}$$

and maximize (13) using the Lagrangean formalism

$$\delta(i - \underset{\sim}{\lambda} (\sum_j \underset{\sim}{f}_j p_j - \underset{\sim}{f})) = 0. \tag{17}$$

In the following we shall denote the set of Lagrange multipliers without the one which is connected with the normalization condition where $f_k = f^{(k)}_i = 1$ by λ'_k. Then the following well known results are obtained

$$Z(\underset{\sim}{\lambda}') = \sum_j \exp(-\underset{\sim}{\lambda}' \, \underset{\sim}{f}_j) \tag{18}$$

which means a generalization of the partition function and

$$f_k = - \frac{\partial}{\partial \lambda_k} \ln Z(\underset{\sim}{\lambda}') \tag{19}$$

$$p_j = Z^{-1}(\underset{\sim}{\lambda}') \exp(-\underset{\sim}{\lambda}' \underset{\sim}{f}_j). \tag{20}$$

(19) serves for a determination of the Lagrange multipliers.

12.4 Relation between the Slaving Principle and Jaynes' Principle

In the following we shall identify the index j with u, s, i.e. with respect to the indices of the unstable and stable modes used in the slaving principle [6]. We therefore make the identification

$$j \longrightarrow u, s \qquad (21)$$

$$p_j \longrightarrow p(u, s). \qquad (22)$$

The information then reads explicitly

$$i = - \sum_{u, s} p(u, s) \ln p(u, s). \qquad (23)$$

The slaving principle implies that the joint probability distribution for u and s can be split into the form

$$p(u, s) = P(s|u) f(u) \qquad (24)$$

where P can be split into a product over the individual indices s. Obviously

$$\sum_{s} P(s|u) = 1 \qquad (25)$$

must hold. By means of (24) we may cast (23) into the form

$$i = - \sum_{u} f(u) \ln f(u)$$

$$- \sum_{u} f(u) \sum_{s} P(s|u) \ln P(s|u) \qquad (26)$$

or in short

$$i = i_f + \sum_{u} f(u) i_s(u) \qquad (27)$$

where i_f and i_s refer to the information of the order parameters and slaved modes, respectively. To demonstrate the significance of this result, we quote a typical result for P(s|u), namely

$$P(s|u) = N \exp \{(s - s(u))^T Q (s - s(u))\}. \qquad (28)$$

By a simple shift of coordinates s one may convince oneself readily that

$$i_s(u) = - \sum_s P(s|u) \ln P(s|u) \qquad (29)$$

becomes independent of u. In other words that (29) does not change in the transition region. Therefore, the only relevant contribution to (27) stems from

$$i_f = - \sum_u f(u) \ln f(u) \qquad (30)$$

or in other words, the macroscopic behaviour is completely determined by the order parameters u . Thus the slaving principle is in accordance with Jaynes' principle or it even provides a microscopic derivation of Jaynes' principle in the region close to nonequilibrium phase transition points.

12.5 The Crucial Problem: How to Find the Adequate Constraints

As is well known, by means of Jaynes' principle the well known relations of thermodynamics such as between free energy, entropy and internal energy can be easily derived provided the energy is used as a constraint (e.g. [1]). More generally speaking, in thermodynamics we may use the conserved quantities such as energy, particle number, angular momentum, etc., as adequate constraints. In view of the strong coupling of open systems to the surrounding, it is doubtful whether these constraints can be used for nonequlibrium systems. Indeed, any effort so far to derive their distribution functions by means of these constraints has been in vain.

Here we arrive at a crucial problem in the whole formulation of the maximum entropy or maximum calibre principle, namely there are no general recipes known for finding constraints for general systems. Therefore we need a guideline how to obtain such constraints [7]. Our guidelines will be:

(a) In the maximum entropy principle it is stated that those quantities should be used as constraints which are observed experimentally. We will come back to this approach in a minute.
(b) We may be guided by the microscopic theory where at least in a number of cases the distribution functions are known to us explicitly and which may then serve as a check whether we have

used the correct constraints.   Let us turn to approach No. 1 and consider an explicit example, namely the single mode laser.
Its field strength can be written in the form of a standing wave

$$E(x,t) = E(t) \, \sin kx \tag{31}$$

The time-dependent amplitude can be split into a rapidly oscillating factor with the optical transition frequency ω and a slowly varying amplitude B which is a random variable because of quantum noise

$$E(t) = (Be^{-i\omega t} + B^* e^{i\omega t}). \tag{32}$$

It is well known from the laser that the measurable quantities are there the first and second moment of the intensity which can be expressed by B, B^* in the form

$$f_1 = <B^* B>, \tag{33}$$

$$f_2 = <B^{*2}B^2> \tag{34}$$

respectively.   Identifying the index j of p_j with B^*, B and denoting the Lagrange multipliers belonging to (33) and (34) by λ_1 and λ_2 respectively, we obtain the following distribution function

$$p(B,B^*) = N \exp (\alpha \, |B|^2 - \beta \, |B|^4) \tag{35}$$

where α and β are directly related to λ_1 and λ_2.   This is the well-known laser light distribution function derived in laser theory [8].   The result tells us that we have chosen the correct constraints by (33) and (34).   This result can be generalized to the multimode laser with and without phase locking [7].   In order to discuss the meaning of such a result, let us consider a somewhat simpler case, namely in which B is replaced by a real variable q and the distribution function p_j is replaced by f(q) [9].

12.6 Numerical Results

One can readily convince oneself that the information becomes infinite for a continuously distributed variable.   Therefore to be realistic we have to introduce an accuracy interval of measurements which we shall call ε [9].   Then the expression for the information reads

$$i = - \int dq \, f(q) \, \ln f(q) - \ln \varepsilon. \tag{36}$$

We shall evaluate (36) for the distribution function

$$f(q) = N \exp [\alpha q^2 - \beta q^4] \tag{37}$$

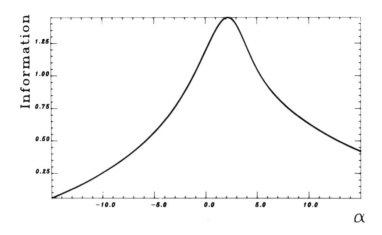

Figure 2. The information as a function of the control parameter α.

which occurs in numerous nonequilibrium phase transitions for the order parameter q. The change of information with increasing control parameter α is exhibited in Fig. 2. As we see the information goes across a maximum and then it drops again. When we choose the same accuracy interval below and above threshold, above it the information levels down at one bit, whereas below threshold it was zero bit. We thus see that when a system goes from its disordered into its ordered state, the information entropy increases instead of decreasing as was usually expected by people using thermodynamic entropy in nonequilibrium phase transitions.

This result can easily be visualized. As was shown on many occasions, the behaviour of the order parameter q can be interpreted as that of a particle moving in an overdamped fashion in a potential well which is exhibited in the left upper corner of Fig. 3 for a negative control parameter α. The distribution function belonging to this potential is shown one step below. With the absolute value of α increasing this distribution function becomes narrower and eventually fills up more or less the accuracy interval so that only a single state can be realized. On the other hand, when the control parameter α acquires positive values the potential of the right upper corner of Fig. 3 applies. In such a case the distribution function f has acquired two minima which become narrower with increasing control parameter α until the two parts of f fill the accuracy intervals. Quite evidently, here two different states can be occupied, i.e. one bit of information can be stored by such a system.

12.7 Determination of Order Parameters by the Maximum Information Principle (General Procedure)

In this section we wish to show how the order parameters and the slaved modes and their distribution functions can be determined by

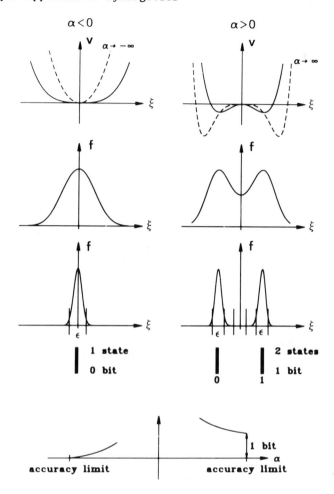

Figure 3. Compare text.

means of the maximum information entropy principle provided we use the adequate constraints [10]. The results of Section 12.5 suggest that the moments are the adequate constraints. But now we do not consider the moments of the order parameters but all the moments of the observed quantities of the system up to fourth order. As we shall see at the end of our calculation, the resulting distribution functions coincide with those known from the microscopic theory. And therefore we feel that this is a second way of justifying the choice of the moments as constraints.

We denote the state vector of the observed quantities by

$$q^T = (q_1, q_2, \ldots, q_N) \qquad (38)$$

and call the moments

$$f_1 = <q_i>; \quad f_1^{(1)} = q_i$$

$$f_{ij} = <q_i \, q_j>, \quad f_{ij}^{(2)} = q_i q_j \, \cdots \qquad (39)$$

The method of Lagrange multipliers then provides us with the result

$$p = \exp \{V(\underset{\sim}{\lambda}, \underset{\sim}{q})\} \qquad (40)$$

and

$$V(\underset{\sim}{\lambda}, \underset{\sim}{q}) = \lambda + \sum_i \lambda_i q_i + \ldots.$$

$$+ \sum_{ijkl} \lambda_{ijkl} q_i q_j q_k q_l \qquad (41)$$

where we first introduce new coordinates $\underset{\sim}{q}$ so that the extrema of V lie at the point of highest symmetry

$$\frac{\partial V}{\partial q_i} = 0 \text{ for } \underset{\sim}{q} = 0. \qquad (42)$$

We now study the neighbourhood of $\underset{\sim}{q}_o$ by making the replacement

$$\underset{\sim}{q} = \underset{\sim}{q}_o + \underset{\sim}{w}. \qquad (43)$$

This transforms V into $\tilde{V}$ where $\tilde{V}$ reads

$$\tilde{V} = \tilde{\lambda} + 0 + \sum_{ik} \tilde{\lambda}_{ik} w_i w_k + \ldots \qquad (44)$$

In order to find the order parameters and slaved modes we diagonalize the bilinear part of (44) by means of a transformation

$$w_i = \sum_k a_{ik} \xi_k \qquad (45)$$

Actually (44) may serve for a procedure in analogy to catastrophe theory where specific normal forms of (44) are established by suitable coordinates. By means of (45), (44) is transformed into

$$\tilde{V}(\underset{\sim}{\tilde{\lambda}}, w) \dashrightarrow \bar{V} \, (\underset{\sim}{\lambda}, \underset{\sim}{\xi}) \qquad (46)$$

and

179

$$\bar{V} = \lambda + \sum_k \lambda_k \xi^2_k + \ldots \tag{47}$$

We now make the following identification for the order parameters and slaved mode amplitudes

$$\lambda_k \geq 0, \ k \dashrightarrow u, \ \lambda_k < 0, \ k \dashrightarrow s \tag{48}$$

Accordingly we may decompose $\bar{V}$ into V_u and V_s

$$\bar{V} = V_u + V_s \tag{49}$$

where V_u contains the order parameters only

$$V_u = \sum_u \lambda_u \xi^2_u + \ldots$$

$$\sum_{uu'u''\ u'''} \lambda_{uu'u''\ u'''} \xi_u \xi_{u'} \xi_{u''} \xi_{u'''} . \tag{50}$$

We introduce the abbreviation g according to

$$\int \exp V_s \ d\underset{\sim}{\xi}_s = g(\underset{\sim}{\xi}_u) > 0 \tag{51}$$

and write this result in the form

$$g(\underset{\sim}{\xi}_u) = \exp \ [-h(\underset{\sim}{\xi}_u)]. \tag{52}$$

Lumping these results together we may introduce the conditional probability

$$P(\underset{\sim}{\xi}_s | \underset{\sim}{\xi}_u) = \exp \ [h + V_s] \tag{53}$$

for the slaved mode amplitudes. The total probability distribution can then be written in the form

$$\exp \bar{V} = P(\underset{\sim}{\xi}_s | \underset{\sim}{\xi}_u) \ f(\underset{\sim}{\xi}_u). \tag{54}$$

Because the slaved mode amplitudes $\underset{\sim}{\xi}_s$ are in general still rather small we may approximate the conditional probability by a Gaussian approximation

$$\bar{P}(\underset{\sim}{\xi}_s|\underset{\sim}{\xi}_u) = \exp\{\bar{\lambda}(\underset{\sim}{\xi}_u) + \sum_s \bar{\lambda}_s(\underset{\sim}{\xi}_u)\xi_s$$

$$+ \sum_{ss} \bar{\lambda}_{ss}(\underset{\sim}{\xi}_u)\xi^2_s\}. \qquad (55)$$

We have chosen (55) in such a way that $\bar{P}$ is normalized and possesses the same first and second moments as the original $P(\underset{\sim}{\xi}_s|\underset{\sim}{\xi}_u)$. Quite evidently our method allows us to determine the

distribution functions both for the order parameters and the slaved mode amplitudes. The distribution function found in this way is in agreement with those which result from the microscopic theory based on the slaving principle in the lowest approximation [1,2] as was exhibited in the preceding paragraphs. The whole procedure can be used to determine patterns which develop in systems close to their points of nonequilibrium. When we go far away from the nonequilibrium phase transition, the distribution functions can be approximated totally by Gaussian functions and then conventional methods of pattern recognition theory can be applied. Here a more elaborate procedure is necessary because of the critical fluctuations which imply that symmetry breaking has not yet taken place.

12.8 Derivation of the Fokker-Planck Equation by Means of the Maximum Calibre Principle

The preceding paragraphs were devoted to the determination of a steady state distribution function or at least to distribution functions not too far away from steady state. We now wish to show that we may derive the joint distribution function for a time-dependent process, or in other words, the distribution functions occurring under path integrals.

To this end we write the joint distribution function in the form [11]

$$P_N = P(\underset{\sim}{q}_N, t_N; \underset{\sim}{q}_{N-1}, t_{N-1}; \ldots; \underset{\sim}{q}_1, t_1). \qquad (56)$$

Its information is given by

$$i = - \int Dq \; P_N \; \ln P_N. \qquad (57)$$

In the following we shall assume that we are dealing with a continuous Markov process so that (56) can be written in the form

$$P_N = \Pi \; P(\underset{\sim}{q}_{i+1}, t_{i+1}|\underset{\sim}{q}_i t_i) \; P_o(\underset{\sim}{q}_1, t_1). \qquad (58)$$

We have performed the whole procedure for a many variable problem including non-stationary Markov processes. Here we shall exhibit the whole procedure by means of an example, namely of a stationary Markov process and a single variable. We shall choose as

Macroscopic approach to synergetics

constraints

$$f_1 = \langle q_{i+1} \rangle_{q_i} \tag{59}$$

and

$$f_2 = \langle q_{i+1}^2 \rangle_{q_i} \tag{60}$$

that means the first two conditional moments, i.e. we assume that at time i the state variable q was measured and is measured then again at a later time, whereby the measurement applies to an ensemble and we have to average over the ensemble. In the following we shall make the substitution

$$i + 1 \; \text{---} > i + \tau \tag{61}$$

in order to exhibit the fact that the time interval τ becomes infinitely small. The maximum information entropy principle immediately yields

$$P(q_{i+\tau}|q_i) = \exp\{\lambda + \lambda_1 q_{i+\tau} + \lambda_2 q^2_{i+\tau}\} \tag{62}$$

where we may now impose the following further conditions: It must be normalized and for $\tau \to 0$

$$P \to \delta(q_{i+\tau} - q_i) \tag{63}$$

must hold. Writing (62) in an equivalent form

$$P = N \exp\left\{-\lambda_2 (q_{i+\tau} - \frac{1}{2\lambda_2})^2\right\} \tag{64}$$

the singularity condition (63) implies that in lowest approximation in τ we have

$$\lambda_2 = \frac{Q}{\tau} . \tag{65}$$

Furthermore we must have (for $\tau \text{ --> } 0$)

$$q_{i+\tau} - \frac{\lambda_1}{2\lambda_2} \quad q_{i+\tau} - q_i . \tag{66}$$

Assuming that λ_1 and λ_2 can be expanded into power series in τ we must have

$$\frac{\lambda_1}{2\lambda_s} = q_i + \tau K(q_i) + \ldots \tag{67}$$

Please note that λ_1 and λ_2 may be functions of q_i because the conditional moments were moments under given q_i. The result we obtain for P thus reads

$$P = N \exp \left\{ -\frac{Q}{\tau} (q_{i+\tau} - q_i - \tau K(q_i))^2 \right\} \tag{68}$$

which is the well known short time propagator for the Fokker-Planck equation. It is now a trivial task to derive the explicit form of the Fokker-Planck equation. In conclusion we may state that the maximum information entropy principle in connection with the constraints of the first and second moment allow us to derive explicitly the path integral solution and furthermore to reconstruct a Fokker-Planck equation in which the drift and diffusion terms can be explicitly determined.

12.9 Conclusion

In our above paper we have shown how a macroscopic approach to synergetics can be constructed. This approach now appears on a similar footing to traditional thermodynamics. That field may be based on adequate macroscopic quantities, such as energy, particle numbers a.s.o. which serve as constraints for a maximum entropy principle and the well known relations of thermodynamics follow whereby also the Lagrange parameters acquire a physical meaning, for instance, such as temperature or pressure.

Here, we were able to devise a similar approach at least for the class of nonequilibrium phase transitions where the macroscopic behaviour of a system is governed by a few order parameters. We have shown that we may either introduce these order parameters as macroscopic variables, or may even determine them from measured data on a set of suitably chosen variables which were determined with their moments up to fourth order.

Of course, now one may study the physical meaning of the Lagrange multipliers in these transitions. There are several problems left for future research, for instance, we now know the adequate constraints for systems in thermal equilibrium and for those close to nonequilibrium phase transitions. The question remains open what has to be chosen as a constraint for the region in between.

REFERENCES

[1]   H. Haken (1983). *Synergetics. An Introduction,* 3rd edition, Springer-Verlag, Berlin, Heidelberg, New York.
[2]   H. Haken (1982). *Advanced Synergetics,* Springer-Verlag, Berlin, Heidelberg, New York.
[3]   H. Haken (1982). A. Wunderlin, Z. *Physik* **B47**, 179.
[4]   H. Haken, J.A.S. Kelso and H. Bunz (1982). *Biol. Cybern.*

51, 347.

[5] E. T. Jaynes (1975). *Phys. Rev.* **106**, 4, 620; E. T. Jaynes (1957). *Phys. Rev.* **108**, 171.

[6] H. Haken (1985). Z. *Physik* **B61**, 329.

[7] H. Haken (1985). Z. *Physik* **B61**, 335.

[8] H. Haken (1970) in *Encyclopedia of Physics,* ed. S. Fluegge, Vol. XXV/2c, Springer-Verlag, Berlin, Heidelberg, New York.

[9] H. Haken (1986). Z. *Physik* **B62**, 225.

[10] H. Haken (1986). Z. *Physik* **B63**, 487.

[11] H. Haken (1986). Z. *Physik* **B63**, 505.

13   *A.V. Holden and G. Matsumoto*

PATTERNED AND IRREGULAR ACTIVITY IN EXCITABLE MEDIA

Abstract

The excitable membranes of nerve and muscle may be represented by relatively high order (>3) nonlinear differential systems that can have equilibrium, periodic and chaotic solutions. These membrane equations may be combined with the equations for passive current flow to give a finite, 1-dimensional excitable medium representing a nerve fibre.   Space-time patterns in such a system can be complicated, especially if branching is allowed.

Space-time patterns in nervous tissue do not reflect activity in an excitable medium: they reflect the organized connectivity between the cellular elements of the nervous system.   A preparation that allows visualization of this patterned activity is described.

13.1 Introduction

Classical experiments on excitable tissues - nerve and muscle tissue - used indirect methods of detecting electrical activity in the tissue: the presence of activity in a (motor) nerve fibre was detected by the contraction it evoked in the muscle fibres it innervated.   Such experiments led to the all-or-none principle for propagative activity in excitable tissues, that was first proposed for cardiac muscle, and extended to nerve fibres, and then demonstrated experimentally for motor and sensory fibres.   In the context of a single nerve fibre, the all-or-none principle means that in the absence of a stimulus, there is a spatially uniform, resting state, and that the response to a brief stimulus applied at a point is either local, and decremental (a sub-threshold stimulus fails to produce an action potential) or is a pair of solitary travelling waves (action potentials) that propagate away from their point of initiation, with an amplitude and conduction velocity that is independent of the stimulus current.   Since the usual experimental situation is to stimulate one end of a fibre the response is usually taken as a single action potential.   Such discrete single action potentials were first observed in single sensory fibres using extracellular recording methods.

This identification of the behaviour of an excitable cell or tissue with the image of a single, propagating action potential is rather limited.   A propagating action potential has a duration (of the order of ms) and a conduction velocity (of the order of 1-100 m s^{-1}) and so has a spatial extent from 1-100 mm, which is much larger than the size of most nerve cells. A different way of measuring length is in terms of the space constant, where the space constant for a cylindrical fibre is the distance, measured from the point of current injection, within which a steady voltage decays to 1/e of its starting value. The dendritic processes of even large neurones, such as motoneurones, have an electrotonic length of only a few space constants. Thus there is often not enough space for there to be a clearcut separation between a local and a propagative response.

Even in situations where the fibres are long enough for there to be enough space for propagation, the normal mode of behaviour is an irregular, repetitive discharge: what is of interest is not a single action potential but periodic, and irregular, propagative activity.

Nerve fibres are not uniform cylinders, but are branched, and so their pattern of behaviour is influenced by their geometry.

Here we shall consider periodic and irregular activity in excitable membranes, and some of the kinds of spatio-temporal patterns that are found in an idealized one-dimensional excitable tissue. Patterns of activity that occur in the central nervous system should not be considered as behaviour in an excitable medium, described by a nonlinear partial differential system: the nervous system is an organization of interacting components, where the functional units may be parts of cells, single cells, or assemblies of cells. The patterns of activity in such systems are strongly determined by the connectivities among the components, and so general modelling must be based on a detailed knowledge of actual connectivities and behaviour. It is not possible (yet) to record actual space-time patterns of activity from a neural system with a large number of functional components, with a resolution at the level of a single functional unit. However, the pattern of skin pigmentation in some marine molluscs provides a visual mapping of space-time patterns generated by a neural system, and so provides a suitable preparation for characterizing and quantifying space-time patterns generated by neural systems.

13.2 Membrane Potentials and Currents

Extracellular recordings can show if an action potential has occurred: they cannot give any insight into the dynamics of the processes generating the action potential. Models based on extracellular recordings describe the conditions under which an action potential can be generated: the all-or-none nature of the action potential implies the existence of a threshold, and a simple model for a neurone is the leaky integrator, where the effects of subthreshold inputs sum linearly and decay exponentially:

$$V(t) = \int_0^t x(u) \exp(-(t-u))du; \quad 0 \le V(t) \le V_{th} \quad (1)$$

When $V(t)$ reaches V_{th} an action potential is generated. This

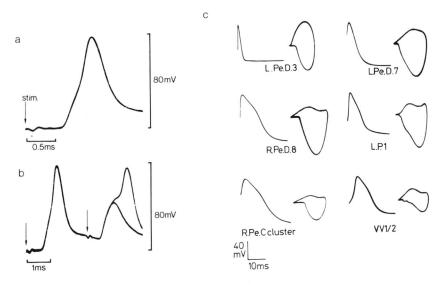

Figure 1. Intracellular recordings of somatic action potentials. (a,b) From rat motoneurone, produced by an electrical stimulus to the motor axon. The action potential propagates back and invades the initial segment and soma, the second action potential in (b) sometimes fails to invade the soma. Both the initial segment and soma action potentials are all-or-none. (c) Spontaneously generated action potentials from different identified neurones in the isolated brain of the pond snail. The different shapes of the action potentials are illustrated using phase plane displays (Holden & Winlow, 1982; Winlow et al., 1982).

simple model gives a strength-duration relation

$$I = I_{rh}/(1 - \exp(-t/\tau))$$ (2)

where a pulse of intensity I and duration t excites, and I_{rh} is the rheobase. This is close to the shape of experimentally obtained strength-duration curves. If the input x(t) is a constant current, the model gives a periodic discharge of action potentials; if the input is sinusoidally modulated, the response shows phase-locking and entrainment [1,2], if the input is stochastic, random walk and diffusion models of neuronal variability result (see the monographs [3,4]). Thus a very simple model, with a threshold and a single time constant, can account for a wide range of observed behaviours.

The introduction of intracellular recording methods has shown that an action potential is not just an event, which may or may not occur, but that action potentials have a characteristic waveform. Different cells have different shaped action potentials, and the shape of action potentials can be rate-dependent.

The recorded waveform V(t) is the result of ionic currents I_{ion}

flowing through the cell membrane; for an isopotential area of membrane, where $dV/dx = 0$

$$C_m dV/dt = -I_{ion}. \tag{3}$$

Such an isopotential preparation (space clamp) is used in experiments designed to characterize $I_{ion}(V)$. The ionic currents are the sum of currents flowing through different (in terms of their ionic selectivities, pharmacology and voltage dependence) ionic conductances; the richness in behaviours seen in different excitable membranes results from **quantitative** differences in the density of a relatively small number of distinct **qualitatively** similar conductances.

These conductances, and their kinetics, have been characterized by voltage clamp experiments, and the first preparation to be analyzed in detail was the membrane of the giant axon of the squid. Biologically, this is a very peculiar axon, and its normal mode of behaviour is a single isolated propagating action potential that triggers the escape response of the squid. However, its description, by the Hodgkin-Huxley equations, is the text book example of a biophysical excitation system, and these equations have been examined by a wide range of methods: see [5] for a review.

In the Hodgkin-Huxley equations the ionic current is the product of a conductance and a potential difference. The membrane conductance is the sum of voltage dependent conductances (g_{Na} and g_K) and a voltage dependent conductance. In other excitable membranes the empirical relation between current and ionic permeability is more complicated, there can be different types of voltage dependent conductance, and there is also the possibility of a charge-dependent conductance (such as $g_{K,Ca}$, a K^+ selective conductance that is activated by a change in the intracellular Ca^{2+} activity, produced by Ca^{2+} entry through a Ca^{2+} selective pathway).

13.3 The Hodgkin-Huxley (H-H) Membrane and Cable System

In the H-H membrane equations the ionic current density is the sum:

$$I_{ion} = I_{Na} + I_K + I_L$$

of the sodium I_{Na}, potassium I_K and leakage I_L current densities (μAcm^{-2}). Each current is a product of a conductance and a driving potential (mV), and the voltage dependent conductances are the products of a maximal conductance $\bar{g}$ ($mScm^{-2}$) and voltage dependent gating variables m, h and n;

$$I_{ion} = m^3 h \bar{g}_{Na}(V-V_{Na}) + n^4 \bar{g}_K(V-V_K) + g_L(V-V_L). \tag{4}$$

The activation (m,n) and inactivation (h) gating variables may be considered as probabilities

$$0 \leq m, n, h \leq 1$$

and obey first order kinetics, with voltage dependent rate coefficients - for the activation variable m

$$dm/dt = \phi[\alpha_m(1-m) - \beta_m m], \qquad (5)$$

where $\phi = 3^{(T - 6.3)/10}$ gives the temperature dependence.

Empirical expression for the voltage dependence of the rate coefficients were obtained from an analysis of the results of voltage clamp experiments.

The H-H membrane equations may be combined with the partial differential equation for electric current flow in a cable (see [6]) to give the H-H cable equation for propagating activity in an axon:

$$\partial^2 V/\partial x^2 - C_m \partial V/\partial t - I_{ion} = 0. \qquad (6)$$

13.3.1 *Equilibrium and threshold*

The standard H-H membrane equations have a single, stable equilibrium solution; the numerical value of the reversal potential for the leakage pathway is fixed to three decimal places at 10.598 to give the equilibrium solution at 0 mV. There is a sharp threshold for instantaneous voltage perturbations from the equilibrium solution: if V is taken instantaneously to a value > 6.3751 an action potential is generated, with the latency dependent on the amplitude of the voltage perturbation.

This all-or-none behaviour of the H-H membrane equations is not due to a saddle point: if V is controlled to an unreasonable 10^{-15} mV there are graded responses [7], and these in fact may be seen experimentally at high temperatures, where the all-or-none law fails [8].

Changes in parameters (such as V_K, $\bar{g}_K$, extracellular Ca^{2+} activity) and an applied hyperpolarizing current density can take the H-H membrane equations into a region of parameter space where there are multiple equilibria [9-11], two of which can be stable. The stable equilibrium solutions lose their stability either at a fold or at a Hopf bifurcation: bifurcations between equilibrium points occur when there is a single zero eigenvalue, and a simple Hopf bifurcation occurs when a single complex conjugate pair of eigenvalues crosses the imaginary axis.

13.3.2 *Hopf bifurcation*

The H-H membrane equations are complicated, so the identification of Hopf bifurcation points relies on numerical methods. Bifurcation parameters that have been examined are:

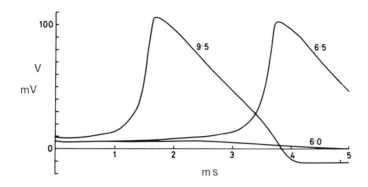

Figure 2.   Numerical solutions of standard Hodgkin-Huxley membrane equations in response to an instantaneous step in V of 6.0, 6.5 and 9.5 mV.

(i)   applied current density I, so that

$$C_m dV/dt = -I_{ion} + I. \qquad (7)$$

Sub- and supercritical bifurcations occur at $6.3°C$ for I = 9.8 and 154 A cm^{-2}.   As the temperature is increased, these points move towards each other, and fuse close to $27.5°C$.

(ii) $\bar{g}_K$.   Sub-critical bifurcations occur at $6.3°C$ for $\bar{g}_K$ close to 3.8 and 19.7 mS cm^{-2}.

(iii) $[Ca^{2+}]_o$.   Sub-critical bifurcations occur at $6.3°C$ at external calcium activities of 3 and 21 mmol dm^{-3}.

Bifurcations in the V_K - I plane are shown in Figure 3.   In all these cases the equilibrium loses its stability at a simple Hopf bifurcation: at   a   sub-critical   bifurcation   unstable   small amplitude periodic solutions emerge, which provide a path to the large amplitude periodic solutions that correspond to a repetitive discharge   of   action   potentials,   and   at   a   supercritical bifurcation, small amplitude periodic solutions emerge.   The periodic solutions emerge at a Hopf bifurcation, and vanish at a Hopf bifurcation.   No complicated periodic, or chaotic, solutions are   seen.   If   the   Hodgkin-Huxley   equations   are   taken   as   the classical   biophysical   excitation   equation,   then   an   excitable medium may be expected to be quiescent, or to show autowaves.   It could also respond to appropriate perturbations by travelling waves or wave trains.

13.4 Solitary Travelling Waves

The Hodgkin-Huxley membrane equations may be combined with the cable equation to give a reaction diffusion equation

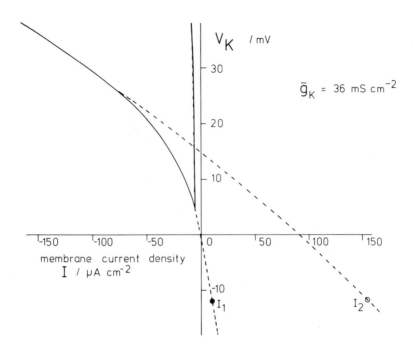

Figure 3.   Bifurcation curves in the V_K - I plane. I_1 and I_2 are the Hopf bifurcation points for the standard Hodgkin-Huxley membrane equations when I is the bifurcation parameter, the dashed lines are the Hopf bifurcation curves and the solid curves enclose a region where there are three equilibrium solutions.

$$\partial^2 V/\partial x^2 - C_m \, \partial V/\partial t - f(V,m,n,h) = 0 \qquad (8)$$

that has a solitary travelling wave solution that corresponds to a propagating action potential. Assuming such a solution, with a constant velocity θ, gives the autonomous system of equations

$$d^2/d\xi^2 - C_m dV/d\xi - f(V,m,n,h) = 0.$$
$$dm/d\xi = \alpha_m(1 - m) - \beta_m m, \qquad (9)$$

similarly for h, n. The problem is to find a value of θ for which V remains bounded and that generates a trajectory from the singular point at which all derivatives w.r.t. ξ are zero. This ordinary differential system for the travelling wave solution is very sensitive to the parameter value. θ = 18.74 m s^{-1} gives a large amplitude, stable solution that corresponds to a propagating action potential and the conduction velocity is reasonable. There is also an unstable travelling wave solution: these two travelling wave solutions coalesce as a parameter (temperature, or maximal cation conductance) is changed.

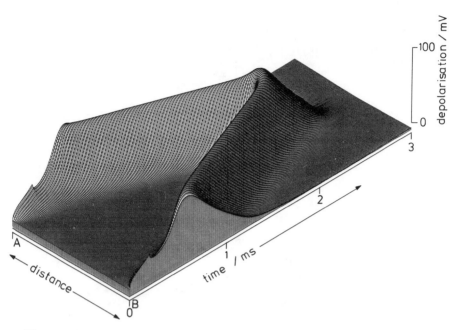

Figure 4.   Numerical solutions of standard Hodgkin-Huxley cable equations: the solitary travelling waves initiated by brief depolarizing currents at A and B propagate towards each other and annihilate each other.   Although the voltage quickly returns close to its resting value the K^+ selective conductance decays slowly, so that there are after-effects to collision.

The travelling wave solution is **not** a soliton: it is annihilated by collision.   Such collisions are likely to occur in branched axonal systems, when action potentials may be initiated at a variety of sites.   Action potential initiation in branched sensory axons is usually at the terminals, or at branch points, and so the temporal pattern of sensory action potential can be complicated by the results of collision: see chapter 9 of [3].

Collision does not just annihilate the colliding action potentials: it leaves the membrane in a refractory state, which transiently alters the local conduction velocity for further nerve impulses.   Such activity-dependent changes in conduction velocity are also seen after a single propagating action potential [12] and so influence the pattern of pulse train solutions.

13.5 **Pulse Train Solutions**

The after-effects of an action potential (a slowly decaying K^+ conductance) may be seen as after-potentials, or demonstrated by changes in excitability or in conduction velocity: for the standard Hodgkin-Huxley system they persist for up to 30 ms.   For a pulse train of two impulses, the second action potential travels in the after-effects of the first, and so at a changed velocity: this will alter the intervals between the action potentials.   If the interval between the propagating action potentials starts at

Figure 5.   Numerical solution of Hodgkin-Huxley cable equations showing reduced conduction velocity of the action potential of a pair initiated 2 ms apart.

less than 10 ms, the second action potential will travel slowly, lengthening the interval; if it is about 10 - 20 ms the second pulse will travel faster, shortening the interval. Thus there is a tendency for an interval of 10 ms, or a rate of 100 s^{-1}. These

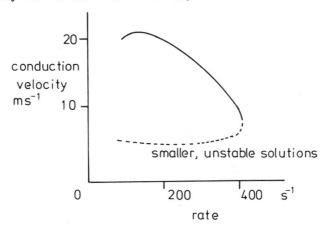

Figure 6.   The conduction velocity for pulse train solutions of the Hodgkin-Huxley cable equations as a function of discharge rate.

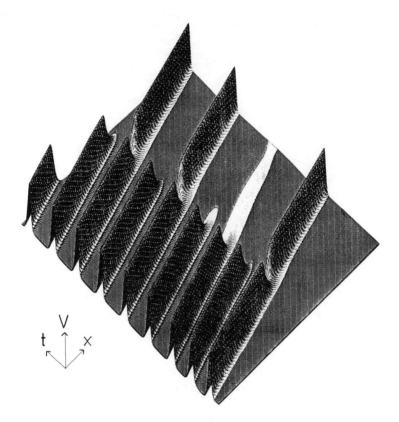

Figure 7. The smaller amplitude, slower, unstable solitary or periodic travelling wave solutions are not seen in numerical integrations: however, this appears to be an example. It has been produced by repetitive activity propagating through a region of reduced excitability that simulates the effect of local anaesthesia.

dispersive effects have been seen in squid axons. The dependence of conduction velocity on rate of travelling waves shown in Figure 6 is probably not significant in normal functioning in molluscs, since it occurs at high rates that are rarely met; however, it may be important in mammals.

Nerve axons are usually less than, or of the same order as, the 'wave length' (the product of conduction velocity and duration) of an action potential, and so although they can sustain high (up to about 100 s^{-1} discharge rates, there is rarely more than one action potential on the fibre at any one time. However, spatial nonuniformities (a change in axon diameter, or a local change in membrane conductances) can radically alter conduction velocities, and enhance interactions between subsequent action potentials.

13.6 Finite Cables

Although 'text book' nerve cells have long axons, in the central nervous system there are many more small neurones with relatively short axons (local circuit neurones) than large neurones with long axons - spatial patterns of activity within a nerve cell should not be thought of as analogous to propagation in a semi-infinite cable, but as propagation in a finite cable. In a finite cable with sealed ends there can be spatially uniform or nonuniform solutions, and the spatially uniform solutions can be time-varying.

If the maximal K^+ conductance of the Hodgkin-Huxley membrane equations is reduced, the membrane becomes autorhythmic: in a finite cable a spatially uniform reduction of g_K can give a spatially uniform, time periodic response. If travelling waves are introduced into the oscillating cable the travelling wave solutions dominate the spatially uniform solutions.

In a real axonal system, with membrane properties that allowed periodic or chaotic autorhythmicity, the region that first generated propagating action potentials would dominate and act as a 'pacemaker'. This pacemaker locus would suddenly switch to another region as soon as an action potential from this new locus escaped. Such switching between pacemaker sites, with discrete changes in the discharge pattern, is in fact seen in some sensory receptors, such as muscle spindle afferents, where action potentials can be initiated at different axonal terminals.

13.7 2- and 3-Dimensional Excitable Media

The close correspondence between propagating activity in a nerve axon and travelling wave phenomena in a 1-dimensional excitable medium could suggest that neural tissue, such as the roughly 2-dimensional, sheet-like structure of the cortex, or the 3-dimensional structure of neural nuclei, might be treated as excitable media. Such excitable media are well known in chemistry and physics and perhaps such a description is appropriate for cardiac tissue - see the monographs of Winfree [13] and Zykov [14]. However,the cardiac cells in a piece of cardiac tissue are coupled by low resistance pathways, and activity propagates through the tissue. The resultant behaviour - travelling and interacting wavefronts - are the result of excitation and propagation, or reaction and diffusion.

Although there are low resistance pathways between some nerve cells, the nervous system behaves as a system of connected, interacting elements, rather than as a continuous, excitable medium. The patterns of connectivity, their strengths and signs, determine the possible behaviours. It is only under pathological conditions, as seen for example in spreading depression, that parts of the nervous system can behave as excitable media. Under normal conditions the spatio-temporal patterns of activity will be more complicated than the spirals and scrolls described for excitable media.

Although these patterns may be imagined - Sherrington (1940) described the pattern of electrical activity within part of the brain as "an enchanted loom where millions of flashing shuttles weave a dissolving pattern, always a meaningful pattern though never an abiding one; a shifting harmony of subpatterns" [15] - they may not be directly visualized, as one cannot directly record the individual electrical activities of a large number of functionally related neurones during a biologically relevant behaviour. What one would like to do would be to represent on a video-display unit (and hence also in the memory of a computer) a signal that locally coded the temporal activity of a single neurone, and that globally recorded the spatial pattern of these activities, in such a way that close points on the screen represented the activities of correspondingly close neurones in a functionally coherent part of the nervous system.

13.8 Space-time Recordings from the Squid

Just as it was possible to idenfity and analyze the ionic current mechanisms underlying excitation in the squid giant axon before experimental techniques permitted such a voltage-clamp analysis of mammalian axons and nerve cell bodies, the squid also provides a preparation where an indirect visualization of space-time patterns modulated by nervous activity is possible. The skin of the squid and other marine molluscs, such as the octopus and cuttlefish, contain pigment spots whose activity may be modulated by neural activity. An increase in the discharge rate to a (cluster of) pigment spots leads to their contraction, and so a loss of colour at that point on the skin corresponds to an increase in the firing rate of the driving neurone. The relation between size of colour and discharge rate may be described by a decreasing sigmoid function. The relation between neurones on the surface of the ganglion and points on the skin will show a somato-topic mapping.

The types of behaviour that are seen in the wild - camouflage patterns, attack patterns - are under neural control and so represent a mapping on to the skin surface of patterns generated within the nervous system. Thus a video-tape of the patterns seen on the surface of the skin of the squid may be treated as a mapping of space-time patterns generated by a sheet-like array of nerve cells.

Examples of such recordings are shown in Plates 4 and 5: these plates show the granularity underlying the pattern, and illustrate how the size of the pigment spots can oscillate. Viewing such a videotape has the same fascination as gazing into a fire - one attempts to impose some recognisable order on the continually changing pattern.

Although it is feasible to record and digitize such patterns, a suitable method for analyzing these patterns has still to emerge. What is required are intuitively satisfactory measures of order or complexity that change in apropriate ways as biologically relevant changes in the patterns occur.

References

[1]   R.B. Stein, A.S. French and A.V. Holden (1972). *Biophys. J.* **12**, 295.

[2]   J.P. Keener, F.C. Hoppensteadt and J. Rinzel (1981). *SIAM J. Appl. Math.* **41**, 503.

[3]   A.V. Holden (1976). *Models of the Stochastic Activity of Neurones,* Springer-Verlag, Berlin.

[4]   F.C. Hoppensteadt (1986). *An Introduction to the Mathematics of Neurones,* Cambridge University Press, Cambridge.

[5]   A.V. Holden (1982) in *Biomathematics in 1980,* ed. L.M. Ricciardi and A.C. Scott, North-Holland, Amsterdam.

[6]   J.J.B. Jack, D. Noble and R.W. Tsien (1975). *Electric Current Flow in Excitable Cells,* Clarendon Press, Oxford.

[7]   R. FitzHugh and H.A. Antosiewicz (1959). *J. Soc. Industr. Appl. Math.* **7**.

[8]   K.S. Cole, R. Guttman and F. Bezanilla, *Proc. Natl. Acad. Sci. USA* 65, 884, 1970.

[9]   K. Aihara and G. Matsumoto (1983). *Biophys. J.* **41**, 87.

[10]   A.V. Holden and W. Winlow (1983). *IEEE Trans. SMC* **13**, 711.

[11]   A.V. Holden, P.G. Haydon and W. Winlow (1983). *Biol. Cybern.* **46**, 167.

[12]   A.C. Scott and U.V. Pinardi (1982). *J. Theoret. Neurobiol.* **1**, 150-172, 173-195.

[13]   A.T. Winfree (1987). *When time breaks down: the three dimensional dynamics of electrochemical waves and cardiac arrhythmias,* Princeton University Press.

[14]   V.I. Zykov (1988). *Simulation of Wave Processes in Excitable Media,* Manchester University Press.

[15]   C.S. Sherrington (1940). *Man and his Nature,* Cambridge University Press.

DYNAMICS AND PHASE TRANSITIONS IN ARTIFICIAL INTELLIGENCE SYSTEMS

Abstract

Large artificial intelligence systems and distributed computational networks pose interesting challenges when trying to describe their dynamical behaviour. I review some results of a theory which we have recently formulated, and which has concrete predictions for the dynamics of such systems from knowledge of their generic features. Novel phenomena that we have discovered include the existence of phase transitions in spreading activation networks, and critical slowing down of their collective dynamics at the phase boundaries.

14.1 Introduction

The existence of machines capable of computing with thousands of processors [1], of distributed computational networks [2], and of large artificial intelligence systems elicits a set of interesting questions about their behaviour. Foremost is the question of predicting their global dynamics from knowledge of their generic features, a complex and difficult problem since total specification of the underlying processes is an impossible task for systems whose sizes range from moderate to large.

The application of a dynamical perspective to artificial intelligence and large scale computational systems is hindered by several problems. An important one is the fact that the systems of interest are composed of very many interacting parts, thus preventing the use of notions associated with the evolution of a few degrees of freedom. Another problem is that it is not clear how to extrapolate from the behaviour of small systems to predict the dynamics of larger ones. A possible solution to the above, and the one that I will describe below, is to resort to statistical mechanics methods in order to study the very large systems and to explore in that limit the appearance of qualitative new regimes. If successful, such an approach can be made relevant to more realistic cases by considering finite size corrections to the phenomena encountered in the infinite case.

Another important issue is that of the proper level of granularity at which the system is to be studied. This is a non-trivial task for such complex structures, and involves problems such as their parametrization into lower dimensional dynamics, as well as the kind of answers that one is seeking. Just as a description of a large computational system in terms of electron motion (or of the dynamics of logic gates) would quickly fail to convey relevant information about higher levels of software dynamics, a consideration of the whole system as a one parameter input-output device might not yield much insight into its time evolution either.

In what follows, I will report results that we have obtained with Tad Hogg [3], and which show promise for the understanding of the global behaviour of a class of interacting **processes** which go under the name of spreading activation networks. These networks, used in various applications of artificial intelligence such as semantic nets [4], disambiguation strategies, and machine learning [5], also provide a simplified description of highly interconnected computing environments. The phenomena that we have discovered range from phase transitions to critical slowing down of the network as a function of its parameters, and they provide a useful paradigm with which to study those systems.

14.2 Spreading Activation Networks

Spreading activation networks consist of a set of nodes representing various potentially active states, with weighted links between them. These weights determine how much the activation of a given node directly affects others. The activity of a node is used to encode either the current belief in the validity of its contents, or its importance for further analysis. The behaviour of these networks is thus controlled by three parameters. The first, specifying their topology, is the average number of links per node; μ. The second one, α, describes the weights and controls the relative amount of activity that flows from a node to its neighbours per unit time. The third parameter is the relaxation rate: γ, and measure the time constant with which the activity of a given node decays.

In a typical application, some nodes are initially activated by external inputs. These nodes in turn cause others to become active with varying intensities, leading to complicated dynamics characterized by a spatio-temporal modulation of the total net activity. This raises two general questions concerning the behaviour of the net. The first one concerns its general dynamical behaviour and its approach to equilibrium. In this context one important consideration is the rate at which equilibrium is reached and how it compares with the characteristic times with which inputs change at the sources. The second question deals with the fraction of the net that partakes in its activity, which determines the extent to which the far regions of the net influence the parts under consideration. It turns out that there exist several operating regimes separated by sharp boundaries. These phase transitions, which depend on both the topological connectivity and the ratio of excitation to relaxation in the networks, lead to a rich phase diagram which we now elucidate.

Consider the simple case shown in Fig. 1. As illustrated, the net consists of n nodes connected by a set of undirected links and

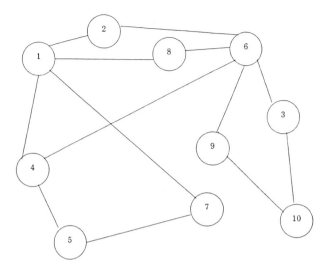

Figure 1.   Schematic network with undirected links between the nodes.

it is characterized by three parameters.   The first one is the average number of links per node, μ, which specifies its topology. The second is the relaxation rate with which the activity of an isolated node decays to zero, denoted by γ, which ranges between 0 and 1.   Finally, there is a coupling constant, α, which controls the relative amount of activity that flows from a node to all of its neighbours per unit time.   As implemented in computers, we have to specify the dynamics in discrete time steps.   Let $A(N)$ be a vector whose i^{th} element is the activation of the ith node at time step N, and let $C(N)$ be a vector whose elements specify the external source at the same time step.   Using the standard model of activation plus relaxation, one can write the time evolution of the net as

$$A(N) = A(N) + M \, A(N-1) \qquad (1)$$

where **M** is a matrix determined by the net conectivity.   It can be written as

$$M = (1-\gamma)I + \alpha R \qquad (2)$$

where **I** is the identity matrix and **R** is a matrix with zero diagonal elements and whose off-diagonal element R_{ij} is the weight of the link from node j to node i, which is defined to be zero for non-existing links.   This weight determines how much of the activation at node j spreads to node i in a single step.   Since the links are undirected, R_{ij} is nonzero if and only if R_{ji} is nonzero.   We can also define the total activation of the net, $T(N)$, by the sum of the elements of the vector $A(N)$.

    In general, the net will consist of a number of distinct

connected components and the activity in each component will be independent of the others. This implies that by suitably relabelling the nodes the matrix **M** can be cast into block-diagaonal form, where each block $m^{(k)}$ corresponds to a single cluster, with all elements outside the blocks equal to zero. Notice that the matrix **R** will also have this form with blocks $r^{(k)}$. An important constraint of this model, as used in artificial intelligence applications [4], is that the activation from any node is divided among the attached nodes according to the weight of their connections. For those blocks $r^{(k)}$ which have more than one element, this implies that the sum of the elements in each column equals one, i.e.

$$\sum_i r^{(k)}_{ij} = 1 \qquad (3)$$

for every j, where the sum extends over all elements of the jth column.

Eq. (1) completely determines the dynamics of the spreading activation net if the sources and the initial conditions are specified. In order to answer the two questions posed above, we will consider the specific case of a constant source, $C(N) = C$, applied to a single source node (e.g. node 1 so only the 1st element of **C** in nonzero) and determine the consequent spread of activity through the net. In doing so, the two relevant questions are: a) the existence of an asymptotic regime and the time it takes to reach it, and b) the number of nodes that are significantly activated in the long time limit.

The asymptotic behaviour of the net is determined by the fixed point A^* of Eq. corresponding to the component of the net which contains the source. This is obtained when $A(N) = A(N-1)$ and leads to a linear equation for A^*, namely

$$A^* = C + MA^*. \qquad (4)$$

Since the vector C has only one nonzero entry, elements of A^* will be nonzero only for nodes which are in the same cluster as the source node. Whether the net actually settles down to the fixed point depends on its stability. Specifically, if the activation at a given time is near its fixed point value, the dynamics of the net can either drive it towards the fixed point or away from it, the latter leading to unbounded growth of the activation. From Eqs. (1) and (3), the total activation corresponding to a source applied to a nonisolated node evolves according to

$$T(N) = C + (1-\gamma + \alpha)T(N-1) \qquad (5)$$

where C is the total source activation. Note that when γ is larger than α this aproaches a fixed limit $T^* = C/(\gamma-\alpha)$. Conversely, if γ is smaller than α the total activation grows without bound and the fixed point A^* is unstable. Eq. (5) also determines the relaxation time, τ, with which this component of the net relaxes towards its asymptotic state like $\exp(-N/\tau)$ after N time steps. It is given by

$$t = -1/\ln(1-\gamma + \alpha). \qquad (6)$$

When the system is unstable, the exponential growth of the activity will also be determined by this equation, but with the opposite sign.

When dealing with large systems (i.e. the number of nodes, n, becomes infinite) the law of large numbers implies that they will almost always be very close to the average. Thus the relevant consideration is their average generic behaviour. In this context we consider the dynamics of a **typical** network rather than any **particular** one, and say that the net will **almost surely** have a given behaviour if its likelihood approaches one as the size of the net grows. A typical network can be simply modelled as a random graph with the constraint that μ, the average number of edges leaving a node, remains constant as the size of the net increases, i.e. it is a finite quantity. For simplicity we will consider the uniform case in which the weights on all the links leaving a node are equal. From Eq. (3) this implies that $R_{ij} = 1/\deg(j)$ whenever the nodes i and j are linked and where $\deg(j)$ is the number of links leaving node j. In this case the matrix R entering the dynamical equations is a random one whose off-diagonal elements are non-zero with probability p, which is given by $\mu/(n-1)$. The blocks $m^{(s)}$ will correspond to the connected components of the random graph. Note that in this model any two nodes are equally likely to be connected, which contrasts with analogous physical situations where the strength of their connection weakens as the euclidean distance between them increases.

14.3 TRANSITIONS

The phase transitions that occur in these networks are shown in Fig. 2, which we now discuss. From Eq. (5) one sees that the dynamics undergoes a sharp transition at the line $\alpha/\gamma = 1$. Above this line, the fixed point is unstable and the activation grows without bound, whereas below it the activity always reaches a stable fixed point. A simple and intuitive way of understanding this result is provided by the expectation that the fixed point of the net dynamics will become unstable when a typical node receives slightly more input than it loses due to its relaxation in a given time step. Since a node is on the average connected to μ other typical nodes, i.e. the average degree of a node is μ, the fraction of activation it receives in a given time step is $\mu(\alpha/\mu)$. Furthermore, the fraction it loses in the same amount of time is γ. This will balance when $\gamma = \alpha$, leading to the above equation.

As in the case of random graphs [6], there is a phase transition at $\mu = 1$ where the topology of the network suddenly changes from small isolated clusters to a giant one containing very many nodes. The consequent spread of activation can be defined as the number

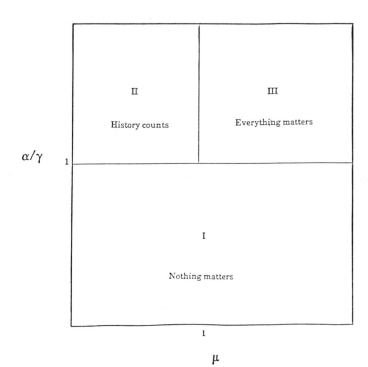

Figure 2. Phase diagram of a spreading activation network. The vertical axis represents the dynamical parameter, whereas the horizontal one depicts the connectivity parameter.

of nodes whose activity is above some specified positive threshold. The existence of these giant clusters allows the activation to reach arbitrarily remote regions of the network.

Several regions of this phase diagram, shown in Fig. 2, are worth pointing out. First consider the case where α/γ are small (i.e. region I). In this regime, the activation of the net relaxes quickly to its asymptotic value and it is localized in space. Furthermore, the topological transition at $\mu = 1$ has no observable effect on the spread of activation, since even for $\mu > 1$ the number of nodes activated above threshold in the infinite cluster remains finite. This implies that events at the source node do not significantly affect the activation of the far regions of the net. Moreover, they will have little effect on the source itself at later times. Because of its limited extent in time, this dynamics can be thought of as taking place in a finite temporal cluster.

For computational purposes, the existence of such a phase implies that one can effectively ignore a *priori* both far regions of the net as well as its ancient history. Thus, in this regime the event horizon affecting a given node at a given time is localized in both space and time. This leads to an effective equilibration process so that a net with time varying sources will always remain near the corresponding fixed point, provided they do not change much in time intervals comparable to τ.

As α/γ increases, the overall relaxation of the net towards its stable fixed point becomes increasingly sluggish, with its characteristic relaxation time diverging to infinity as the transition is approach. This divergence, determined by Eq. (6), is given by

$$\tau = 1/\gamma\varepsilon \qquad (7)$$

where the parameter ε, given by $1-\alpha/\gamma$, measures the relative distance below the transition line. This leads to sharp phase transitions into phases II and III. In phase II, one encounters a regime where the event horizon grows indefinitely in time but remains localized in space. This means that unlike phase I, ancient history matters in determining the activation of any node, and that since the net reverberations never settle down (i.e. the activity keeps increasing) the assumption of an equilibrium between the net and the time variations at the source no longer holds.

Finally in region III not only is the event horizon extended in time but also in space. This means that the amount of spreading grows indefinitely and therefore far regions of the net can significantly affect each other. In other words, everything matters. This region is separated from phase II by a sharp transition in which the number of nodes with activation values above a given positive threshold, grows explosively. This can be seen by a mean field theory argument which shows that although the activation decays away exponentially with distance from the source, the individual activities of the other nodes grow like $(1-\gamma + \alpha)^N$, showing that arbitrarily remote nodes eventually exceed threshold.

We should point out that the finite size of any real network will introduce corrections analogous to the ones encountered in physical systems. In addition to the spatial smoothing out of the transition due to the finite number of nodes in the net, one also observes [7] a smoothing of the dynamical transition due to observations over a finite number of time steps. In particular, this means that slightly below the transition, observations might not last long enough to distinguish between continued activity growth and the eventual reaching of the fixed point.

14.4 CONCLUSION

The results that we have obtained and described in this paper show that the methods of statistical mechanics and dynamical systems theory can be used to obtain generic information about very large computational systems. One can then expect AI to pass a threshold into simpler domains as systems do indeed become extremely large.

We have also shown that large scale AI systems can behave very differently from their small scale counterparts. In particular, as local parameters are varied they may show sudden changes in overall performance which cannot be generally inferred from investigations of corresponding small scale systems. This aproach is to be contrasted with canonical bottom-up methods, which predict the behaviour of the larger system as an extrapolation from its smaller version. Although an explicit analytic determination of the precise location of phase boundaries may be very difficult for real systems, the generic qualitative behaviour we have predicted will be pervasive in many systems and observables. This is another manifestation of the well known phenomenon of universality, in which deep underlying similarities are masked by diversity at the surface level.

Finally, the apearance of more complicated dynamics at the process level can lead to much richer behaviour than the one discussed here. For example, the existence of delays, as well as nonlinear dynamics, along with the presence of inhibitory links, can lead to complicated oscillatory cycles and moving wavefronts, as well as chaotic reverberations of the whole network. These behaviours, associated with the appearance of limit cycles and strange attractors, have time-dependent asymptotic dynamics, and as a consequence, the commonly used criterion which infers equilibrium from an unchanging activity is no longer valid [8].

I can safely conclude by stating that the study of very large computational systems, as well as the dynamics of symbolic computation and searches in problem spaces, can uncover novel phenomena and also provide a useful framework with which to study their behaviour.

ACKNOWLEDGEMENTS

I want to thank Tad Hogg and Jeff Shrager for their insightful remarks and enthusiasm throughout this collaboration. Part of this research was supported by ONR contract No. N00014-82-0699.

REFERENCES

[1]   For an example of such machines, see D. Hillis (1985). *The Connection Machine,* MIT Press, Cambridge, Mass.

[2]   Large computing networks are discussed in a number of places. For a description of the Cedar Project, see L.P. Deutsch and E.A. Taft (1980). Xerox PARC technical report CSL-80-10.

[3]   B.A. Huberman and T. Hogg (1987). *Artificial Intelligence* **33**, 155.

[4]   A review of spreading activation networks in cognitive science appears in J.R. Anderson (1983). *The Architecture of Cognition,* Harvard University Press, Cambridge, Mass.

[5]   See for example, *Machine Learning, An Artificial Intelligence Approach,* ed. R.s. Michalsky, J.G. Carbonell and T.M. Mitchell, Tioga Press, Palo Alto, CA., (1983).

[6]   B. Bollobas (1985). *Random Graphs,* Academic Press, Orlando, Florida.

[7]   The observation of such phase transitions is reported in J. Shrager, T. Hogg, and B.A. Huberman (1987). *Science* **236**, 1092.

[8]   M.Y. Choi and B.A. Humberman (1985). *Phys. Rev.* **B31**, 2862.

15   *R.L. Kautz*

GLOBAL STABILITY OF THE CHAOTIC STATE NEAR AN INTERIOR CRISIS

Abstract

In dissipative systems, chaotic trajectories are locally unstable in that they show a sensitive dependence on initial conditions but globally stable in that they are represented by attractors.   We explore   the   global   stability   of   the   chaotic state   in   the neighbourhood of an interior crisis using the rf-biased Josephson junction   as   an   example.   A   measure   of   global   stability   is developed by considering the response of the system to thermal noise.   In the presence of noise the system occasionally escapes from the region of state space containing the attractor and the temperature dependence of the average escape time is characterized by   an   activation   energy.   This   activation   energy   is   a   useful measure of the global stability of the chaotic state, being analogous to the barrier height for a particle in a potential well.

15.1 Introduction

Chaos   in   nonlinear   systems   is   paradoxical   in   that   the   observed pseudo-random behaviour results from a deterministic equation of motion.   A   second,   related   paradox   of   the   dissipative   chaotic state is its simultaneous possession of global stability and local instability.   Global   stability   is   implied   by   the   fact   that   the chaotic   state   is   represented   by   an   attractor   and   an   associated basin   of   attraction   in   state   space.   If   initial   conditions   are chosen   within   the   basin   of   attraction,   then   the   solution approaches   the   attractor   in   the   limit   of   large   times.   Global stability   is   the   result   of   dissipation   which   assures   that   a collection   of   points   in   state   space   will   evolve   such   that   the collection   occupies   an   ever   decreasing   volume.   In   spite   of   its global   stability,   a   chaotic   solution   is   locally   unstable   in   that an   infinitesimal   perturbation   produces   a   new   trajectory   which diverges   from   the   original   trajectory   at   an   exponential   rate. Local   instability   gives   rise   to   the   property   of   extreme sensitivity   to   initial   conditions,   a   hallmark   of   chaotic behaviour.   The   apparent   contradiction   between   global   stability and   local   instability   is   resolved   through   the   fractal   nature   of the   chaotic   attractor.   The   dimension   of   the   attractor   is   large enough to include diverging trajectories but small enough that it does not occupy a volume of state space.

Local and global stability are distinct in that the former concerns deviations from a particular trajectory produced by infinitesimal perturbations and the latter concerns deviations from an attractor produced by possibly finite perturbations. Quantitative measures of both the local stability and global stability of a solution derive from the Liapunov chatacteristic exponents λ_i. Local stability is related to the maximum exponent λ. A solution is locally stable if λ is negative and locally unstable if λ is positive. Global stability is measured by the average divergence of the flow, that is, the average rate of expansion Λ of state-space volumes in the vicinity of the attractor. In terms of the Liapunov exponents [1],

$$\Lambda = \sum_i \lambda_i. \tag{1}$$

The more negative Λ is, the more rapidly trajectories beginning within a basin of attraction converge towards the attractor. Although Λ is a useful measure of global stability in some circumstances, it can also be misleading. Because state-space volumes necessarily contract with time in dissipative systems, Λ is negative for all trajectories including unstable periodic orbits. Thus, while Λ quantifies the attraction of attractors, it does not distinguish attractors from other fixed orbits and in this sense fails as a measure of global stability.

In the present work we develop an alternative measure of global stability by studying the response of a system to thermal noise. In general terms, we propose defining the global stability of a steady-state solution by measuring the average time required for thermal noise to produce an occurrence of some event that is not characteristic of the unperturbed solution. This average time will be called the lifetime of the state with respect to the uncharacteristic event. The degree to which a lifetime represents a fundamental property of the steady-state solution depends on the nature of the event which defines it. The event might be specified, for example, by an arbitrary boundary surrounding the attractor in state space. The lifetime for remaining within the boundary will depend on the position of the boundary and, while such a lifetime represents a potentially useful measure of global stability, it is not a particularly fundamental one. In many cases, however, there are natural state-space boundaries which can be used to define a lifetime. If the basin of attraction does not include all of state space, then it is natural to choose as a stability parameter the lifetime defined by escape from the basin of attraction. In this case, we expect the lifetime to measure a fundamental property of the attractor.

Another natural definition of lifetime, the one of direct concern in this paper, applies in the vicinity of an interior crisis. As discussed in a review by Grebogi et al. [2], a crisis results when increasing some system parameter α to a critical value α_c causes a chaotic attractor to collide in state space with an unstable periodic orbit. The crisis is said to be a boundary crisis if the chaotic attractor disappears at α_c and an interior crisis if the attractor merely undergoes a sudden expansion in state space. In both cases, for α less than α_c the attractor is contained within a region of state space bounded by the unstable periodic orbit and its stable manifold. For a boundary crisis, this region is identical to the basin of attraction. For an interior crisis, the unstable periodic orbit defines a region, to be called an interior basin, which is smaller than the basin of attraction and contained within it. The difference between a boundary crisis and an interior crisis follows directly from the difference between a basin of attraction and an interior basin. In each case the crisis occurs when the attractor expands beyond the bounds of the unstable periodic orbit. When this orbit defines a basin of attraction, increasing α beyond α_c destroys the basin since trajectories beginning in that region of state space can escape into other regions. In contrast, when the unstable periodic orbit defines an interior basin, increasing α beyond α_c simply allows the attractor to expand into other parts of the basin of attraction.

In the vicinity of an interior crisis the lifetime with respect to escape from the interior basin is a natural global-stability parameter. For situations in which escape from the interior basin is a physically significant event, this lifetime can be a useful parameter. The interior crisis to be discussed here is a case in point. The example chosen for study derives from the rf-biased Josephson junction as modelled by a damped, driven pendulum [3]. This system, described by a second-order differential equation with a time-dependent drive, is among the simplest physical systems to show chaotic behaviour and has often served as a model for the study of chaos [4]. A common and sometimes useful condition observed in the rf-biased junction is a state of phase lock between the junction phase and the rf bias. At the interior crisis to be considered, this phase-locked condition is destroyed with the sudden expansion of the attractor. Thus, calculating the lifetime defined by escape from the interior basin is equivalent in this case to calculating the lifetime of the phase-locked state and leads to a measure of the global stability of phase lock.

As has been emphasized by Landauer [5], global stability is generally difficult to assess for systems far from equilibrium like an rf-biased junction: there is no general formalism for evaluating the effect of thermal noise on steady-state solutions in driven, dissipative systems. The lifetimes calculated here are obtained by Monte Carlo simulations in which thermal noise is modelled using a random-number generator. Although the results are necessarily empirical, they indicate that loss of phase lock is typical of thermally activated processes in that the temperature dependence of the lifetime takes the form

$$\tau = \tau_o e^{\varepsilon/kT}, \tag{2}$$

in the limit of low temperatures. The parameters ε and τ_o, called the activation energy and the attempt time, conveniently summarize the effect of thermal noise on phase lock. The activation energy is of particular interest as it represents a useful measure of global stability.

It is important to note that there is no *a priori* reason to expect that the asymptotic low-temperature lifetime of a chaotic state will be of the form given by Eq. (2). This form can be derived for the simpler case of the thermally activated escape of a particle from a potential well in the presence of damping [6]. In this case, which excludes time dependent forces, the steady-state solutions at zero temperature all correspond to situations in which the particle is at rest at a potential minimum and chaotic behaviour is impossible. The simplicity of the steady-state solution for a particle in a well allows Eq. (2) to be derived by construction. The complexity of chaotic solutions suggests that Eq. (2), if generally valid in the chaotic case, will be difficult to prove. However, if Eq. (2) is valid for some chaotic systems, as our simulations suggest, then for these systems ε and τ_o are natural parameters that can be used to characterize the system's response to thermal noise.

The physical significance of ε and τ_o is suggested by the roles they play in the escape of a particle from a potential well. In the potential-well problem, the activation energy ε is equal to the depth of the well, that is, the minimum energy required for escape. While it is generally impossible to define a well depth for an interior basin or a basin of attraction, reasoning by analogy suggests that for the chaotic case the activation energy is similarly the minimum energy required for escape. The attempt time τ_o in the potential-well problem is, for a lightly damped particle, equal to the inverse of the natural oscillation frequency at the bottom of the well. In more physical terms, the attempt time is the interval between the times at which the particle makes a close approach to the energy barrier. As will be discussed in section 15.5, the attempt time for the chaotic case is expected to be of a similar nature.

In the example to be considered here, the stability of phase lock is studied as a function of a system parameter, the dc bias i_o, over a range which takes the system from a region of harmonic motion, through a period-doubling cascade, and into a region of phase-locked chaotic motion which is terminated by an interior crisis at $i_o = i_{oC}$. Calculations of the activation energy and attempt time associated with the lifetime of phase lock indicate that ε goes to 0 and τ_o goes to ∞ as i_o approaches i_{oC}. The asymptotic behaviours of ε and τ_o are aparently of the form $(i_{oC} - i_o)^{\gamma_\varepsilon}$ and $(i_{oC} - i_o)^{-\gamma_\tau}$ respectively. By analogy with the patterns of divergence of other quantitities at the crisis point, the critical exponents γ_ε and γ_τ, which are evaluated here numerically, may prove to be universal constants. Because γ_ε and

γ_τ determine how stability is lost near an interior crisis, they are of potential importance in applications where the optimum operating point of a device lies near a chaotic region. If these critical exponents are in fact universal constants, they may be of use in situations where it is necessary to design around chaos.

In the remainder of this paper we focus on the particular interior crisis mentioned above. Section 15.2 develops the nature of the solutions in the neighbourhood of the crisis point and section 15.3 considers the local stability of these solutions. Section 15.4 describes the calculation of the lifetime of phase lock and its associated activation energy and attempt time. Section 15.4 explores the behaviour of the latter quantities near the crisis point.

15.2 Interior Crisis

The equation of motion for the phase ϕ of a dc- and rf-biased Josephson junction can be written as [4]

$$\beta\ddot{\phi} + \dot{\phi} + \sin\phi = i_o + i_1 \sin\Omega_1 t, \qquad (3)$$

where all parameters are reduced to dimensionless quantities. Here β is the hysteresis parameter, i_o is the dc bias, i_1 is the amplitude of the rf bias, Ω_1 is the rf frequency, dots indicate derivatives with respect to the time t, and $\dot{\phi}$ is the junction voltage . Eq. (3) is also the equation of motion for a damped pendulum driven by a torque which includes a constant component i_o and a sinusoidal component $i_1 \sin\Omega_1 t$. In this case ϕ is the angle of the pendulum from the downward vertical position.

The presence of a periodic drive suggests that steady-state solutions of Eq. (3) will commonly be periodic with a period equal to the drive period or possibly some multiple of it. Periodic solutions necessarily have the form

$$\phi(t + 2\pi m/\Omega_1) = \phi(t) + 2\pi l, \qquad (4)$$

for all t with m and l integers. The smallest integer m for which Eq. (4) is satisfied is the period of the solution measured in rf cycles. All periodic solutions are phase locked because the phase advances by exactly l revolutions during m rf cycles. The average voltage of a periodic solution is $\langle\dot{\phi}\rangle = (l/m)\Omega_1$.

Periodic solutions corresponding to a given l and m often exist over a range of dc bias. This gives rise to what are called rf-induced constant-voltage steps, that is, regions of dc bias where the average voltage is fixed at the value $(l/m)\Omega_1$. Steps for which l/m is an integer n are called principal steps and those for which l/m is not an integer are called subharmonic steps. For the principal steps, which concern us here, it can be shown that in the limit $\Omega_1 \gg 1$ or in the limit $\Omega_1^2 \beta \gg 1$, the dc-bias range of the nth-order step is [7]

$$n\Omega_1 - |J_n(\tilde{i}_1)| < i_o < n\Omega_1 + |J_n(\tilde{i}_1)|, \qquad (5)$$

where J_n is the nth-order Bessel function and $\tilde{i}_1$ is an alternative measure of the rf amplitude defined by

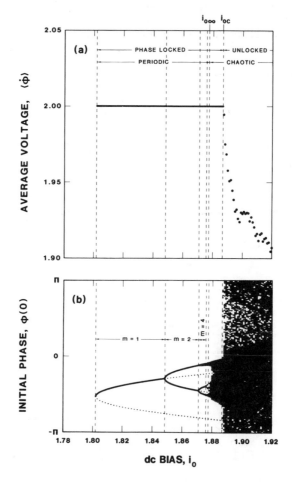

Figure 1. (a) *Average voltage and* (b) *initial phase as a function of dc bias for* $\beta = 25$, $\Omega_1 = 0.2$, *and* $\tilde{i}_1 = 10$. *In* (a) *the average voltage was computed using an averaging time of* 10^4 *rf cycles. For each dc bias level in* (b), *the value of the phase at the beginning of the rf cycle is plotted for 500 successive rf cycles. The initial phases for unstable periodic solutions having periods of 1 and 2 rf cycles are indicated by dotted lines.*

$$\tilde{i}_1 = \frac{i_1}{\Omega_1 \sqrt{\Omega_1^2 B^2} + 1}. \tag{6}$$

Over the range of validity of Eq. (5), solutions on the principal steps are period-1 solutions and chaotic behaviour is not observed.

Chaos is commonly observed in the rf-biased junction when $\beta > 1$, $\Omega_1 < 1$ and $\Omega_1^2 \beta$ is less than or comparable to 1 [8]. In this paper we focus on chaotic behaviour associated with the 10th principal step for $\beta = 25$, $\Omega_1 = 0.2$, and $\tilde{i}_1 = 10$ ($i_1 = 10.198$). This parameter set is also explored in two papers [9, 10] written before the existence of chaotic crises had been recognized. As it happens, the dc bias range of the 10th step includes, in this case, a good example of loss of phase lock through an interior crisis.

The nature of the solution at the lower dc-bias range of the 10th step is shown in Fig. 1. Figure 1(a) plots the average voltage as a function of dc bias. For $\Omega_1 = 0.2$ the value of the average voltage on the 10th step is 2. According to Eq. (5), which is not strictly applicable in the present case since $\Omega_1^2 \beta = 1$, the voltage is expected to maintain this value from $i_o = 1.7925$ to 2.2075. In fact, the step begins at $i_o = 1.8021$, reasonably close to the predicted lower limit, but ends prematurely at $i_{oC} = 1.8867$. The loss of phase lock at i_{oC} is caused by the interior crisis which is central to this paper.

The sequence of changes in the solution which lead to the crisis are revealed by the bifurcation diagram shown in Fig. 1(b). In this diagram solutions are characterized by values of the phase at the beginning of the rf cycle. The initial phases for each of 500 successive cycles of the steady-state solution are plotted for each value of dc bias. At the lower end of the step the solution is periodic with a period of 1 rf cycle and all 500 values of initial phase are coincident. Above $i_o = 1.8483$, however, the period-1 solution becomes unstable and a stable period-2 solution develops through a pitchfork bifurcation. The period-2 solution alternates between two values of initial phase and is represented on the diagram by two points at each value of dc bias. The initial phase of the unstable period-1 solution is indicated by a dotted line, the middle tine of the pitchfork. A second period-doubling bifurcation occurs at $i_o = 1.8708$ where the period-2 solution becomes unstable and a stable period-4 solution develops. As expected on general grounds [1], this period-doubling process repeats ad infinitum with increasing i_o and the dc-bias ranges of successively higher-period solutions

approach 0 in a geometric series. Based on the stability range of
the period-64 solution, the accumulation point of this bifurcation
cascade is predicted to be at $i_{O\infty}$ = 1.8773.

Above $i_{O\infty}$ the steady-state solution is chaotic and the possible
values of initial phase span continuous intervals. In the chaotic
region, the 500 values of $\phi(0)$ plotted for each bias point in Fig.
1(b) merely sample the range of possibilities. For i_O between $i_{O\infty}$
and i_{oC}, however, the values of $\phi(0)$ span only a fraction of the
interval from $-\pi$ to π. In this range of i_o, the solution is
chaotic but phase locked. Phase-locked chaos gives rise to
pseudo-random values of initial phase but this randomness is
restricted such that the phase advances by very nearly 10
revolutions during each rf cycle. Above i_{oC}, the range of initial
phases suddenly expands to fill the entire interval from $-\pi$ to π
and here the chaotic motion includes 2π phase slips. These random
2π slips cause the average voltage to deviate from 2 for i_o above
i_{oC}.

Poincaré sections of the chaotic solution in the neighbourhood

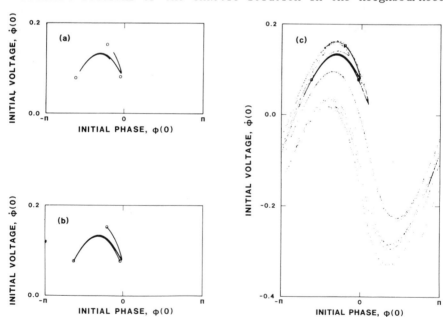

Figure 2. *Poincaré sections of chaotic solutions for β = 25, Ω_1*
= 0.2, $\tilde{i}_1$ = 10 and three values of dc bias: (a) i_o = 1.882, (b)
i_o = 1.8867, and (c) i_o = 1.8868. Each section includes 5 x 10^4
points corresponding to the location of the system in state
space at the beginning of as many successive rf cycles. Also
included are circles which locate the three points corresponding
to the section of an unstable period-3 solution.

of the crisis point are shown in Fig. 2. These Poincaré sections are formed by plotting a point corresponding to the location of the system in state space at the beginning of the rf cycle, $(\phi(0), \dot{\phi}(0))$, for each of 5×10^4 successive rf cycles. As shown in Fig. 2(a), the points of the section for the chaotic solution at $i_o = 1.883$ fall on a folded arc. This folded arc is the chaotic attractor. Figure 2(a) also shows circles which locate the three values of $(\phi(0), \dot{\phi}(0))$ corresponding to an unstable period-3 solution which is the cause of the interior crisis. These three points are saddle points and might be thought of as marking the location of mountain passes. The saddle points together with their stable manifold (not shown) surround the chaotic attractor and define the boundary of the interior basin. As i_o approaches i_{oC}, the ends of the attractor expand until they reach the saddle points, as shown in Fig. 2(b). A further small increase in i_o from 1.8867 to 1.8868 allows the chaotic attractor to expand beyond the interior basin into distant parts of the surrounding basin of attraction, as shown in Fig. 2(c).

The expansion of the attractor to fill the entire $-\pi$ to π range of possible initial phases occurs discontinuously at i_{oC}. However, for i_o just above i_{oC}, the system is most often found within the interior basin. This fact is suggested by the relatively few points falling outside of the interior basin in Fig. 2(c). While spontaneous 2π phase slips are always possible for i_o greater than i_{oC}, their frequency approaches 0 as i_o approaches i_{oC}. This limiting behaviour, typical of interior crises [2], explains why the average voltage is continuous at the crisis point, as Fig. 1(a) suggests. An interior crisis thus produces discontinuities in some properties of the solution while preserving continuity in others.

15.3 Local Stability

The local stability of a solution is determined by the response of the system to infinitesimal perturbations. To evaluate the local stability we consider the effect of a small displacement in the state vector at time t_o on the state vector at later times t.

When the displacement is infinitesimal, such effects are determined by the Jacobian matrix which is defined in terms of the state vector $(x_1, x_2) \equiv (\phi, \dot{\phi})$ by,

$$J_{jk}(t, t_o) = \frac{\partial x_j(t)}{\partial x_k(t_o)}. \tag{7}$$

Depending on the stability of a solution, the components of the Jacobian matrix tend to increase or decrease exponentially in time. The Liapunov exponents, defined by

$$\lambda_i = \lim_{t \to \infty} \frac{1}{t-t_o} \ln|i^{th} \text{ eigenvalue of } J(t,t_o)|, \qquad (8)$$

characterize the long-term growth of infinitesimal perturbations. The maximum exponent, to be denoted λ, determines the local stability of a solution. If λ is positive then infinitesimal perturbations grow exponentially and the solution is locally unstable. If λ is negative then infinitesimal perturbations decay exponentially and the solution is locally stable.

Liapunov exponents have been computed previously for the rf-biased junction by a number of authors [4,8-13]. Results for the present case, shown in Fig. 3, follow the expected pattern. The maximum exponent is negative throughout the period-doubling cascade except at the bifurcation points where λ goes to 0 as stability is transferred between solutions having different periods. For i_o greater than $i_{o\infty}$, λ is positive, confirming the chaotic nature of the solutions in this region, whether phase locked or unlocked. We also note that there is an apparent discontinuity in the derivative of λ at the interior crisis. Thus, the occurrence of bifurcations, chaos, and crises are all reflected in the behaviour of λ.

As was noted in the introduction, the sum of the Liapunov exponents is equal to Λ, the divergence of the flow. Because Λ is

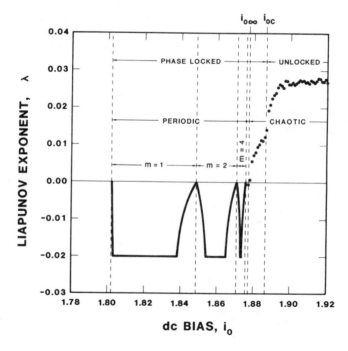

Figure 3. Maximum Liapunov exponent as a function of dc bias for $\beta = 25$, $\Omega_1 = 0.2$, and $\tilde{i}_1 = 10$, calculated using an averaging time of 10^4 rf cycles.

related to the rate at which trajectories converge toward an attractor, it represents a potentially useful measure of global stability. For the rf-biased junction, it can be shown rigorously that [8]

$$\Lambda = \lambda_1 + \lambda_2 = 1/\beta. \tag{9}$$

The fact that Λ is a constant implies that any set of trajectories having initial conditions within a state-space region of volume V_o will occupy a volume $V_o e^{-t/\beta}$ at time t, regardless of the nature of the trajectories. Thus, for the rf-biased junction, Λ provides no information about the relative global stability of attractors.

15.4 Global Stability

The global-stability parameter to be discussed here, the average time for escape from an interior basin in the presence of thermal noise, is a natural physical parameter. For the case at hand, this lifetime corresponds to the lifetime of phase lock, a quantity which is in principle accessible to experiment. This quantity is, however, inaccessible to theory in parameter regions near the onset of chaos. In these regions it can only be calculated through time-consuming Monte Carlo simulations.

The equation of motion for the rf-biased junction including the Johnson noise associated with the junction resistance is [14]

$$\beta\ddot{\phi} + \dot{\phi} + \sin\phi = i_o + i_1\sin\Omega_1 t + i_N(t), \tag{10}$$

where the noise current i_N has an impulsive autocorrelation function,

$$\langle i_N(t)i_N(t')\rangle = 4\Gamma\delta(t - t'), \tag{11}$$

and a white power spectrum. Here Γ is a reduced temperature, the ratio of kT to the Josephson coupling energy (the difference in potential energy between the $\phi = \pi$ and $\phi = 0$ states of the junction). In integrating Eq. (10), i_N is evaluated as its average over each integration interval Δt. That is, i_N is represented by a series of random numbers having a Gaussian distribution with a mean of 0 and a variance of $4\Gamma/\Delta t$. The lifetime of phase lock on the nth step is calculated through a series of trials, each beginning with initial conditions, $\phi(0)$ and $\dot{\phi}(0)$, corresponding to the unperturbed phase-locked state. A trial is terminated after the jth rf cycle if the phase at the end of the cycle differs from the phase-locked value, $\phi(0) + 2\pi jn$, by more than 2π. This procedure yields an approximate value for the lifetime given by $\tau = 2\pi\langle j\rangle/\Omega_1$ where $\langle j\rangle$ is the average of j over the series of trials.

The calculated lifetime of phase lock on the 10th step is

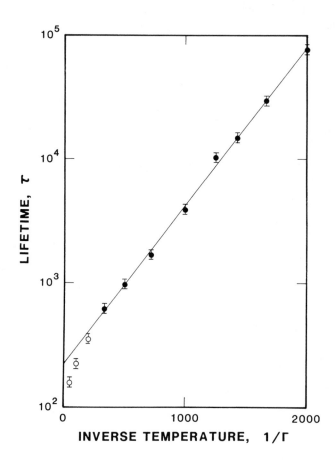

Figure 4. Lifetime of phase lock on the 10th step as a function of inverse temperature for $\beta = 25$, $\Omega_1 = 0.2$, $\tilde{i}_1 = 10$, and $i_o = 1.83$. Each data point is the average of 100 trials. Error bars show the expected standard deviation of the average value. The solid line is a least squares fit to the eight points at lower temperatures indicated by filled circles.

plotted in Fig. 4 as a function of inverse temperature for a dc bias of $i_o = 1.83$ where the unperturbed solution has a period of 1 rf cycle. The most important feature of this graph is the fact that the lifetime increases exponentially with $1/\Gamma$ in the limit of large $1/\Gamma$ (low temperature). The existence of an exponential asymptote implies that the temperature dependence of τ can be characterized in terms of an activation energy and an attempt time, defined as ε and τ_o by Eq. (2). A similar conclusion is reached for other dc bias levels along the range of phase lock shown in Fig. 1(a), including points within the phase-locked

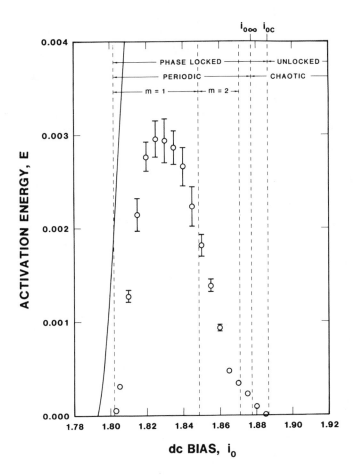

Figure 5. Activation energy for loss of phase lock on the 10th step as a function of dc bias for $\beta = 25$, $\Omega_1 = 0.2$, and $\tilde{i}_1 = 10$. Monte Carlo results are plotted as circles and a solid line shows the approximation given by Eq. (12).

chaotic region. Our simulations thus provide strong empirical evidence that the stability of phase lock in the rf-biased junction can be characterized in general by an activation energy and an attempt time.

To determine ε and τ_0 for a set of data such as that shown in Fig. 4, we use the method of least squares to fit a straight line to a plot of $\ln\tau$ versus $1/\Gamma$. The slope of this line is E, the activation energy measured in units of the Josephson coupling energy, and the $\ln\tau$-intercept is $\ln\tau_0$. To ensure that only the exponential portion of the data set is used, high-temperature points are successively eliminated until the included values of τ all exceed the calculated τ_0 by a factor of 2. We also require that the included points span at least a decade of lifetimes. On this basis, the eight data points indicated by filled circles in Fig. 4 were selected for computing the low-temperature asymptote which is shown by a solid line.

The computed activation energy and attempt time are plotted as a a function of dc bias in Figs. 5 and 6. Error bars associated with these results are defined by straight lines fit to the $\ln \tau$ versus $1/T$ plots that yield an rms error twice that of the least squares fit. The Monte Carlo results are compared here with analytic expressions for E and τ_o that have been derived in the limit $\Omega_1^2 \beta \gg 1$. In this limit the dynamics of phase lock can be modelled by a particle in a potential well and Fokker-Planck methods used to calculate the lifetime of phase lock [7,15-18]. For a. bias point on the nth step, this model yields a low-temperature asymptote described by [7]

$$E = |J_n(\tilde{i}_1)|[\tilde{i}_o \sin^{-1}\tilde{i}_o - \frac{\pi}{2}|\tilde{i}_o| + \sqrt{1 - \tilde{i}_o^2}], \qquad (12)$$

$$\tau_o = \frac{4\pi\beta}{\sqrt{4\beta|J_n(i_1)|\sqrt{1-1_o^2} + 1} - 1}, \qquad$$

where $\tilde{i}_o$ is a measure of the dc bias appropriate to the nth step [cf. Eq. (5)],

$$\tilde{i}_o = \frac{i_o - n\Omega_1}{|J_n(\tilde{i}_1)|}. \qquad (14)$$

While Eqs. (12) and (13), plotted in Figs. 5 and 6 as solid lines, are not strictly applicable in the present case, for which $\Omega_1^2 \beta = 1$, there is some agreement with the Monte Carlo results at the low dc-bias end of the step. In fact, a reasonably good fit is obtained if the theoretical curves are displaced upward in dc bias so that the theoretical endpoint of the step matches the actual ' endpoint. This agreement provides evidence for the accuracy of the Monte Carlo simulations. Because the theory does not include period doubling or chaos, the observed divergence between the theoretical and Monte Carlo results at higher dc-bias levels is to be expected.

Both the activation energy and the attempt time show singular behaviour at the ends of the dc-bias range of phase lock, E going to 0 and τ_o going to ∞. It is not surprising that near the end of the phase-locked range only a small amount of energy is required to break phase lock since a small change in dc bias is sufficient to move the system into a region where phase lock is unstable.

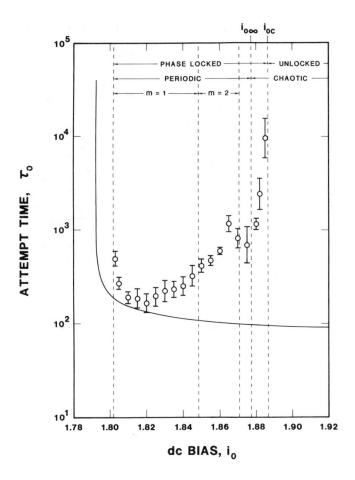

Figure 6. Attempt time for loss of phase lock on the 10th step as a function of dc bias for $\beta = 25$, $\Omega_1 = 0.2$, and $\tilde{i}_1 = 10$. Monte Carlo results are plotted as circles and a solid line shows the approximation given by Eq. (13).

However, it is perhaps surprising that the attempt time approaches ∞. The divergence in τ_o at the lower end of the step is accounted for by the theory that gives rise to Eq. (13). In this theory, the attempt time is directly related to the natural response time of the phase-locked state and this response time diverges as i_o approaches the lower end of the step. The situation at the upper end of the step, where the phase-locked state is chaotic, is discussed in section 15.5.

It is of interest to compare the local stability of the phase-locked state as measured by λ with the global stability as measured by E. Comparing Figs. 3 and 5 reveals that these quantities are almost totally unrelated. In particular, between $i_{o\infty}$ and i_{oC}, the phase-locked solution is locally unstable

221

according to λ but globally stable according to E, quantitatively verifying one of the essential paradoxes of chaotic behaviour. It is also apparent that while the local stability goes to 0 at each period-doubling bifurcation, the global stability remains high at these points. Some understanding of this situation can be gained by considering an analogy between the phase-locked junction and a particle in a potential well. Three potential functions analogous to situations near a period-doubling bifurcation are shown in Fig. 7. The period-1 phase-locked state corresponds to a particle at rest at the centre of the potential shown in Fig. 7(a). In the analogy, loss of phase lock corresponds to activation over the potential barrier and the global stability of phase lock is measured by the barrier height ε. The local-stability parameter is the curvature of the potential at its minimum. In Fig. 7(a), the curvature is positive and the minimum is locally stable but in Fig. 7(b), which represents the situation at the bifurcation point, the curvature is 0 and local stability is marginal. Figure 7(c) corresponds to a situation just beyond the bifurcation point. Here there are two locally stable minima representing the two period-2 phase-locked states and a central unstable maximum representing the unstable period-1 phase-locked state [cf. Fig. 1(b)]. The important point to note is that the barrier height ε remains large during the process of bifurcation even though the local stability goes to 0 at the bifurcation point. Thus, although the dynamics of phase lock cannot generally be modelled

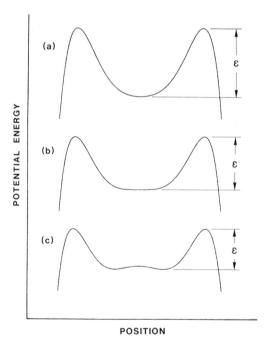

Figure 7. Symmetric sixth-order potential functions having (a) a positive quadratic term, (b) no quadratic term, and (c) a negative quadratic term.

between λ and E at a period-doubling bifurcation.

by the behaviour of an autonomous particle in a potential well, the analogy helps us to understand the difference in behaviour

For i_o greater than i_{oC}, phase lock is globally unstable and the chaotic solutions observed here show spontaneous 2π phase slips. This fact does not imply that unlocked chaotic states are globally unstable in any general sense, only that their stability cannot be measured with respect to loss of phase lock because phase lock is not one of their attributes. In the present instance, the unlocked chaotic solution coexists with a stable period-1 solution on the 8th step so its global stability can be measured with respect to escape from its basin of attraction. Thus, while the lifetime of phase lock is an important parameter to the rf-biased junction, it is just one of a class of measures that can be used to quantify global stability.

15.5 Critical Exponents

In order to examine the asymptotic behaviour of E and τ_o near the crisis point, these quantities are plotted again in Fig. 8 as a function of $(i_{oC} - i_o)$ using logarithmic scales for both axes. The nearly linear relationships apparent in Fig. 12 within the chaotic region suggest that the asymptotic forms of E and τ_o are

$$E \propto (i_{oC} - i_o)^{\gamma_E}, \tag{15}$$

$$\tau_o \propto (i_{oC} - i_o)^{-\gamma_\tau}. \tag{16}$$

Such forms are typical of quantities which show singular behaviour at a crisis point. Fitting straight lines to the five data points between $i_{o\infty}$ and i_{oC} using the method of least squares yields for the critical exponents $\gamma_E = 2.3$ and $\gamma_\tau = 1.4$. The accuracy of these values for γ_E and γ_τ is difficult to assess because it depends on whether the data points are sufficiently close to i_{oC} to reveal the true asymptotic behaviour.

Ultimately, the asymptotic behaviour of E and τ_o at a crisis point must be established by analysis rather than simulation. As a step in this direction we consider the situation qualitatively by picturing the crisis as a collision between a chaotic attractor and a set of saddle points which define an interior basin. Because E is defined by a low-temperature asymptote, we infer that it is related to the minimum energy required for escaping the interior basin. The fact that E goes to 0 as i_o approaches i_{oC} reflects the fact that the outer reaches of the chaotic attractor approach the saddle points more closely as i_o aproaches i_{oC} and less energy is required to produce a deviation in the trajectory which will take the system over the saddle points and into remote parts of the basin of attraction.

The fact that τ_o goes to ∞ as i_o approaches i_{oC} is more difficult to explain. For the case of a lightly damped particle in a potential well, the attempt time is the interval between the times at which the particle makes a close approach to the energy barrier. By analogy we expect that near a crisis point the

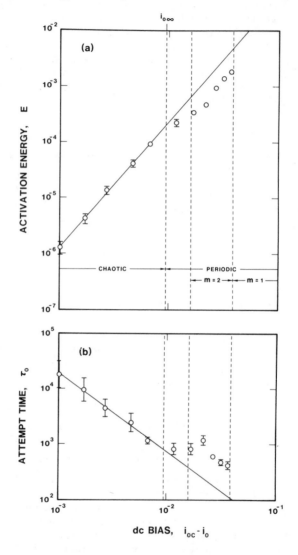

Figure 8. Log-log plot of (a) activation energy and (b) attempt time for loss of phase lock on the 10th step as a function of $i_{oC} - i_o$ for $\beta = 25$, $\Omega_1 = 0.2$, and $\tilde{i}_1 = 10$. Straight lines are least-squares fits to the four data points in the chaotic region.

attempt time of a chaotic system will be related to the time between events in which the system visits portions of the attractor from which escape is possible with thermal energies comparable to the minimum required energy, say within a factor of 2. In the limit of low temperatures escape is probable only from these portions of the attractor because the probability of obtaining thermal excitations greatly in excess of the minimum is small relative to the probability of obtaining the minimum. As i_o aproaches i_{oC} the minimum escape energy decreases and the fraction of the attractor from which escape is possible with twice the minimum energy also decreases. The attempt time thus increases as i_o approaches i_{oC} because near i_{oC} the system spends most of its time on portions of the attractor from which escape is virtually impossible given an amount of thermal energy comparable to the minimum energy required for escape. These qualitative ideas not only suggest why E and τ_o show singular behaviour at the crisis point but may prove useful as a starting point for quantitative analysis of the asymptotic behaviour.

In conclusion, the simulations presented here demonstrate that the lifetime for thermally activated escape from an interior basin is a natural and useful measure of global stability for chaotic systems. The simulations suggest that the low-temperature asymptote of the lifetime can be characterized by an activation energy and an attempt time and that the behaviour of these parameters at an interior crisis is described by critical exponents. We also speculate that these critical exponents are universal constants but this possibility and indeed the existence of the activation energy and attempt time as well defined parameters remains to be rigorously demonstrated for chaotic systems.

Acknowledgement

This work was supported in part by the Office of Naval Research under contract number N00014-85-F-0085. (Contribution of the U.S. Government, not subject to copyright.)

References

[1]   A.J. Lichtenberg and M.A. Lieberman (1983). *Regular and Stochastic Motion,* Chap. 7, Springer-Verlag, New York.

[2]   C. Grebogi, E. Ott and J.A. Yorke (1983). *Physica* 7D, 181.

[3]   W.C. Stewart (1968). *Appl. Phys. Lett.* 12, 277;   D.E. McCumber (1968). *J. Appl. Phys.* 39, 3113.

[4]   R.L. Kautz and J.C. Macfarlane (1986). *Phys. Rev.* A33, 498, and references therein.

[5]   R. Landauer (1978). *Phys. Today* 31(11), 23.

[6]   H.A. Kramers (1940). *Physica* 7, 284.

[7]   R.L. Kautz (1981). *J. Appl. Phys.* 52, 3528.

[8]   R.L. Kautz and R. Monaco (1985). *J. Appl. Phys.* 57, 875.

[9]   R.L. Kautz (1981). *J. Appl. Phys.* 52, 6241.

[10]   R.L. Kautz (1983). *IEEE Trans. Magn.* MAG-19, 465.

[11]   W.-H. Steeb and A. Kunick (1982).   *Phys. Rev.* **A25**, 2889.
[12]   V.N. Gubankov, S.L. Ziglin, K.I. Konstantinyan, V.P. Koshelets and G.A. Ovanikov (1984).   *Zh. Eksp. Teor. Fiz.* **86**,   343 [*Sov. Phys.-JETP* 59, 198, 1984].
[13]   Z.-D. Wang and X.-X. Yau (1985).   *Acta Phys. Sin.* 34, 1149[*Chin. Phys.* 6, 326, 1986].
[14]   R.L. Kautz (1985).   *J. Appl. Phys.* **58**, 424.
[15]   M.J. Stephen (1969).   *Phys. Rev.* **186**, 393.
[16]   J.R. Waldram, A.B. Pipard and J. Clarke (1970).   *Phil. Trans. R. Soc. London* **A286**, 265.
[17]   P.A. Lee (1971).   *J. Appl. Phys.* **42**, 325.
[18]   E. Ben-Jacob and D.J. Berman (1984).   *Phys. Rev.,* **A29**, 2021.

MODIFIED PAINLEVE TEST FOR ANALYTIC INTEGRABILITY

Abstract

The characteristic feature of the so-called Painlevé test for integrability of an ordinary (or partial) analytic differential equation, as usually carried out, is to determine whether all its solutions are single-valued by local analysis near individual singular points of solutions.   This test, interpreted flexibly, has been quite successful in spite of various evident flaws.   A more robust and generally more appropriate definition is proposed: a multi-valued function is accepted as an integral if its possible values (at any given point in phase space) are not dense.   This definition is illustrated and justified by examples, and a widely applicable method of testing for it is presented, based on asymptotic analysis covering several singularities simultaneously.

16.1 Introduction

In the last quarter century a great many nonlinear equations, and systems of equations, have been discovered to have the so-called "soliton property".   In brief, and necessarily oversimplified, this is the property of supporting essentially localized (extending over a limited region) steady travelling waves that can plunge into interaction with an arbitrary (nonsteady) localized wave and emerge completely unchanged in form and velocity;   such steady travelling waves are called "solitons".   For equations that are linear, or trivially transformable into linear, such behaviour is a simple consequence of the superposition principle, but for genuinely nonlinear ones it is striking.

Most of the celebrated soliton systems are partial differential equations (p.d.e.'s) in two independent variables (often representing space and time dimensions and denoted by x and t). They include the prototypical Korteweg-de Vries equation (KDV) in which solitons were first discovered, the sine-Gordon (SG) with solitons (called kinks and antikinks) and breathers, and the nonlinear Schrödinger equation (NLS), all of which are of wide physical applicability, but also many interesting (systems of) ordinary differential equations (o.d.e.'s) like those describing the Toda lattice, integrodifferential equations such as the Benjamin-Ono equation, equations with more independent variables, and so on and on.

It is in fact suprising how fast the number of equations in this only recently visible class has been growing, so that one might almost think that soliton equations are destined to become preponderant. But of course there are also a great many "nonsoliton" equations, and one would like a way to distinguish between them.

16.2 **The Painlevé Test**

The soliton property is closely allied to "complete integrability", a concept usually though of in connection with Hamiltonian systems but, properly interpreted, of much wider applicability. Ablowitz, Segur and Romani, among others, have proposed reducing p.d.e.'s by imposing invariance conditions on the unknown function (e.g. similarity under some scaling), and applying the so-called Painlevé test to the resulting o.d.e.'s. (Weiss, Tabor and Carnevale have shown how the Painlevé test can be applied directly to the p.d.e.'s, a very substantial improvement.)

The Painlevé test, as usually applied in practice (and this is a weakening of the true proper test, but the main point to be made applies nevertheless; Painlevé had a different goal in mind in devising it, anyway) consists of determining all possible structures of isolated movable singularities in the finite complex plane of the solutions of o.d.e. under investigation; only poles are allowed if it is to "pass" and be called integrable. This condition must be interpreted liberally, since many equations one would like to call integrable have first to be transformed in some simple (or not so simple) way in order to pass the test. But it is not apparent why poles are allowed, and other singularities not; nor why fixed singularities are ignored. Nevertheless, this test has had remarkable success in sifting out integrable systems. (Of course only "mostly analytic" equations are under consideration here.)

16.3 **Single-valuedness**

It is proposed that the essential ingredient for the success of the common Painlevé test is that it picks out equations whose solutions are all single-valued. That is true if the solution has only poles; but why not allow a solution with an essential singularity, e.g. a two-way infinite Laurent series? The only practical reason is that it is harder to discover if there are such solutions. Now single-valuedness is too severe a condition anyway, as one can easily construct equations with multivalued solutions that must be considered integrable by any reasonable standard, in fact that permit single-valued integral even though the solution itself is multivalued. Take, for instance, a form of the second of Painlevé's six equations, $u'' = 6u^3 + xu$, which has solutions with simple movable poles $(x-a)^{-1}$ and $-(x-a)^{-1}$. It has the Painlevé property par excellence, and is to be accounted integrable even though we cannot write its integral in any explicit way. Make the substitution $u = v^2$ and obtain an equation for v that also has a single-valued integral and so must be accounted integrable, yet which has branched solutions such as start their expansions with $(x-a)^{-1/2}$.

So single-valuedness is not really the answer. It works well, so we do not want to abandon it, but just generalize it somewhat.

16.4 Nondense Multivaluedness

We cannot just abandon single-valuedness completely, as that would open the floodgates to too many systems not integrable in any reasonable sense. The basic concept can be seen in a simple, amost trivial, real system:

$$dx/dt = a, \quad dy/dt = b, \qquad (1)$$

where a, b are constants, and the unknowns x, y are angle variables (taken modulo unity). Then (1) represents uniform flow on a torus, and one can ask, when does this flow possess a reasonble time-independent integral (constant of motion)? By "reasonable" all we need mean is continuous (and not identically constant: not totally independent of x and y, of course). It is well known and easy to see that the answer depends on whether the ratio r = a/b is rational. For x − at and y − bt are two time-dependent integrals, and t can be eliminated by forming the combination k = (x − at) − r(y − bt) = x − ry. However, this latter is in general not single-valued on the torus, even when taken modulo unity. But if r = p/q, where p, q are relatively prime integers, then K takes on only a finite number (q) of distinct values modulo unity at a given point (x,y), and multiplying K by q gives qx − py, all of whose values at a given point are the same (modulo unity); indeed to form a genuinely single-valued integral one need only apply an appropriate periodic function, thus: sin 2π (qx − py). On the other hand, if r is irrational, then the possible values of K at a given point are not only infinite in number but dense; any trajectory covers the entire torus densely, so any continuous integral must be identically a constant and of no use (in distinguishing one orbit from another).

To sum up, (1) is integrable if, and only if a/b is rational, and if and only if there is a continuous nontrivial integral that is (possibly but) not densely multivalued. This illustrates in the simplest conceivable case the new concept of integrability proposed.

Naturally, in higher order systems there is a concept of degree of integrability, counting how many independent continuous integrals exist. And these notions also have to be adapted to the special circumstances of Hamiltonian theory. But instead, on this occasion, let me elaborate on what may seem too elementary a type of question but is actually quite illuminating.

Consider a single (nonautonomous) "mostly analytic" (i.e. analytic except possibly for some isolated lower-dimensional manifolds) first order ordinary differential equation in the complex plane. What do we mean for it to be integrable, and how do we tell whether it is? To focus our attention and ensure that we are not getting lost in the abstractions of pure mathematics, we choose a quite concrete example. And consonant with good scientific methodology we start with the simplest possible case that is not too trivial to be interesting.

Now linear first differential equations are integrable by quadrature, and although even that does not necessarily render copletely trivial the question of integrability in the sense we are discussing, it does simplify it substantially. So let us set du/dz equal to a nonlinear function of u. The simplest nonlinearity is quadratic, but that would lead to a Riccati equation, which can be transformed to a linear second order equation by a standard transformation. We are determined to choose a genuinely nonlinear equation for illustration, so take a cubic right side, as simple as possible while still retaining some dependence on z, say

$$du/dz = u^3 + z. \qquad (2)$$

Is this equation integrable?

Now (2) immediately fails the usual Painlevé test, since comparison of the first two terms shows that the leading behaviour of a solution at a movable singular point a is proportional to $(z - a)^{-1/2}$. However, we know that (2) may nevertheless be integrable, though that is in some sense unlikely. (There are some quite nontrivial examples of just such a phenomenon, for instance Dym's equation.) How can we be sure?

If denseness of multivaluedness is the issue, we cannot tell for sure from the behaviour around just one singular point. It may take the winding of a path of integration in and out around several singular points to generate enough values for denseness.

16.5 The Poly-Painlevé Test

Since the Painlevé test looks at one singular point (at a time), and the proposed new test looks at several, we call it the poly-Painlevé test. All such tests are perturbation expansion tests, but what are we to expand around here? In the Painlevé test we expand around the singular point, but now there will be several and we cannot prefer any single one of them. Instead, we introduce a parameter into the equation by transformation and expand with respect to that parameter.

The method can only be sketched here, even for this very simple case. In (2), replace z by z + b, where b is a constant; this does not affect the character of the equation. Now let b get large, and scale u and z by appropriate powers of b for "maximal balance", keeping as many terms of comparable dominant order as possible. This requires u^3 to scale like b, so u like $b^{1/3}$, and then for du/dz to participate in the dominant order requires z to scale like $b^{-2/3}$. Denoting the rescaled variables by capital letters leads to

$$dU/dZ = U^3 + 1 + cZ. \qquad (3)$$

where $c = b^{-5/3}$ is a small parameter in which we can expand. Equations (2) and (3) are equivalent as to integrability.

To lowest order, i.e. without the cZ term, (3) is autonomous and can be integrated in terms of quadratures. The solution has many singular points, in fact an infinite lattice of them, as is best seen by inverting the variables, thinking of Z as a function of U. The natural integral is multivalued, in fact infinitely so, but since its values, for given (Z,U), form a regular double array or lattice, they are not dense and do not rule out integrability.

Continuing the expansion, we find, although we cannot give the calculation here, that in the first order the equation remains integrable. In the next (second) order, however, the traversal of one "period parallelogram" of the zero order lattice does not close but in effect introduces a third period, and since the new period is proportional to c^2 it is essentially arbitrary and generically leads to denseness. This shows that (3), hence also (2), is not integrable.

16.6 Further Discussion

The same method can be and has been applied to higher order equations. For instance, the second order equation

$$u'' = u^2 + f(z), \qquad (4)$$

which is known to have the Painlevé property only if the arbitrary analytic function f is linear (or constant), might nonetheless be completely integrable for some other functions f; however the poly-Painlevé test shows that it is not. The test should be capable of determining whether (4) is partly integrable, i.e. whether it has one but not two independent integrals, but that calculation has not yet been carried out.

Even for first order equations, one serious limitation of the new test should be pointed out, a limitation which is seemingly more severe for first than for higher order equations. The test requires finding, or creating, a suitable expansion procedure, one which can be carried out farily explicitly order by order, and which retains several singularities in the lowest order. This is not always possible. Perhaps the simplest counterexample is a modification of (2):

$$du/dz = u^3 + k/(z^2 - 1). \qquad (5)$$

For which values of the constant k is (5) integrable? It does not appear possible to devise an expansion to answer that question. If, as seems most unlikely, (5) happens to be integrable for all k, then an expansion in k could show that to be the case.

The reason for the difficulty in treating (5) is that the decay of the last term like z^{-2} as z approaches infinity is too fast for a scaling procedure to work like before (z^{-1} would be all right; the critical cut-off power is $-3/2$), while the simple-pole blow-up of the last term at z = 1 (or -1) is too slow for a scaling procedure to work in that vicinity (again -3/2 is the critical power).

16.7 Conclusion

It has only been possible here to indicate the basic idea of the new method, which has very considerable potential for wide application. For higher order systems there seems generally to be much more freedom for devising appropriate expansions, therefore the difficulty discussed in the preceding section is not as serious as might appear. Some further examples have been worked out, but many more must be done to explore the possibilities.

If the method could show the integrability of some systems not already known about, that would be a major triumph. Actually, it is easier to use the method to show nonintegrability to all orders of the expansion and also proving the convergence of the expansion. However, in any particular example it would be fairly convincing evidence of integrability if it holds for several orders, say six or better eight.

17 *M. Marek, I. Schreiber and L. Vroblová*

COMPLEX AND CHAOTIC WAVES IN REACTION-DIFFUSION SYSTEMS AND ON THE EFFECTS OF ELECTRIC FIELD ON THEM

17.1 Introduction

Reaction-diffusion systems can have properties of distributed excitable media and conduct excitations (e.g. pulses, fronts) if reactions with suitable properties are considered. Reaction-diffusion models are therefore widely used in the description of wave propagation in physical, chemical and biological systems ranging from ecology to an electron-hole plasmas in semiconductors [1-5]. Homogenous liquid reacting systems often contain ionic components. As the spatial nonhomogenous concentration profiles of ionic components give rise to electric potential gradients (it is particularly common in biological systems), it is also of interest to study reaction-diffusion-electric field interactions. We have earlier reported on detailed experimental studies of pulse and front waves and on the effect of the imposed electric fields on them. Nonlinear effects such as pulse annihilation, splitting, reversal of the direction of the wave movement and a formation of stationary dissipative structures were observed [6-8].

The results of modelling of the above experiments by a reaction-diffusion-convection (ionic migration) model with the SH-kinetics were discussed in [9]. The SH model describes kinetics of the metabolism of low-molecular thiols [25]. The pulse waves were generated either by a permanent perturbation at the left hand boundary of the system. A periodic perturbation excited simple periodic, complex periodic and chaotic wave trains, depending on the frequency and amplitude of the perturbation.

Here we discuss the process of generation of pulse waves in detail. The observed (periodic or aperiodic) regimes will be characterized by a "firing number" ν, which can be formally defined as $\nu = \lim_{q \to \infty} \sup p/q$, where p is a number of travelling waves generated in the course of q external perturbations. First we briefly discuss the results of experiments showing that a dependence of the "firing number" on a parameter (an amplitude or a period of the perturbation) is a step-wise function, similar to a devil's staircase [10], i.e. to a continuous function which is constant on intervals (where it takes on rational values) and increasing at points (where it takes on irrational values). Then results of the modelling will be used to demonstrate that the parametric dependence of the "firing number" has also a devil's staircase-like structure and can be understood on the basis of the study of low-dimensional models describing the situation close to the point where the pulse waves are generated.

17.2 Experiments

A modification of the Belousov-Zhabotinski reaction mixture [7] (composition 0.286M $HBrO_3$, 0.007M KBr, 0.05M $CH_2(COOH)_2$ and 0.00375M ferroin) was used in experiments, which were conducted in a thermostated shallow layer of liquid (the height was aproximately equal to 1 mm). The waves were initiated by a periodically switched voltage applied between the silver and platinum electrodes immersed in the reaction mixture [11]. No generation of waves is observed when the silver electrode is negatively biased while a travelling wave may be formed when the silver electrode is positively biased [12]. The positive voltage difference determines the dissolution of silver ions which decrease the concentration of the bromide ions below a critical value and a wave can be generated [11,12].

The length LP of the positive voltage pulse and the period T between two successive voltage pulses were chosen as variable parameters. The firing number v is determined as the ratio of the number of waves generated to the number of voltage pulses applied within an experimental run. The dependence of the firing number v on the pulse "amplitude" LP is shown in Fig. 1a and the dependence on the forcing period T in Fig. 1b. The full circles denote periodic regimes and the empty ones irregular patterns of the wave trains. We can observe a typical stepwise dependence similar to a devil's staircase.

17.3 Mathematical Model

We consider a periodically perturbed, spatially one-dimensional

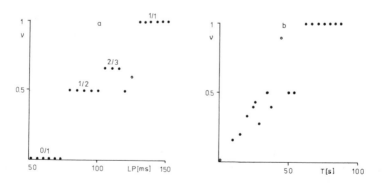

Figure 1. Dependence of the firing number v: (a) on the perturbation amplitude (LP - length of the pulse) for T = 50 s; (b) on the perturbation period T for LP = 90 ms. Experiments with the Belousov-Zhabotinski reaction.

two-component reaction-diffusion-convection system with the mass transport occurring by a molecular diffusion and an ionic migration in the constant electric field. The mass balance equations can be written in the form [7]

$$\frac{\partial X}{\partial t} = D_X \frac{\partial^2 X}{\partial z^2} + Z_X D_X U \frac{\partial X}{\partial z} + f(X,Y)$$

$$ (1) $$

$$\frac{\partial Y}{\partial t} = D_Y \frac{\partial^2 Y}{\partial z^2} + Z_Y D_Y U \frac{\partial Y}{\partial z} + g(X,Y), \quad z \in (0,L).$$

Here X, Y are concentrations of the active reaction components, D_X, D_Y their diffusion coefficients, Z_X, Z_Y corresponding electric charges, U is the intensity of the imposed electric field, t time and and z the spatial coordinate [9]. The excitable kinetics is modelled by the SH model

$$f(X,Y) = \alpha \frac{\nu_0 + X^\gamma}{1 + X^\gamma} - X(1 + Y)$$

$$g(X,Y) = X(\beta + Y) - \delta Y, \quad (2)$$

where α, δ, $\nu_0 > 0$ and γ, $\beta > 1$ are kinetic parameters. We choose zero flux boundary conditions for both components at z = L:

$$\frac{\partial X}{\partial z}\Big|_{z=L} = \frac{\partial Y}{\partial z}\Big|_{z=L} = 0. \quad (3)$$

Mixed boundary conditions for X and Y at z = 0 describe the periodic perturbation:

$$(1 - p(t)) \frac{\partial W}{\partial z}\Big|_{z=0} + p(t)(W\Big|_{z=0} - A_W)) = 0 \quad (4)$$

$$W = X, Y$$

Here p(t) is a T-periodic function consisting of rectangular pulses

$$p(t) = \begin{array}{l} 0 \text{ for } t \in [t_k, t_k + T_1) \\ 1 \text{ for } t \in [t_k + T_1, t_k + T), \quad t_k = kT, \end{array} \quad (5)$$

$$k = 0, 1 \ldots,$$

The function p(t) switches between two types of boundary conditions for X and Y at z = 0 and thus defines the periodic

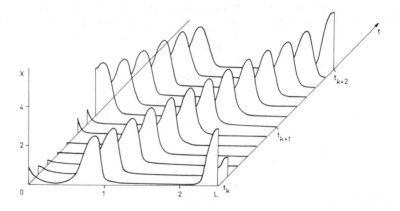

Figure 2. Pulse wave generation, spatio-temporal periodic pattern of the component X, $v = 1/2$, SH-model, Eqs. (1-5), A = 0.8, T = 3.1.

perturbation of the system with the period T. The "amplitude" of the perturbation is defined by two numbers A_X and A_Y. We have set $A_Y = 0.19$ and $\Delta T = T - T_1 = 0.6$ for the reported numerical computations; forcing is then described by two parameters, the amplitude $A = A_X$ and the period T.

Our basic assumption is that the unperturbed system (1) with zero flux boundary conditions on both ends has for $U = 0$ a unique stable spatio-temporally constant solution. Numerical simulations suggest that this happens for $\alpha = 12$, $\beta = 1.5$, $\gamma = 3$, $\delta = 1$, $v_0 = 0.01$, $D_X = 0.008$, $D_Y = 0.004$ and $L = 2.5$.

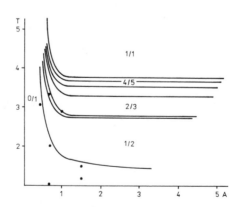

Figure 3. Approximate regions of main observed resonances in the parameteric plane (A, T), SH-model, Eqs. (1-5). Higher resonances and aperiodic patterns are scattered among the main resonances; 0 denote some locations of chaotic regimes.

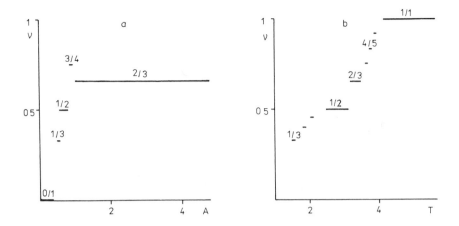

Figure 4.   Dependence of the firing number ν: (a) on the perturbation amplitude A for T = 3.1;   (b) on the perturbation period T for A = 0.8 SH-model, Eqs. (1-5).

17.4 Dynamics in the Absence of External Electric Field

First, we assume U = 0 and study the dynamics of the periodically perturbed system of Eqs. (1-5).   A typical periodic pattern generated by Eqs. (1-5) is shown in Fig. 2. This regime has the firing number ν = 1/2 which corresponds to one initiated travelling wave within two forcing periods.   The ranges of low order resonances observed in simulations are depicted in Fig. 3 (the boundaries are approximate).   Between two simplest resonances 0/1 and 1/1 are located other resonances arranged according to the Farey tree.   However, it is not clear, whether the overall structure of the higher order resonances satisfies also the Farey sequences.

The dependence of the firing number ν on A and T is shown in Fig. 4a,b.   The similarity of these step-wise plots to the devil's staircase is evident.   More detailed studies of the dynamics between the depicted resonances reveal patterns which do not periodically repeat after some integer multiple of the forcing period T.   An example of such an aperiodic dynamics is given in the form of the Poincaré maps in Fig. 5a,b,c.   The Figures were generated by taking three locations along the z axis and plotting X and Y at times immediately before the switching from the zero flux boundary conditions to the constant concentration boundary conditions (i.e. at successive times $t_k + T_1$) after transients converged close to an asymptotic behaviour.   A small scale folding can be observed in Figs. 5a,b,c, thus the dynamics can be considered as chaotic though it seems to be close to a quasiperiodic one.

The presence of the chaotic dynamics prevents a direct comparison with the classical devil's staircase known from the studies of invertible or critical circle maps, as is, for example, the map

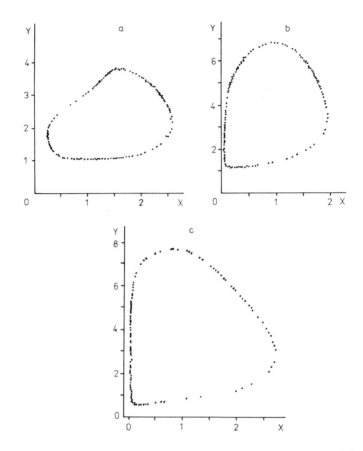

Figure 5.   Poincaré sections of a chaotic regime: (a) z = 0;
(b) z = 1.25;   (c) z = 2.5, A = 0.55, T = 3.1.

$$X_{k+1} = T + X_k + (A/2\pi) \sin 2\pi X_k \pmod 1, \text{ for } A \leq 1. \qquad (6)$$

The resonance regimes of (6) form cusp shaped regions in the parametric plane A, T – the Arnold's tongues [14] and the dynamics is quasiperiodic (not chaotic) among the resonances.   This structure cannot be identified in   Fig. 3.   Hence we cannot directly apply the results of the studies of the dynamics of circle maps to the phenomena observed in periodically perturbed excitable distributed systems.

An interesting aspect of the dependence of ν on A is the observation that below a critical value T_C of the forcing period T we cannot reach the 1/1 resonance even with arbitrary large values of the amplitude A, cf. Fig. 3, 4a.   The critical value of the period ($T_C \approx 3.8$ in Fig. 3) can be decreased if the diffusion coefficients are increased.   This is illustrated in Fig. 6a, which corresponds to Fig. 4a but with the diffusion interaction 80 times higher.

The increase of the diffusion interaction causes a "spreading out" of the travelling pulse waves and the concentration profiles are flattened. However, even if the pulse wave ceases to be a well defined object, it is still possible to differentiate between the "excited" state (high levels of concentrations of the components X, Y along the spatial axis) and the "non-excited" state (a relaxation to a stationary homogeneous regime). This enables one again to define the firing number ν.

The above behaviour suggests that the dynamics of the pulse wave excitation is controlled rather by the form of the kinetics relations than by the diffusion transport. To simplify the analysis we shall disregard the fact that a strong diffusion interaction means that both the shape and the velocity of the wave are not well defined. We shall assume that the excitation as a result of the interaction of diffusion and kinetics can occur only close to the left boundary of the system.

This strong simplification enables us to describe the system by two nonautonomous ordinary differential equations

$$\dot{X} = f\ (X, Y) + D_1\ p(t)\ (A_X - X)$$

$$\dot{Y} = g\ (X, Y) + D_2\ p(t)\ (A_Y - Y). \tag{7}$$

Here, similarly to the distributed system, A_X and A_Y are "perturbation amplitudes" and $p(t)$ is defined by eq. (5). However the diffusion coefficients D_1 and D_2 are not directly related to D_X and D_Y.

The firing number ν can be again defined as an asymptotic ratio of the number of successful external perturbations in the course

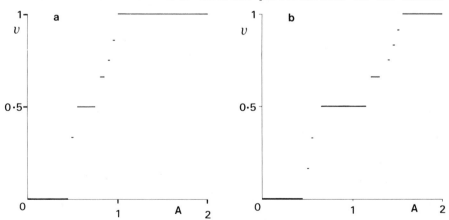

Figure 6. Dependence of the firing number ν on the amplitude A of perturbations for $T = 3.1$: (a) distributed system, Eqs. (1-5), $D_X = 0.64$, $D_Y = 0.32$; (b) two-dimensional model, Eqs. (5,7), $D_1 = 1$, $D_2 = 0.5$.

of which the system reaches an "excited state" to an overall number of external perturbations. In the present computations the excited state was determined from the condition that the value of the concentration Y must pass through Y = 5 from the left to the right. The concept of a rotation number can be used in the refining of the definition of ν [15]. The resulting dependence of ν on A is given in Fig. 6b. The similarity to the spatially distributed system (cf. Fig. 6a) as well as a devil's staircase-like structure can be observed.

The equations (5,7) define a two-dimensional Poincaré mapping. The devil's staircase-like dependence could be more easily studied if an appropriate one-dimensional mapping (possibly noninvertible) could be connected with the system. This connection can be made, for example, by an assumption that the system is controlled purely by kinetics, i.e. if we shall study the system of Eqs. (5,7) in the limit D_1, $D_2 \to \infty$ (or, equivalently Eqs. (1-5) for D_x, $D_y \to \infty$). However, we obtain a system defined by a one-periodic orbit $X_k = A_X$, $Y_k = A_Y$, $k = 0, 1, \ldots$, which does not have any degree of freedom. But if the zero-flux boundary condition $\frac{\partial Y}{\partial z}\big|_{z=0}$ = 0 is introduced instead of Eq. (5) for the component Y in the case of the distributed system or the condition $A_Y = Y$ for the system (5,7), we obtain under the assumption of an infinitely fast diffusion a one-dimensional noninvertible mapping [16]. This mapping can generate both periodic and chaotic solutions and a devil's staircase-like behaviour which is associated with the intermittency [17].

17.5 The Influence of External Electric Field

The inclusion of the terms describing the effects of the imposed external field on the mass transport transforms the pure reaction-diffusion problem into a reaction-diffusion-convection problem. Let us assume that the intensity U of the electric field is nonzero. Then the convection terms in Eq. (1) describing an ionic migration of the reaction components can be zero, positive or negative, depending on the charge Z of the component in question. The presence of charged components may have a profound effect on the dynamics of Eqs. (1-5). In particular "convective instabilities" [18,19] and a sensitive dependence on infinitesimal random perturbations [20] may occur.

The question of existence of travelling waves in an infinite system [21] can be answered by an application of the theorems by Kopell and Howard [22]. Examples of the effects of the intensity of the electric field U on pulse and front waves were dsicussed in [9,23].

Here we study the dependence of the firing number ν on U. In addition to fixed values of parameters of the Eqs. (1-5) defined earlier in the text we set $A_X = A = 0.8$, T = 3.1 and the charges $Z_X = Z_Y = -1$. We have found that the positively (negatively,

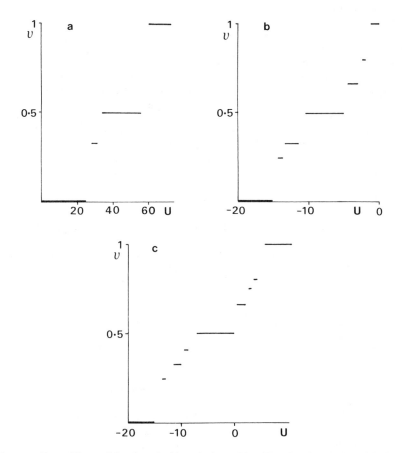

*Figure 7. The effect of the intensity U of electric field on
the firing number v; SH-model, Eqs. (1-5):*
(a)   resonance 0/1 for U = 0;   A = 0.3, T = 3.1;
(b)   resonance 1/1 for U = 0;   A = 0.8, T = 4.1;
(c)   resonance 1/2 for u = 0;   A = 0.8, T = 3.1.

resp.) directed electric field increases (decreases, resp.) the
firing number. Three examples are presented in Figs. 7a,b,c. The
1/1 regime can be reached from the 0/1 regime on applying a
positive U, cf. Fig. 7a, and vice versa, cf. Fig. 7b. The 1/2
resonance can be changed to the 1/1 resonance or to the 0/1
resonance on applying a positive or a negative U, respectively,
cf. Fig. 7c. All the dependences again have a devil's
staircase-like structure. Thus a positively (negatively, resp.)
oriented electric field has the same effect on the firing number
as an increased (decreased, resp.) diffusion interaction in the
absence of an electric field.

17.6 Conclusions

The phenomena observed both in the first round of experiments described above and in the modelling of a spatially one-dimensional system with SH-kinetics demonstrate the need for new refined experiments, directed to detailed observations of temporal and spatial patterns. Also the effects of electric field (representing a component selective type of convection) deserve a more detailed study. Spatially two-dimensional systems are also of interest; for example a spatial and temporal organization of biological systems, which is connected with the transfer of information via concentration pulses of specific substances [24] often occurs on quasi-two-dimensional surfaces.

References

[1]   R.J. Field and M. Burger (eds.) (1985). *Oscillations and Travelling Waves in Chemical Systems*, Wiley, New York.
[2]   V.S. Markin, V.F. Pastushenko and Ju.A. Chizmadzkev (1981). *Theory of Excitable Media (in Russian)*, Nauka, Moscow.
[3]   V.S. Zykov (1984). *Modelling of wave processes in excitable media (in Russian)*, Nauka, Moscow.
[4]   M.T. Grechova (ed.) (1981). *Autowave Processes in Systems with Diffusion (in Russian)*, IPF AN SSSR, Gorkii.
[5]   V.I. Krinski (1984). *Selforganization, Autowaves and Structures Far from Equilibrium*, Springer Verlag, Heidelberg.
[6]   H. Sevciková and M. Marek (1984). *J. Phys. Chem.* **88**, 2183.
[7]   H. Sevciková and M. Marek (1983). *Physica* **9D**, 140.
[8]   H. Sevciková and M. Marek (1984). *Physica* **13D**, 379.
[9]   H. Sevciková and M. Marek (1986). *Physica* **21D**, 61.
[10] M. Herman (1977) in *Geometry and Topology*, Vol. 597 of *Lecture Notes in Mathematics*, J. Palis,(ed.), Springer-Verlag, Berlin.
[11] L. Vroblová (1986). *M.Sc. Thesis*, Dept. of Chemical Eng., Prague Institute of Chemical Technology.
[12] K. Showalter, R.M. Noyes and H. Turner (1979). *J. Am. Chem. Soc.* **101**, 7463.
[13] G.M. Hardy and E.M. Wright (1965). *An introduction to the theory of numbers*, Clarendon Press, Oxford.
[14] V.I. Arnold (1961). *Izv. Akad. Nauk. SSSR, Ser. Mat.* **25**, 21.
[15] M. Marek and I. Schreiber, in preparation.
[16] M. Marek and I. Schreiber (1987). *Formation of periodic and aperiodic waves in reaction-diffusion systems*, Proc. of Conference "Bifurcation - Analysis, Algorithms, Applications", University of Dortmund, August, 18-22, 1986 R. Seydel and H. Tröger (eds.), Birkhäuser, Basel.
[17] P. Manneville and Y. Pomeau (1980). *Commun. Math Phys.* **74**, 189.
[18] E.M. Lifsic and L.P. Pitajevskij (1979). *Physical Kinetics (in Russian)*, Nauka, Moscow.
[19] M.I. Rabinovich and D.I. Trubeckov (1984). *An introduction to the theory of oscillations and waves (in Russian)*, Nauka, Moscow.

[20] R. J. Deissler (1985). *J. Stat. Phys.* **40**, 371.
[21]   A. Klic, M. Holodniok, M. Kubicek and M. Marek, *On the existence of travelling waves in reaction-diffusion-ionic migration systems,* to be published.
[22] N. Kopell and L. Howard (1975). *Advances in Math.* **18**, 306.
[23] H. Sevciková, M. Kubicek and M. Marek (1983). *Concentration waves - effects of an electric field,* Proc. of IV Int. Conf. on Math. Modelling, Zürich, August 1983, P. Avula (ed.).
[24]   A. V. Holden (1976). *Biol. Cybernetics* 21, 1; (1983). *Bull. Math. Biol.* **45**, 443.
[25] E. E. Sel'kov (1976). *Biofizika* **15**, 1076 (in Russian).

GROWTH AND DECAY OF DENDRITIC MICROSCTRUCTURES

Abstract

The effect of capillarity-driven diffusion on the recalescence of mushy zones is studied in pure materials. A global description of this phenomenon is presented which relates a characteristic lengthscale of the microstructure, $\bar{R}$, to the mean undercooling of the mixture, $<\Delta T>$. Experiments on mushy zones in succinonitrile, ethylene carbonate and ice/water indicate that the mean undercooling decays as $t^{-1/3}$, implying that the characteristic lengthscale of the system grows as $t^{1/3}$. This result is in agreement with the interpretation of coarsening as a statistical ripening phenomenon.

18.1 Introduction

Solidification of a material from the melt often produces a partially solidified region, consisting of a solid microstructure embedded in a continuous liquid phase. This region can be formed just ahead of the advancing solid front, as in dendritic growth, or further away in the supercooled liquid by local nucleation and growth of the solid phase. The two-phase region is geometrically complex, consisting of many lengthscales in varying spatial orientations, and is known as a "mushy zone".

Mushy zones are dynamic regions in which there occur many interactions resulting from local gradients in free energy and composition. These gradients arise primarily from interfacial curvature (capillarity) effects in the two-phase system. The presence of curved solid/liquid interfaces throughout the fine solid-liquid microstructure leads to changes in the local equilibrium concentrations in alloys, and to variations in the local equilibrium temperature in pure materials. The magnitudes of these local interfacial fluctuations are given by the Thomson-Freundlich and Gibbs-Thomson equations [1] and vary as the mean local curvature, K. The dispersion of interfacial curvature tends to excite fluxes of solute or enthalpy among the different interfacial regions. These fluxes reduce the local mean curvatures through remelting of highly-curved regions, while further solidification occurs at flatter regions. Such interactions have been observed both during growth of the solid

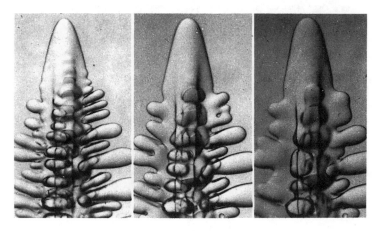

Figure 1.

phase and during isothermal holding of the two-phase region [2], and are similar to Ostwald ripening phenomena which occur among isolated domains embedded in a diffusing matrix [3].   An example of isothermal coarsening is shown in Fig. 1 for the case of a single dendrite of pure succinonitrile, in which the blunting of highly curved features due to local thermal fluxes is clearly visible.

The melting of highly-curved regions and reduction of the mean interfacial curvature leads to a coarser microstructure composed of relatively longer lengthscales.   The dynamics of this coarsening process are of particular interest because they markedly affect the resultant compositional segregation and microstructure of the completely solidified material [2,4,5]. This phenomenon has been modelled locally in a binary system by Reeves and Kattamis [6], and by Whisler and Kattamis [7].   This study will focus on the coarsening phenomenon in mushy zones of pure materials (maintained at constant solid-phase volume fraction) where compositional effects are absent and the lengthscale evolution is driven only by a reduction in the total interfacial energy of the two-phase system.   A description of the coarsening process will be presented here which avoids simplifying geometric assumptions, and instead characterizes the dynamics in terms of global parameters of the mushy zone such as volume fraction of the solid phase, V_v, and the total interfacial area per unit volume, S_v.

18.2 Global Description of the Coarsening Process

Coarsening in the mushy zone of a pure material is driven primarily by fluxes of heat excited by temperature gradients between interfacial regions of differing curvature.   The relationship between the curvature at a point on the interface and the local equilibrium temperature is given by the Gibbs-Thomson equation, viz.

$$T_\infty - T_e = \Delta T_e = \Gamma K \qquad (1)$$

where T_∞ is the nominal (bulk) melting point of the material, T_e is the local equilibrium temperature of a curved region, Γ is a material-dependent constant, and K is the local total curvature, defined as the sum of $1/r_1$ and $1/r_2$, where r_1 and r_2 are the principal radii of curvature. The constant Γ is equal to $\Omega\gamma/\Delta S_f$, where γ is the solid/liquid interfacial energy, Ω is the molar volume, and ΔS_f is the molar entropy of fusion. The Gibbs-Thomson equation expresses the equlibrium condition for freezing or melting at a curved solid-liquid interface. This equilibrium condition requires that the energy needed to change a differential volume of solid to liquid, given by $(\Delta T_e S_f/\Omega)dV_s$, is balanced by the corresponding change in stored interfacial energy, given by γdA. The ratio (dA/dV_s) at a point on a curved interface is simply K, the local total curvature.

The local Gibbs-Thomson equation applies at each point on the solid/liquid interface, which leads to temperature gradients throughout the two-phase system. Scaling arguments indicate that diffusion of heat induced by these gradients occurs on a much faster timescale than the rate of change of temperature gradients caused by the slowly varying interfacial curvatures. The diffusion of enthalpy through the slowly changing temperature field is therefore quasi-static, and this temperature field can be described by Laplace's equation,

$$\nabla^2 T = 0. \qquad (2)$$

The liquid-phase temperature field is fully specified at any time t through the application of the Gibbs-Thomson equilibrium condition at each point of the solid/liquid interface.

If the volume fraction of the solid phase, V_v, is held constant, then the diffusion-controlled coarsening in a mushy zone is analogous to an Ostwald ripening process, with the particle size distribution of a ripening system replaced by an equivalent curvature distribution [8]. The range of interfacial temperatures can be considered to interact in a statistical sense with an intermediate, mean temperature determined by the mean curvature of the entire microstructure. This mean temperature can be expressed by the relation

$$\langle \Delta T \rangle \propto 1/\bar{R}, \qquad (3)$$

where $\langle \Delta T \rangle$ is the global mean undercooling of the mushy zone and is proportional to some mean curvature, $1/\bar{R}$.

An alternative derivation of this proportionality can be had through a direct application of the Gibbs-Thomson relation, Eq.

(1), to the entire mushy zone. For the case of constant volume fraction of solid, the global value of (dA/dV_s) is simply S_v, the specific interfacial area of the mushy zone. The global Gibbs-Thomson relationship can then be written as

$$<\Delta T> = \Gamma S_v, \tag{4}$$

where S_v, like K, has units of $(length)^{-1}$. The global parameter S_v for steady-state scaled curvature distribution of a collection of spheres is given by DeHoff and Iswaran [9] as

$$S_v = (3V_v S_2/S_3)(1/\bar{R}), \tag{5}$$

where S_2 and S_3 are functions of the scaled curvature distribution, and the factor of 3 is a shape factor specific to spheres and would vary somewhat for more complex geometries [10]. Eq. (5) indicates that S_v is proportional to $1/\bar{R}$ for the case of constant volume fraction. This result may be combined with the global Gibbs-Thomson condition, Eq. (4), to yield the proportionality expressed by (3).

The mean undercooling of the mushy zone, $<\Delta T>$, which arises from capillary effects at the curved solid/liquid interfaces, is seen to be proportional to $1/\bar{R}$, where $\bar{R}$ is a characteristic lengthscale of the microstructure at a given time t. The growth of $\bar{R}$ in a complex microstructure at constant volume fraction of solid (or, equivalently, the decrease in S_v) has been found to obey a power law of the form

$$\bar{R}^3(t) = \bar{R}^3(0) + kt, \tag{6}$$

where $\bar{R}(0)$ is the initial mean lengthscale at t = 0, t is time, and k is a volume-fraction dependent rate constant. This result has been obtained by Voorhees [11] in the numerical simulations of Ostwald ripening between spheres, and has been suggested by Mullins [12] based solely on the assumption of self-similar lengthscale distribution, regardless of the specific geometry of the system.

The $\bar{R}^3(0)$ term in Eq. (6) will be rapidly dominated by the $\bar{R}^3(t)$ term, because an initially fine microstructure in a mushy zone will coarsen rapidly. Thus, equation (6) can be well-approximated by the asymptotic form

$$\bar{R}(t) \stackrel{\sim}{=} (kt)^{1/3}. \tag{7}$$

Inserting this dynamic behaviour of $\bar{R}$ into Eq. (3) yields the relationship

$$<\Delta T> = C\, t^{-1/3}, \tag{8}$$

Inserting this dynamic behaviour of $\bar{R}$ into Eq. (3) yields the relationship

$$<\Delta T> = C \, t^{-1/3}, \tag{8}$$

where C is a rate constant which depends on both volume-fraction of solid and the material being studied. Equation (8) is a global relation characteristic of the mushy zone as a whole. This equation describes a thermodynamic effect resulting from the mean lengthscale of the microstructure, which grows as $t^{1/3}$.

Scaling arguments indicate that the rate constant C of Eq. (8) is of the form

$$C = \left[\frac{1}{C_P \kappa}\right]^{1/3} \left[\frac{\gamma \Omega}{\Delta S_f}\right]^{2/3} H(V_v), \tag{9}$$

where L is the latent heat of fusion, C_p is the heat capacity of the liquid, and κ is the thermal diffusivity. $H(V_v)$ is a statistical factor of unit order which accounts for the volume fraction effects which influence both the local diffusion rates and scaled curvature distributions in the microstructure. In the next section, an experimental technique will be described which enables this slow recalescence of the mean temperature toward the nominal melting point to be measured.

18.3 Experimental

18.3.1 *Equipment*

A method has been developed by the authors to create and maintain a large, uniform mushy zone in a glass ampoule. Coarsening studies were then performed on three different pure materials. These materials are succinonitrile (SCN), ethylene carbonate (ETC), and ice/water (H_2O). SCN is a BCC material which forms a highly-branched dendritic microstructure, and melts at about $58.08^{\circ}C$. ETC has a melting point of about $34.34^{\circ}C$, and forms long, branchless blades or needles when solidifying from the melt. Ice forms hexagonal planar dendrites and melts at $0.01^{\circ}C$ under its own vapour pressure.

Purification of SCN and ETC was accomplished by distillation under vacuum followed by a two-stage refining process. The purified materials were then directly transferred into glass ampoules containing reentrant thermometer wells, and sealed under vacuum. This process is described in more detail elsewhere [13]. The H_2O sample used distilled, deionized water which was slowly frozen and thawed three times under open vacuum (to drive out trapped gases), and then distilled into a glass ampoule and sealed hermetically under its own vapour pressure.

The thermometric elements used to measure the mean mushy zone

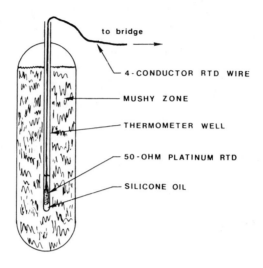

Figure 2.

resistance to the International Practical Temperature Scale of 1968 (IPTS-68) [14]. These thermometer elements were placed near the bottom of the thermometer well in each ampoule. A drop of silicone oil was then added to improve thermal contact between each element and the surrounding glass well. The design of these ampoules is illustrated in Fig. 2.

Thermometer resistance is measured with a Tinsley Automatic Thermometer Bridge. This digital resistance bridge has a resolution of 10^{-5} ohm, and was chosen because of its remarkable stability and negligible long-term zero-drift. The conversion of measured resistance values to temperature readings results in a temperature resolution of about 5×10^{-5} K. The resistance bridge is interfaced with a microcomputer which records resistance measurements at pre-set time intervals. Resistance readings are taken at a rate of 6 per second over short time intervals about each data point and averaged. The duration of the averaging interval becomes longer as the experiment progresses and the time between data points increases. This procedure allows some interpolation of the sixth decimal place in resistance values, and improves the temperature resolution to about 3×10^{-5} K at a mean temperature of approximately 300 K, which is an overall resolution of 1 part in 10^{7}.

18.3.2 *Procedure*

To provide a uniform mushy zone in an ampoule for a coarsening experiment, the specimen ampoule is initially immersed in a water bath that is about 15 k above the melting point of the test material. After complete melting of the material, the ampoule is transferred to a controlled water bath which is set at some specified initial supercooled temperature. Once the melt has reached a uniform supercooled temperature (as measured by the platinum element), the ampoule is removed from the bath. The solid phase is immediately nucleated by applying a short burst of cold aerosol to the bottom of the ampoule. The solid phase then grows rapidly upward through the supercooled melt, forming a fine, uniform microstructure embedded in the liquid phase. The entire mixture recalesces rapidly to within a few mK of the melting point. The initial volume fraction of solid formed in this rapid phase transformation is given by the relation

$$V_v = \frac{\Delta T_o}{(L/C_p)},\qquad (10)$$

where ΔT_o is the initial uniform undercooling of the melt. Thus, the volume fraction of solid in the mushy zone can be controlled conveniently through the initial temperature of the melt. This volume fraction is maintained constant during the course of an experiment by immediately immersing the solid/liquid mixture into a second controlled-temperature bath which is set within 2 mK of the melting point. This provides an essentially isothermal environment around the ampoule, and the resulting minimization of external heat flow prevents any further net solidification or melting.

At the moment of solid-phase nucleation, the data-recording routine on the microcomputer is started. Time-temperature data are automatically recorded at equal intervals of 0.0025 sec$^{-1/3}$ on a $t^{-1/3}$ time scale, beginning at 125 seconds after nucleation. After sufficient time has elapsed, the experimental run is terminated. The sample may then be removed from the thermostatted bath and re-melted, and the entire procedure can then be repeated

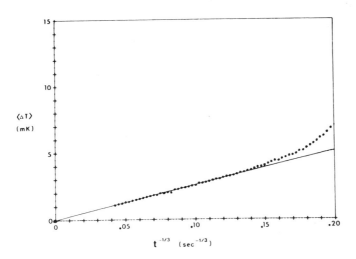

Figure 3.

either at the same or at a different solid volume fraction.

18.4 Discussion of Results

A number of experimental coarsening runs have been performed on SCN and ETC mushy zones, as well as on ice/water mixtures. The solid volume fractions studied ranged from about 5% to 20%, and are limited on the high end by the amount of initial undercooling that can be achieved without spontaneously nucleating the solid phase. Data for a typical ETC run is plotted in Fig. 3, and similar data acquired for an SCN mushy zone is shown in Fig. 4. time increases non-linearly from right to left on the horizontal axis, with the origin representing the asymptotic equilibrium of zero undercooling at infinite time ($S_v \to 0$). The straight line through each plot represents the capillary-controlled recalescence in the mushy zone, which shows excellent agreement with the theoretically predicted $t^{-1/3}$ dependence discussed earlier. Similar behaviour has also been observed in the ice/water system. The mean undercooling $<\Delta T>$ was shown in Section 18.2 to be proportional to $1/\bar{R}$, where $\bar{R}$ is a characteristic lengthscale of the microstructure (related to S_v^{-1}). Figures 3 and 4 also demonstrate that there is a characteristic lengthscale in the complex microstructure which grows as $t^{1/3}$.

The undercooling is seen to be larger at early times in both of these data plots than would be predicted by the straight-line power-law fit. This initial behaviour is caused by a transient that arises from the slow thermal response of the glass thermometer well and platinum element. In the inital undercooled state, both the melt and thermometer structure are at a uniform temperature, $T_\infty - \Delta T_0$. After the nucleation and rapid growth of the solid phase, the temperature of the mushy zone quickly rises to

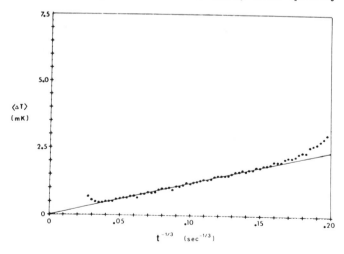

Figure 4.

within a few mK of the equilibrium temperature T_∞ because of the released heat of fusion.   However, to track this recalescence, some additional heat must be transferred from the two-phase region to the thermometer well to raise its temperature close to T_∞. This transient lasts about 300 seconds in Figs. 3 and 4.   Other data,   not   presented   here,   show   that   this   transient   thermal response becomes somewhat larger in magnitude and longer in duration for higher initial undercoolings, and less significant at smaller undercoolings.   This transient also seems to depend upon the thermal properties of the two-phase mixture, being shorter in materials having larger thermal diffusivities.

The temperature resolution required to monitor microstructural coarsening thermally is illustrated by some numerical data for the SCN coarsening experiment of Fig. 4 in Table 1.   The upper set of points   occur   during   the   initial   transient   response   of   the thermometer, while the lower set shows the asymptotic $t^{-1/3}$ power-law coarsening.   The use of thermometer structures with smaller thermal mass is being investigated to improve on the thermal response and extend the range of useful data.   It is also possible to correct for the transient mathematically (based on a step-function response of the thermometer element) in a manner outlined recently by Dantzig [15].   Voorhees and Glicksman [16] have concluded from numerical simulations that in order to track the mean temperature in a coarsening solid-liquid mixture, the diameter of the thermal sensor must be larger than about 20 R̄. (Otherwise,   local   fluctuations   become   significant.)   This condition is easily met with the present apparatus, and the smoothness of the data confirm this.   Because of the extremely fine size scale of the rapidly-formed mushy zone, much smaller thermometer elements could be used without introducing excessive statistical noise from the thermal fluctuations occurring in the partially solidified mixture.

The slopes of the lines in Figs. 3 and 4 can be viewed as coarsening rate constants, whereby the increasing lengthscales in the microstructure manifest themselves by reducing the global undercooling $\langle\Delta T\rangle$.   It has been observed in all three materials that this slope increases with increasing volume fraction, V_v.

This effect is physically reasonable as the mean interfacial separation is reduced at higher values of V_v, resulting in steeper gradients   of   temperature   and   a   faster   diffusion-controlled coarsening   process.   More   detailed   statistical   analysis   is currently underway to quantify the relationship between the observed rate constants and the volume fraction of solid.

The data in Fig. 4 indicate a deviation from linearity in the $t^{-1/3}$ plot at late times.   This effect is caused by gravitational settling of the denser solid phase towards the bottom of the ampoule.   Settling increases the amount of solid near the thermometer element, thereby increasing the interfacial density S_v.   This effect tends to increase the local mean undercooling $\langle\Delta T\rangle$, which can slow down or even reverse the diffusion-controlled recalescence at late times.   Significant settling of the solid was noticeable in the SCN ampoule after about 5,000 seconds and was less visible in the ETC mushy zones. In ice/water mixtures, the

Table 1: Time-Temperature Data for a Typical SCN Coarsening Experiment

Data Point #	Time (secs)	Temperature ($^{\circ}$C)
1	125	58.09534
2	129	58.09554
3	134	58.0957
4	140	58.0958
5	145	58.0959
6	151	58.09596
7	157	58.09601
.	.	.
.	.	.
.	.	.
35	657	58.09724
36	702	58.09727
37	751	58.09725
38	805	58.0973
39	863	58.09733
40	928	58.09737
41	1000	58.0974

less dense ice tended to float upward at long times, leading to a reverse effect as the solid floats away from the element.

Small variations of less than 1 mK in the extrapolated values of T_{∞} were observed in each material between different runs. These variations arise primarily from thermally-induced microstrains in the platinum resistor elements which probably occur during the rapid initial recalescence. These strains tend to increase the nominal resistance R_o of the platinum resistors. Such small changes in the R_o values produce a uniform offset in the calculated temperature over the very small experimental temperature ranges encountered. The relative precision of the data points remains unaffected by these microstrains. In this sense, the observation of capillary induced recalescence involves relative temperature measurements with high resolution, rather than absolute temperature measurements with high accuracy.

18.5 Conclusions

The following conclusions can be drawn from the results presented here:

(a) The progress of microstructural coarsening in rapidly solidified pure materials can be followed in situ via recalescence effects using coarse grained, i.e. macroscopically averaged, thermometric measurements of the two-phase mixture.

(b) The measured mean temperature in the two-phase mushy zones can be theoretically related to the interfacial area per unit volume through the Gibbs-Thomson equation and the statistical distribution of length scales in the microstructure.

(c)   The diffusion-controlled recalescence displays $t^{-1/3}$ kinetics after the initial transient response of the thermal sensor. This time dependence is in agreement with theoretical predictions based on coarsening kinetics of discrete spherical particles. These systems are related by the analogy between the size distribution of ripening spheres and the statistical lengthscale distribution of the more complex solid-liquid microstructures examined.

(d)   Identical diffusion-controlled kinetics were observed for the disparate morphologies obtained with SCN, ETC and ice/water indicating that it is the statistical lengthscale distributions, and not the details of the microstructure, that govern the coarsening. In each of these materials, the rate constant is observed to increase monotonically with increasing volume fraction of the solid phase.

Acknowledgement

The authors wish to acknowledge the support provided for this work by the National Science Foundation under research grant DMR83-08052.

References

[1]   R.K. Trivedi (1975) in *Lectures on the Theory of Phase Transformations,* H. Aaronson (ed.), The Metallurgical Society/AIME, New York, 51-81.

[2]   T.Z. Kattamis, J..C. Coughlin and M.C. Flemings (1967). *Tran. AIME* **234**, 1504-1511.

[3]   P.W. Voohees (1985). *J. Stat. Phys.* **38**, 231-252.

[4]   T.Z. Kattamis and M.C. Flemings (1966). *Tran. AIME* 236, 1523-1532.

[5]   M. Basaran (1981). *Met. Trans.* **12A**, 1235-1243.

[6]   J.J. Reeves and T.Z. Kattamis (1971). *Scripta met.* **5**, 223-230.

[7]   N.H. Whisler and T.Z. Kattamis (1972). *J. Crystal Growth* **15**, 20-24.

[8]   M.E. Glicksman and P.W. Voorhees (1984). *Met. Trans.* **15A**, 995-1001.

[9]   R.T. DeHoff and C.V. Iswaran (1982). *Met. Trans.* **13A**, 1384-1395.

[10]   R.T. DeHoff, private communication.

[11]   P.W. Voorhees (1982). *Ostwald Ripening in Two-Phase Mixtures,* Ph.D. Dissertation, Rensselaer Polytechnic Institute.

[12]   W.W. Mullins (1986). *J. Appl. Phys.* **59**, 1341-1349.

[13]   M.E. Glicksman, R.J. Schaefer and J.D. Ayers (1976). *Met. Trans.* **7A**, 1747-1759.

[14]   The International Practical Temperature Scale of 1968 (1969). *Metrologia* 5, 35-44.

[15]   J.A. Dantzig (1985). *Rev. Sci. Instrum.* **56**, 723-725.

[16]   P.W. Voorhees and M.E. Glicksman (1985). *J. Crystal Growth* **72**, 599-615.

19 *L.F. Mollenauer*

SOLITONS IN OPTICAL FIBRES: AN EXPERIMENTAL ACCOUNT

Abstract

This paper reviews the experimental study of solitons in optical fibres.   In addition to describing the first experimental observation of fibre solitons, it relates invention of the soliton laser, discovery of the soliton self frequency shift, the study of interaction forces between solitons, and the possible use of solitons in telecommunications.   Throughout, it is shown how experiment has provided important correction to predictions of the nonlinear Schrödinger equation, and has stimulated new theoretical development.

19.1 Introduction

In optical fibres, solitons are nondispersive light pulses based on nonlinearity of the .fibre's refractive index.   Such fibre solitons have already found exciting use in the precisely controlled generation of ultrashort pulses, and they promise to revolutionize telecommunications.   In this paper, I shall describe those developments, and the experimental studies they have stimulated or have helped to make possible.   Thus, besides the first experimental observation of fibre solitons, I shall describe invention of the soliton laser, discovery of a steady down shift in optical frequency of the soliton, or the "soliton self frequency shift", and the experimental study of interaction forces between solitons.

As early as 1973, Hasegawa and Tappert pointed out [1] that "single-mode" fibres - fibres admitting only one transverse variation in the light fields - should be able to support stable solitons.   Such fibres eliminate the problems of transverse instability and multiple group velocities from the outset, and their nonlinear and dispersive characteristics are stable and well-defined.   The first experiments [2], however, had to wait a while, for two key developments of the late '70s.   The first was fibres having low loss in the wavelength region where solitons are possible, and the second was a suitable source of picosecond pulses, the mode-locked colour centre laser.

But the first experiments led almost immediately to further developments.   For one, the manifest abilities of single mode fibres to compress and shape pulses led to invention and development of the soliton laser [3]. The precisely shaped pulses from that laser in turn provided the key to the further experimental work mentioned above. Additionally, by demonstrating that there were no hidden problems in the generation of fibre solitons, the first experiments stimulated much thought about the possibility of creating an all optical, soliton-based communications system [4]. Such ideas have been further amplified and tested by computer simulation [5,6] and by real world experiment [7].

Many ideas about solitons were first worked out (and in some cases discovered or conceived) in highly abstract and purely mathematical terms. But the existence of simple and clear experimental results has stimulated the quest for, and has often generated more direct explanations. Such simple physical models often yield valuable insight and stimulate further invention. In the following, I shall refer to such models wherever possible.

19.2 Dispersion, Nonlinearity and Pulse Narrowing in Fibres

The dispersive qualities of quartz glass and the loss per unit length of the best single-mode fibres presently available are both shown in Fig. 1. As will be demonstrated shortly, soliton effects are possible only in the region of "negative" group velocity

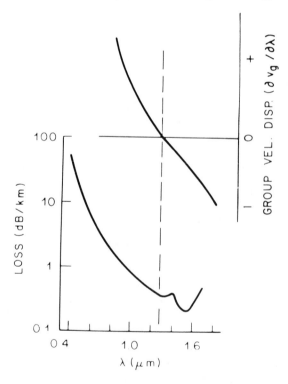

Figure 1. Loss of best single-mode fibres and group velocity dispersion of quartz glass as functions of wavelength.

dispersion ($\partial V_g/\partial\lambda<0$).

As shown by Fig. 1, such dispersion occurs only for wavelengths greater than ≈ 1.3 μm. (The net dispersion of a given fibre is also determined by the ratio of the core diameter to the wavelength, but such waveguide contributions can only push the zero of dispersion to longer wavelengths. In so-called "normal" fibre, the net dispersion is little different from that shown in the figure.) Note that the region of negative dispersion includes the region ($\lambda \sim 1.5$ μm) of lowest energy loss (the minimum loss can be as low as 0.16 db/km). In the experiments to be described, the central frequency of the pulse has usually been at or near that loss minimum.

Dispersion alone - regardless of sign - always causes the higher and lower frequency components of a pulse to separate, and thus always serves only to broaden the pulse. Pulse narrowing, and by extension, solitons, are made possible by the fact that the medium is nonlinear, that is, the index of refraction is a function of the light intensity:

$$n = n_0 + n_2 I \tag{1}$$

For quartz glass, n_2 has the numerical value 3.2×10^{-16} cm^2/W and I is the light intensity in compatible units.

The effect of the index nonlinearity is to produce "self phase modulation" [8]. That is, a wave will experience phase retardation in direct proportion to the intensity-induced change in index and (temporarily assuming negligible pulse reshaping) to the length of fibre traversed:

$$\Delta\phi = \frac{2\pi}{\lambda} Ln_2 I \tag{2}$$

In a pulse such as that shown in Fig. 2a, the phase retardation will be greatest at its peak. Thus there will be a crowding together and spreading apart of waves in the trailing and leading halves of the pulse, respectively, resulting in the frequency distribution shown in Fig. 2b. When such a "chirped" pulse is acted upon by the fibre's negative group velocity dispersion, the leading half of the pulse, containing the lowered frequencies, will be retarded, while the trailing half, containing the higher frequencies, will be advanced, and the pulse will tend to collapse upon itself as shown in Fig. 2c. If the peak pulse intensity is high enough, such that the chirp is large compared to that produced by dispersion, the degree of pulse narrowing can be substantial.

19.3 **The Nonlinear Schrödinger Equation and Solitons**

In writing down a differential equation to describe nonlinear pulse propagation in fibres, it is convenient to eliminate terms containing the central optical frequency from the wave equation. One thereby obtains an equation involving only the pulse envelope function u. The process involves expanding dispersive terms about

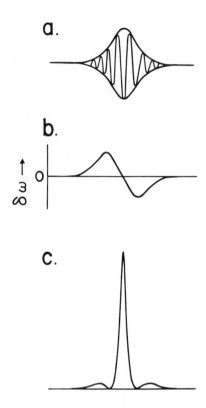

a.

b.

$\uparrow \dfrac{\partial \phi}{\partial t}$ 0

c.

Figure 2. (a) Optical pulse that has experienced self phase modulation. (b) Corresponding frequency chirp. (c) Resultant compressed pulse in a fibre with negative group velocity dispersion. (See text.)

the central frequency. When only the leading terms are kept, the result is the nonlinear Schrödinger equation:

$$-i\frac{\partial u}{\partial \xi} = \frac{1}{2}\frac{\partial^2 u}{\partial s^2} + |u|^2 u \qquad (3)$$

where $\xi = z/z_c$ and $s = (t - z/v_g)/t_c$, represent distance of progagation along the fibre and (retarded) time measured in the soliton units z_c and t_c, respectively.

Experimental discovery of the soliton self frequency shift (see Section 19.6) implies that the nonlinear term of (3) is actually time-dependent. The appropriate correction (see (10)) becomes most important in the "femtosecond" regime, and its effects are often much greater than those of higher order dispersive terms. Thus, (3) is fairly accurate for pulses of at least a few ps duration, and when the central frequency is well away from that of zero dispersion. The representation clearly becomes worse for very short pulses, or where the dispersion varies rapidly with

frequency.
  The fundamental soliton is the following special solution to (3);

$$u(z,t) = A \; sech(s/A)e^{i\xi/2} \qquad (4)$$

where A is often taken to be unity. Note that the fundamental soliton does not change shape as it propagates along the fibre. (Except for the phase term, (4) is independent of ξ). Physically, it represents a condition of exact cancellation between the chirp produced by the nonlinearity, and that produced by dispersion. (In other words, the pulse shape and amplitude are such that the last two terms of (3) cancel, except for that part that generates the phase term.)
  Note that area of the soliton, defined as $S = \int|u|ds$, is independent of A. Thus, S is an important constant of the motion ($S = \pi$) for solitons. The stability of solitons is often best expressed in terms of area concepts. For example, any pulse of reasonable shape, for which $\pi/2 \leq s \leq 3\pi/2$, will eventually turn into a (fundamental) soliton. This is an important concept, as it shows how easy it is to create solitons, or to maintain them once they have been created.

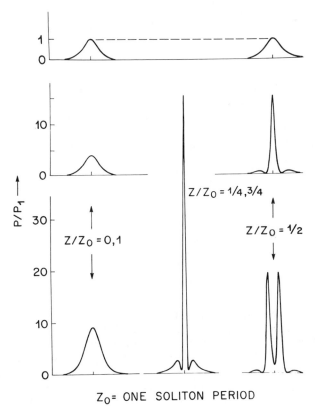

Z_0= ONE SOLITON PERIOD

Figure 3.   Theoretical behaviour of the fundamental (N = 1) and two higher order solitons with propagation.

One usually has $t_c = \tau/1.76$, where τ is the intensity FWHM (full width at half maximum) of the fundamental soliton with $A = 1$. Then one has

$$z_c = 0.322 \; \frac{2\pi c}{\lambda^2} \frac{\tau^2}{D} \tag{5}$$

where λ is the vacuum wavelength, and where the dispersion parameter D reflects the change in pulse delay with change in wavelength, normalized to the fibre length. (For "normal" fibre, $D \sim 15$ ps/nm/km at 1.5 μm.) Finally, the peak power corresponding to the fundamental soliton is given by the expression:

$$P_1 = 0.776 \; \frac{\lambda^3}{\pi^2 c n_2} \frac{D}{\tau^2} A_{eff} \tag{6}$$

where n_2 has the numerical value given earlier, where A_{eff} refers to the fibre core area, and where c is the speed of light in vacuum.

An important set of solutions to (3) result from an input function of the form:

$$u(0,s) = N \, \text{sech}(s) \tag{7}$$

where N is an integer, the "soliton number". A few of those solutions are shown graphically in Fig. 3. Note that $N = 1$ yields the fundamental soliton, already discussed.

For integers $N \geq 2$, sech input pulses always lead to pulse shaping that is periodic with period $\xi = \pi/2$. (The periodicity sonner or later breaks down, however, mainly because of the time-dependence (10) of the nonlinear term.) In real space, the period is

$$z_0 = \frac{\pi}{2} z_c \tag{8}$$

The peak input powers are, of course,

$$P_N = N^2 P_1 \tag{9}$$

For $N = 2$, the behaviour is particularly simple: the pulse alternately narrows and broadens, achieving minimum width at the half period. For greater N, the behaviour becomes more complex, but always consists of a sequence of pulse narrowings and splittings (see Fig. 3).

Although the $N = 1$ soliton is unique, the $N = 2$ and $N = 3$ solitons shown in Fig. 3 each represent but one member of a continuum. For example, the $N = 2$ soliton can be looked upon as

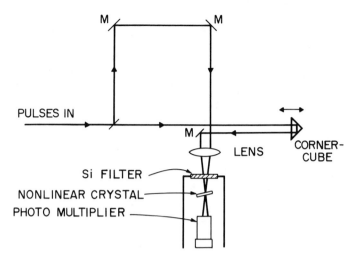

Figure 4. Schematic of apparatus for observing pulse shapes in autocorrelation. Variable delay is accomplished through translation of the corner cube. The Si filter passes 1.5 μm light but keeps out visible room light.

a nonlinear superposition of two fundamental solitons; the continuum of solutions is obtained by varying the relative amplitudes and widths of the two components. (The particular N = 2 soliton shown in Fig. 3 corresponds to components with amplitude and width ratios of 3:1 and 1:3, respectively, and it is the only N = 2 soliton to pass through the sech shape at any point in its period.)

Finally, we note that z_0 (or z_c) has a physical interpretation independent of solitons or other nonlinear effects: it represents the distance an initially minimum bandwidth pulse of width τ must travel to be dispersively broadened by a factor on the order of two. Thus, z_0 is of universal importance as a scaling factor. It is the most important and all-pervasive parameter of pulse propagation in optical fibres.

19.4 Experimental Verification

To verify the predicted soliton effects, it is necessary to observe the shapes of pulses as they emerge from a length of fibre. The pulse shapes can be observed by autocorrelation. In that technique, the beam is divided into two roughly equal parts, which, after travelling separate paths, are brought together in a nonlinear crystal; see Fig. 4.

In the arrangement shown there, second harmonic light is generated only if pulses from both beams are simultaneously present in the crystal. Thus the strength of the second harmonic (registered by the photomultiplier) reflects the temporal overlap of the pulses in the two converging beams. A measurement of second harmonic intensity as a function of relative delay then yields the pulse shape in autocorrelation.

As can be seen from Fig. 3, the most varied changes in pulse shape (with changing input power) are to be observed at the output end of a fibre whose length is one-half the soliton period. The first experiments [2] were conducted with such a half-period fibre; Fig. 5 summarizes the results. The autocorrelation trace labelled "laser" describes the colour centre laser output (fibre input) pulses, and corresponds to $\tau = 7$ psec and an approximately sech^2 shape.

The lower row of Fig. 5 shows the experimentally determined fibre output pulse shapes at certain critical power levels, where one sees, respectively, the expected half-period behaviour of the fundamental and several higher-order solitons.

The actual length (700 m) of fibre used in the above experiment agrees rather well with the half-period ($z_0/2 = 675$ m) calculated from (5) and (8) for $\tau = 7$ psec and for the fibre's dispersion, D = 15 psec/nm/km. The calculated value of P_1 also agrees well with experiment. From (6), $P_1 = 1.0$ W for the parameters just cited and for an effective core area $A_{eff} \approx 1 \times 10^{-6}$ cm^2, whereas the average of P/N^2 for the first three solitons yields $P_1 = 1.2$ W.

In a similar experiment [9] carried out with a full-period length of fibre, it has been possible to demonstrate directly the periodicity of the higher order solitons. In that experiment, at the critical power levels for solitons, both the pulse shapes and the pulse frequency spectra were observed to return to the input

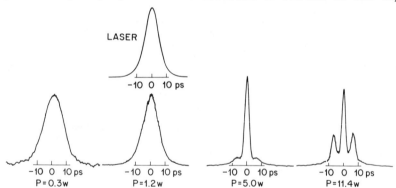

Figure 5. *Results of experiments with fibre whose length is one-half the soliton period. Above: Autocorrelation shape of the laser pulses launched into the fibre. Below: Autocorrelation shapes of pulses emerging from the fibre for various input powers. P = 0.3 W: negligible nonlinear effect; only dispersive broadening is seen. P = 1.2 W: return to input pulse width; corresponds to the fundamental soliton. P = 5 W: Pulse narrowed to minimum width corresponds to half-period behaviour of the N = 2 soliton. P = 11.4 W: first well-resolved splitting; corresponds to the N = 3 soliton. (Note that the three-fold splittng in autocorrelation corresponds to a two-fold splitting of the pulse itself.)*

values, whereas for intermediate powers, the pulses were narrowed and the frequency spectra correspondingly broadened.

19.5 The Soliton Laser

As may be inferred from the behaviour shown in Fig. 3, extreme pulse narrowing should be obtained at high soliton number in a judiciously chosen length of fibre. Indeed, compression by factors as great as 30x have been obtained experimentally [10]. But one pays a penalty for such extreme compression: a large fraction of the pulse energy (for N > 6, more than half) remains in uncompressed "wings" surrounding the central spike. The soliton laser [3] represents a way to obtain ultrashort pulses without having to pay that penalty. Instead, feedback from a pulse shaping fibre enables the laser to produce minimum bandwidth, sech^2 shaped pulses of controlled width. As indicated earlier, the precisely controlled pulses from the soliton laser have been vital to further experiments on solitons.

As shown in Fig. 6, the soliton laser consists of two cavities, the main cavity, and the cavity containing the fibre, called the control cavity. The two cavities are coupled together through their common mirror, M_0. The main cavity belongs to a synchronously pumped, mode locked colour centre laser, continuously tunable over a broad band centred at $\lambda \sim 1.5 \mu m$. (It

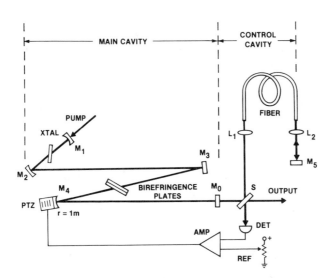

Figure 6. Schematic representation of the soliton laser, including stabilization loop. Beam splitter S has reflectivity R = ~ 30-50%. L: microscope objectives PZT: piezoceramic translator. (This is the very latest version, which differs from that described in [12] only in that the piezoceramic translator has been moved from the control cavity to the main cavity; this reduces the translator motion required when the control fibre is long.)

is thus very similar to the laser used in the first experiments.)
Without the control cavity, the laser produces mode locked pulses
of ≥8 ps width [11]. A controlled fraction of its ouput pulse
energy travels around the control cavity which is matched in round
trip time to the main cavity. The fibre compresses the pulses and
sends them back into the main cavity, thereby stimulating the
colour centre laser to produce narrower pulses. Thus, there is a
successive compression until the fibre pulses become solitons, at
which point the laser operation reaches a stationary state.

It was expected [3] that the laser would operate on an N = 2
soliton (see Fig. 3). If so, and if there is a constant relation
between z_0 and the control fibre length L, the width of the pulses
formed by the soliton laser should scale with the square root of
L. This dependence of the pulse widths on L has indeed been
confirmed experimentally in [3]. That is, the data there fit the
relation $L \approx z_0/2$ (see Fig. 7), with the implication that in the

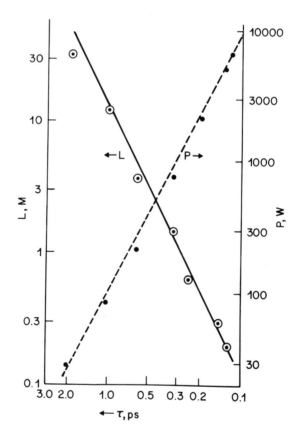

*Figure 7. Control fibre lengths (encircled points) and peak
powers at input to control fibre (solid points) as functions of
the obtained laser output pulse width. Solid line:* $\frac{1}{2} z_0(\tau)$
from (5) and (8). Dashed line: $P_2(\tau)$ *from (6) and (9).*

stationary state, pulses returned from the fibre have substantially the same width as those launched into it. Nevertheless, it was later learned [12] that stable solutions also exist for which L is considerably less than $z_0/2$.

Note that the system of coupled cavities allows for a high degree of indpendence between the laser and control cavity powers. This flexibility is vital, as the ratio of laser output to fibre soliton power is too great to be spanned by the available optical gain, as would be required in a single cavity. Successful operation of the two cavity laser requires, however, that the pulses returned from the control cavity have the correct optical phase with respect to the pulses circulating in the main cavity. The servo loop shown in Fig. 6 maintains [12] that phase by constant correction of the relative cavity lengths against effects of vibration and thermal drift. The necessary error signal is easily obtained, based on the following. First, the power in the control cavity varies with ϕ, the round trip optical phase shift in the control cavity about as shown in Fig. 8. Second, soliton laser action is correlated with a well defined level lying somewhere within the middle of the range of cavity powers.

Thus, at least in the neighbourhood of the soliton level, the control cavity power is a good measure of ϕ. The appropriate error signal for the control of ϕ is then generated, simply by taking the difference between the detector signal and a reference voltage empirically set to the soliton level. The op-amp magnifies that difference and drives the piezoelectric translator of the cavity mirror M_4. Thus, assuming correct choice of signal polarity, a deviation of ϕ from the soliton value produces a change in detector signal that results in corrective displacement of M_4.

The stabilized soliton laser emits a stream of pulses that are very uniform in width and height. The noise on the pulses can be as low as ~1 percent of the full intensity. Figure 9 shows a time exposure of a typical autocorrelation trace as seen on an

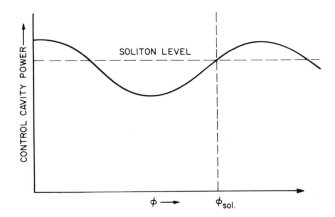

Figure 8. Variation of control cavity power with round trip optical phase shift ϕ.

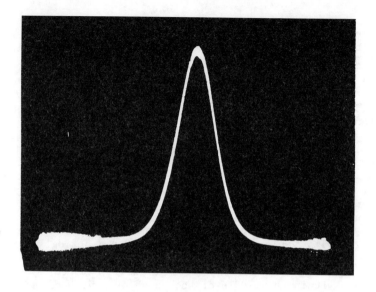

Figure 9.   A typical autocorrelation trace of the laser output pulses.   In this example, the actual FWHM of the pulses is 580 fsec (L = 1.6 m).

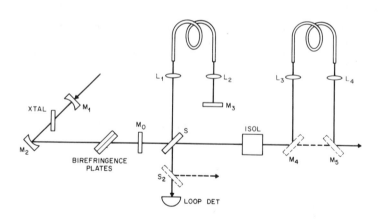

Figure 10.   Set-up to duplicate the control fibre with a "test" fibre.   Mirrors M_4 and M_5 can slide out of the beam to allow the laser output to be measured directly.   (The beam from S_2 is used for direct confirmation of the datum point at L' = 2L.   See text.)

oscilloscope screen, where the sharp definition of the trace
indicates the absence of serious noise.

To analyze the soliton laser more completely, conditions (input
pulse shape, power) in the control fibre were accurately
duplicated [12] in a second length of fibre cut from the same
spool; see Fig. 10. By gradually cutting back the length L' of
this "test" fibre, and by observing its output pulse shapes in
autocorrelation, it was possible to infer what was happening in
the control fibre itself.

Figure 11 summarizes the results of a typical experiment. Note
that the pulse shapes and widths make good fit to the expected
behaviour of sech N = 2 solitons. But also note that, at least in
this particular example, the round trip distance in the control
fibre (2L) is considerably less than the soliton period (z_0).

Thus the pulses returned from the control cavity are considerably
narrower than those launched into it.

The above result yields additional insight into operation of the
soliton laser. Without feedback from the control fibre,
subpicosecond pulses will tend, with repeated round trips,
gradually to broaden in time and to collapse in the frequency

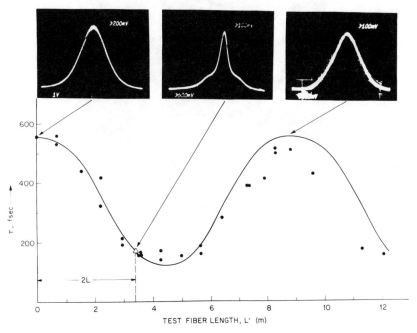

Figure 11. FWHM of pulses emerging from test fibre of length L'
when $sech^2$-shaped, laser output pulses of 560 fsec FWHM were
launched into it. The solid line is the theoretically expected
value for a sech N = 2 soliton calculated for D = 14.5 psec
$nm^{-1} km^{-1}$. The power level (in both control and test fibers) was
equal to the N = 2 soliton power as calculated from n_2 =
$3.2x10^{-16} cm^2/W$ and A_{eff} = 86 μm^2 for 560 fsec pulses. The datum
point at 2L was found directly from the control fibre.

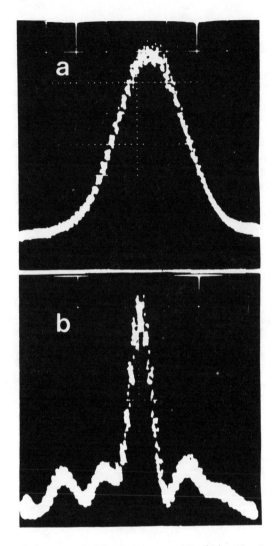

Figure 12. Autocorrelation traces of (1) 60 fs FWHM pulses produced directly by the soliton laser and (b) the same, but after external compression to 19 fs FWHM. Note: the regular ripples seen on the traces are not noise, but represent interference fringes between the two nearly parallel beams. Hence, the distance between bumps is the optical period (5 fs).

domain. The collapse would be brought about by dispersion-induced phase shifts and attenuation, primarily in the far wings of the pulse's frequency spectrum.

Feedback from the fibre provides the correction necessary to sustain mode locking, and thus to prevent collapse. The need for some degree of pulse narrowing is now easily understood. That is, amplitudes of the pulses returned from the control cavity are often just a few percent or less of those in the main cavity.

Nevertheless, the feedback is strong where it is most needed, at the extreme fringes of the pulse spectrum, as the narrowed pulses contain relatively much more power there than do those circulating in the main cavity. Of the several recent attempts at theoretical analysis of the soliton laser [13-15], perhaps [14] and [15b] come closest to this model derived from experiment. On the other hand, the experiment of Figs. 10 and 11 shows unequivocally that the soliton laser cannot be modelled as a single cavity, as was attempted in [13].

To date, the shortest pulse produced directly by the soliton laser [16] has τ = 60 fs FWHM (see Fig. 12a). That pulse was produced by using a special "dispersion flattened" fibre [17] in the control cavity, one with D = 0 at two wavelengths, 1.37 and 1.62 μm.

The 60 fs pulses were compressed in an $L' \sim z_0/2$ external piece of the same fibre to about 19 fs FWHM, when launched at a power level corresponding to N = 2 (see Fig. 12b). Incidentally, the 19 fs pulse width corresponds to a bit less than 4 optical cycles at 1.5 μm, or the same number of cycles as the shortest known light pulse (8 fs, produced by a dye laser in the visible). (Note that once again, this experiment, besides producing the shortest possible pulse, shows what the pulses must look like at one point in the control fibre itself.) The measured frequency spectrum of the 19 fs pulse just filled the space between the two D = 0 points of the fibre. In the meantime, the bandwidth of 60 fs pulse is only about 25% of the gain bandwidth of the colour centre. Thus, one can conclude that the limit to ultrashort pulse production in the soliton laser is set primarily by the dispersion properties of the control fibre. In principle, the above result could be improved upon by a more suitable fibre. It is considered possible, for example, to produce dispersion flattened fibres with no second wavelength for which D = 0.

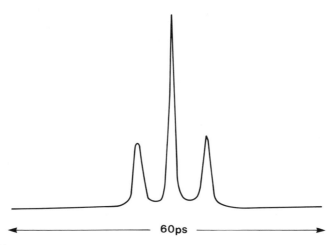

← ——————————— **60ps** ——————————— →

Figure 13. An autocorrelation trace of the pulses at the end of the test fibre (L' = 392 m). Parameters are: $\tau_0 \approx$ *500 fs, P* $\approx$ *1.5 P_1; full range of scan 60 ps.*

19.6 Discovery of the Soliton Self Frequency Shift

When the experiments of Fig. 11 were completed, the test fibre was made much longer (L' = 392 m, or ~46 z_0 for τ = 560 fs). The object was to see how well N = 2 solitons could survive such a long fibre. But when pulses of P considerably greater than P_1 were launched into the fibre, the pulses displayed an unexpectedly large temporal splitting at the fibre output (see Fig. 13).

The corresponding frequency spectrum (Fig. 14) provided the first clue to this strange behaviour.

It was clear that the pulse had split into a narrow, intense, fundamental (N = 1) soliton, and a weaker, non-soliton part that broadened as it travelled down the fibre. The spectrum of the weaker pulse remained more or less centred at the original laser frequency, but the soliton had undergone a strong shift to lower frequencies. The large temporal splitting could thus be explained in terms of the combined effects of group velocity dispersion and the large splitting in frequency.

The frequency shift is made possible by the fact that in glasses, the Raman gain [18] extends right down to zero frequency difference between pump and signal (see Fig. 15).

Thus, the higher frequency components of the pulse can act as a pump to provide Raman gain for the lower frequency components, and by this means, there can be a steady flow of energy to lower frequencies. Clearly, the increasing Raman gain with increasing frequency would make the effect increase rapidly with decreasing pulse width (increasing pulse bandwidth). Experiment showed just such a scaling. For example, for a 260 fs soliton at output of the 392 m fibre, the down shift was 8 THz, or 4%. But for a 120 fs soliton, the shift was 20 THz (10%), and in just 52 m of fibre.

Immediately following these experiments [19], Gordon produced a proper theory [20] of the effect. First, he noted that existence of the Raman effect requires that the nonlinear term in (3) be modified to include a delayed response, as follows:

$$|u|^2 u \rightarrow u(t) \int ds\ f(s)|u(t - s)|^2 \qquad (10)$$

where f(s) is real if there are no losses other than the Raman type, f(-|s|) = 0 to ensure causality, and $\int f(s)ds = 1$ to recover (3) for sufficiently short delays.

But as the modified equation (3) is difficult to work with directly, Gordon took its Fourier transform. The empirically known Raman gain (Fig. 15) could then be introduced in a natural way, and its effects could be treated as a perturbation on the soliton. The final result is a rate of frequency shift

$$\frac{d\nu_0}{dz} = 0.0436\ h(\tau)/\tau^4 \qquad (11)$$

where $h(\tau)$, a slowly varying function on the order of unity, reflects details of the Raman spectrum. The much stronger τ^{-4}

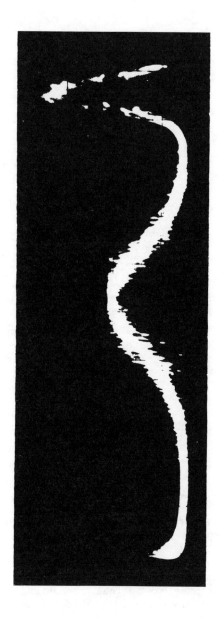

Figure 14. Typical spectrum of the pulses at the output end of the 392 m test fibre, under conditions similar to those of Fig. 13. Scan width is one free spectral range of the scanning Fabry Perot etalon (4.6 THz); frequency increases towards the right. The narrow peak near the right margin is at ν_0.

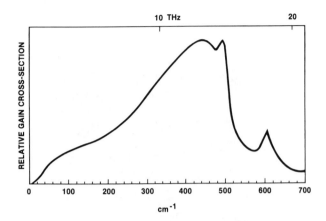

Figure 15. Relative Raman gain versus frequency difference between pump and signal (from Ref. 18).

dependence can be understood as a result of the τ^{-2} dependence of the soliton power (P_1, eq. (6)), and the τ^{-1} dependence of its spectral width, which proportionally increases both the effective Raman gain and the resulting frequency displacements. The power of the τ^{-4} dependence can perhaps be best appreciated from the log-log plot of Fig. 16. Note that while the effect can be very large in just a few metres of fibre for pulse widths of a few tens of femtoseconds, it becomes negligibly small for pulse widths of several tens of picoseconds. The latter extreme is fortunate for telecommunications.

19.7 Experimental Observation of Interaction Forces Between Solitons

Several papers have predicted the existence of interaction forces between copropagating solitons. In 1981, Karpman and Solov'ev [21] analyzed the interaction using a perturbation theory, and later, Gordon [22] derived the soliton interactions directly from the exact two-soliton function. Nevertheless, in both papers, the main prediction was of a force which decreases exponentially with the initial distance of separation of equal amplitude solitons, attractive when the waves under the pulses are in phase, and repulsive for waves out of phase. The existence of such forces is of obvious significance to optical communications, and may have important implications for optical computing and other applications as well.

Recently, pulses from the soliton laser have made possible the first experimental observation [23] of such forces. The essentially isolated (separated by 10 ns) pulses from that laser were first passed through a Michelson interferometer to turn them into pulse pairs. A micrometer screw adjustment of the length of one interferometer arm allowed the pulse separation to be varied continuously from zero to many picoseconds, while piezoelectric

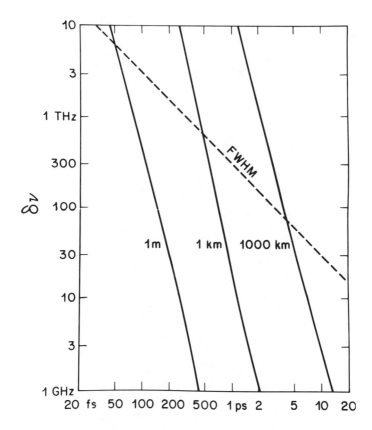

Figure 16. Soliton frequency shifts $\delta\nu_0$ versus τ, for various unit lengths of lossless silica glass fibre having D = 15 ps/nm/km, and for λ = 1.5 μm. The pulse bandwidth (FWHM) is shown for comparison. (From [20].)

transducer control of the same allowed for precise adjustment of the relative optical phase. The pulse pairs were then observed in autocorrelation both before and after they had passed through a length of single mode fibre. Solitons were obtained by adjusting coupling into the fibre until the emerging pulses had exactly the same width as at input. Also, by alternately blocking the two arms of the interferometer, each member of the pulse pair could be adjusted for equal intensity as measured at the fibre output.

The first experiments were carried out in a 340 m length of polarization preserving, low loss (>0.3 dB/km) fibre. The pulse widths were $\tau \sim 1$ ps FWHM, which made the fibre of order 10 soliton periods long. This length would theoretically allow convergence of attractive pulses with a well-resolved initial separation (a little over 3 pulse widths).

The results of a typical measurement are shown in Fig. 17, for both in phase (attractive) and opposite phase (repulsive) pulses.

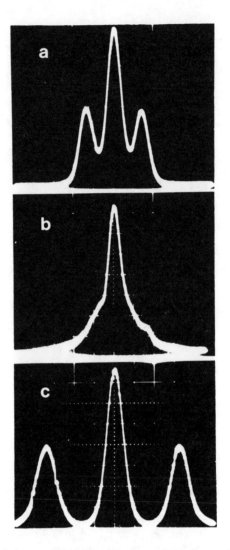

Figure 17.   Autocorrelation traces of pairs of τ = 0.90 pulses (a) at input of 340 m fibre and (b,c) at fibre output, for pulses in phase (attractive) and of opposite phase (repulsive), respectively.   Pulse separation at input, 2.33 ps.   (All traces are to a common time scale.)

Figure 18 is a plot of final pulse separation σ_{out} as a function of initial pulse separation σ_{in}, for both the repulsive and attractive cases.   The theoretical curves shown there for repulsion, and for attraction up to the first overlap, are based on Eq. (18) of [22].   (The cosine function given there is valid only for the attractive case (ϕ = 0.)   For the repulsive case (ϕ = π), that function must be replaced by the hyperbolic cosine [24].) Note the close agreement between theory and experiment for the repulsive case and for the attractive case up to the point of first pulse overlap (for which $\sigma_{in} = \sigma_c$).

But, for the attractive branch and for smaller initial pulse separation ($\sigma_{in} < \sigma_c$), the observed behaviour (again, see Fig. 18) is different from prediction. In theory [22], the two pulses reach a bound orbit around their common centre, so that they periodically "pass through each other". (Note that this is merely a convenient manner of speaking, as the pulses are indistiguishable.) This orbital motion is reflected in the oscillatory structure of the theoretical curve, which the experimental curve clearly does not follow. Rather, the experiment shows that the pulses never completely overlap. This is a consequence of the instability of the attractive phase [22], and the fact that perturbations exist to trigger the instability. For small angles, and further assuming constant pulse separation, Eq. (13) of [22] has the aproximate solution

$$\phi = \phi_0 \exp (\alpha z/z_0) \tag{12}$$

where

$$\alpha = \pi\sqrt{2} \exp(-0.88\sigma/\tau) \tag{13}$$

Although it has been shown numerically [25] that higher order dispersion can trigger the instability, for the conditions of the experiments, it is more likely that the soliton self frequency shift provides the effective perturbation. The phase shift caused by a common frequency shift $\Delta\nu$ of pulses separated by time interval σ is

$$\Delta\phi = 2\pi \, \Delta\nu \, \sigma \tag{14}$$

According to (11) each 1.17 ps pulse alone should experience a 3.4 GHz shift over the 340 m fibre length. Thus, for $\sigma \sim \sigma_c = 3.79$ ps, (14) yields $\Delta\phi \approx 0.08$ radians at the end of the fibre, and proportionally less for some fraction thereof.

When the pulses merge, α becomes large ($\alpha \approx 4.4$ for $\sigma = 0$), and the exponential growth, accumulated over the last few soliton periods of the fibre, becomes more than large enough to amplify the $\Delta\phi$ estimated above (or some reasonable fraction thereof) to a value on the order of π. In this way, the unstable attractive phase will be converted to the stable repulsive one. On the other hand, for σ_{in} considerably greater than σ_c, the pulses always remain apart, α correspondingly remains small, and the initial phase perturbation cannot grow more than a few times. Thus, the net phase shift remains well within the domain of attraction.

Once the laser was accidentally maladjusted (the polarization axis of the control fibre was misaligned) such that the laser produced a weak (relative intensity $\sim$ 2%) satellite pulse about 7 ps away from the main pulse. Then in a measurement like that of Fig. 18, there were extra bumps in the curves near $\sigma_{in} = 7$ ps,

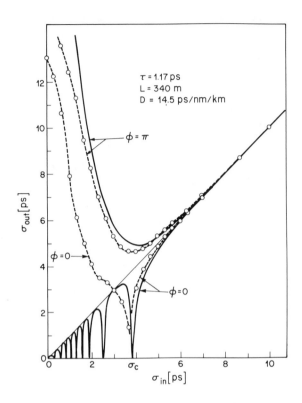

Figure 18. Pulse separation at output (σ_{out}) versus same at input (σ_{in}) of 340 m fibre. Solid curve, theory; dashed line, experiment.

corresponding to an approximately ± 20% change in σ_{out}. This shows how even very small coherent satellites can act as emissaries of the main pulse, greatly extending the effective range of interaction. There is an obvious lesson here for telecommunications, namely, that the space between the pulses must be kept clean to avoid serious interactions. It is also clear that small deviations from the perfect pulse shape can result in large ʻchanges of the pulse interaction. The close fit of experimental results to theory (Fig. 18) is therefore testament to the accuracy of the sech amplitude envelope shape of the soliton laser pulses.

The derivations in [21] and [22] are based on the highly abstract inverse scattering theory, and other analyses [25,26,27], while differently based, are still purely mathematical. But the attraction and repulsion can be simply understood in terms of the chirp each pulse induces on the other. For the isolated, symmetrical pulse shown in Fig. 2, the chirp is correspondingly symmetric, and (negative) dispersion can only compress the pulse. For a pulse partially overlapped by a second pulse, however, its intensity distribution, and hence its chirp, will be unbalanced.

(The fact that chirp in the isolated soliton is exactly cancelled out by dispersion in no way changes the argument.) If the waves under the pulses are in phase, the intensity will be increased, and the chirp **decreased**, in the region of overlap. (The chirp is proportional to the slope of intensity.) Hence, there will be a net acceleration of the pulses toward each other (attraction). By a similar argument, if the waves are out of phase, just the opposite will happen. Note that by the argument given here, the existence of a force does not require solitons. For solitons, however, the interaction is not complicated by simultaneous decay of the pulses; hence, solitons are the simpest pulses for study. Note also that a force will exist between pulses of different optical frequencies, but then, the intensities simply add, and the force will always be attractive.

19.8 Solitons in Telecommunications

Conventional fibre optic telecommunications systems use but a tiny fraction of the potential information carrying capacity of optical fibres. That is, the optical signals are detected and electronically regenerated every 20-100 km before continuing along the next span of fibre. But electronic repeaters limit rates to a few Gbits/sec or less per channel. Furthermore, the use of multiple channels, or wavelength multiplexing, is difficult and cumbersome, as the demultiplexing/multiplexing must be performed at each repeater. Thus, the only sensible way to achieve higher bit rates is by allowing the signals to remain strictly optical in nature.

Figure 19 shows a proposed all-optical system, where the fibre losses are compensated by Raman gain, and the information is transmitted as trains of solitons. To provide for the gain, cw pump power is injected at every distance L (the "amplification period") along the fibre, by means of directional couplers. (The wavelength-dependent couplers provide for efficient injection of the pump power, but allow the signal pulses to continue down the main fibre with little loss.) A system might contain as many as

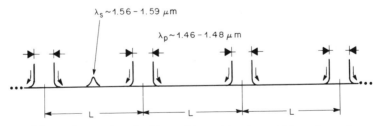

$\lambda_s \sim 1.56 - 1.59 \, \mu m$

$\lambda_p \sim 1.46 - 1.48 \, \mu m$

Figure 19. Segment of the all optical soliton based system. Single laser diodes are shown here at each coupler, but the required pump power, ~50 to 100 mW, would best be supplied by a battery of, say, a dozen lasers, each tuned to a slightly different wavelength, their outputs combined through a diffraction grating. In this way, stimulated Brillouin back scattering can be avoided. The multiplicity of pump lasers would also provide a built-in, fail-safe redundancy.

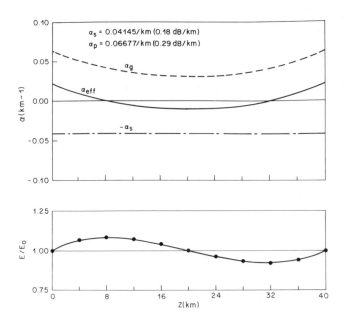

Figure 20. Above: Coefficients of loss $(-\alpha_g)$ and Raman gain (α_g) and their algebraic sum (α_{eff}) for an amplification period $L = 40$ km. Below: The corresponding normalized pulse energy.

100 or more amplification periods.

Loss at the pump wavelength makes the Raman gain nonuniform within each amplification period. Nevertheless, the signal pulse energy fluctuations can be surprisingly small. Figure 20 shows gain and signal energy for $L = 40$ km and low loss single mode fibre. The result of bidirectional pumping, the Raman gain is the sum of two decaying exponentials, and is adjusted (through control of the pump intensity) such that there is no net signal gain or loss over the period. Note that with the Raman gain, the signal pulse energy varies by no more than about ±8%, whereas without it, more than 80% of the signal energy would be lost in just one period.

The stable transmission of fibre solitons over a gain compensated fibre of 10 km has already been demonstrated experimentally [7]. Nevertheless, the question remains of just how well the solitons will stand up to the periodically varying pulse energy in the system of Fig. 19. Here, numerical studies [5,6] have provided some important answers. In [6], a gain/loss term $-i\Gamma u$ (where Γ corresponds to the α_{eff} of Fig. 20), was added to the right hand side of (3). Solutions were then obtained on a Cray.

Figure 21 shows a principal result of the calculations of [6], the pulse distortion obtaining at the end of one amplification period, graphed as a function of the parameter L/z_0. Here the change δS, of pulse area is used as a measure of the pulse distortion from a true soliton. (For convenience, S has been renormalized to unity.) The peak in δS, occurring at $z_0 \approx L/8$, corresponds to resonance between the soliton's phase term (see (4)) and the periodic pulse energy variation. Note the excellent recovery of the soliton for z_0 both long and short with respect to the resonance value.

Nevertheless, the region of "long" z_0 is the one of practical interest. First, it allows P_1 to be just a few mW, as opposed to the greater power required for short z_0. (From (5), (6), and (8), P_1 is inversely proportional to z_0.) Depletion of the Raman pump power (typically ~100 mW) by the signals themselves then becomes negligible, as required for stable gain, independent of the signals. Second, the soliton pulses are exceptionally stable for long z_0. For example, the calculations of [5] and [6] show negligible increase in pulse distortion after 50 or more amplification periods, and such stable behaviour is expected to continue indefinitely. Furthermore, modest lumped losses, such as from couplers, should have negligible effect as long as they, too, are cancelled out by Raman gain.

The required values of z_0, on the order of 40 km or more, would be obtained by using pulses of several tens of ps width at low values of D; see (5) and (8). (In "dispersion-shifted" fibres, λ_0, the wavelength where the dispersion passes through zero, can

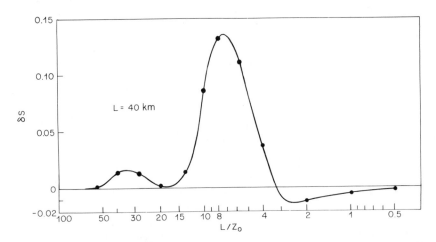

Figure 21. Computed change in pulse area (a measure of pulse distortion) obtaining at the end of one amplification period L = 40 km for a perfect soliton launched at input, versus the quantity L/z_0. (For other values of L, only the resonance peak height changes; its location remains the same [6].) Note that the region of "large" z_0 is to the far right, while the region of "small" z_0 is to the far left.

be pushed as far as desired toward the wavelength of minimum loss. Thus, in principle, D can be made arbitrarily small at that wavelength.)

The question is often raised, why not simply propagate the signals at λ_0? For one reason, that scheme would not allow for wavelength multiplexing, as λ_0 corresponds to just one wavelength. More fundamentally, for pulses of reasonable power, the combination of dispersion terms of higher order and index nonlinearity leads to severe pulse distortion and broadening in long fibres [28]. Thus, the soliton is *the* stable pulse.

The maximum propagation distance will therefore be limited not by instability of the solitons, but by the accumulated effects of noise. The most serious of these effects arises from random modulation of the pulse frequencies (and hence of their velocities) by Raman spontaneous emission [29,6]. The corresponding random (Gaussian) distribution in pulse arrival times can lead to significant error if the path is long enough. (Provided the gain is always held close to unity, the Raman spontaneous emission itself increases only in direct proportion to the total path length, and remains negligibly small for many thousands of kilometres. Also, of course, it is assumed that the pulses are initially of minimum bandwidth, that is, that the laser source is without chirp. Chirped pulses would produce additional problems [30].)

In a typical arrangement, the pulses are initially spaced apart by about 10 pulse widths (see Fig. 22), to avoid the interaction forces discussed in the previous section. But it also helps to provide for a spread in arrival times; as shown in the figure, the detection window can be nearly as large as the pulse separation.

It can be shown that for fixed fibre characteristics, and a given error rate, the bit-rate system-length product is a constant. For example, for an assumed error rate of 10^{-9}, for D = 2 psec/nm/km, and for reasonable values of the other pertinent fibre parameters, the rate-length product is ~29,000 GHz-km. Thus, for example, for a 4 GHz bit rate (250 psec spacing between 25 psec wide pulses), the maximum allowable distance of transmission would be Z = 7250 km.

In contrast to systems involving electronic repeaters, the all optical system is highly compatible with wavelength multiplexing. By virtue of the fibre's dispersion, however, pulses at different wavelengths will have different velocities and will pass through each other. Hence, it is necessary to consider the possible effects of soliton-soliton collisions. Computer simulation has shown [6] that when pulses collide they modulate each other's frequencies (hence velocities), the sign of the effect depending on whether the collision takes place in a region of net gain or loss. Thus, a statistically large number of collisions can add a second source of random pulse arrival times.

The effect can be understood in terms of a cross phase modulation, or mutually induced chirping, similar to that giving rise to the interaction forces described in the previous section. To be sure, were the collision to take place under conditions of

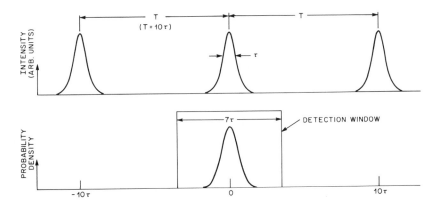

Figure 22. *Above:* Train of pulses in their initial relative positions. *Below:* Detection window and corresponding Gaussian probability density of pulse arrival times for an error rate of 10^{-9}.

no energy gain or loss, the chirp each pulse induces on the other in the second half of the collision would exactly cancel the chirp induced in the first half, and there would be no net effect. But the cancellation is not complete when the collision takes place in a region of net gain or loss. Hence each pulse suffers a net frequency (velocity) shift.

The effect scales inversely with the square of the frequency (velocity) difference between the pulse streams. (The effect is, of course, proportional to the distance the pulses travel during the collision. But it is also proportional to the change of pulse energy during the collision; for constant energy gain (or loss) rate, this results in a second factor of proportionality to the length of the collision.) Thus, the variance in arrival times can be made comparable to or smaller than that produced by the Raman noise, just by making the wavelength separation between adjacent channels great enough.

A few examples [6] of system design, with and without multiplexing, are summed up in Table 1. Note that the values of z_0 fall well to the right of the resonance peak in Fig. 21. Note also that the best approach is to use the combination of many channels at a modest rate per channel. (Compare 2b or 2c with 1b.) In addition to producing the highest overall rate-length product, that approach puts the major burden for separating signals on simple optics, rather than on ultrafast electronics. But above all, note the tremendous overall rates made possible by the all-optical nature of the systems: in one instance, the rate is ~106 GHz, or about two orders of magnitude greater than possible with conventional systems!

Recently, L.F. Mollenauer and K. Smith have performed an experimental test of the above ideas. By recirculating 55 ps soliton pulses (at λ_s = 1.60 μm) many times around a 42 km loop with loss exactly compensated by Raman gain (λ_p = 1.50 μm), they have observed nearly distortionless transmission to distances in excess of 4000 km. The experiment is fully described in a paper submitted to Optics Letters in April, 1988.

Table 1: Design examples of high bit rate, soliton based, all optical systems

L = 40 km and D = 2 psec/nm/km

Design No.	z_0(km)	τ(psec)	P_1(mW)	Z(km)	N	NR(GHz)
1a	30	12.3	10	3600	1	8.1
1b				2860	5	40.5
2a	100	22.6	3	6600	1	4.4
2b				5200	10	44
2c				3000	24	106

The numbers in this table were derived as follows: Choice of D and z_0 serve to fix τ, through (5) and (8), and also P_1. The single channel bit rate R = 1/(10τ), consistent with a spacing of 10τ between pulses. The system overall length Z = 29000/R (see text). The total wavelength span for multiplexing is somewhat arbitrarily set at ~20 nm, and the algorithm for determining N, the number of channels, is detailed in [6]. In examples 1b, 2b, and 2c, Z is reduced, to reduce the variance in pulse arrival times caused by the Raman noise effect; this allows room for the variance caused by soliton-soliton collisions.

19.9 Conclusion

The work reviewed here demonstrates, once again, the truism that "physics is an experimental science". To be sure, we have seen how many properties of light pulses in optical fibres, such as the existence of a stable, fundamental soliton, are described rather well by the simple nonlinear Schrödinger equation. Yet we have also seen how other important features are not. The soliton self frequency shift, for example, came as a complete surprise. Still other features are well predicted in first approximation, but not in fine details. Higher order solitons, for example, exhibit the expected behaviour over at least a few periods, but tend eventually to break apart. The attractive force between solitons, otherwise well predicted by (3), has been shown unable to survive the large phase instability of close contact and the effects of self modulation.

In spite of the complications, however, we have seen that there is a common and simple conceptual thread binding together almost all of the experimental results. That is, most of the effects observed so far (except for the soliton self frequency shift) can be understood in terms of phase modulation resulting from an instantaneous index nonlinearity. Thus, pulse compression and solitons result from self phase modulation and a resultant self chirping, while cross phase modulation (mutually induced chirping) explains the interaction forces between solitons and the result of soliton-soilton collisions. Both the experimental results and this simple physical model give the effects a reality and an existence independent of more abstract models. Such concepts are vital to further invention and development.

Finally, the soliton laser, so important to the other experiments, itself provides another dramatic example of the need for experiment. The soliton laser began with a simple concept, but just how well that concept would work out could not possibly be foreseen without empirical test; the device is just too complex. Thus experiment has also been required for generation of appropriate models of the soliton laser, and has been vital for the development of practical details. Happily, here the final result has surpassed the wildest dreams of the inventor!

References

[1]   A. Hasegawa and F. Tappert (1973). *Appl. Phys. Lett.* **23**, 142.

[2]   L.F. Mollenaeur, R.H. Stolen and J.P. Gordon (1980). *Phys. Rev. Lett.* **45**, 1095.

[3]   L.F. Mollenauer and R.H. Stolen (1984). *Opt. Lett.* **9**, 13.

[4]   A. Hasegawa (1983). *Opt. Lett.* **8**, 650.

[5]   A. Hasegawa (1984). *Appl. Opt.* **23**, 3302.

[6]   L.F. Mollenauer, J.P. Gordon and M.N. Islam (1986). *IEEE J. Quantum Electron.* **QE-22**, 157.

[7]   L.F. Mollenauer, R.H. Stolen and M.N. Islam (1985). *Opt. Lett.* **10**, 229.

[8]   R.H. Stolen and C. Lin (1978). *Phys. Rev.* **A17**, 1448.

[9]   R.H. Stolen, L.F. Mollenauer and W.J. Tomlinson (1983). *Opt. Lett.* **8**, 186.

[10]   L.F. Mollenauer, R.H. Stolen, J.P. Gordon and W.J. Tomlinson (1983). *Opt. Lett.* **8**, 289.

[11]   L.F. Mollenauer, N.D. Vieira and L. Szeto (1982). *Opt. Lett.* **7**, 414.

[12]   F.M. Mitschke and L.F. Mollenauer (1986). *IEEE J. Quantum Electron.* **QE-22**, 2242.

[13]   H.A. Haus and M.N. Islam (1985). *IEEE J. Quantum Electron.* **QE-21**, 1172.

[14]   K.J. Blow and D. Wood (1986). *IEEE J. Quantum Electron.* **QE-22**, 1109.

[15]   F. If, P.L. Christiansen, J.N. Elgin, J.D. Gibbon and O. Skovgaard (1986). *Opt. Commun.* 57, 356; P. Berg, F. If, P.L. Christiansen and O. Skovgaard, to be published in *Phys. Rev. A*.

[16]   F.M. Mitschke and L.F. Mollenauer (1987). *Opt. Lett.* **12**, 407.

[17]   L.G. Cohen and A.D. Pearson (1983). *Proc. SPIE* **425**, 28.

[18]   R.H. Stolen, C. Lee and R.K. Jain (1984). *J. Opt. Soc. Am.* **B1**, 652.

[19]   F.M. Mitschke and L.F. Mollenauer (1986). *Opt. Lett.* **11**, 659.

[20]   J.P. Gordon (1986). *Opt. Lett.* **11**, 662.

[21]   V.I. Karpman and V.V. Solov'ev (1981). *Physica* **3D**, 487.

[22] J.P. Gordon, *Opt. Lett.* 8, 596, 1983.

[23] F.M. Mitschke and L.F. Mollenauer (1987). *Opt. Lett.* **12**, 355.

[24] J.P. Gordon, private communication.

[25] P.L. Chu and C. Desem, *Electron. Lett.* 21, 228, 1985.

[26] D. Anderson and M. Lisak, *Opt. Lett.* 11, 174, 1986.

[27] K.J. Blow and N.J. Doran, *Electron. Lett.* 19, 429, 1983.

[28] G.P. Agrawal and M.J. Potasek, *Phys. Rev.* A33, 1765, 1986.

[29] J.P. Gordon and H.A. Haus, *Opt. Lett.* 11, 665, 1986.

[30] J. Mork, P.L. Christiansen, H.E. Lassen and B. Tromborg, *Evolution of Soliton from Chirped and Noisy Pulses in Single-Mode Fibres,* to appear in *IEE Proc. J. Optoelectron.*

GENERATION OF SPATIALLY ASYMMETRIC, INFORMATION-RICH STRUCTURES IN FAR FROM EQUILIBRIUM SYSTEMS

Abstract

It is shown that one-dimensional sequences which are spatially asymmetric and information-rich may be generated from a time irreversible dynamics possessing a chaotic attractor. An algorithm is suggested, whereby the time irreversibility is manifested as a spatial asymmetry in a one-dimensional space. The statistical properties of the sequences generated in the particular case of Rössler's model are analyzed. The biological significance of these results is discussed.

20.1 Introduction

In this Chapter we present some ideas on the origin of asymmetrical patterns. In particular, we show that there is a mechanism based on self-organization phenomena that can lead to a systematic selection of such patterns against their mirror images.

One powerful motivation behind such an investigation is the fact that much of what we call "information" is carried in nature by a one-dimensional spatial structure which displays an absolute asymmetry in the form of a preferred polarity. This asymmetrical structure is, as the reader may have already guessed, the DNA and the information carried is the genetic code. Indeed, the genetic code depends essentially on two elements: a particular sequence of codons along the genetic material; and the possibility of "reading" this sequence unidirectionally. Although perfectly reproducible from one generation to the next, the codon sequence is basically unpredictable in the sense that its global structure cannot be inferred from the knowledge of a part of it, however large. It can thus be regarded as a stochastic process, and it is this possibility that allows us to speak of "information". Significantly, in all known biosynthetic reactions in which the information is revealed, there are start signals and the reading proceeds down the messenger, three nucleotides by three in a fixed direction from the start point, corresponding to the juxtaposition of the $5'$-$3'$ ends of the individual nucleotides. The primordial importance of this asymmetry is illustrated by the inability of synthetic molecules lacking polarity to act as efficient messengers [1]. Similarly, amino-acids display an absolute three-dimensional asymmetry in the form of preferred chirality, which is of crucial importance in the structural and functional properties of the proteins and enzymes. In what follows, however, we shall focus entirely on the selection of one-dimensional asymmetric patterns.

The careful reader will probably be wondering by now whether this question has not been settled 20 years or so ago, when the possibility of symmetry-breaking transitions in far-from-equilibrium systems was discovered [2-4]. Let us stop therefore for a moment and examine critically to what extent such transitions provide us with the solution of the problem of asymmetry. To fix ideas we consider a reaction-diffusion system,

$$\frac{\partial X}{\partial t} = v(X, \lambda) + D \cdot \nabla^2 X \qquad (1)$$

where $X = X_1, \ldots, X_n$ is the set of concentration variables, v the reaction rate vector, λ a set of characteristic parameters, and D the diffusion coefficient matrix. We assume that the system operates in an isotropic medium free of external fields: v is thus space and time-independent, and the boundary conditions are symmetrical.

The study of Eq. (1) under the conditions just defined has attracted a great deal of interest in the last two decades. In particular it has been established [3] that beyond a critical value λ_c of the parameter λ the uniform steady-state solution loses its stability. This gives rise to the bifurcation of space-dependent steady-state solutions, generally asymmetric and characterized by a well-defined intrinsic wavelength, or of wave fronts corresponding to various kinds of spatio-temporal organization within the system. There is at present ample experimental evidence, reviewed for instance in [5], that real world chemical systems like the Belousov-Zhabotinski reaction give rise to these phenomena.

At this point, however, it is important to notice that in an isotropic medium, in which only scalar processes (here chemical reactions) take place, for every non-uniform solution $X(r, t)$ there exists a solution $X' = X(r', t)$, where

$$r' = G \cdot r$$

$\underline{G}$ being a group of transformations that leave Eq. (1) invariant.

All members of the multiplet obtained by the action of such a symmetry group, including any given solution and its mirror image, have identical stability properties. None of them can therefore dominate in a selection process. In other words, we arrive at an **extended Curie principle** [6], whereby the symmetry broken at the level of an individual solution is restored when the full set of available solutions is considered. This is in complete agreement with experiment. For instance spiral waves in the Belousov-Zhabotinski reaction always arise by pairs, the two members of which are rotating in opposite directions.

Our principal goal in the present work is to see how one can come to terms with this fundamental restriction. An additional feature we require is that the asymmetric pattern arising in the process of selection to be explained presently should be an information carrier.

20.2 A Mechanism of Selection of One-dimensional Asymmetric Patterns

Consider a dissipative dynamical system like, for instance, a set of chemical reactions giving rise to the synthesis of species whose concentrations we denote by X_i. Suppose that we record the set of values of some representative variables X_i obtained for increasingly large times. We can regard this sequence as a one-dimensional string of digits. Now, because of the irreversibility inherent in the evolution of any dissipative system, the "direct" sequence so generated will be asymmetric in the sense that typically, it will be different from the sequence that one would read in the "reverse" direction: the latter would, in fact, be generated by a physically unacceptable dynamical system obtained from the initial one upon time reversal. We therefore see that the irreversibility can induce an absolute asymmetry in the abstract space of one-dimensional sequences of symbols. The question we raise now is how can this asymmetry of the symbolic dynamics be transferred to a one-dimensional system embedded in the physical space.

Let us propose a particular algorithm by which this goal can be achieved. Assume that the chemical system under consideration is capable of producing sustained oscillations in time, not necessarily periodic. Assume furthermore that when X_i crosses a certain level L_i with a positive slope (Fig. 1) a new process is switched on as a result of which substance i is rapidly precipitating or diffusing outside the reaction space, being subsequently collected on a "tape". The space pattern that will result in this way will carry the signature of the various thresholds encountered sequentially by the various concentrations. As pointed out earlier this sequence will typically be asymmetric because of the time symmetry-breaking afforded by irreversibility [7]. We have thus succeeded in mapping irreversibility on to a one-dimensional spatial pattern. More generally, it can be shown that in a reaction-diffusion system coupled to a polar vector (such as the gravitational field or an electrostatic field), space patterns carrying the signature of the sequence of values reached irreversibly by the concentration variables in the course of time can be generated. This spatial mapping of irreversibility cannot be achieved by a coupling with an axial vector, since such a coupling cannot appear in the balance equation of a scalar variable like a concentration. This is the reason why the mechanism discussed in the present work gives no information on the selection of chirality, the second major form of asymmetry observed in nature.

Variable

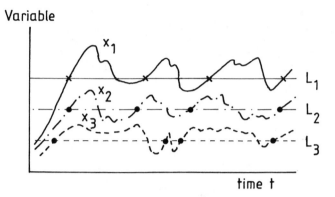

Figure 1.   An asymmetric space sequence of X_1, X_2, X_3 can be generated as these substances precipitate when their respective concentrations cross the threshold values L_1, L_2, L_3 with a positive slope.

Let us classify the patterns that can be generated by different chemical processes. Supppose first that X_i tend asymptotically to a limit cycle.   It is well known that this kind of behaviour requires at least two coupled variables, and that the phase difference between any two of them will be constant for all times. It follows that the level crossings will at best generate patterns defined by the following symbolic dynamics:

$$\ldots xyxyxy\ldots \qquad\qquad (a)$$

if 2 variables manage to cross the assigned levels,

$$\ldots xyzxyzxyz\ldots \qquad\qquad (b)$$

if 3 variables manage to cross the assigned levels, etc.
    Sequence (a) is reversible, as it reads the same in either direction.   Sequence (b) describes, on the other hand, an absolutely asymmetrical object.   Still, by introducing an abbreviated symbol (the analogue of a "codon") through

$$\alpha = \{xyz\}$$

sequence (b) reads

$$\ldots \alpha\alpha\alpha\ldots \qquad\qquad (c)$$

and loses in this form its space asymmetry.
    Whether the structure is to be understood in terms of (b) or (c), one can hardly speak of information, since the system is completely predictable.   For instance, by applying Shannon's definition of information, $I = -\sum_i p_i \ln p_i$, one would find I identically equal to zero [8].

Suppose now that the system runs on a chaotic attractor. It is well-known that under appropriate conditions the time evolution of such a dynamical system can be mapped to a stochastic process [9]. In this sense, therefore, the transcript of chaotic dynamics according to the mechanism presented in this section may potentially give rise to structures that are: (i) asymmetric because of irreversibility, and (ii) information-rich, since the information theoretic entropy of a stochastic process is in general different from zero. For this to be really the case, however, some additional conditions must be satisfied as we show presently.

Let K be the number of states of the process, $i_1 \ldots i_K$, K symbols characterizing these states in a unique manner. Referring again to our level crossing mechanism, if two variables X, Y manage to cross the assigned levels then K = 2, i_1 = X and i_2 = Y, whereas if this happens for three variables X, Y, Z then K = 3, i_1 = X, i_2 = Y and i_3 = Z. Consider now a sequence of symbols released in the course of time, $i_{\alpha_1} \ldots i_{\alpha_n} \ldots$ One can immediately see that if the process is zeroth order Markov-like, for instance a Bernouilli process [10], the asymptotic probability (in the sense of long time limit on the stable attractor) for any given sequence of symbols is equal to the asymptotic probability for the reverse sequence:

$$P(i_{\alpha_1} \ldots i_{\alpha_n}) = p(i_{\alpha_1}) \ldots p(i_{\alpha_n}) = P(i_{a_n} \ldots i_{\alpha_1}). \quad (2)$$

In this case, therefore, the time irreversibility giving rise to the attractor cannot lead to the selection of an absolute asymmetry. A well-known example of a chaotic attractor giving rise to a Bernouilli process is the logistic map or the tent map, when the probability of being in the left or in the right half of the unit interval is considered [11]. This kind of symbolic representation is therefore inadequate for coding.

Consider next a first order Markov chain $i_{\alpha_1} \ldots i_{a_n}$, where n can be as large as desired. The probability of a given finite "word" can then be computed in terms of an initial probability vector and of the conditional probability matrix. If as before attention is limited to the asymptotic behaviour on the stable attractor the initial probability will be the invariant steady-state probability p_s, therefore

$$P(i_{\alpha_1} \ldots i_{\alpha_1}) = p_s(i_{\alpha_1}) \, P(i_{\alpha_2}|i_{\alpha_1}) \ldots P(i_{\alpha_1}|i_{\alpha_{1-1}}), \quad (3)$$

We want now to evaluate the probability that this Markov chain generates the reverse word. Suppose first that one deals with a two-state process, involving the two symbols X and Y. The probability of a word XY and of its reverse are respectively

$$P(XY) = p_s(X)P(Y|X) \qquad (4a)$$

$$P(YX) = p_s(Y)P(X|Y). \qquad (4b)$$

On the other hand the absolute probabilities p_s are stationary solutions of the Chapman-Kolmogorov equation [10]

$$p_s(X) = P(X|Y)p_s(Y) + P(X|X)p_s(X)$$

or, using $P(X|X) + P(Y|X) = 1$,

$$P(Y|X)p_s(X) = P(X|Y)p_s(Y). \qquad (5)$$

It follows that $P(XY) = P(YX)$, in other words, in the framework of a first order Markov process by using only two symbols one cannot satisfy the property of asymmetry required by a reasonable coding.

Consider now a three-state process involving the three symbols X, Y and Z. The probabilities of a word XYZ and of its reverse are

$$P(XYZ) = p_s(X)P(Y|X)P(Z|Y) \qquad (6a)$$

$$P(ZYX) = p_s(Z)P(X|Y)P(Y|Z). \qquad (6b)$$

(We emphasize that we enquire about the frequency of the reverse word generated by the original Markov chain. It can be shown [12] that by using the invariant probability vector as initial state vector, the reverse chain is also a Markov chain. This, however, is not of interest in the present context.)

These two probabilities are not identical: one can easily find a counter example corresponding to the following (stochastic) conditional probability matrix $\mathbf{P} = \{P(j|i)\} = \{P_{ij}\}$, $i,j = 1,2,3$:

$$\mathbf{P} = \begin{bmatrix} \frac{1}{2} & \frac{1}{4} & \frac{1}{4} \\[2mm] \frac{3}{4} & 0 & \frac{1}{4} \\[2mm] \frac{1}{4} & \frac{1}{4} & \frac{1}{2} \end{bmatrix}$$

The invariant state vector is found to be $\underline{p} = (7/15,\ 1/5,\ 1/3)$ It follows that, for instance,

$$P(123) = (7/15).(1/4).(1/4) = 7/240$$

whereas

$$P(321) = (1/3).(1/4).(3/4) = 1/16.$$

On the other hand, there exist classes of systems for which the equality between (6a) and (6b) is fulfilled. The most obvious class is given by systems satisfying detailed balance in the strict sense [13], $P(j|i) = P(i|j)$, p_s = const. = 1/3. A wider class is constituted by systems enjoying cyclical symmetry, like for instance random walk with reflecting barriers.

In summary, we arrive at the conclusion that a system that can be mapped to a first order Markov process involving at least three symbols generates asymmetric sequences and can therefore be used for coding. In the next section we examine the difference between this "minimal" coding and the coding associated with a higher order Markov process, using a particular example.

20.3 A Simple Model of Chaotic Dynamics Generating A High Order Stochastic Process

We consider a 3-variable system described by the equations [14]:

$$\frac{dX}{dt} = - Y - Z$$

$$\frac{dY}{dt} = X + aY$$

$$\frac{dZ}{dt} = bX - cZ + XZ. \qquad (7)$$

For the parameter values a = 0.38, b = 0.3 and c = 4.5, this model gives rise to a chaotic attractor [15], depicted in Fig. 2. A reasonable set of threshold values turns out to be $L_x = L_y = L_z =$ 3.0. Using the initial conditions $X_0 = Y_0 = Z_0 = 1.0$ one can then generate through the level-crossing mechanism described above a sequence of symbols of the form (printing starts only after a sufficiently long time for transients to die out has elapsed):

ZYX ZXYX ZXYX ZYX ZXYX ZYX ZYX ZX ZYX ZYX ZXYX ZYX.... (8)

It can be checked that this sequence can be entirely reformulated by introducing the **hypersymbols**

$$\alpha = ZYX, \quad \beta = ZXYX, \quad \gamma = ZX \qquad (9)$$

The result reads:

$$\alpha\ \beta\ \beta\ \alpha\ \beta\ \alpha\ \alpha\ \gamma\ \alpha\ \alpha\ \beta\ \alpha\ \ldots \qquad (10)$$

This interesting property has to be attributed to the deterministic origin of the mechanism giving rise to the sequences. It suggests the existence of strong correlations in the succession of the symbols X, Y, Z, that is to say, a high

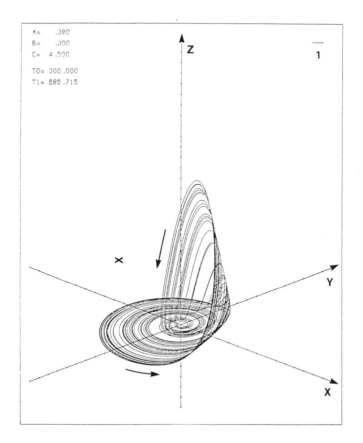

Figure 2. Chaotic attractor in Rossler's model obtained for the parameter values a = 0.38, b = 0.30, c = 4.50. Initial condition: X = Y = Z = 1; x: fixed point not at the origin; |———|: unit interval along the axes.

order Markov process. To check this we evaluate the statistical properties of the sequences as follows. Eqs. (7) are integrated up to about 40,000 time units, generating in this way about 10,000 symbols and 3,000 hypersymbols. The results are recorded starting at t = 300, in order to have all transients die out. The numbers of observed singlets (X, Y, or Z), doublets (XX, XY, etc.), triplets, etc. are counted. The conditional probabilities are then deduced through such relations as

$$P(B|A) = \frac{P(AB)}{p_s(A)}$$

$$P(C|AB) = \frac{P(ABC)}{P(AB)} \tag{11}$$

where A, B, C can be either of the X, Y and Z. (By increasing the integration time one can verify that the probability of finding a given symbol or a succession of symbols in a given order tends to a definite limit. It seems therefore that one is dealing with a stationary stochastic process.) The fact that doublets like XX never occur shows that the sequence is not completely random. Moreover, the absence of the triplet YXY rules out the possibility that the sequence defines a simple Markovian process since in this case

$$P(YXY) = p_s(Y)P(X|Y)P(Y|X)$$

all of which are different from zero. We have to look therefore for a higher order Markov process whereby the memory of an event extends several steps backward. As shown in Tables 1 and 2, good agreement with the numerically computed frequencies of septuplets of symbols is reached by assuming a fifth order Markov chain (that the chain should be of at least fourth order could be inferred from the existence of the longest hypersymbol and from the fact that Z is always preceded by X.):

$$P(ABCDEFG) = P(ABCDEF) \ P(G|BCDEF)$$

$$= P(ABCDE) \ P(F|ABCDE) \ P(G|BCDEF) \qquad (12)$$

where $P(G|BCDEF)$ represents the numerically computed conditional probability of the last symbol given the quintuplet BCDEF. As shown in the preceding section, this guarantees automatically the asymmetry of the generated sequences.

A close inspection of Table 1 reveals some surprising properties. For instance of all possible 3^7 sequences long of 7 symbols that can be constructed from X, Y and Z, only 21 are realized by the dynamics. Moreover, for about half of them the conditional probablity of the last symbol given the five preceding ones turns out to be equal to one. Everything happens therefore as if the system is endowed with a set of "grammatical" rules followed automatically as a result of the dynamics.

The statistical properties of the hypersymbols can be studied in a similar manner. One finds that their sequences involve a much larger number of combinations compared to the sequences of symbols. This suggests that the hypersymbol sequence is a Markov chain of lower order than the one for the symbols.

The above procedure has been repeated by changing the value of parameter a of model (7) to 0.40. The results are qualitatively similar, including the possibility to express the sequences in terms of hypersymbols, the only change being in the numerical values of the probabilities of the different multiplets. Similarly, a small change in the threshold values leaves the main results unchanged. On the other hand, a large change in the value of the thresholds has drastic consequences. For instance, using as threshold for X, Y and Z the time averages of these quantities one finds sequences which cannot be expressed entirely in terms of hypersymbols. Moreover, a Markov chain model of order up to six is unable to fit the data adequately.

The information content of the sequence of X, Y and Z can be computed from an extension of Shannon's definition [8,16,17] where

Table 1:

Sequence	Probability deduced from counting	Probability inferred from a fifth order Markov chain (Eq. (12))	Conditional probability	No. of occur-rences
XYXZXYX	.04233	.04233	1.000	423
XYXZXZY	.01871	.01871	1.000	187
XYXZYXZ	.06484	.06484	1.000	648
XZXYXZX	.06104	.06104	.4849	610
XZXYXZY	.06484	.06484	.5152	648
XZXZYXZ	.03552	.03552	1.000	355
XZYXZXY	.08347	.08347	.8324	834
XZYXZXZ	.01681	.01679	.1677	168
XZYXZYX	.04152	.04152	1.000	415
YXZXYXZ	.12590	.12590	1.000	1258
YXZXZYX	.03552	.03552	1.000	355
YXZYXZX	.07565	.07516	.7113	756
YXZYXZY	.03062	.03113	.2879	306
ZXYXZXY	.04233	.04232	.6935	423
ZXYXZXZ	.01871	.01871	.3066	187
ZXYXZYX	.06484	.06484	1.000	648
ZXZYXZX	.02461	.02510	.6930	246
ZXZYXZY	.01091	.01039	.3071	109
ZYXZXYX	.0855	.08355	1.000	835
ZYXZXZY	.01681	.01681	1.000	168
ZYXZYXZ	.04152	.04152	1.000	415

Table 2: Statistical test of the Markovian character of the sequences as deduced from the χ^2-distribution

Orders compared	Degrees of freedom	Values of χ^2
1st - 0th	1	9166
2nd - 1st	2	2954
3rd - 2nd	4	457.6
4th - 3rd	8	137.8
5th - 4th	16	42.73
6th - 5th	32	.4599

the state i is now given by a quintuplet of symbols. It is found to be finite, but much less than the maximum corresponding to equiprobability. By producing these sequences as a natural outcome of its nonequilibrium dynamics, out of the much more numerous random sequences that can be envisaged, the system has thus become a generator of information [17]. Notice, however, that each individual message remains unpredictable as witnessed by the stochastic character of the sequence. On the other hand, the degree of unpredictability is tempered by the existence of correlations extending over five nearest neighbours of the chain. To borrow a term used in communication theory [8, 17], we can say that this enables the system to **compress** the information contained in the sequences of X, Y, Z by reproducing them more succinctly in terms of the hypersymbols α, β, γ. We thus find, on our simple model, one of the essential aspects of biological information pointed out in the Introduction.

20.4 Discussion

The origin of information raises automatically two issues, which appear to be mutually contradictory. On the one side information must be associated with some kind of randomness – whence the numerous attempts to relate it to "ignorance" and to "incomplete specification" of a system. But on the other side, the astronomically large number of random sequences makes it extremely improbable to select, on *a priori* grounds, the particular class of sequences that is likely to play the major role in an observed phenomenon. For instance, as pointed out by Eigen [18], the number of random polypeptides of a moderate length, say 100 units, which have to be scanned to select a biologically meaningful protein is of the order of 10^{130}. On these grounds, the spontaneous origin of information as a result of a purely probabilistic game must be ruled out.

In this paper we have suggested that informationally meaningful structures can be generated from an underlying mechanism which is nonlinear, time irreversible, and operating in the far-from-equilibrium chaotic region. In this way randomness and asymmetry – two prerequisites of information – are incorporated from the outset in the resulting structure. In addition, being the result of a mechanism – which has ultimately to do with the existence of molecules endowed with suitable catalytic properties – these structures automatically overcome the difficulty of the tremendous "thermodynamic improbability" characterizing random sequences.

Because of the dissipative character of the dynamics, the system can run on an asymptotically stable attractor, a situation which would be impossible had the dynamics been conservative. As a result, the sequences generated by the mechanism discussed in sections 2 and 3 are structurally stable towards small changes of parameters, including the threshold values L_i assigned.

Specifically, although the particular succession of symbols may be modified considerably, their statistical properties remain essentially unchanged. This provides us with a dynamical analogue of a mutation. It also suggests that it is more meaningful to speak of families of sequences rather than of a preferred one, a situation which is somewhat reminiscent of the quasispecies concept of Eigen and Schuster [18].

Naturally in order that the scheme described in this paper be tested against real-world biopolymers carrying biological information, it will be necessary to identify the series of reactions which may result in such polymers or precursors thereof, for instance in the context of prebiotic chemistry. In this respect, it is worth noting that the distribution of configurational sequences in vinyl polymers obtained from polymerizing monomers like CH_2 = CHF or even the symmetrically substituted monomers such as CHF = CHF, can be accounted for by a third order irreversible Markov chain [19,20]. More to the point, the analysis of DNA sequences of various organisms [21] suggests that there exist correlations between the nucleotides. Finally, evidence has recently accumulated that tRNA and rRNA share deep structural affinities [22], suggesting that these two species may have derived from a common group of ancestral molecules. A primordial chaotic dynamics of a suitable set of chemical reactions would provide a mechanism for such an ancestor. In any case, the generality of the arguments invoked in our analysis, suggests that whatever the details of these reactions might be, biological information was bound to have a one-dimensional character, and to arise in a world of broken time symmetry.

Acknowledgements

We express our gratitude to Professors I. Prigogine and H. Frisch for stimulating discussions. J.S.R. and G.S.R. would like to thank Prof. I. Prigogine and their respective institutions (Jawaharlal Nehru University, New Delhi and All India Institute of Medical Sciences, New Delhi) for providing them with the opportunity to work at Brussels. This research was supported, in part, by the U.S. Energy Department under contract number DE-AS05-81ER10947.

REFERENCES

[1]   S. Luria (1975). *36 Lectures in Biology,* MIT Press, Cambridge, Mass.

[2]   P. Glansdorff and I. Prigogine (1971). *Thermodynamic theory of structure, stability and fluctuations,* Wiley, London.

[3]   G. Nicolis and I. Prigogine (1977). *Self-organization in non-equilibrium systems,* Wiley, New York.

[4]   A. Turing (1952). *Phil. Trans. Roy. Soc.* **B237,** 37.

[5]   C. Vidal and A. Pacault (1984). *Nonequilibrium dynamics in chemical systems,* Springer, Berlin.

[6]   G. Nicolis and I. Prigogine (1981). *Proc. Nat. Acad. Sci. (USA),* **78,** 659.

[7]   I. Prigogine (1980). *From Being to Becoming,* Freeman, San Francisco.

[8]   A. Khinchin (1957). *Mathematical Foundations of Information Theory,* Dover, New York.

[9]   J. Guckenheimer and Ph. Holmes (1983). *Nonlinear oscillations, dynamical systems, and bifurcations of vector fields,* Springer, Berlin.

[10]   W. Feller (1957). *An introduction to probability theory and its applications,* Wiley, New York.

[11]   H.G. Schuster (1984). *Deterministic chaos,* Physik-Verlag, Weinheim.

[12]   J. Kemeny, J. Snell and A. Knaap (1976). *Denumerable Markov Chains,* Springer, Berlin.

[13]   N. Van Kampen (1981). *Stochastic processes in physics and chemistry,* North-Holland, Amsterdam.

[14]   O. Rössler (1979). *Ann. New York Acad. Sci.* **316**, 376.

[15]   P. Gaspard and G. Nicolis (1983). *J. Stat. Phys.* **31**, 449.

[16]   W. Ebeling and R. Feistel (1982). *Physik der Selbst-organisation and Evolution,* Akademie-Verlag, Berlin.

[17]   J. Nicolis (1986). *Dynamics of hierarchical systems,* Springer, Berlin.

[18]   M. Eigen and P. Schuster (1979). *The Hypercyle,* Springer, Berlin.

[19]   H. Frisch (1984). *Adv. Chem. Phys.* **55**, 201.

[20]   H. Frisch, C. Mallows and F. Bovey (1966). *J. Chem. Phys.* **45**, 1565.

[21]   D. Lipman and W. Wilbur (1983). *J. Mol. Biology* **163**, 363.

[22]   A. Nazarea, D. Bloch and A. Semrau (1985. *Proc. Nat. Acad. Sci. (USA)* **82**, 5537.

PROTOTYPE LONG-SCALE EQUATIONS OF DISSIPATIVE DYNAMICS

Abstract

Long-scale equations valid in the vicinity of singular bifurcation manifolds are simplest forms retaining the variety of behaviour of underlying systems, and are analogous in this respect to normal forms of the singularity theory. Among different forms of long-scale equations, dependent on the type of the bifurcation and on the symmetry of the problem, of special interest are forms with built-in multiple scaling that combine non-dissipative dynamics on a relatively fast time scale with slower dissipation.

21.1 Introduction

Non-equilibrium systems exhibit universal behaviour near bifurcation points, that are analogous in this respect to critical points of equilibrium systems [1]. More complex behaviour, that still remains model-independent, can be observed near multicritical points or near degenerate (singular) bifurcations. Dynamics in the proximity of singular bifurcation points of different types is characterized by long-scale prototype equations (normal forms) that are uniquely defined by the type of the singularity. Hierarchies of equations of this kind, growing increasingly complex with increasing codimension of the singularity, have been derived for dynamical systems where state variables depend on time only [2,3]. Adding a weak spatial modulation leads to prototype partial differential equations that can describe both relaxation to stationary or time-dependent homogeneous states and spatial symmetry breaking phenomena, such as formation of stationary structures, waves and spatio-temporal turbulence.

In the following, we shall derive long-scale equations of several types taking as a starting point the reaction-diffusion system. This system has an advantage of combining a simple general structure with qualitatively rich behaviour and inexhaustible stock of available parameters. Long-scale equations associated with either bifurcation at simple zero eigenvalue (section 21.2) or Hopf bifurcation (section 21.3) can be obtained by straightforward expansion, leading to equations of Landau-Ginzburg type for either real or complex order parameter (amplitude). Additional parametric restrictions, that either increase multiplicity of solutions, or cancel effective amplitude diffusivity, can be imposed to arrive at equations with more elaborate structure and richer variety of behaviour.

We shall pay most attention to the analysis of long-scale equations that possess a double-scale structure, such that approximate symmetries (not inherent to the underlying system) appear on an intermediate scale. The examples of highly symmetric intermediate forms, suggesting a general structure of families of solutions, are given by the nonlinear Schrödinger equation at a singular Hopf bifurcation (section 21.2) and the nonlinear Klein-Gordon equation in the vicinity of bifurcation at double zero eigenvalue (section 21.4).

21.2 Simple Zero Eigenvalue

Consider a reaction-diffusion system with a constant matrix of diffusivities D and a nonlinear source function $f(u;p)$ dependent on the state vector u and the parametric vector p:

$$du/dt = D \cdot \nabla^2 u + f(u;p) \qquad (1)$$

Let $u = u_0(p)$ be a homogeneous stationary solution satisfying $f(u_0;p_0) = 0$, and p_0 be fixed at a codimension one bifurcation manifold where the Jacobi matrix $f_u(u_0;p_0)$ has a simple zero eigenvalue; it is presumed that all other eigenvalues of f_u have negative real parts. Then normal forms can be obtained by straightforward expansion of variables and parameters in powers of a dummy small parameter ϵ. The source function in (1) is expanded as

$$f(u;p) = f_u \cdot u + f_p \cdot p + \frac{1}{2} f_{uu} : \tilde{u}\tilde{u} + f_{up} : \tilde{u}\tilde{p} + \frac{1}{6} f_{uuu} \therefore \tilde{u}\tilde{u}\tilde{u} + \ldots \qquad (2)$$

where $\tilde{u} = u - u_0$, $\tilde{p} = p - p_0$ and f_u, f_p, f_{uu}, etc. are matrices of derivatives with respect to the variables and parameters computed at $u = u_0$, $p = p_0$. To make the forthcoming expressions more transparent, we restrict to a generic case when all eigenvalues λ_i of f_u are distinct, setting

$$\tilde{u} = \phi U + \Sigma \chi_i V_i \qquad (3)$$

with ϕ, $\chi_i = O(\epsilon)$, and project the Taylor expansion of (1) on to the basis (U, V_i) formed by eigenvectors of f_u ($f_u \cdot U = 0$, $f_u \cdot V_i = \lambda_i V_i$, $\text{Re}\lambda_i < 0$). Since amplitude χ_i corresponding to vectors V_i relax to quasistationary values on the rapid time scale, their time dependence can be expressed through the long-scale derivative $\dot{\phi}$ of a single amplitude ϕ:

$$\dot{\phi}=\delta\nabla^2\phi+K+\Sigma L_i\chi_i+\Sigma\delta_i\nabla^2\chi_i+\frac{1}{2}M\phi^2+\phi\Sigma M_i\chi_i+\frac{1}{2}\Sigma M_{ij}\chi_i\chi_j+\frac{1}{6}N\phi^3+\ldots \quad (4)$$

$$\chi_i=-\lambda_i^{-1}(-\phi d\chi_i/d\phi+K^i+\Sigma L_j^i\chi_j+\delta^i\nabla^2\phi+\Sigma\delta_j^i\nabla^2\chi_j+\frac{1}{2}M^i\phi^2). \quad (5)$$

Projections of matrices of derivatives are defined as

$$K=U^\dagger\cdot f_p\cdot\tilde{p}, \quad L=U^\dagger\cdot f_{up}:U\tilde{p}, \quad M=U^\dagger\cdot f_{uu}:UU, \quad N=U^\dagger\cdot f_{uuu}\therefore UUU,$$

$$K^i=V_i^\dagger\cdot f_p\cdot\tilde{p}, \quad M^i=V_i^\dagger\cdot f_{uu}:UU, \quad M_i=U^\dagger\cdot f_{uu}:UV_i, \quad \text{etc.} \quad (6)$$

i.e. indices in subscript and superscript positions are added, respectively, when U is changed to V_i or $U^\dagger$ to $V_i^\dagger$, the latter being eigenvectors of the transposed matrix f_u. The amplitude diffusivities in (5) are defined in a similar way:

$$\delta=U^\dagger\cdot D\cdot U, \quad \delta_i=U^\dagger\cdot D\cdot V_i, \quad \text{etc.} \quad (7)$$

The equation of ϕ is obtained from (4) after amplitudes of "slaved" modes χ_i have been found to a desired order in ϵ using (5). Parametric conditions, required to suppress evolution on successive time scales t_k extended by factors $\epsilon^{-k}(k=1,2\ldots)$ relative to the original scale of (1), coincide with well-known conditions of cusp, swallowtail, etc. singularities:

$$M(p_0)=0, \quad \hat{N}(p_0)=N-3\Sigma\lambda_i^{-1}M^iM_i=0, \quad \text{etc.} \quad (8)$$

Changes on a spatial scale extended by a factor $\epsilon^{-1/2}$ have to be excluded whenever changes on a temporal scale t_k are suppressed. Conditions restricting parametric deviations are obtained alongside (8) in respective orders. This procedure leads to a sequence of normal forms

$$\dot{\phi}=\delta\nabla^2\phi+F_k(\phi) \quad (9)$$

where $F(\phi)$ is a kth degree polynomial

$$F(\phi) = \sum_{i=0}^{k-2} a_i\phi^i - \phi^k. \quad (10)$$

All coefficients of $F(\phi)$ depend on parametric deviations, except the coefficient at the leading term that can be typically rescaled to -1 without reversing time. The time and spatial coordinates in (9) are extended, respectively, by the factors ϵ^{k-1} and $\epsilon^{(k-1)/2}$.

Equation (9) is applicable only at $\delta > 0$; otherwise, developing short-scale inhomogeneities invalidate the assumed spatial scaling. Note that (9) possesses gradient structure (absent in the original system (1)), and its stationary states are extrema of the functional

$$\Phi = \int [\tfrac{1}{2}\delta |\nabla\phi|^2 + V(\phi)]dx; \quad V(\phi) = -\int F_k(\phi)d\phi. \quad (11)$$

The simplest form (9) does not exhibit spontaneous symmetry breaking, but can describe propagating waves that effect transitions between its homogeneous stationary states satisfying $F_k(\phi)=0$. This equation underscores a far-reaching analogy between this kind of transition involving alternative nonequilibrium stationary states and phase transitions in equilibrium systems. Indeed, (9) has the same form as the space-dependent Landau-Ginzburg equation of phase transitions [4], that also can include higher degree polynomials at multicritical points. Another related form (symmetric in ϕ but spatially anisotropic) is the amplitude equation of Segel-Newell-Whitehead [5].

The singular point $\delta=0$ marks the borderline between systems stable to inhomogeneous perturbations and those exhibiting spontaneous spatial symmetry breaking. When δ is negative but small, spontaneous inhomogeneities develop on an extended scale, and can be described by equations of universal structure.

Substituting χ_i from (5) into (4) yields

$$\dot{\phi} = \delta\nabla^2\phi - \eta_0(\nabla^2)^2\phi - \eta_1|\nabla\phi|^2 - (\eta_1+\eta_2)\phi\nabla^2\phi + \tfrac{1}{2}M\phi^2 + \tfrac{1}{6}\hat{N}\phi^3 + \ldots \quad (12)$$

with

$$\eta_0 = \Sigma\lambda^{-1}_i\delta_i\delta^i, \quad \eta_1 = \Sigma\lambda_i^{-1}\delta_i M^i, \quad \eta_2 = \Sigma\lambda_i^{-1}\delta^i M_i \quad (13)$$

and $\hat{N}$ given by (8). Near a generic (tangent) bifurcation the appropriate scaling for (12) is $\phi=O(\epsilon)$, $\partial_t=O(\epsilon)$, $\nabla=O(\epsilon^{1/4})$. Allowing for parametric deviations of a suitable order of magnitude we obtain in the leading order (after rescaling)

$$\dot{\phi} = \delta\nabla^2\phi - \eta_0(\nabla^2)^2\phi + a_0 - \phi^2. \quad (14)$$

The biharmonic term stabilizes inhomogeneities on short scales, provided $\eta_0 > 0$. Nonlinear differential terms turn out to be of a smaller order of magnitude here, but they appear in the leading order in the proximity of the cusp singularity at $M=0$. The scaling is changed then to $\phi=O(\epsilon)$, $\partial_t=O(\epsilon^2)$, $\nabla=O(\epsilon^{1/2})$. After rescaling both dependent and independent variables (assuming $\delta=O(\epsilon)<0$ and $N<0$) we obtain the equation

$$\dot{\phi} = -\nabla^2\phi - \eta_0(\nabla^2)^2\phi - \eta_1|\nabla\phi|^2 - (\eta_1 + \eta_2)\phi\nabla^2\phi + a_0 + a_1\phi - \phi^3. \qquad (15)$$

The spatial differential terms coincide, not unexpectedly, with those from the theory of long-scale phase inhomogeneities of Sivashinsky and Kuramoto [6], but the effective diffusivity becomes amplitude-dependent. Other terms containing the amplitude are the same as in (9). Equation (15) does not need to be connected with the reaction-diffusion system (1); in particular, it has been derived and studied in connection with a problem of Marangoni convection with chemical reaction [7].

The only symmetry-breaking bifurcation from homogeneous stationary states that is observed in (14), (15) leads to formation of ordered stationary patterns. Equation (14) still retains gradient structure with the potential

$$\Phi = \int \left[\frac{1}{2}\delta|\nabla\phi|^2 - \frac{1}{2}\eta_0(\nabla^2\phi)^2 + a_0\phi - \frac{1}{3}\phi^3\right]dx. \qquad (16)$$

This potential, however, reaches a local minimum at an inhomogeneous state when the amplitude diffusivity is sufficiently negative. As follows from the linear stability analysis of (14), formation of stationary patterns with the incipient wavenumber $k=|\delta/2\eta_0|^{1/2}$ is favoured at $\delta < -2\eta_0^{1/2}a_0^{1/4}$. Equation (15) lacks gradient structure and can generate complex nonstationary behaviour away from the primary symmetry-breaking bifurcation point. The likelihood of turbulent behaviour under suitable conditions is indicated by the similarity between (15) and the Kuramoto-Sivashinsky equation, which can be obtained just by removing amplitude-dependent terms (that would violate the symmetry to translations of the phase variable) from (15).

21.2 Hopf Bifurcation

At the locus of the Hopf bifurcation $p=p_0$, the Jacobi matrix f_u has a pair of imaginary eigenvalues $\pm i\omega$, while its other eigenvalues have negative real parts. In the vicinity of $p = p_0$, the solution is presented in the leading order as

$$\tilde{u} = \epsilon\phi e^{i\omega t}U + c.c. \qquad (17)$$

where $f_u \cdot U = i\omega U$, and ϕ is a complex amplitude. Standard bifurcation analysis yields the amplitude equation

$$\dot{\phi} = \delta\nabla^2\phi + \mu\phi - \nu|\phi|^2\phi. \qquad (18)$$

The linear growth coefficient μ depends on parametric deviations from the locus of the Hopf bifurcation, while the amplitude diffusivity δ and the nonlinear interaction parameter ν are defined on the locus $p=p_0$ itself. All parameters of (18) are complex, $\mu = \bar{\mu} + i\tilde{\mu}$, etc. The temporal and spatial scales are extended, respectively, by the factors ϵ^{-2} and ϵ^{-1}.

Equation (18) has stable nontrivial solutions only at $\bar{\nu}>0$, $\bar{\delta}>0$. Conditions $\bar{\nu}=0$, $\bar{\delta}=0$ define loci of singular bifurcations on the hypersurface $p=p_0$. Equations describing dynamics in the vicinity of these higher-codimension manifolds in the parametric space can be derived by continuing the bifurcation analysis of Eq. (1) to higher orders. These equations should imply different scaling of extended spatial and temporal variables, and contain higher-order terms necessary to stablize small-amplitude solutions.

At the singular point $\bar{\nu}=0$, the scaling has to be changed to $\partial_t=0(\epsilon^4)$, $\nabla=0(\epsilon^2)$, and a fifth-order nonlinearity is brought in to ensure the stabilization at large amplitudes. Further parametric restriction, analogous to singularity conditions involved in higher-order equations (10), can be imposed to arrive at the hierarchy of equations yielding up to n periodic orbits surrounding a single stationary state:

$$\dot{\phi} = \delta\nabla^2\phi + \sum_{k=0}^{n} \nu_k|\phi|^{2k}\phi. \qquad (19)$$

The condition $\bar{\delta}=0$ marks the borderline between bifurcation of homogeneous oscillations and propagating waves. Near this point, it is necessary to use two time scales, $T=\epsilon t$ and $\tau=\epsilon^2 t$, extended by factors of different order of magnitude; the appropriate spatial scaling is $\nabla=\epsilon^{1/2}$. The equation obtained on the intermediate time scale is the Schrödinger equation $i\phi_T+\tilde{\delta}\nabla^2\phi=0$. The amplitude modulation on a slower time scale is described by an evolution equation containing a stabilizing biharmonic term:

$$\phi_\tau = \mu'\phi + \delta'\nabla^2\phi - \sigma(\nabla^2)^2\phi - \nu|\phi|^2\phi. \qquad (20)$$

A new feature introduced by this equation is a primary bifurcation of wave patterns at $\bar{\delta}'<0$.

Equations (19) and (20), though generating more complex dynamics than the generic amplitude equation (18), still do not help to elucidate the most interesting phenomenon of self-focussing, or Benjamin-Feir instability [8] that occurs in (18) at $\mathrm{Re}(\nu\delta^*)<0$. If the complex amplitude is presented as $\phi=\rho e^{i\theta}$, and the phase θ is allowed to vary on a spatial scale long compared to that of (18), its evolution can be described by the Burgers equation containing the phase diffusivity $\mathrm{Re}(\nu\delta^*)$. If the latter is negative but small, the incipient phase turbulence under conditions of weak self-focussing can be described by the Kuramoto-Sivashinsky equation [6] containing a stabilizing biharmonic term. Chaotic behaviour of solutions of (18) under conditions of strong self-focussing can be explored, however, only numerically [9]. In order to reduce this kind of behaviour to a form amenable to analytical studies, one has to ascend one more step on the ladder of singular bifurcations.

If both $\bar{\delta}=0$ and $\bar{\nu}=0$ hold simultaneously, the intermediate scale amplitude equation is the nonlinear Schrödinger equation (NLS)

$$i\phi_T + \tilde{\delta}\nabla^2\phi - \tilde{\nu}|\phi|^2\phi = 0. \tag{21}$$

The scaling is the same as in Eq. (18), i.e., $\nabla=O(\epsilon)$, and the extended time variable is $T=\epsilon^2 t$. Parametric deviations from the locus of Hopf bifurcation are restricted here to $O(\epsilon^4)$, thereby eliminating the linear growth term. Equation (21) has soliton solutions when $\tilde{\delta}$ and $\tilde{\nu}$ have opposite signs [10]. Note that this is the case when the phase stability condition $\mathrm{Re}(\nu\delta^*) = \tilde{\nu}\tilde{\delta} > 0$ is violated. Thus, soliton dynamics in the vicinity of this singular bifurcation point can give an insight into elusive nonlinear development of self-focussing instability.

NLS equation has an infinite number of conservation laws that are not inherent to the underlying system (1). These symmetries can be therefore only approximate, and are broken when higher-order corrections are taken into account. Clearly (21) is not sufficient for description of bifurcating patterns, and has to be complemented by an evolution equation on a slower time scale. In the sequel, we restrict to a one-dimensional system (thus avoiding the problem of lateral stability) and show how the slow evolution selects solitons with a certain amplitude and wavelength out of the two-parametric family of one-soliton solutions of (21).

At $\tilde{\delta}>0$, $\tilde{\nu}<0$, (21) is rescaled to the standard form

$$i\phi_T + \phi_{xx} + 2|\phi|^2\phi = 0. \tag{21}$$

The one-soliton solutions of (21a) can be presented as

$$\phi = q(\zeta)\,\exp[i(k\zeta+\Omega T)] \tag{22}$$

where $\zeta=x-cT$ is a moving coordinate, and

$$c=2k, \quad \Omega=\eta^2+k^2, \quad q = \eta\ \mathrm{sech}\ \eta\zeta. \tag{23}$$

The amplitude η and the wavenumber k are allowed to vary on the slow time scale $\tau = \epsilon^2 T = \epsilon^4 t$. The evolution equation for these parameters can be obtained from higher-order solvability conditions of the bifurcation expansion. In the course of ϵ-expansion in the vicinity of a Hopf bifurcation, solvabiilty conditions apepar only in odd orders, since even-order equations do not contain secular terms. Introducing a $O(\epsilon^2)$ correction ψ to the amplitude ϕ, we obtain in the fifth order of the bifurcation expansion an equation of ψ in the form

$$L\psi = iF(\phi) \qquad . \tag{24}$$

where the operator L is a linearization of NLS and the inhomogeneity $F(\phi)$ is

$$F(\phi) = \mu'\phi + \delta'\phi_{xx} - \nu'|\phi|^2\phi - \sigma\phi_{4x} + \kappa_1|\phi|^2\phi_{xx} + \kappa_2\phi^2\phi_{xx}^*$$

$$- \kappa_3\phi_x\phi_x^{*2}\phi + \kappa_4\phi_x^2\phi^* - \lambda|\phi|^4\phi - \phi_\tau. \tag{25}$$

The first term is contributed by fourth-order deviations of parameters affecting the Jacobi matrix f_u. The second term is due to $O(\epsilon^2)$ deviations of components of the diffusivity matrix D. The third term is due to $O(\epsilon^2)$ deviations of parameters affecting the nonlinearities in $f(u)$. Coefficients of other terms are evaluated at the singular bifurcation point itself; the notation in (25) is chosen in such a way that higher-order terms are stabilizing when their coefficients are positive.

Both (21a) and (24) can be combined into a single equation

$$i\phi_T + \phi_{xx} + 2|\phi|^2\phi = i\epsilon F(\phi)$$

with $F(\phi)$ given by (25), excluding the last term. This places the multiscale bifurcation expansion into a more familiar context of studies of perturbed soliton systems [10,11].

Presenting ψ, analogous to (22), as

$$\psi = [\bar{\rho}(\zeta,\tau)+\tilde{\rho}(\zeta,\tau)]\exp i[k(\tau)\zeta+\Omega(\tau)T] \tag{26}$$

and separating the real and imaginary parts in (24) yields

$$L_1\bar{\rho} + \tilde{F}(\phi) = 0, \quad L_2\tilde{\rho} - \bar{F}(\phi) = 0 \tag{27}$$

where $\bar{F}(\phi)=\text{Re}[Fe^{-i(k\zeta+\Omega T)}]$, $\tilde{F}(\phi) = \text{Im}[Fe^{-i(k\zeta+\Omega T)}]$, and L_1, L_2 are self-adjoint operators

$$L_1 = \partial_\zeta^2-\eta^2+6q^2, \quad L_2 = \partial_\zeta^2-\eta^2+2q^2, \tag{28}$$

The homogeneous equations $L_1\bar{\rho}=0$, $L_2\tilde{\rho}=0$ are satisfied, respectively, by $\bar{\rho}=q_\zeta$, $\tilde{\rho}=q$. By Fredholm alternative, the solvability conditions of the inhomogeneous equations (27) are

$$\langle qF(\phi)\rangle = 0, \quad \langle q_\zeta\tilde{F}(\phi\rangle = 0 \tag{29}$$

where $\langle...\rangle = \int_{-\infty}^{\infty}...d\zeta$. Since $q(\zeta)$ is even and q_ζ is odd, only even part of $\bar{F}(\phi)$ and odd part of $\tilde{F}(\phi)$ bear upon the solvability conditions. Using (22), (25) in (29) yields evolution equations for the soliton parameters

$$\frac{1}{2}\eta_\tau = \eta(\bar\mu - \bar\delta'k^2 - \sigma k^4) - \eta^3(\frac{1}{3}\bar\delta' + \frac{2}{3}\bar\nu' + C_1 k^2) - C_2\eta^5$$

$$\frac{1}{2}k_\tau = -\frac{2}{3}k\eta^2(\bar\delta' + 2k^2\bar\sigma - C_3\eta^2) \tag{30}$$

where constants C_i are expressed through real part of parameters of (25).

If both "superviscosity" $\bar\sigma$ and the coefficient C_2 in (30) are positive, both runaway to high amplitudes and unlimited acceleration of solitons are ruled out. Under these conditions, stable stationary solutions (30) correspond, in different parametric domains, to (a) trivial solution ($\eta=0$, k indefinite), (b) a standing soliton ($\eta\neq0$, k=0), or (c) a soliton propagating with a certain amplitude-dependent speed in either direction ($\eta\neq0$, k≠0). Both standing and propagating solutions can bifurcate from the trivial state either supercritically or subcritically. In the latter case, solitons are formed via a first-order transition, and hysteresis is observed. The propagating solution can also undergo a Hopf bifurcation leading to solitons with the amplitude and speed changing periodically on the slow time scale.

Further analysis is required to incorporate radiation and interaction between solitons. At sufficiently low densities (not exceeding η^{-1}), pair interaction in arrays of solitons (either standing or propagating in the same direction) can be taken into account by evaluating overlap integrals between one-soliton solutions. Interaction can be either repulsive or attractive, depending on imaginary parts of parameters appearing in (25). Repulsive interaction would stabilize uniformly spaced arrays or trains, while attractive interaction might bring about "condensation" of the soliton gas. The soliton interaction is likely to chaotize motion of solitons with periodically varying amplitudes and speeds.

21.3 Double Zero Eigenvalue

Equations admitting a still larger variety of primary bifurcations can be obtained in the proximity of bifurcation at double zero eigenvalue (degenerate geometrically as well as algebraically). As a starting point, one can take the reaction-diffusion system (1) expanded in Taylor series and projected on to an appropriate Jordan basis (U, V):

$$\dot\phi = \delta^*\Delta^2\phi + \delta_*^*\nabla^2\psi + \psi + F^*(\phi,\psi)$$

$$\dot\psi = \delta\nabla^2\phi + \delta_*\nabla^2\psi + F(\phi,\psi). \tag{31}$$

Since the amplitudes ϕ and ψ have to be scaled in a different way, straightforward expansion can be misleading, and an alternative method of nonlinear near-identity transformations is preferrable.

A suitable nonlinear transfomation is

$$\phi \to \phi, \quad \psi \to \dot{\phi} - \delta^* \nabla^2 \phi - \delta_*^* \nabla^2 \psi - F^*(\phi, \psi). \tag{32}$$

This transformation differs from one suggested for perturbative reduction of dynamical systems to normal forms [3] by added terms containing spatial derivatives. Since spatial variations are supposed to occur on an extended scale, the differential terms are small, and (32) remains a near-identity transformation that can be inverted perturbatively to the needed order. A variety of normal forms can be obtained at singular bifurcation points of higher codimension belonging to the two-dimensional hierarchy normal forms of dynamical systems [3]. This variety is further increased by the possibility to impose conditions on diffusivities, like $\delta=0$.

The lowest-order form obtained at the generic (tangent) bifurcation is

$$\dot{\phi} = \nabla^2 \phi + a_0 - \phi^2 + \epsilon[(b_0 + b_1\phi)\dot{\phi} + \eta\nabla^2\dot{\phi}] \tag{33}$$

where the amplitude diffusivity δ, presumed positive, is rescaled to unity, and $\eta = (\delta^* + \delta_*)/\delta$. The implied scaling is $\phi=O(\epsilon^2)$, $\partial_t=O(\epsilon)$, $\nabla=O(\epsilon)$; the parameters $a_0=O(\epsilon^4)$, $b_0=O(\epsilon^2)$ depend on parametric deviations.

Equation (33) truncated at $\epsilon=0$ is the nonlinear Klein-Gordon equation (NKG), that is conservative and Lorenz-invariant. Waveform solutions of the truncated equation can be expressed as functions of the eikonal $\Theta-\omega t-\underline{k}\cdot\underline{x}$, where ω is frequency and $\underline{k}$ is the wavenumber vector. These solutions satisfy

$$m^2\phi_{\Theta\Theta}+V'(\Theta)=0 \tag{34}$$

where $m^2=\omega^2-k^2$ is the "squared mass". This equation is integrated to

$$\phi_\Theta = (\sqrt{2}/m)[E-V(\phi)]^{1/2}. \tag{35}$$

The integration constant E has a meaning of energy, as becomes clear when (35) is rearranged as

$$E = m^2\phi_\Theta^2/2+V(\phi). \tag{35a}$$

The relationship between energy and mass is obtained by presenting the half-period as

$$\pi = \int_{\phi_-}^{\phi_+} \frac{d\phi}{\phi_\Theta} = \frac{m}{\sqrt{2}} \int_{\phi_-}^{\phi_+}[E-V(\phi)]^{-1/2}d\phi. \tag{36}$$

The integration limits $\phi_\pm$ correspond to points where $E-V(\phi)$

changes its sign. The integration interval includes the stable stationary point $\phi=1$ for "timelike" solutions (m real, E–V positive) and the unstable stationary point $\phi=-1$ for "spacelike" solutions (m imaginary, E–V negative). Periodic solutions of (34) exist within the interval $-2/3<E<2/3$. Due to the Lorentz invariance, these two one-parametric families expand into two-parametric families of timelike and spacelike waveforms parametrized by energy and wavenumber k.

Both Lorenz invariance and the conservation law are broken when $O(\epsilon)$ terms in (33) are taken into account. These "dissipative" terms cause a drift of energy on an extended time scale. A waveform satisfying (34) can remain, to the leading order, a steady solution of the untruncated equation (33) if the dissipation averaged over the period vanishes.

The stationary condition can be derived in a simple way by differentiating (35a) with respect to Θ and using (33) to obtain

$$E_\Theta = \phi_\Theta (m^2 \phi_{\Theta\Theta} + V') = \epsilon(\omega/m)\phi_\Theta \Phi(\phi);$$

$$\Phi(\phi) = m[(b_0 + b_1\phi)\phi_\Theta + \eta k^2 \phi_{\Theta 3}] = [b_0 + b_1\phi - \eta k^2 V''/m^2]m\phi_\Theta. \tag{37}$$

It is convenient to transform to the coordinate frame where the solution is either uniform in space (when it is timelike) or time-independent (when it is spacelike). The transformation to this "comoving" frame is effected by a Lorenz boost with a speed $v<1$ ($v=k/\omega$ for timelike, and $v=\omega/k$, for spacelike waves). The Lorentz transformation affects only the dissipative part of (33), transforming (37) to

$$\Phi(\phi) = [b_0 + b_1\phi - \eta v^2 V''\sqrt{1-v^2}]m\phi_\Theta. \tag{37a}$$

Introducing an extended comoving coordinate $\theta=\epsilon\Theta$ and averaging over the period on the original short scale yields

$$E_\theta = \omega F(E, v) \tag{38}$$

where $F(E, v)$ is the average dissipation:

$$F(E, v) = \langle \Phi(\phi) \rangle = \langle (b_0 + b_1\phi) - \eta v^2 V''/\sqrt{1-v^2}]m\phi_\Theta \rangle;$$

$$\langle \ldots \rangle = \frac{1}{2\pi}\int_0^{2\pi} \ldots \phi_\Theta d\Theta = \frac{1}{\pi}\int_{\phi_-}^{\phi_+} \ldots d\phi. \tag{39}$$

The condition $F(E, v)=0$ selects a one-parametric family of steady waveforms out of the two-parametric family of solutions dependent on energy and wavenumber or speed. It is clear from (38) that a necessary condition for stability of these solutions is $\partial F/\partial E<0$.

A convenient explicit form of the stationarity condition is

$$F/J_1 = Rb_0 + b_1 - 2\eta k^2/m^2 = 0 \qquad (40)$$

where $R = J_0/J_1$ and

$$J_n(E) = \frac{m}{\pi} \int_{\phi_-}^{\phi_+} \phi^n \phi_\Theta d\Theta = \frac{\sqrt{2}}{\pi} \int_{\phi_-(E)}^{\phi_+(E)} \phi^n \sqrt{E-V(\phi)} \, d\phi. \qquad (41)$$

Recall that the parameter b_1 is defined on the bifurcation manifold, while b_0 depends on parametric deviations and can be varied at will. Only the sign of b_1 is relevant qualitatively, and this coefficient can be reduced by rescaling to ± 1. Both J_0 and J_1 increase with E from 0 at E = -2/3 to 12/5 and 12/7 respectively at E = 2/3, while the ratio R increases from 1 to 7/5.

If $b_1=1$, the oscillatory solution (k=0) emerges via a Hopf bifurcation at $b_0=-1$, and disappears via a saddle-loop bifurcation when this parameter increases to -5/7. Within this interval, the oscillatory solution exists at some energy level E* satisfying $b_0=-1/R(E^*)$ within the interval -2/3<E*<2/3.

A stationarily propagating timelike wave with a suitable k exists at $b_0>-1$ at any E from the interval -2/3<E<E*. As b_0 increases, stationary waves at each given energy level shift towards shorter wavelengths. At $b_0>-5/7$, waves exist in the entire interval -2/3<E*<2/3.

At $b_1=-1$, the oscillatory solution exists within the parametric interval 5/7<b_0<1, at the energy level E* satisfying $b_0=1/R(E^*)$. Waves with k≠0 exist at energy levels from the interval E*<E<2/3, and fill the entire interval -2/3<E*<2/3 when b_0 exceeds unity.

The stability condition $\partial F/\partial E<0$ can be reduced using (40) to

$$J_1^{-1}\partial F/\partial E = -(b_1 - 2\eta k^2/m^2)\partial \ln R/\partial E = b_0 \partial \ln R/\partial E < 0 \qquad (42)$$

Since R increases with E, it follows from (42) that at $b_1 = -1$ all solutions are unstable. At $b_1 = 1$, timelike waves are stable, provided $k^2 < m^2/2\eta$. Further analysis taking into account instabilities associated with changes of ω and k on an extended scale, will be reported elsewhere.

Several types of higher singularities can be obtained by annulling some coefficients of equation (33). Singularities common with corresponding dynamical systems [3] occur at either $a_2=0$ or $b_1=0$, and a two-dimensional hierarchy of singularities can be further ascended by annulling coefficients or higher-order terms. The corresponding hierarchy of normal form (cf. 11) is

$$\phi = \nabla^2 \phi + F_k(\phi) + \epsilon^{m-(k-1)/2} G_m(\phi)\dot{\phi} + \epsilon^{(k-1)/2}\eta\nabla^2\dot{\phi} \qquad (43)$$

where F_k, G_m are polynomials of degree k and m, and both time and spatial coordinates are extended by the factor $\epsilon^{(k-1)/2}$ when $\phi=0(\epsilon)$. At $m>(k-1)/2$, equations from this hierarchy have the same approximate symmetries as (33), and can be analysed in the same manner.

Singularities specific for distributed system correspond to two kinds of diffusional instabilities in (33) at $\delta=0$ and $\eta=0$. Catastrophic short-scale instability occurs when either of these coefficients is negative; the fastest growing mode corresponds to stationary patterns near $\delta=0$, and to "luminar" waves with the speed $\sqrt{\delta}$ near $\eta=0$. Higher-order differential terms have to be introduced in the vicinity of these points.

The lowest-order equation obtained near $\delta=0$ is (cf. 14)

$$\phi = \delta\nabla^2\phi-\delta^*\delta_*(\nabla^2)^2\phi+(\delta^*+\delta_*)\nabla^2\phi+a_0-\phi^2 \qquad (44)$$

The implied scaling is $\phi=0(\epsilon^2)$, $\partial_t=0(\epsilon)$, $\nabla=0(\epsilon^{1/2})$, $\delta=0(\epsilon)$. This equation does not have useful intermediate symmetries. The double-scale structure of (33) can be, however, retained in a suitably defined vicinity of another singular point $\eta=0$. At $\eta<0$, the catastrophic short-scale instability in (33) manifests itself in the unlimited growth of energy of "nearly luminar" waves ($k\to\infty$, or $v\to1$), as follows from (37).

Higher-order differential terms can stabilize the system before the dissipative part becomes of the same order of magnitude as the conservative part when $\eta=-\epsilon^{2/3}\eta'$, $1-v=0(\epsilon^{2/3})$. The stabilizing term making its way into the dissipative function under these conditions is

$$-\epsilon^4[\delta^*\delta_*(\nabla^2)^2\phi-\delta_*^*\nabla^2\phi] = -\epsilon^4 k^2[\delta^*\delta_* k^2-\delta_*^*\omega^2]\phi_\ominus 4.$$

The fourth derivative is expressed as

$$\phi_{\ominus 4} = \partial_\ominus^2(V'/m^2) =$$

$$(V'''\phi_\ominus^2+V''\phi_{\ominus\ominus})/m^2 = [V'V''-2(E-V)V''']/m^4 = W)\phi)/m^4.$$

The Lorenz boost with the speed $v=1-\epsilon^{2/3}w/2$ transforms the dissipative function (39) to

$$F(E,w) = \epsilon^{2/3}[b_0 J_0+(b_1+2\eta/w)J_1 + (\delta_*^*-\delta^*\delta_*)w^{-3/2}<W(\phi)>] \qquad (45)$$

This expression has to vanish at values of w, E corresponding to a steady "near-luminar" wave.

Acknowledgement

This work has been supported by the US-Israel Binational Science Foundation.

References

[1]   H. Haken (1983). *Synergetics*, Springer; G. Nicolis, C. Van den Broeck (1984) in *Nonequilibrium Cooperative Phenomena in Physics and Related Fields,* p. 473, ed. M. G. Velarde, Plenum.

[2]   V. I. Arnold (1983). *Geometrical Methods in the Theory of Ordinary Differential Equations,* Springer; J. Guckenheimer, and P. Holmes (1983). *Nonlinear Oscillations, Dynamical Systems and Bifurcations of Vector Fields,* Springer.

[3]   L. M. Pismen (1986). *Lect. Appl. Math.* **24**, 175.

[4]   L. D. Landau and E. M. Lifshits (1980). *Statistical Physics* (3rd ed.), Part 2, 178, Pergamon Press.

[5]   L. A. Segel (1969). *J. Fluid Mech.* **38**, 203; A. C. Newell and Y. A. Whitehead (1969) *J. Fluid Mech.* **38**, 279.

[6]   G. I. Sivashinsky (1977) *Acta Astronautica* **4**, 1177; Y. Kuramoto (1978). *Progr. Theor. Phys. Suppl.* **64**, 346.

[7]   L. M. Pismen (1984). *J. Coll. Int. Sci.* **102**, 237.

[8]   A. C. Newell (1980) in *Proceedings of International Symposium on Synergetics,* p. 244, ed. H. Haken.

[9]   Y. Kuramoto and S. Koga (1982). *Phys. Lett,* **A92**, 1; L. Sirovich and P. K. Newton (1986). *Physica* **21D**, 115.

[10]   M. J. Ablowitz and H. Segur (1981). *Solitons and Inverse Scattering Transform,* SIAM, Philadelphia.

[11]   D. J. Kaup and A. C. Newell (1978). *Proc. Roy. Soc. London* **A361**, 413.

TIME OF EMERGENCE AND DYNAMICS OF COOPERATIVE GENE NETWORKS

Abstract

Presuming the existence of a vast number of relatively short, self-replicating ribonucleotide strands, this paper establishes an explicit expression for the time of emergence of the first catalytic RNA-feedback cycles in terms of conditions characteristic of prebiotic Earth.   Our theory is based on an extension of results on phase transitions in random graphs.   The theory also determines the types of hypercycles which are most likely to be formed.   Evaluated in light of recent experimental results on RNA self-catalysis, using very conservative assumptions the theory predicts a time of emergence which in order of magnitude is comparable to or less than the lifetime of the Universe.   Chaotic behaviour and other highly nonlinear dynamic phenomena resulting from catalytic feedback between the first genes are described.

22.1 Introduction

The emergence of life out of simple organic and inorganic compounds of primordial world is one of the most fascinating of all self-organizing processes.   Exactly how things happened when life began some 3 billion years ago is not known, but analyses performed particularly by Eigen and coworkers [1-4] have provided general conceptions about the kind of processes which most likely took place.   It is quite clear that evolution has faced a number of severe crises in which new principles of organization had to develop to protect the information already accumulated and to make continuation of the information build-up possible.   One such crisis was solved through the development of cells, but even before that stage, the evolution of information rich nucleotide strands had to find a way.

Even at this molecular level, the evolutionary process can be interpreted in terms of natural selection.   It is a trial and error process in which a wide spectrum of randomly produced structures are tested with respect to their rate of formation and their resistance to decomposition.

A basic step in the evolution of life was the development of self-replicating nucleotide strands.   But even this step is associated with problems of chemical nature [5,6].   Many questions are still unanswered with respect to the synthesis of nucleic acid-like components under prebiotic conditions. It is possible that the first steps have involved mechanisms and materials different from what we find in today's organisms [7].

The self-replicating nucleotide strands introduced a first order autocatalytic process and thus gave rise to exponential growth in the concentrations of nucleic acids. With resources in the form of mononucleotides the net rate of increase in the concentration of a particular nucleic acid x may then be expressed as:

$$\frac{dx}{dt} = k \cdot (1 - \alpha)^S \cdot x \cdot r - x/\tau \tag{1}$$

r is here the nucleotide concentration, and k is the rate constant for the self-replication process. α is the error probability per base, and s the number of bases in the strand. τ is the lifetime of nucleic acid molecules as limited by hydrolysis and other decomposition processes.

Since the binding energies associated with base pairing are limited, and since replication takes place at finite temperatures, there is a certain probability of error. Without the catalyzing and error correcting effects of specific proteins, thermodynamic calculations show that the error probability is of the order of α = 0.02 [8]. This is a fundamental parameter in determining the maximum strand length.

The remaining parameters may be estimated for conditions thought to be representative of prebiotic Earth. Allowing for possible weak catalyzing effects from various inorganic compounds and from miscellaneous primitive proteins, the maximum sustainable nucleotide strand is found to be of the order of 50-100 bases [8], corresponding approximately to today's transfer RNA-molecules. At this stage the information build-up is terminated by inevitable errors. The information carried by such nucleotide strands is not sufficient to code for more specific enzymatic properties, and without the enzymatic effects, the error suppression in the replication process required to produce long nucleotide strands cannot be attained.

One could imagine, perhaps [9] that by accident a nucleotide string was produced which carried sufficient information to code for the proteins required for its own reproduction and to control the metabolic processes of a primitive life form. However, even the simplest known organism capable of providing its own metabolism is an extremely intricate system, which functions through a complicated interplay between a DNA-string of more than 10^6 bases, 1000 or more different RNA-molecules, and 2-3000 different, highly specific proteins. Simple combinatorial arguments show [8] that the chance of generating the information accumulated in such a system in a random process is so small that the emergence of life in average would take more than 10^{100} times the lifetime of the Universe. The "frozen accident hypothesis" thus makes the evolution of life an extraordinary unprobable event.

As an alternative mechanism, Eigen [1-4] has suggested that the

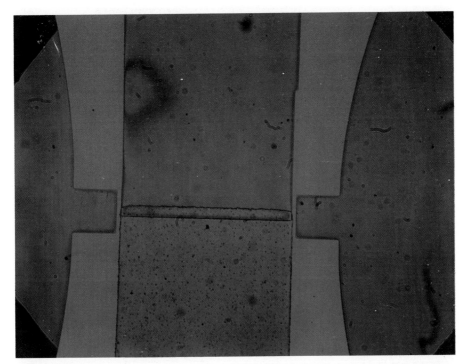

Plate 1 Nb-NbO$_x$-Pb thin film tunnel junction (green border) produced at Physics Laboratory I, The Technical University of Denmark on glass substrate (brown areas). Junction dimensions: length 400μ, width 16μ. Micro strip antennas for high frequency connections to the junction (red areas). See p. 93.

Plate 2 Computer graphics produced at the Laboratory of Applied Mathematical Physics, The Technical University of Denmark. The x-derivative of the solution, $\Phi(x,t)$, to Eq. (1) with parameters $\alpha = 0.252$, $\beta = 0$, $\Gamma = 0.71$, $\eta = 1.35$, and $\lambda = 5$. The solution exhibits solitons corresponding to current singularities in the I-V characteristic for the Josephson junction. See p. 93.

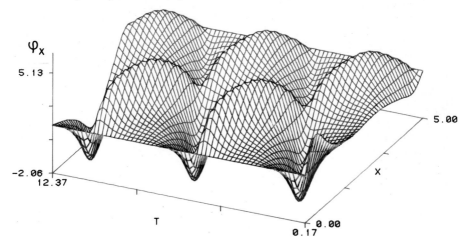

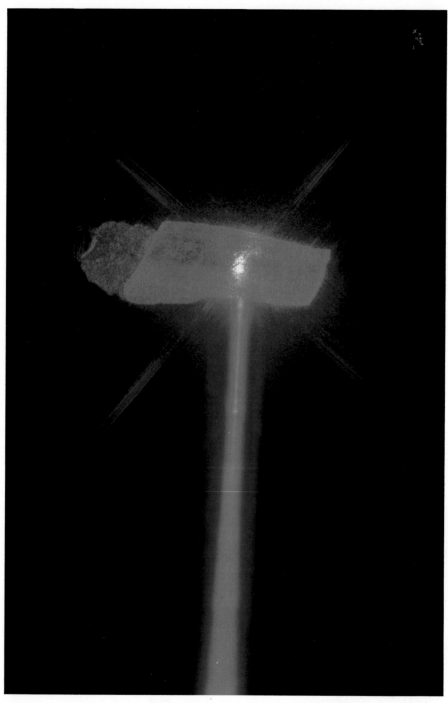

Plate 3 A single crystal of ACN (acetanilide) scattering laser light in a laser Raman experiment at Los Alamos. (Courtesy of Irving J. Bigio, Los Alamos National Laboratory). See p. 139.

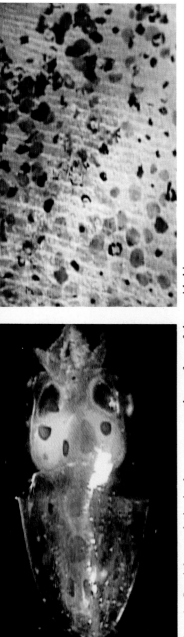

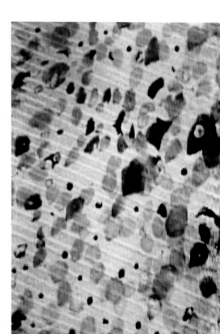

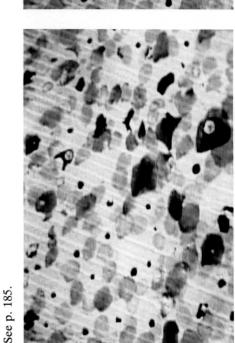

Plate 4 Spatial granularity in the pattern on the surface of the squid skin.

(a) The newly hatched squid has only a few pigment spots: these oscillate independently, at rates about 0.2-0.5 Hz.

(b) In a fragment of adult squid skin the spatial pattern (colour above, lack of colour below) is clearly due to the constriction of the pigment spots. See p. 185.

Plate 5 Temporal changes in the pattern on the squid surface are produced by temporal changes in the pigments: panels a and b are from the same area of isolated skin viewed about a second apart: close examination of individual corresponding spots show changes in their size. See p. 185.

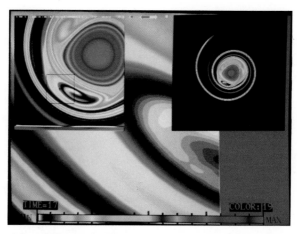

Plate 6 The vorticity of the run in Figure 1 at t=17 in N. J. Zabusky, "Model Complementarity, Visualisation and Numerical Dignostics" (three different magnifications are shown).

Plate 7 The vorticity transport term, u·∇ω, in the laboratory frame at t=4 and t=17. See p. 465.

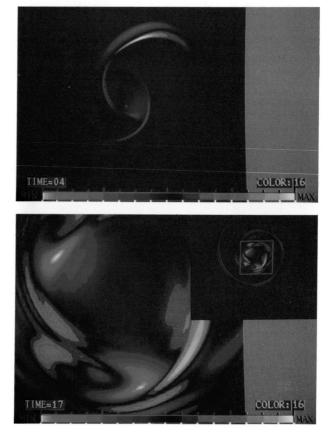

production of longer nucleotide strings was accomplished through the formation of cooperative structures (hypercycles) between different types of nucleotide strands. A hypercycle would result, for instance, if one type of nucleotide strand by error replication happened to code for a protein which could slightly facilitate and stabilize the production of another type of nucleotide string, and if at the same time this second type of nucleotide strand happened to code for a protein which could assist the replication of the first nucleic acid.

In the beginning, such specific catalytic effects might not occur. Rather, the simple proteins produced by a given type of nucleic acid could be relatively unspecific and catalyse the production of a great number of other nucleic acids. Other types of nucleic acids might also start to produce relatively unspecific, slightly enzymatic proteins, and at a certain time, a reinforcing positive loop involving a number of different nucleotide strands could be established. The formation of such a catalytic feedback would give the involved nucleic acids an advantage over other types of nucleotide strands with respect to the rate of formation and with respect to the replication accuracy. This would stabilize the information carried by the collaborating nucleotide strands, and it would permit the gradual development - hand in hand - of longer and longer strands and of more and more specific proteins.

The advantages of being a member of a catalytic feedback structure are quite significant. With no resource limitation, the rate of nucleic acid synthesis is now essentially given by

$$\frac{dx}{dt} = k_c x^\beta \tag{2}$$

where $\beta > 1$ is the order of the catalytic process. As before, x denotes the concentration of a particular nucleic acid, and k_c is the characteristic reaction constant. The dynamics of (2) has hyperbolic properties [1,2], and information carriers in a hypercycle structure will therefore rapidly outgrow isolated nucleotide strings.

The nucleic acids first involved in these processes are usually considered to be RNA [1,2], because RNA molecules are simpler than DNA, because RNA is directly involved in the protein synthesis of today's organisms, and because RNA molecules can develop complex tertiary structure with different functions. In this connection it is interesting to note that it is believed that not only the protein part but also the ribosomal RNA in ribosomes can have catalytic activity [10]. In addition, recent experimental results indicate that ribosomal RNA precursors can perform self-splicing processes [11]. Bass and Cech [12] have shown that rRNA splicing in Tetrahymena Thermophyla is catalyzed by RNA. Ribosomal RNA precursor is processed such that an intermediate sequence (the intron) is cut out and the two flanking sequences (the exons) are joined to form the functional RNA molecule. It is the folded structure of the intron that mediates the specific cleavage-ligation reaction. A similar type of reaction is found in other organisms such as yeast and bacteriophage T4.

There are a number of reasons to suggest that self-splicing could have played a role in prebiotic hypercycle formation:

(a) the catalytic activity is restricted to a relatively small fragment of the RNA molecule,

(b) a number of very primitive viroids with between 250 and 400 bases of RNA are known to have sequence homology with the self-splicing introns from Tetrahymena Thermophyla, and

(c) self-splicing can facilitate recombination by cutting and splicing different pieces of RNA.

It is also worth noticing that while self-splicing is thermodynamically neutral, translation is a rather energy-consuming process. In addition, the amount of information needed for replication in today's biological systems is usually less than the information needed for translation. The number of bases defining the genes necessary for replication in the bacterium E. coli, for instance, is about 10-20000 bases. For comparison, the minimal translational machinery needed for protein synthesis in E. coli is of the order of 80000 bases.

In the next section we shall sketch how random graph theory can be applied to determine both the time of emergence and the type of the first RNA feedback structures to develop in a system of a vast number of different, relatively short, isolated RNA molecules. A more detailed exposition of this theory will be given elsewhere. We shall thereafter try to evaluate the time of emergence for the first hypercycles in terms of conditions characteristic of prebiotic Earth. Finally, biological clocks and chaotic phenomena resulting from interactions between the first genes will be described.

G

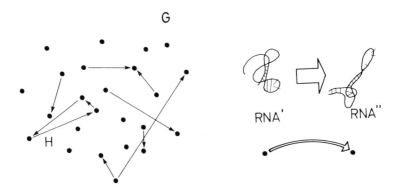

Figure 1. *(a)* *A random graph G is defined as a set of vertices* $V_n = \{1, 2, \ldots, n\}$ *with a number of randomly distributed edges between the vertices. We study the appearance of subgraphs, such as the 3-vertex cycle H, as G acquires more and more edges.*

(b) *As a model for prebiotic evolution, the vertices of the random graph are interpreted as specific types of information carriers, and the oriented edges as catalytic interactions.*

22.2 Random Graph Models of Prebiotic Evolution

The study of random graphs was initiated by Erdös and Rényi [13] who discovered that if a graph defined over a very large set of vertices evolves by randomly acquiring more and more edges, then most properties of the graph will appear rather suddenly, almost like a phase transition in a physical system. Figure 1a shows an example of a graph G defined over a set of vertices $V_n = \{1, 2, ..., n\}$, and with a number of randomly distributed oriented edges. Note that G contains an oriented 3-vertex cycle. To contain a particular subgraph is a property of a random graph.

Random graph theory appears to offer an appropriate framework for studying the emergence of catalytic RNA feedback structures in a system of a vast number of isolated RNA species. As illustrated in Figure 1b, the vertices of the graph are taken to represent particular species of information carriers, and the oriented edges represent directed catalytic effects. The random graph theory has recently been extended particularly by Bollobás [14,15]. His results do not apply directly to our problem. We have generalized the graph theoretical treatment to allow for double edges and loops [16].

A random graph $\mathbb{G}_n$ represents the set of possible outcomes of a process in which edges are randomly thrown on to a vertex set V_n. We can denote the random graph either by the number M of edges it has acquired: $\mathbb{G}(n, M)$ or by the probability p that a given edge exists: $\mathbb{G}(n, p)$. Assuming that each vertex point in principle can communicate with any other point, and including possible self-interacting loops, the total number of potential edges is $N = 2\binom{n}{2} + n = n^2$. The $\mathbb{G}(n, M)$ description has $\binom{N}{M}$ different realizations, each with equal probability $\binom{N}{M}^{-1}$. A particular realization of $\mathbb{G}(n, M)$ is denoted $\mathbb{G}_M$. In the $G(n, p)$ description, $0 \leq p \leq 1$, we have 2^N different realizations, and the probability of finding a particular realization $\mathbb{G}_M$ with M edges is $p^M (1-p)^{N-M}$.

A subset Q of $\mathbb{G}_n$ is said to define a property Q for the realization $G \in \mathbb{G}_n$, if $G \in Q$. As an example, it is a property of a graph to be connected, in which case Q represents the subset of all connected realizations of $\mathbb{G}_n$. It is also a property of a graph to contain a specific subgraph H. A property is said to be monotone, if $G_1 \in Q$ and $G_1 \subset G_2$ implies $G_2 \in Q$. A property is convex if $G_1 \subset G_2 \subset G_3$ and $G_1 \in Q$, $G_3 \in Q$ imply that $G_2 \in Q$. The property of containing a specific subgraph is monotone and convex.

A graph process $(G_M)_0^N$ on the vertex set V_n may be defined as a sequence of graphs G_M where $G_0 \subset G_1 \subset G_2 \subset --- \subset G_N$. $(G_M)_0^N$ thus represents the process of adding edges to V_n one by one while maintaining already established communications. As an example, Figure 2 illustrates the set of possible processes through which the complete 2-graph can be generated. The heavy arrows define one particular process out of the 24 possible processes. If edges are thrown on to V_n at random at a given rate (number per unit time), the evolution of the graph can be described as a Markov process whose states are particular graphs on V_n.

If, for a random graph in $\mathbb{G}(n,M)$, the probability $P_M(Q)$ of possessing a certain property Q asymptotically approached 1 for n → ∞, then almost every graph $G_M \in \mathbb{G}(n,M)$ has the property Q. On the other hand, almost no graph $G_M \in \mathbb{G}(n,M)$ has the property Q if $P_M(Q) \to 0$ for n → ∞. Most monotone properties in large random graphs appear rather suddenly. Thus, if almost no realization of $\mathbb{G}(n,M)$ for a given M has the property Q, for a slightly higher number of edges then almost all realizations have Q. This behaviour is illustrated in Figure 3 where $P(H \subset G)$ measures the probability that a realization of $\mathbb{G}(n,M)$ contains the subgraph H as a function of M. M_c is the critical edge density above which almost all realizations of G will contain H.

The abrupt change in the probability of containing a certain subgraph can be expressed in terms of a threshold function. The function $M^*(n)$ is said to be a threshold function for the monotone property Q if

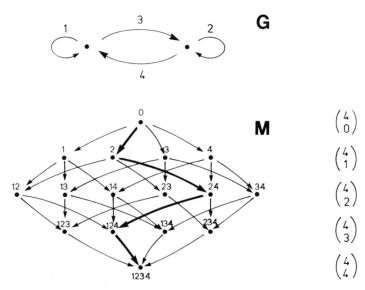

Figure 2. The complete 2-graph (top) together with a graphical representation of the 24 different processes through which the complete 2-graphs can be generated by adding one edge at a time. The complete n-graph can be generated in $(n^2)!$ different ways.

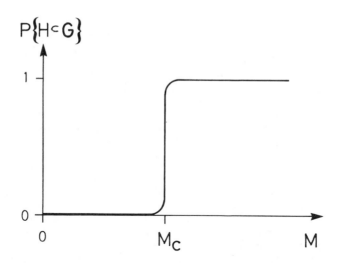

Figure 3.   The probability $P(H \subset G_M)$ that a realization of the random graph $G(n, M)$ contains a given subgraph H changes abruptly as the number M of edges in G pass a critical value M_c.

$$\lim_{n \to \infty} P_M(Q) = \begin{cases} 0 & \text{if } M(n)/M^*(n) \to 0 \\ 1 & \text{if } M(n)/M^*(n) \to \infty \end{cases} \qquad (3)$$

M(n) is here a function which specifies how the number of edges is assumed to vary with the number of vertex points.   To determine the threshold function for the property of containing a particular subgraph, the subgraph must be balanced.   Loosely speaking this means that the subgraph should not itself contain subgraphs of higher edge density.   The edge density of a subgraph is defined as the number of edges relative to the number of vertex points, i.e. $d = s/k$.

Bollobás [14, 15] and Rasmussen [16] have shown that the threshold function for containing a particular subgraph H depends on the maximal edge density of any of its subgraphs.   If H is balanced, one has

$$M^*(n) = cn^{2-k/s}. \qquad (4)$$

c is here a proportionality factor which is independent of n.

With these introductory remarks we can now return to our problem of hypercyle formation.   Let us envisage the manifold of uncoupled RNA molecules existing in primordial soup at a certain stage of the evolutionary process.   From our discussion in the previous section it follows that these molecules typically had lengths up to about 100 nucleotides.   It can also be shown [8] that the longer species must have had a predominant cytidine-guanosine backbone.   Since we are looking for the appearance of cooperative structures, the interesting subset of RNA species are those which through error replication (or self-splicing) can acquire both catalytic effects and the ability to be catalysed by other RNA molecules. At this stage it is not necessary to specify whether the catalytic interactions are direct or via simple proteins. Presumably both types of processes have been involved.

With present knowledge it is not possible to specify which RNA molecules are to be counted as members of our subset of interesting species, and which are not. It is clear, though, that the longer molecules are more likely to be candidates than the shorter ones. Let n denote the total number of different RNA species in the subset, and let each of these species be represented by a point in the labelled vertex set $V_n\{1,2,\ldots,n\}$.

We assume that each point in principle can communicate with any other point through the establishment of a directed edge between the two points. Further, as a first approach we consider the case in which each mutation causing the appearance of a catalytic interaction corresponds to the generation of a single random connection between two RNA species.

Let us now consider what happens as we randomly start to throw edges on to the vertex set V_n. In the beginning, the graph will consist of isolated points with a few scattered edges. In accordance with the threshold function

$$M^*(n) = cn^{2-k/s} = cn^{1/2}, \tag{5}$$

the appearance of the first 2-trees ($k = 3$ and $s = 2$) takes place when the number of edges becomes comparable to the square root of the number of vertex points. For $M < cn^{1/2}$, the graph almost certainly only contains isolated edges, and for $M > cn^{1/2}$ the graph almost certainly contains 2-trees. The first 3-trees ($k = 4$ and $s = 3$) appear for $M \sim cn^{2/3}$, and the first k-trees appear for $M \sim cn^{(k-1)/k}$. The appearance of these trees does not depend on whether the edges are oriented or not.

For trees, the maximal edge density $d = s/k$ is always less than 1, while for cycles the density is identically 1. All cycles therefore occur more or less at the same density, namely when the number of edges becomes comparable with the number of vertex points $M \sim cn$.

The expectation value for the number X_k of oriented k-cycles in the random graph $\mathbb{G}(n,M)$ can be obtained as

$$E(X_k) = \binom{n}{k}(k-1)!\binom{M}{k}/\binom{N}{k} \sim \frac{1}{k}\left(\frac{M}{n}\right)^k \tag{6}$$

with $k = 1, 2, \ldots$

Here $\binom{n}{k}$ gives the number of ways in which k vertices can be chosen out of V_n. For each of these subsets of vertices, there are $(k-1)!$ possible ways of choosing the edges of a directed k-cycle. The factor $\binom{n}{k}(k-1)!$ therefore determines the total number of k-cycles in the complete n-graph. $\binom{M}{k}$ is the number of ways in which k edges can be selected from the existing M edges, and $\binom{N}{k}$ is the number of ways in which k edges can be chosen out of the $N = n^2$ edges of the complete n-graph. The factor $\binom{M}{k}/\binom{N}{k}$ therefore expresses the probability that a k-cycle involving k specific vertex points exists with M edges on V_n. The approximation in the final expression for E(X) involves replacing $\binom{M}{k}$ by $M^k/k!$ for $k \ll M$.

The expression for $E(X_k)$ shows that as long as the edge density is relatively low $M/n \ll 1$, the number of small cycles will dominate over the number of large cycles. Cycles involving a single or a few RNA species are therefore the first to appear, and only later, as more catalytic interactions have been established, will larger cycles start to become abundant.

Furthermore, it has been shown by Bollobás [14,15] and Rasmussen [16] that X_k, $k = 1,2,3....$ asymptotically are independent Poisson distributed random variables, if the subgraph is strictly balanced. A strictly balanced subgraph is a graph where the densest part is the graph itself. A directed cycle and a tree are strictly balanced.

Because X_k, $k = 1,2.....$ are independent, Poisson distributed random variables, we can determine the probability that $\mathbb{G}(n,M)$ contains any cycle. Inserting $M = cn$ in (6) we get:

$$P(H \subset \mathbb{G}(n,M), \text{ H is a cycle}) =$$

$$1 - \exp(-\sum_{k=1}^{\infty} c^k/k) = c, \quad c < 1 \tag{7}$$

This means that the graph with probability c contains a directed cycle for $c < 1$ and the graph with probability 1 contains a directed cycle for $M \geq n$. In this case we do not have a sharp threshold function as indicated by Figure 3. However, the threshold function becomes sharper and sharper the longer cycles we are asking for.

From a biochemical point of view it appears unlikely that the first catalytic interactions should be very specific as assumed in the above analytical derivation. We have therefore considered the problem of non-specific catalysis. Through simulations on random graphs with n = 10,000 vertex points and with up to 100 edges created by mutation of a given species at a time we have found that the time of emergence for the first directed cycles is lowered by a factor approximately equal to the valency of the catalytic interactions. Even with this extension, the small cycles are the first to appear. These results will be discussed in more detail elsewhere.

We have also examined the effect of allowing the strength of the catalytic interactions to vary. This seems only to increase the tendency for the small cycles to be formed first. We emphasize this result because it appears to contradict expectations by Eigen and coworkers [3], according to which relatively large hypercycles initially could play a significant role in the evolutionary process.

In the further development of the graph, the number of small cycles increases gradually while at the same time more and more large cycles are formed. For M > n, a transition occurs in which suddenly the large cycles start to dominate. This part of the graph evolution is not so important, however, if we believe that catalytic effects are crucial in the selection between the different RNA subsystems. Here the emergence of the first hypercycles and their structure are the most interesting findings, because the further evolution so heavily depends on them.

22.3 Time of Emergence for the First Catalytic Feedback Cycles

To quantitatively estimate the emergence time for the first catalytic feedback cycles is one of the outstanding problems of the theory of prebiotic evolution. Unfortunately, present knowledge of the physico-chemical and biochemical conditions of premordial Earth do not suffice for this purpose. It is therefore of interest to invert the problem, knowing that life did evolve, and try to estimate the total number of RNA-molecules which must have existed in order for the formation of catalytic feedback cycles to take place within a reasonable time span. The oldest definite trace of living organisms dates back about 3.1 billion years, at which time life seems to have evolved to the cell level [17]. Since the age of the Earth is about 4.6 billion years, the time of emergence for the first catalytic feedback cycles is limited to $t_{em} < 10^9$ years.

The rate at which new RNA species are produced through error replication of already existing species is given by

$$f = k \cdot m \cdot (1 - (1 - \alpha)^s) \qquad (8)$$

where m is the number of RNA molecules in a given prebiotic niche. As before, k is the rate constant in the RNA self-replication process, s is the typical number of bases for the RNA molecules of interest, and α is the error probability per base. For a more precise interpretation, m must be considered as the number of molecules within a volume small enough for the molecules to come in contact with each other through diffusion, convection and other mixing processes. To independently estimate m we must therefore know both the concentration of RNA molecules in a successful niche of prebiotic Earth, and the volume within which RNA molecules can interact with each other.

A fraction b of the total number of generated RNA species can be assumed to have catalytic activity. An estimate of b may be obtained from the number of prescribed bases required in modern RNA catalysis [11,12], assuming that the remaining bases can be substituted freely. The rate of formation of catalytic RNA species is hereafter

$$\nu = b \cdot f \tag{9}$$

In accordance with our results in the previous section, the number of edges required for the first catalytic feedback cycles to appear is of the order of

$$M \sim cn \tag{10}$$

where the parameter $c < 1$, as determined by (7) controls the certainty with which feedback cycles exist. As before, n is the total number of potentially communicating RNA species.

The time of emergence for the first catalytic feedback cycles can hereafter be obtained from

$$\nu t_{em} \sim cn$$

or

$$t_{em} \sim \frac{cn}{bf} . \tag{11}$$

Implicitly, we have here assumed that once a catalytic RNA species has been produced it will continue to exist until a feedback cycle is formed. This is, of course, a simplification. It is presumably true, however, that catalytic RNA species on average have considerably longer lifetimes than non-catalytic ones, since the folding of the former usually makes them more persistent to hydrolysis.

By definition the fraction b of catalytic RNA molecules determines the total number of different catalytic RNA species. Furthermore, the graph theoretical model is valid only for large n. Therefore, the emergence time of catalytic feedbacks is independent of the actual value of b, as long as b is high enough to yield a large number of catalytic RNA families.

To proceed with our estimate we assume that the sequence length for the RNA species of interest, like modern tRNA, has been of the order of $s \simeq 80$, perhaps with a minimum length $s - \Delta s \simeq 60$. For strings with a predominant cytidine-guanosine backbone, we shall assume that $\gamma = 90\%$ of the nucleotides have been either cytidine or guanosine, and that the remaining $(1-\gamma) = 10\%$ have been randomly chosen between the 4 different types of ribonucleotides. This gives a total of

$$n_{tot} \sim \Delta s \cdot 2^{\gamma s} \cdot 4^{(1-\gamma)s} \tag{12}$$

for the number of possible strings.

If, as a first approach we take the number of bases which have to be specific in order for an RNA molecule to acquire catalytic effects to be $\sigma \simeq 10$, the fraction of RNA molecules which are produced with catalytic effects is large enough to guarantee asymptotic results.

Substituting (8) and (11) into (10), we obtain for the time of emergence of the first catalytic feedback cycles

$$t_{em} \sim \frac{c \cdot \Delta s \cdot 2^{\gamma s} \cdot 4^{(1-\gamma)s}}{k \cdot m \cdot (1-(1-\alpha)^S)} \qquad (13)$$

or

$$m \sim \frac{c \cdot \Delta s \cdot 2^{\gamma s} \cdot 4^{(1-\gamma)s}}{k \cdot (1-(1-\alpha)^S) \cdot t_{em}} \qquad (14)$$

Inserting finally

$$k \; \tilde{=} \; 10^{-6}/sec \; (12),$$

$$\alpha \; \tilde{=} \; 10^{-2} \; (8),$$

$$t_{em} \; \tilde{=} \; 10^9 \; years \; \tilde{=} \; 3 \cdot 10^{16} \; sec,$$

and the probability that we have a catalytic feedback

$$c = 0.5$$

we obtain

$$m \; \tilde{=} \; 1.4 \cdot 10^{17} \; molecules.$$

Altogether we thus need of the order of 10^{17} RNA molecules within a volume of prebiotic soup which is small enough for the molecules to come in contact with each other through diffusion, convection and other mixing processes.

If we compare this number with estimates of the effective reaction volume (~ one ml) we find a minimal prebiotic RNA concentration of around $2 \cdot 10^{-4}$ moles/litre. This is the same orderof magnitude as than the concentration we find for tRNA in modern bacteria, such as E. coli [18]. The above estimate relates to our analytical results and therefore gives a very conservative estimate. With non-specific catalysis, the RNA concentration required for hypercycle formation is much smaller. Furthermore, our estimate only considers a micro niche of one ml. The oceans and the lakes of the premordial earth probably had space for many parallel evolutionary processes [19].

We therefore conclude that the proposed evolutionary scheme in principle is capable of explaining the formation of catalytic RNA cycles within a time span considerably less than 1 billion years.

22.4 Dynamics of Catalytic Feedback Cycles

Eigen and Schuster [3] have studied the dynamics of elementary catalytic feedback cycles. They have found that for hypercycles involving 2 RNA species only, the concentration variables tend toward constant equilibrium values. For higher order hypercycles the typical behaviour is limit cycles, where the oscillation period is controlled by the catalytic reaction constants and by

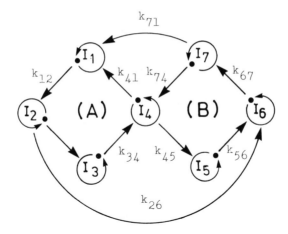

Figure 4. Two interacting 4-cycles. For a wide set of parameter values, this structure produces chaotic behaviour.

the lifetime of the produced information carriers. With increasing difference between the various reaction constants, the oscillations tend to become sharper to finally approach a typical relaxation type oscillation.

In the present paper we would like to demonstrate how behaviour is likely to be found in networks where two or more cycles with different periods interact. The selection mechanisms involved in hypercycle evolution are so strong, however, that coexistence of hypercycles over longer periods of time only can be realized if the cycles support one another. This can occur, for instance, if the information carriers of one cycle by error replication produce information carriers of the other cycle, and vice versa. It can also occur if (relatively weak) catalytic interactions exist between molecules in the two main hypercycles. Let us first present results for the latter of these possibilities.

As illustrated in Figure 4, the system which we have studied consists of n = 7 interacting RNA species. Species 4 constitutes the common vertex of the two main hypercycles (A) and (B). In addition to these hypercyles it is assumed that species 7 catalyses the production of species 1, and that species 2 similarly catalyses production of species 6. Error replication has so far been neglected. The equations of motion for the system are given by

$$\frac{dx_i}{dt} = k_i x_i r + \sum_j^n k_{ji} x_i x_j r - x_i/\tau_i \qquad (15)$$

with $i,j, = 1,2,\ldots,n$. Here x_i is the concentration of RNA species i, k_i is the rate constant in the self-replication of this species, and τ_i is the corresponding molecular lifetime. k_{ji} is the catalytic rate constant for the formation of species i by

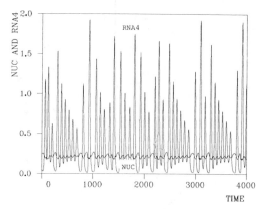

ALFA=0, K41=.9, K45=1.1, K67=.5, K26=.15, K71=.24

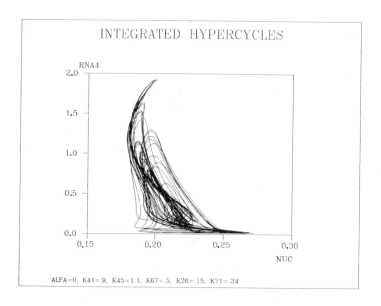

ALFA=0, K41=.9, K45=1.1, K67=.5, K26=.15, K71=.24

Figure 5.   Temporal variation of the nucleotide concentration and the concentration of RNA species 4 for a particular set of rate constants (a).   Corresponding phase projection (b).   Both figures indicate the occurrence of chaotic behaviour.

species j.

Figure 5 shows an example of the obtained results.   We have here taken k_{41} = 0.9, k_{45} = 1.1, k_{67} = 0.5, k_{26} = 0.1 and k_{71} = 0.24. The remaining rate constants are all k_i = k_{ij} = 1.0, and the molecular lifetimes are τ_i = 0.3.   With these parameters, the two two main cycles are about equally strong and have different periods.   Figure 5a displays the variation with time of the nucleotide concentration and of the concentration of RNA species 4.   Figure 5b shows the corresponding phase plot.   Both figures

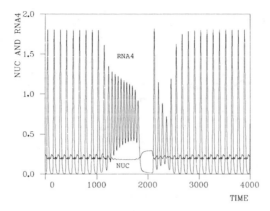

ALFA=.0, K41=.9, K45=1.1, K67=.5, K26=.90, K71=.24

Figure 6.   Shift in dominance between 3 different cycles is observed for a somewhat higher value of the catalytic coupling constant k_{71}. All other parameter values are as in Figure 5.

indicate the occurrence of chaos.

By changing the relative magnitudes of the various rate constants a great variety of different modes of behaviour can be observed.   As a further example Figure 6 shows the variation of the nucleotide concentration and the concentration of RNA species 4 for k_{41} = 0.9, k_{45} = 1.1, k_{67} = 0.5, k_{26} = 0.9, and k_{71} = 0.24. In this case one can observe a shift in dominance between 3 different cycles.

We have also studied the effect of error replication.   By adding terms to the equations of motion (15) which account for

(i)   the loss of copies of species i through errors in the self-replication and in the catalytic production of this species, and

(ii)   the additional production of copies of species i through errors in the replication and catalytic production of other species.

We have found that even with error probabilities as small as 10^{-6} there is a clear tendency for this mechanism to make the behaviour of the system more regular.

The possible evolutionary advantage of these coupled hypercycles is a better adaptability, because the efficiencies of RNA synthesis in the feedbacks probably vary differently with the environmental conditions.

22.5 Conclusion

An extension of random graph theory has allowed us to calculate both the threshold edge density and the distribution of directed cycles on a very large set of vertex points.   By interpreting the cycles as catalytic feedback loops in a prebiotic system of a vast number of different RNA species we have obtained an explicit expression for the time of emergence t_{em} of the first catalytic feedback networks, and we have found that t_{em} with conservative parameter values can be less than 1 billion years. The production of these feedback structures is a fundamental step towards the evolution of life because cooperative structures are essential in order to create and maintain a pool of more complicated polymers. Through numerical simulations we have extended the analytical results in various directions, and we have also observed that chaotic behaviour can occur in more complex gene networks.

Acknowledgements

Thanks are due to Carsten Thomassen, Mathematical Institute, The Technical University of Denmark, for enlightening discussions on random graphs and Henrik Nielsen, Biochemical Laboratory B, Panum Institute, University of Copenhagen, for his help on understanding RNA self-splicing in modern life. We are also indebted to Jan Holst, Institute of Mathematical Statistics and Operational Research, The Technical University of Denmark and Torben Smith Sorensen, Physico Chemical Institute, The Technical University of Denmark for critical discussions on our model.

References

[1] M. Eigen (1971). *Naturwissenschaften* **58**, 465.
[2] M. Eigen (1973). in *The Physicist's Conception of Nature,* ed. J. Mehra, Reidel, Dordrecht.
[3] M. Eigen and P. Schuster (1979). *The Hypercycle - A Principle of Natural Self-Organization,* Springer-Verlag, Heidelberg.
[4] M. Eigen, W.C. Gardiner, P. Schuster and R. Winkler-Oswatitsch (1981). *Scientific American* **244(4)**, 88.
[5] R. Lohrmann, P.K. Bridson and L.E. Orgel (1980). *Science* **208**, 1464.
[6] G.F. Joyce, G.M. Vissen, C.A.a. van Boeckel, J.H. van Boom, L.E. Orgel and Y. van Mestrenen (1984). *Nature* **310**, 602.
[7] A.G. Cairns-Smith (1985). *Scientific American* 74.
[8] For a general overview of these problems see e.g. B.-O. Küppers (1983). *Molecular Theory of Evolution,* Springer-Verlag, Berlin.
[9] J. Moned (1970). *Le Hasard et la Necessite,* Editions du Seuil, Paris. English edition, *Chance and necessity,* Knopf (1972) New York.
[10] F.H. Westheimer (1986). *Nature* **319**, 534.
[11] T.R. Cech (1985). *Int. Rev. on Cytology* **93**, 3.
[12] B.L. Bass and T.R. Cech (1984). *Nature* **308**, 820.
[13] P. Erdös and A. Renyi (1960). *Publ. Math. Inst. Hungar. Acad. Sci.* 5.
[14] B. Bollobás (1984). *Lectures in Random Graphs,* Progress in Graph Theory, Academic Press.
[15] B. Bollobás (1985). *Random Graphs,* Academic Press.
[16] S. Rasmussen (1985). *Aspects of Instabilities and Self-Organizing Processes (in Danish),* Ph.D. Thesis, Physics

Laboratory III, The Technical University of Denmark.

[17]   S.L. Miller and L.E. Orgel (1974).   *The Origin of Life on the Earth,* Prentice-Hall.

[18]   J.L. Ingraham, O. Maaloe and F.C. Neidhardt (1983).   *Growth of Bacteria Cells,* Sunderland.

[19]   S. Rasmussen, B. Bollóbas and E. Mosckilde, *Elements of a Quantitative Theory of Prebiotic Evoultion,* preprint, Oct. 1987.

RELATIONSHIP BETWEEN CLASSICAL DIFFUSION IN RANDOM MEDIA AND QUANTUM LOCALIZATION

Abstract

The problem of diffusion in random media, as described by the Langevin or Fokker-Planck equation is reduced to an imaginary-time Schrödinger equation, describing the motion of a quantum particle in a random potential. We sketch and review the implications of this mapping.

23.1 Introduction

Much effort has been devoted to the subjects of classical diffusion in random media [1,2,3] and quantum localization [4]. These research fields, however, developed rather independently. Recently, remarkable connections have been established. This was achieved by using the well-known equivalence between the Langevin equation in d-dimensions and the Fokker-Planck equation. The latter is then reduced to an imaginary-time Schrödinger equation [5-7], denoting the Hamiltonian of the associated quantum system. Clearly, the mapping of the Langevin equation on to an associated Schrödinger problem is not restricted to establishing the connection between diffusion phenomena and the motion of a quantum particle. Other important applications of this mapping include: **Critical phenomena** - here, the Langevin equation corresponds to the time-dependent Ginzburg-Landau equation (TDGL) in d-dimensions, describing critical dynamics. The associated Fokker-Planck equation is then reduced to an imaginary-time Schrödinger equation, defining the Hamiltonian of the quantum system, which in turn can be mapped on to a (d+1)-dimensional static and classical counterpart. Up to now, and as far as critical phenomena are concerned, these mappings have been used to derive the following results: (i) Dynamic scaling [8] was traced back to anisotropic static scaling in an associated (d+1)-dimensional classical model [6,7,9]. (ii) Dynamic critical exponents were calculated with conventional renormalization-group techniques from the (d+1)-dimensional classical and static counterpart [9]. (iii) The equivalence of real-space renormalization groups for critical dynamics and quantum systems was established [10]. (iv) The (d+1)-dimensional static and classical model resulting from the TDGL system was shown to exhibit a tricritical Lifshitz point (TLP), belonging to a novel class of TLP's which result from a relevant, nonlocal quartic field interaction, previously ignored [11].

23.1.1 *Simulation of quantum systems in terms of Langevin equations [6,7,12]*

23.1.1.a *Construction of quantum systems with soluble ground state properties [13]*. In this review, we concentrate on the relationship between diffusion phenomena in random media and the motion of a quantum particle. The problem of diffusion in random or modulated media, as described by the Langevin or Fokker-Planck equation, is reduced to the imaginary-time Schrödinger equation, defining the Hamiltonian of the associated quantum system. The quantum analog describes the motion of a quantum particle in a random or modulated potential [14,15].

In the random case, the mapping then connects the hitherto independent fields of classical diffusion in random media [1-3] and localization [4]. Results, as derived from this mapping from one-particle and random systems, include: (i) Establishing of the connection between long-time diffusive behaviour and the low-frequency properties of the density of states and the inverse-exponential localization length [14-16]; (ii) extension of the approach to spatially discrete diffusion [15-16]; (iii) numerical evidence for the low-frequency behaviour of the power spectrum, the density of states and the inverse-exponential localization length for a model exhibiting sublinear diffusion of the Sinai type [16]; and (iv) the low-frequency scaling behaviour has been related to intermittency in the associated one-dimensional map [15].

In this review, we first sketch the mapping of Langevin processes on to an associated quantum system. Here, we sketch the formal structure of the mapping, and discuss the relationship between classical diffusion, $1/\omega$ noise and the motion of a quantum particle in random media.

We distinguish two classes of diffusion models: (A) Diffusion associated with a random drift potential, yielding linear long-time behaviour of the time-dependent mean square displacement and (B) diffusion resulting from a random drift force, leading to sublinear diffusion.

In section 22.3, we extend the mapping to spatially discrete systems, and present numerical results for models (A) and (B).

23.2 **Sketch of the Basic Formalism**

In this section, we sketch the relationship between the Langevin, Fokker-Planck, and imaginary-time Schrödinger equations. The Schrödinger equation defines the Hamiltonian of the associated quantum system and its classical counterpart. We are then prepared to discuss the implications of these connections.

To simplify the notation, we first consider a one-particle system with time evolution defined by the Langevin equation

$$\dot{x} = F(x) + \eta(t). \tag{1}$$

We assume that the drift force F can be derived from the potential W,

$$F(x) = -\frac{\partial W}{\partial x} \; . \tag{2}$$

η is a Gaussian random force with

$$\langle \eta(t) \rangle = 0, \langle \eta(t)\eta(t') \rangle = \sigma\delta(t - t'). \tag{3}$$

Two cases can be distinguished:

A: random drift force F

B: random drift potential W $\Big\}$ $\tag{4}$

The associated Fokker-Planck equation for the probability density is [17]

$$\frac{\partial P}{\partial t} = \frac{\partial}{\partial x} \left(\frac{\partial W}{\partial x} P + \frac{\sigma}{2} \frac{\partial}{\partial x} P \right). \tag{5}$$

It admits the stationary solution

$$P_{eq} \sim \exp \left(-\frac{2W}{\sigma} \right). \tag{6}$$

By invoking the transformation [5]

$$P(x,t) = (P_{eq}(x))^{1/2} \psi(x,t), \tag{7}$$

it reduces to the imaginary-time Schrödinger equation

$$-\frac{\partial \psi}{\partial t} = \left(\frac{\sigma}{2} \frac{\partial^2}{\partial x^2} + V(x) \right)\psi = \mathcal{H}\psi. \tag{8}$$

The potentials W and V are related by the Riccati equation

$$V = \frac{1}{2\sigma} \left(\frac{\partial W}{\partial x} \right)^2 - \frac{1}{2} \frac{\partial^2 W}{\partial x^2} = \frac{1}{2\sigma} F^2(x) + \frac{1}{2} \frac{\partial F}{\partial x} \; . \tag{9}$$

The associated eigenvalue problem,

$$\mathcal{H}\varphi_m = \lambda_m \varphi_m, \tag{10}$$

yields a non-negative energy spectrum, with

$$\lambda_0 = 0, \quad \varphi_0 = (P_{eq})^{1/2} \sim \exp \left(-\frac{1}{\sigma}W \right). \tag{11}$$

The general solution of the Fokker-Planck equation can then be expressed as [7,17]

$$P(x,t) = \varphi_0^2 + \varphi_0 \sum_{m=1}^{\infty} C_m \varphi_m e^{-\lambda_m t}. \tag{12}$$

Invoking the initial condition

$$P(x,0) = \delta(x - x'), \tag{13}$$

and denoting the solution with this initial condition, $P(x,t|x',0)$, we obtain

$$P(x,t|x',0) = \varphi_0(x) \sum_{m=0}^{\infty} \frac{\varphi_m(x')}{\varphi_0(x')} \varphi_m(x) e^{-\lambda_m t} \tag{14}$$

reducing for $t = 0$ to Eq. (12). Correlation functions can then be expressed as

$$S_{xx}(t) = <<x(t)x(0)>> = \int dx \int dx' \; xx' \; <P_{eq}(x') \; P(x,t|x',0)>$$

$$= \sum_m <<0|x|m>^2 e^{-\lambda_m t} > \tag{15}$$

and

$$<<x^2(t)>> = \int \phi x x^2 <P(x,t|0,0)>. \tag{16}$$

$<>$ denotes averages with respect to the random force and the random drift force F or the random drift potential W.

In case B, where the drift potential W is random ergodic (eq. (4)) and bounded, the long-term behaviour of time-dependent mean square displacement $<<x^2(t)>>$ is exactly known, namely

$$<<x^2(t)>> \underset{t \to \infty}{\to} 2Dt. \tag{17}$$

The diffusion coefficient D is given by [3]

$$D = \frac{\sigma}{2} (<\exp -\frac{2W}{\sigma}> <\exp +\frac{2W}{\sigma}>)^{-1}, \tag{18}$$

where

$$<A> = \lim_{L \to \infty} \frac{1}{2L} \int_{-L}^{L} A(x)dx. \tag{19}$$

To establish the connection between the density of states of the quantum analog and the diffusive behaviour, we next introduce the probability that at time t, the particle is still or back at its origin,

$$P(t) = \int dx <P(x,t|x,0)>. \tag{20}$$

336

Using the definition the density of states

$$\rho(\omega) = \Sigma_n \langle \delta(\omega - \lambda_n) \rangle, \tag{21}$$

we obtain, with the aid of Eqs. (14) and (20),

$$\rho(\omega) = \Sigma_n \langle \delta(\omega - \lambda_n) \rangle = \frac{1}{\pi} \text{Im} \, P(\Omega = -\omega + 10^+), \tag{22}$$

where

$$P(\Omega) = \int_0^\infty dt \, e^{-\Omega t} P(t). \tag{23}$$

Finally, to characterize and connect the localized nature of the wave functions with the diffusive behaviour, we introduce the characteristic function [14, 16, 18, 19]

$$\Gamma(\Omega) = \int_0^\infty dz \, \ln(\Omega + z)\rho(z) = \lim_{L \to \infty} \frac{1}{L} \int_0^L R(x)dx \tag{24}$$

and its derivative

$$\frac{d\Gamma}{d\Omega} = P(\Omega), \quad P(\Omega) = \int_0^\infty dt \, P(t) \, e^{-\Omega t}. \tag{25}$$

$P(t)$ is the probability that the classical particle is still at or back to its initial position (eq. (20)). Moreover,

$$R(x) = \frac{d\varphi}{dx}/\varphi. \tag{26}$$

φ is a solution of the quantum problem

$$\mathcal{H}\varphi = \omega\varphi \tag{27}$$

with $\mathcal{H}$ given by Eq. ( ). The inverse-exponential localization length $\gamma(\omega)$ and the integrated density of states $N(\omega)$ then follow from

$$\Gamma(-\omega + 10^+) = \gamma(\omega) + i \, \pi \, \gamma''(\omega), \tag{28}$$

where, by invoking Eq. (23),

$$\gamma'(\omega) = \int_0^\infty \rho(\omega)\ln|z - \omega|dz = \lim_{L \to \infty} \frac{1}{L}\int_0^L |R(x)|dz$$

$$\gamma''(\omega) = N(\omega) = \int_0^\infty \rho(z)dz. \tag{29}$$

The integrated density of states corresponds to the density of nodes of the solution with energy ω. To establish the connection between long-time diffusive behaviour and low-frequency properties in the quantum analog, such as density of states and inverse localization length, more explicitly, we next invoke dynamic scaling [1,2]

$$<P(x,t|0,0)> = P(t)g(x/\xi(t)).$$ (30)

$P(x,t|0,0)$ is the solution of the Fokker-Planck Eq. (5) with initial condition $P(x,0) = \delta(x)$. ξ is the correlation length and

$$P(t) = <P(x,t|x,0)>.$$ (31)

Using expression (16) for the mean-square displacement, namely,

$$<<x^2(t)>> = \int dx\ x^2 <P(x,t|0,0)>$$ (32)

and

$$\int <P(x,t|0,0)>\ dx = 1,$$ (33)

scaling implies

$$<<x^2(t)>> \sim \xi^2(t),\ P(t) \sim 1/\xi(t).$$ (34)

Combining Eqs. (23)-(24), the long-time diffusive behaviour can now be related to the low-frequency properties of the quantum analog. These properties are density ofstates and inverse-exponential localization length. In doing so, we recall that diffusion can be subdivided into two classes (Eq. 4): A - the potential W is stationary random, and B - the drift force F is stationary random. In view of the linear long-time dependence of the mean-square displacement, (Eq. (17)) for class-A diffusion, dynamic scaling, (34) implies

$$<<x^2(t)>> \sim \xi^2(t) \sim t,\ P(t) \sim \frac{1}{\xi(t)} \sim t^{-1/2}.$$ (35)

This information is sufficient to determine the low-frequency behaviour of the quantum analog. In fact, from Eqs. (25), (28) and (29) for the long-time behaviour given by (35), we obtain

$$P(\Omega) \sim \Omega^{-1/2}$$ (36)

for $\Omega \to 0$, and

$$Im\ \frac{d\Gamma}{d\Omega}|_{\Omega=-\omega+10}^{+} = \rho(\omega) \sim \omega^{-1/2}.$$ (37)

An extension of the method developed in [20] to calculate the diffusion coefficient for periodic potentials $W(x)$, for class-A models [16] yields

$$\rho(\omega) = \frac{1}{2\pi}\ (D\omega)^{-1/2},$$ (38)

confirming the dynamic scaling prediction. Here, the diffusion coefficient D is given by Eq. (18). Noting that $N(\omega) \sim \omega^{1/2}$, the characteristic function (24) can be expanded in terms of $\Omega^{1/2}$. The second-order term then also yields the low-frequency behaviour of the inverse-exponential localization length (Eq. (28)), namely,

$$\gamma'(\omega) \sim \omega. \tag{39}$$

As expected, the ground state is not exponentially localized [$\gamma'(\omega=0) = 0$]. Indeed, the occurrence of diffusion requires $<<x^2(t)>> \to \infty$ for $t \to \infty$ so that the variance of the ground-state wave function diverges ($<<x^2>> = <<\varphi_0|x^2|\varphi_0>>$) = ∞.

These properties differ considerably from those in the Anderson model for localization [4],

$$(-\frac{\sigma}{2} \frac{\partial^2}{\partial x^2} + V(x)) \; \varphi = \omega\varphi, \tag{40}$$

where $V(x)$ is a random potential. For bounded potentials, such as

$$P(V) = \{ \begin{array}{ll} 1/2\Delta & \text{for } -\Delta \leq V(x) \leq \Delta \\ 0 & \text{otherwise,} \end{array} \tag{41}$$

the density of states is known to exhibit a Lifshitz tail [21]

$$\rho(\omega) \sim \exp(-\frac{\text{const}}{(\Delta\omega)^{1/2}}), \tag{42}$$

where $\Delta\omega = \omega - \omega_0$ and ω_0 is the ground-state energy. Moreover, all states, including the ground state are exponentially localized. This behaviour differs markedly from the power-law singularity (38) in the density of states, the vanishing inverse localization length for $\omega = 0$ (Eq. 39)) in class-A models. It mirrors the distinctly different statistical properties of the potential $V(x)$. In the Anderson model, the values of $V(x)$ are random and independent, while in model A, they are correlated because $W(x)$ adopts random and independent values (Eq. 30).

Next, we turn to the class-B models, where the drift forces are independent random variables. Recently, this class attracted considerable attention [2,22,23]. For a Gaussian-distributed drift force

$$<F(x) \; F(x')> = \Delta\delta(x - x'), \tag{43}$$

one expects sublinear diffusion [3] of the Sinai type [22], namely,

$$<<x^2(t)>> \sim (\ln t)^4. \tag{44}$$

Thus, the diffusion coefficient vanishes (Eq. (17)). This behaviour differs dramatically from that found in the class-A models. It is a consequence of the different nature of potential W, related to the drift force by $F(x) = -\partial W/\partial x$. Indeed, here $W(x)$ corresponds to a Brownian trajectory,

$$W(x) = -\int_0^x F(y)dy + const, \qquad (45)$$

because the values of $F(x)$ are random and independent. Adopting Sinai-type behaviour, dynamic scaling (Eq. (34)) implies

$$<<x^2(t)>> \sim (lnt)^4 \sim \xi^2(t)$$

$$P(t) \sim 1/\xi \sim 1/(lnt)^2. \qquad (46)$$

From Eqs. (23), (25), (28) and (29), we then obtain the leading low-frequency behaviour for the density of states $\rho(\omega)$, the integrated density of states $N(\omega)$, and the inverse-exponential localization length [16] $\gamma'(\omega)$

$$\rho(\omega) = \frac{2A}{\omega|ln\ \omega|^3} \ , \quad N(\omega) = \frac{A}{|ln\ \omega|^2} \ , \quad \gamma'(\omega) = \frac{A}{|ln\ \omega|} \ , \qquad (47)$$

differing markedly from the corresponding properties in class-A systems and the Anderson model.

To establish the connection to $1/\omega$ noise [24], we next consider the power spectrum

$$S_{xx}(\omega) = \int_0^\infty dt\ cos\ \omega t\ <<x(t)x(0)>>$$

$$= -\frac{1}{2}\int_0^\infty dt\ cos\ \omega t\ <<x^2(t)>>. \qquad (48)$$

Invoking the Sinai-type long-time behaviour (46), from (48) we obtain to leading order in ω

$$S_{xx}(\omega) \sim |ln\ \omega|^4/\omega, \qquad (49)$$

confirming the prediction of Maniari et al. [25] and revealing the generation of $1/\omega$ noise by this diffusion process.

23.3 Extension to Spatially-discrete Systems and Examples

To illustrate some of these connections, to sketch the extension to spatially-discrete systems, and to test dynamic scaling, we finally consider the extension to spatially-discrete diffusion. The discrete form of the Fokker-Planck Eq. (5) for $\sigma = 2$ is the master equation [5]

$$\dot{P}_n = T_{nn+1}P_{n+1} + T_{nn-1}P_{n-1} - P_n(T_{n+1n} + T_{n-1n}), \qquad (50)$$

with n denoting the lattice sites, and $P_n(t)$ the probability of finding a particle at site n at time t. The transition rates are chosen according to the relation

$$T_{nn+1} = 1/T_{n+1n} = \exp(\tfrac{1}{2}F_{n+1}), \tag{51}$$

with F representing the discretized drift force. As in the continuous case, we can distinguish two classes:

Class A: $W_n = \sum_{m=1}^{n} F_m$ independent random variables,

Class B: F_n independent random variables.

In analogy to the continuous analogs, linear diffusion is expected in class A, and sublinear behaviour of the Sinai type in class B.

The master equation (50) can now be mapped on to a Schrödinger equation [15,16] by introducing

$$P_n(t) = e^{-\omega t}\sqrt{P_n^0}\,\varphi_n, \quad P_n^0 \sim \exp(-W_n), \tag{52}$$

yielding

$$-\varphi_{n+1} + \varphi_{n-1} - 2\varphi_n) + V_n\varphi_n = \omega\,\varphi_n \tag{53}$$

with quantum potential

$$V_n = \exp(-\tfrac{1}{2}F_{n+1}) + \exp(\tfrac{1}{2}F_n) - 2. \tag{54}$$

Introducing the variable

$$R_n = \frac{\varphi_n}{\varphi_{n+1}}, \tag{55}$$

the discrete Schrödinger Eq. (53) might be further reduced to the nonlinear map

$$R_{n+1} = V_n - \omega + 2 - 1/R_n \tag{56}$$

with the characteristic function

$$\Gamma(\Omega) = \lim_{N\,\infty} \frac{1}{N} \sum_{n=1}^{N} \ln R_n, \quad \Omega = -\omega + 10^+ \tag{57}$$

corresponding to the expression (23) in the continuous case.

Two models will be considered:

Model A:

$$P(W_n) = \begin{cases} \dfrac{1}{2\Delta} & \text{or } -\Delta \le W_n \le \Delta \\ 0 & \text{otherwise} \end{cases} \tag{58}$$

belonging to class A.   The potential V_n is given by Eq. (54) with

$F_{n+1} = W_{n+1} - W_n$ and $F_n = W_n - W_{n-1}$.

Model B:

$$P(F_n) = \begin{cases} \dfrac{1}{2\Delta} & \text{for } -\Delta \leq W_n \leq \Delta \\ 0 & \text{otherwise} \end{cases} \tag{59}$$

belonging to class B and expected to exhibit sublinear diffusion of the Sinai type.

We treated the recursive relation (56) numerically to estimate the real and imaginary parts of the characteristic function $\Gamma(\Omega)$ (Eq. (57)), giving the inverse-exponential localization length $\gamma'(\omega)$ and the integrated density of states $N(\omega)$ (Eq. (28)). Figure 1 shows the results for Model A.   For comparison, we have included $N_0(\omega)$ of the free-particle analog, where $V_n = 0$.

The leading small ω-behaviour is (Eqs. (37) and (38))

$$N(\omega 0 = \frac{1}{\pi} \left(\frac{\omega}{D}\right)^{1/2}, \quad \gamma'(\omega) \sim \omega \tag{60}$$

where in the present discrete case [16]

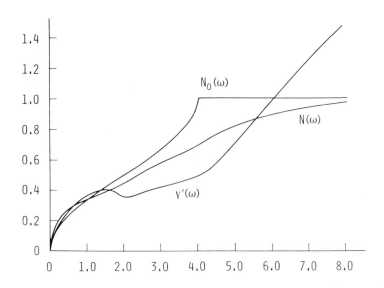

Figure 1.   Numerical results for the inverse-exponential localization length γ' and integrated density of states $N(\omega)$ for Model A with $\Delta = 2$ and $N = 10^5$.   For comparison, we included the free-particle integrated density of states $N_0(\omega)$.   From [26].

$$D = \lim_{N \to \infty} \frac{N^3}{2 \sum\limits_{n=1}^{N} \exp(-W_n) (\sum\limits_{n=1}^{N} \exp\frac{1}{2}W_n)^2} = \frac{\Delta^3}{4} (sh\Delta)^{-1}(sh\frac{\Delta}{2})^{-2}, \quad (61)$$

yielding $D \approx 0.4$ for $\Delta = 2$, and for free diffusion ($\Delta = 0$) $D = 1$. Thus, reduction of the diffusion constant, owing to the random potential W, as shown in Fig. 1, leads to an integrated density of states, which is larger than the free-particle counterpart. At higher frequencies, however, $N(\omega)$ is seen to be smaller than $N_0(\omega)$ and extends to higher frequencies before reaching the limiting value one. The exponentially localized nature of the excited states becomes apparent from the frequency dependence of γ'. We also verified the extension of the asymptotic regime, where $N(\omega)/\omega^{1/2}$ = const and $\gamma'(\omega)/\omega$ = const holds. For $\omega < 10^{-3}$, the deviations are less than 1%.

To explore the validity of dynamic scaling in the case of Sinai-type sublinear diffusion, we also performed a numerical study of model B. Figure 2 shows the results for the integrated density of states and the inverse-exponential localization length.

From the low-frequency behaviour, it is apparent that the expected anomalous behaviour of $N(\omega)$ and $\gamma'(\omega)$, as given by Eq. (46), is not visible on this scale. Comparison of Figs. 2 and 3 also reveals that, except for small ω values, the exponential localization length is much shorter in model A. Numerical results illustrating the approach to the asymptotic regime, where the scaling predictions (47) should hold, are shown in Fig. 3. While

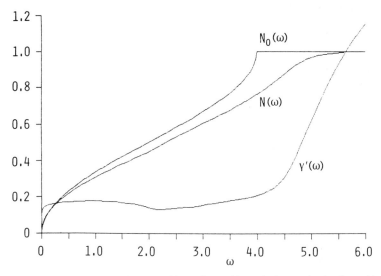

Figure 2. Numerical results for the integrated density of states and the inverse-exponential localization length for Model B, $\Delta = 2$ and $N = 10^5$. For comparison, we included the free-particle integrated density of states. From [16].

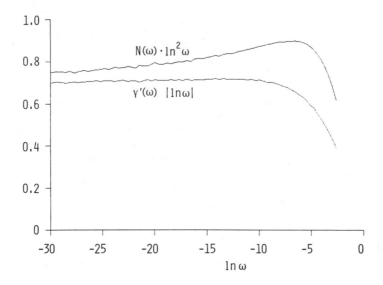

Figure 3. Numerical results for the approach to the asymptotic behaviour of the integrated density of states and the inverse-exponential localization length for $N = 10^7$ and $\Delta = 2$. From [16].

γ' provides strong evidence for the scaling prediction, $N(\omega)$ reaches the asymptotic regime at extraordinarily low frequencies, much smaller than 10^{-14}. This behaviour clearly reveals that the asymptotic Sinai-type regime of sublinear diffusion and for the low-frequency properties of the quantum analog is inaccessibly small. Nevertheless, Fig. 3 shows scaling to be valid within 10% for $\omega \leq 10^{-4}$.

To summarize, we have established the connections between classical diffusion in random media and the motion of a quantum particle in a random potential. Moreover, we have confirmed dynamic scaling. However, it should be emphasized that the quantum analogs of the classical diffusion models differ from the standard Anderson model (Eqs. (40) and (41)), owing to the presence of correlations in the quantum potential, yielding an extended ground state. In other words, we have shown that such correlations suppress exponential localization in the ground state.

Acknowledgements

The author thanks V. Emery, A. Politi, M.P. Sörensen, E. Tosatti and M. Zannetti for extensive discussions.

References

[1]   S. Alexander, J. Bernasconi, W.R. Schneider and R. Orbach (1981). *Rev. Mod. Phys.* **53**, 175.

[2]   D.S. Fisher (1984). *Phys. Rev.* **A30**, 960.

[3]   K. Golden, S.G. Goldstein and J.L. Lebowitz (1985). *Phys. Rev. Lett.* **55**, 2629.

[4]   D.J. Thouless (1974). *Phys. Rep.* **13**, 93.

[5]   N.S. Goel, S.C. Maitra and E.W. Montroll (1971). *Rev. Mod. Phys.* **43**, 231.

[6]   T. Schneider, M. Zannetti, R. Badii and H.R. Jauslin (1984). *Phys. Rev. Lett.* **53**, 2191.

[7]   T. Schneider, M. Zannetti and R. Badii (1985). *Phys. Rev.* **B31**, 2941.

[8]   P.C. Hohenberg and B.I. Halperin (1977). *Rev. Mod. Phys.* **49**, 435.

[9]   T. Schneider and M. Schwartz (1985). *Phys. Rev.* B31, 7484.

[10]   T. Schneider and M. Zannetti (1985). *Phys. Rev.* B32, 6108.

[11]   A. Aharony, E. Domany, R.M. Hornreich, T. Schneider and M. Zannetti (1985). *Phys. Rev.* **B32**, 3358.

[12]   R. Badii, T. Schneider and M.P. Sörensen (1987). *Phys. Rev.* **B35**, 297.

[13]   T. Schneider and R. Badii (1985). *J. Phys.* **A18**, L187.

[14]   T. Schneider, M.P. Sörensen, E. Tosatti and M. Zannetti (1986). *Europhy. Lett.* **2**, 167.

[15]   T. Schneider, M.P. Sörensen, A. Politi and M. Zannetti (1986). *Phys. Rev. Lett.* **56**, 2341.

[16]   T. Schneider, A. Politi and R. Badii (1986). *Phys. Rev.* A34, 2505.

[17]   H. Haken (1977). *Synergetics*, Springer-Verlag, Berlin.

[18]   F.J. Dyson (1953). *Phys. Rev.* 92, 1331.

[19]   T.M. Nieunwenhuizen (1982). *Physica* **113A**, 173.

[20]   R. Festa and E. Galleani d'Agliano (1978). *Physica* 80A, 229.

[21]   B. Simon (1985). *J. Stat. Phys.* 38, 65.

[22]   Ya. Sinai (1982) in *Proc. Berlin Conference on Mathematical Problems in Theoretical Physics,* p. 12, eds. R. Schrader, R. Seiler and D.A. Uhlenbrock, Springer-Verlag, Berlin.

[23]   M. Nauenberg (1985). *J. Stat. Phys.* 41, 803.

[24]   P. Dutta and P. Horn (1981). *Rev. Mod. Phys.* 53, 497.

[25]   E. Marinari, G. Parisi, D. Ruelle and P. Widney (1983). *Phys. Rev. Lett.* **50**, 123.

[26]   T. Schneider (1987) in *Lecture Notes in Physics,* Vol. 268, ed. L. Garrido, Springer-Verlag, Berlin.

STRUCTURE AND DYNAMICS OF REPLICATION–MUTATION SYSTEMS

Abstract

The kinetic equations of polynucleotide replication can be brought
into fairly simple form provided certain environmental conditions
are fulfilled. Two flow reactors, the continuously stirred tank
reactor (CSTR) and a special dialysis reactor are particularly
suitable for the analysis of replication kinetics.   An
experimental set-up to study the chemical reaction network of RNA
synthesis was derived from the bacteriophage $Q\beta$. It consists of a
virus-specific RNA polymerase, $Q\beta$ replicase, the activated
ribonucleosides GTP, ATP, CTP and UTP as well as a template
suitable for replication.

The ordinary differential equations for replication and mutation
under the conditions of the flow reactors were analysed by the
qualitative methods of bifurcation theory as well as by numerical
integration.

The various kinetic equations are classified according to their
dynamical properties: we distinguish "quasilinear systems" which
have uniquely stable point attractors and "nonlinear systems" with
inherent nonlinearities which lead to multiple steady states, Hopf
bifurcations, Feigenbaum-like sequences and chaotic dynamics for
certain parameter ranges.   Some examples which are relevant in
molecular evolution and population genetics are discussed in
detail.

24.1 The Molecular Basis of Polynucleotide Replication

Biological evolution is an exceedingly complicated process which
cannot be analysed on the molecular level yet. It is inevitable
to restrict investigations to some less comprehensive subproblems.
For this goal it is useful to consider the evolutionary process on
several hierarchically ordered time scales (Figure 1).   In this
contribution we shall consider three processes in some detail:

(i) the poplymerization kinetics of polynucleotide replication
and mutation;

(ii)   the dynamics of selection processes leading to
"quasistationary" mutant distributions; and

(iii)   the optimization of replication rates through
evolutionary adaptation to the environmental conditions.

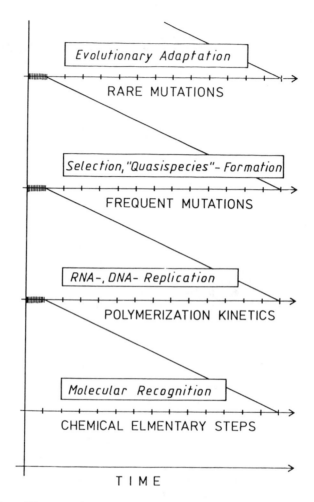

Figure 1. *Time scales in molecular evolution. Recognition of biopolymers is a very fast process and occurs in some ten μsec to msec. Replication of polynucleotides needs more time: the second fastest time scale falls roughly into the range around one minute in a typical testtube experiment depending on chain lengths and enzyme concentrations [1]. Generation times of organisms are longer. Bacteria which replicate fastest require about twenty minutes under optimal conditions. The two slower time scales are appropriately measured in generation times: the formation of quasistationary sequence distributions requires hundreds of generations. Adaptation is still slower by at least two orders of magnitude.*

In most of the intensively studied examples these three hierarchical processes were found to be well separated on the time axes. They interfere with each other only indirectly: the faster processes provide the frame within which the slower processes operate.

Prokaryote evolution, i.e. the evolution of viruses and bacteria is usually dominated by these three processes. The evolution of eukaryotes, in particular that of multicellular, sexually reproducing organisms is much more complicated. It comprises several highly complicated and not yet completely understood processes like morphogenesis, cell differentiation and speciation.

Polynucleotide replication when analysed by the techniques of chemical reaction kinetics turns out to be a complicated reaction network. In essence it represents a template induced, enzyme catalysed polycondensation reaction. The monomers incorporated into the newly synthesized chains are nucleoside triphosphates (GTP, ATP, CTP, UTP in RNA or TTP in DNA, respectively). One pyrophosphate anion per monomer is split off during the formation of the new bond yielding thereby the free energy which drives the polymerization process. The molecular basis of template induction is represented by the complementarity of certain purine and pyrimidine bases in the two base pairs G≡C and A=U in RNA, or A=T in DNA, respectively. RNA replication is by far a simpler process than DNA replication. In nature RNA synthesis is commonly accomplished by a single enzyme, whereas DNA replication usually requires catalysis by at least ten different proteins.

One particular example, RNA synthesis by means of the enzyme Qβ replicase from a simple bacteriophage infected with the bacterium *escherichia coli* has been studied in great detail [1-3]. It represents a case of "complementary replication" (Figure 2). This process leads from a single stranded molecule to two complementary single strands. Double helical structures involving parts of plus and minus strand occur only as intermediates during replication.

DNA replication in nature is an example of "direct replication" (Figure 2). This process leads from one plus-minus complex of double helical structure to two molecules of the same kind.

One of the most important advantages of the molecular approach to biological evolution lies in the fact that mutation is an intrinsic part of the replication mechanism: correct and erroneous replications are parallel reactions of one and the same reaction mechanism. Molecular biology revealed several classes of mutations (Figure 3):

(i)   point mutations or single base exchanges;
(ii)   insertions of bases ranging from one base to many bases;
(iii)   deletions of single bases or entire groups of bases; and
(iv)   rearrangements of polynucleotide sequences which leave the total length (v) essentially unchanged.

Point mutations are the most frequent replication errors, but insertions, deletions and rearrangements are certainly important in evolution as well.

During the last decade a new, quite unexpected property of RNA molecules was discovered: apart from the role transfer- and ribosomal-RNA play in cellular protein synthesis RNA molecules can also act as catalysts in RNA processing. In particular two types of unconventional RNA catalysis were reported:

(i)   self-splicing of messenger-RNA molecules [4] and
(ii)   maturation of RNA by the cleaving enzyme RNAse P [5] which carries an RNA molecule as the catalytically active unit.

Hence, RNA can act in two different catalytic modes in RNA

synthesis: firstly as template and secondly as ordinary biochemical catalyst. Due to the "double function" of RNA reaction networks involving these molecules may lead to highly nonlinear kinetics equations. It is appropriate to distinguish different modes of autocatalysis which reflect the structure of the kinetic equations. Simple template induced polymerization follows a first order "over-all" rate law,

$$I_k + \ldots \to 2I_k: \quad \dot{x}_k \propto x_k.$$

Throughout this paper we denote concentrations by lower case letters: $[I_k]=x_k$ or $[A]=a$; time derivatives are indicated by

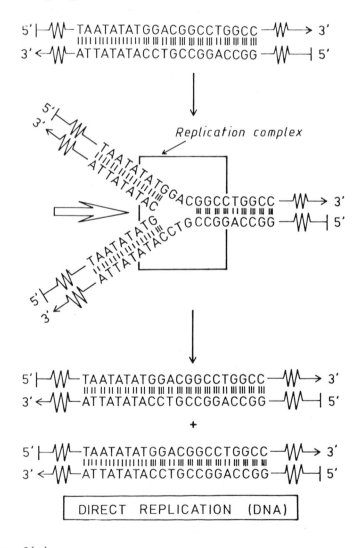

Figure 2(a).

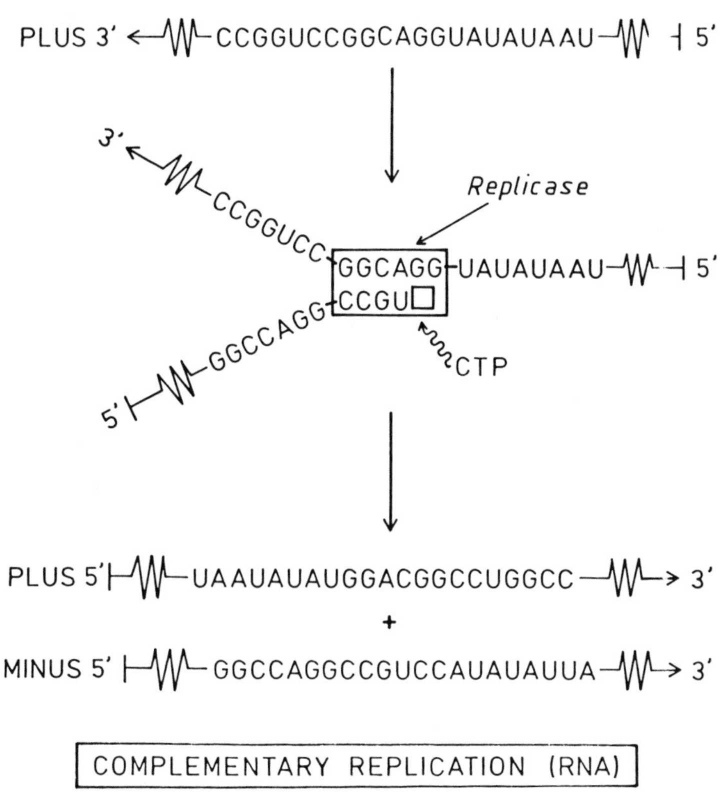

PLUS 3' ←⟍⋀⋀⋀⟋CCGGUCCGGCAGGUAUAUAAU⟍⋀⋀⋀⟋ ⊣ 5'

Replicase

GGCAGG⊦UAUAUAAU⟍⋀⋀⟋ ⊣ 5'
CCGU☐

⟍⟍CTP

PLUS 5'⊦⟍⋀⋀⋀⟋UAAUAUAUGGACGGCCUGGCC ⟍⋀⋀⟋→ 3'

+

MINUS 5' ⊦⟍⋀⋀⟋GGCCAGGCCGUCCAUAUAUUA⟍⋀⋀⟋→ 3'

COMPLEMENTARY REPLICATION (RNA)

Figure 2(b).
Figure 2(a) and (b). Direct and complementary replication of polynucleotides. Direct replication is the common mechanism of DNA synthesis in all organisms (except some viruses). It leads from one double stranded DNA molecule to two double stranded DNA molecules. The process requires a complicated machinery of at least ten enzymes. Viral RNA is commonly multiplied by the mechanism of complementary replication. Plus and minus strands occur as free molecules and are replicated separately. The process has some analogy to the common photographic technique since both use a negative or complementary copy as intermediate. Virus-specific RNA synthesis is usually catalysed by a single enzyme (see e.g. [1-3]).

"dots": $dx_k/dt = \dot{x}_k$, $da/dt = \dot{a}$, etc.

Complementary replication turns out to be a special case of first order autocatalysis. Apart from a transient phase the ensemble consisting of plus- and minus-strands grows according to the rate law given above. The rate constant for the growth of the ensemble turns out to be the square root of the rate constants for the two individual replication steps: $\bar{k} = \sqrt{k_+ . k_-}$.

The corresponding equations of the template induced and catalysed process are of second order in polynucleotide concentrations,

$$I_k + I_j + \ldots \to 2I_k + I_j: \quad \dot{x}_k \propto x_k \cdot x_j.$$

Herein, I_k is the template and I_j the catalyst. Second order autocatalysis may also involve the same type of molecule as template and catalyst.

$$2I_k + \ldots \to 3I_k: \quad \dot{x}_k \propto x_k^2.$$

A reaction of this type represents the essential step of the "Schlögl model" [6]. A closely related reaction step occurs in the Brusselator model [7]. Both processes are at least

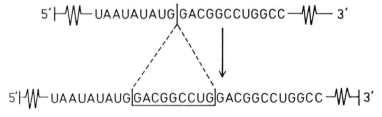

POINTMUTATION

DELETION

INSERTION

Figure 3. The molecular mechanisms of three main classes of mutations: point mutations, deletions and insertions.

trimolecular, since at least one additional molecule is required in the conversion to the new product molecule I_k.

Trimolecular steps are not readily accepted as elementary steps in vapour phase or dilute solution kinetics, because the simultaneous encounter of three molecules is highly improbable. In replication kinetics, however, we are not dealing with elementary steps but with "over-all" rate laws of many thousand individual steps. Then, there is no reason to rule out processes of formally higher orders.

24.2 Flow Reactors and Constraints

Flow reactors are suitable devices for maintaining a controllable flux which keeps chemical reactions far away from equilibrium. A stock solution flows into the tank where chemical reactions take place. The influx is balanced by a compensating, steady outflux of instantaneous tank solution. Intensive stirring is applied to suppress diffusion effects and to achieve spatial homogeneity of the solution. Adjustment of concentrations in the stock solution and control of flow rates allows to run flow reactors under conditions which eventually simplify the kinetic equations.

In the case of "in vitro" RNA replication the stock solution which flows into the reactor, contains the four activated monomers, the ribonucleoside triphosphates GTP, ATP, CTP and UTP, as well as a virus-specific RNA-dependent RNA polymerase, "Qβ replicase".

Two types of flow reactors will be used in the forthcoming discussion:
(i) a continuously stirred tank reactor (CSTR), which is a fairly simple experimental devise; and
(ii) a dialysis reactor, which provides particularly well suited conditions for the mathematical analysis of the rate equations.

Both reactors are sketched in Figure 4. The dialysis reactor requires rather sophisticated mechanical and electronic equipment. It can be simulated with satisfactory accuracy, however, by the technically unpretentious serial transfer technique. The conditions of the dialysis reactor make the analysis of selection processes particularly simple and therefore this reactor has been characterized also as "evolution reactor".

The two reactors impose different constraints on the kinetic equations. In the first case the flow of solution through the reactor is usually kept constant. The second reactor allows control of the total concentration of the polymers in the reactor. The constraint of constant total polynucleotide concentration was found to be particularly useful in theoretical investigations. It is well known as constraint of "constant population size" in theoretical population genetics.

In order to illustrate both constraints we shall analyse the simple case of independent and error-free replication in both reactors. Since the basic features are insensitive to the number of components in the input solution, we shall assume that only one compound, A, and the template, I, are required to synthesize polynucleotides.

The number of possible, different templates is exceedingly large – we have $N=4^\nu$ different polynucleotide sequences of chain length ν. High dimensionality of the kinetic differential equation is characteristic for any reaction network modelling polynucleotide synthesis and hence, we have to use special mathematical techniques in order to be able to analyse systems with arbitrarily large numbers of variables (see e.g. [8]).

The reaction mechanism of independent, error-free polynucleotide replication consists of a large number of parallel reactions:

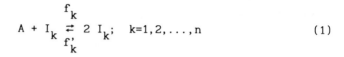

$$A + I_k \underset{f'_k}{\overset{f_k}{\rightleftarrows}} 2\,I_k; \quad k=1,2,\ldots,n \qquad (1)$$

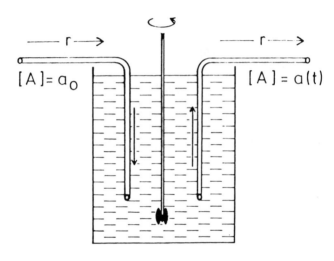

$$[A] = a_0 \qquad\qquad [A] = a(t)$$

A

Figure 4. Two flow reactors to keep chemical reactions away from thermodynamic equilibrium.

Figure 4(a): The continuously stirred flow reactor (CSTR). In this reactor which is known also as "chemostat" in microbiology, the material source is provided by the continuous influx of a solution containing the material which is necessary for polynucleotide replication. In the simplified model systems to be discussed here, we assume only one energy rich input compound, A. The evolutionary constraint is provided by the continuous outflux of solution from the tank reactor. Replicating molecules are injected into the reactor at $t=t_o$.

Then, their concentrations may increase and eventually reach a stationary value, or they may be diluted out of the reactor depending on the concentration of the input solution and the flow rate r which is commonly measured in terms of the reciprocal residence time (τ_R^{-1}) of the solution of the tank reactor.

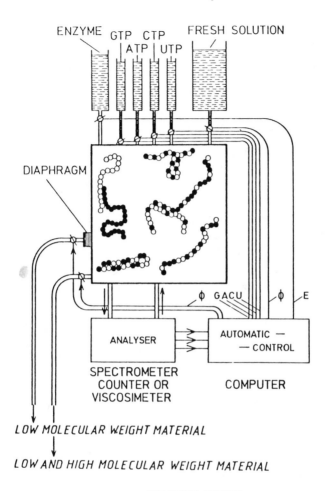

ENZYME GTP CTP FRESH SOLUTION
ATP UTP

DIAPHRAGM

φ GACU φ E

ANALYSER

AUTOMATIC —
— CONTROL

SPECTROMETER
COUNTER OR
VISCOSIMETER

COMPUTER

LOW MOLECULAR WEIGHT MATERIAL

LOW AND HIGH MOLECULAR WEIGHT MATERIAL

●........○●○○●●○ = POLYNUCLEOTIDE

B

*Figure 4(b). The evolution reactor.   This kind of dialysis
reactor consists of a reaction vessel which has walls which are
impermeable to polynucleotides.   Energy rich material is poured
from the environment into the reactor. Degradation products are
removed steadily.   Material transport is adjusted in such a way
that the concentrations of energy rich monomers, the nulceoside
triphosphates, are constant in the reactor.   A dilution flux Φ
is installed in order to remove the exess of polynucleotides
produced by replication. Thus, the sum of the concentrations,*

$$x_1 + x_2 + \ldots + x_n = C_o,$$

*may be controlled by the flux Φ.   Under the constraint of
constant total concentration Φ is adjusted so that C_o =
constant.   The regulation of Φ requires internal control, which*

355

may be achieved by analysis of the solution and data processing by a computer as indicated above. The constraint of constant total concentraation facilitates the analysis of the differential equations for polynucleotide replication considerably, and hence, has been applied frequently [9,10].

In the CSTR this reaction mechanism leads to the following rate equations:

$$\dot{a} = \sum_{k=1}^{n} (f'_k x_k^2 - f_k a x_k) + (a_0 - a)r \qquad (2a)$$

$$\dot{x}_k = x_k(f_k a - f'_k x_k - r); \quad k=1,2,\ldots,n. \qquad (2b)$$

Herein, we denote the concentration of A in the stock solution by a_0. The flow rate r is expressed as the reciprocal mean residence time of the solution in the reactor: $r=(\tau_R)^{-1}$. The dynamics of the parallel reactions (2) is completely determined by two external parameters, the input concnetration a_0 and the flow rate r. We verify easily that the total concentration in the tank reactor, $c=a+\Sigma x_k$, converges asymptotically towards the stationary value $c=a_0$:

$$\dot{c} = (a_0 - c)r \Rightarrow c(t) = a_0 - [a_0 - c(0)]\exp(-rt) \qquad (3)$$

The relaxation time of this process is the mean residence time of the solution in the reactor, τ_R.

The reaction mechanism (2) has 2^n stationary states which fulfil the conditions

$$\bar{x}_k = 0 \text{ or } \bar{x}_k = (f_k \bar{a} - r)/f'_k; \quad k=1,2,\ldots,n$$

and

$$\bar{a} - [a_0 + r \sum_{k=1}^{n} (f'_k)^{-1}]/(1 + \sum_{k=1}^{n} \frac{f_k}{f'_k})$$

in every possible combination. Appropriately, we denote a particular stationary state by $P_{ij\ldots m}$ if $\bar{x}_i \neq 0$, $\bar{x}_j \neq 0, \ldots, \bar{x}_m \neq 0$ and all other $x_k=0$. The "zero state", i.e. the stationary state at which all polynucleotide concentrations vanish, is characterized by P_0. Without losing generality we arrange the individual polynucleotide sequences with decreasing replication rate constants f_k:

$$f_1 > f_2 > \ldots > f_n \qquad (4)$$

Cases of kinetic degeneracy, these are cases in which rate constants are equal, are excluded here. We shall come back to this problem in the last paragraph.

Three lemmas are useful to characterize the properties of the individual stationary states (proofs are found in [8]):

(i) The stable stationary states of the reaction network (2) are unique in the sense that for every pair of external parameters (a_0, r) there exists only one stationary state which is asymptotically stable and which has the entire range of physically meaningful initial concentrations $(a(0) \geq 0; x_k(0) > 0, k=1,2,\ldots,n)$ as basin of attraction.

(ii) For reversible replication, i.e. for nonzero rate constants of the reverse reaction of polynucleotide synthesis $(f'_k \neq 0)$ the sequence of stable stationary states with increasing values of a_0 and/or decreasing values of r is:

$$P_0 \to P_1 \to P_{12} \to P_{123} \to \ldots \to P_{12\ldots n}. \qquad (5)$$

(iii) In the case of irreversibility (for practical purposes) of polynucleotide synthesis $(f'_k = 0)$ there exist only two asymptotically stable stationary states: P_0 and P_1. The two states are mutually exclusive and hence we have either selection of the polynucleotide sequence which replicates fastest or all polynucleotides are diluted out of the CSTR.

In addition it is possible to derive analytical expressions for the lines in the plane of external parameters $(a_0, r0)$ at which the stationary states $P_{12\ldots m}$ become unstable:

$$a_0 = F_m \cdot r \text{ with } F_m = \frac{1}{f_{m+1}} \left(1 + \sum_{i=1}^{m} \frac{f_1 - f_{m+1}}{f'_i} \right). \qquad (6)$$

All lines are straight and pass through the origin of the (a_0, r)-coordinate system. The stationary point $P_{12\ldots m}$ is asymptotically stable in the sector confined by the straight lines $a_0 = F_{m-1} \cdot r$ and $a_0 F_m \cdot r$. In the irreversible case $(f'_k = 0)$ all lines except $a_0 = F_0 \cdot r$ coincide with the a_0-axis and hence, there are only two stable stationary states P_0 and P_1.

The results of this analysis are shown graphically in Figure 5 (for the purpose of comparison with the forthcoming system we have chosen the a_0-axis as abscissa).

The conditions of the dialysis reactor differ from those in the CSTR with respect to two aspects.

The concentration of input material is controlled and kept

357

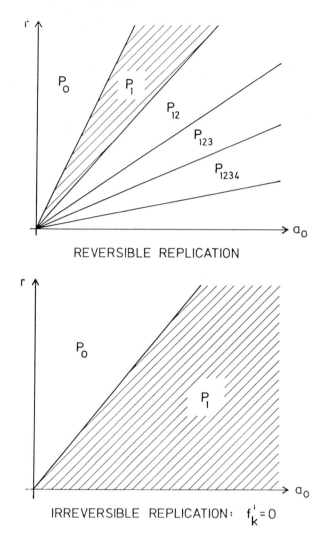

REVERSIBLE REPLICATION

IRREVERSIBLE REPLICATION: $f_k' = 0$

Figure 5. Selection in the CSTR. The asymptotically stable stationary states $P_{12..m}$ of the reaction mechanism (1) are unique in the plane of external parameters (a_0, r) and it can be subdivided into areas within which one particular state is stable. In the reversible case we have n+1 different regions covering all intermediate cases from stability of the zero state P_0 to stability of the state $P_{12..n}$. Selection of the most efficiently replicating type I_1 is observed in the region of stability of P_1. In the irreversible case ($f_k'=0$) only two states P_0 and P_1 are stable. The region within which selection occurs (P_1) is drastically increased.

constant, $[A]=a_0=$const., and therefore we can account for it by proper redefinition of the rate constants: $f_k \cdot a_0 \to f_k$. In addition the total polynucleotide concentration is kept constant by proper adjustment of fluxes: $c_0 = \Sigma x_k$. The mechanism of independent, error-free replication (1) in the dialysis reactor thus leads to the rate equation:

$$\dot{x}_k = x_k(f_k - f'_k x_k - \frac{1}{c_0}\Phi); \quad k=1,2,\ldots,n \qquad (7)$$

with $\Phi = \sum_{\ell=1}^{n} x_\ell(f_\ell - f'_\ell x_\ell)$.

Equation (7) contains only one external parameter, the total concentration c_0. The unspecific dilution flux $\Phi(t)$ is determined by the conservation relation of concentrations, $\Sigma x_k = c_0$. It compensates the net production of polymers in the dialysis reactor. Apart from a concentration factor the flux Φ is identical to the mean net rate of replication $(\bar{E})$. "Net" refers here to the difference between the forward and backward rates of polymerization:

$$\frac{1}{c_0}\Phi(t) = \bar{E}(t) = \frac{\sum_{k=1}^{n} (f_k - f'_k x_k) x_k}{\sum_{k=1}^{n} x_k}$$

The mean net replication rate is optimized under the conditions of the dialysis reactor (see equation (9)).

Equation (7) sustains 2^n-1 different stationary states. They correspond to the states $P_1, P_2, \ldots, P_{12\ldots m}$ discussed above. The zero state P_0 does not exist in the dialysis reactor since it contradicts the conservation relation. In case we order the replication rate constants as shown in equation (4) we observe a succession of unique stationary states which are asymptotically stable for certain ranges of total concentrations. This series, ordered with increasing total concentration c_0, is identical to that shown in (5) apart from the fact that the first state P_0 is missing now. The state $P_{12\ldots m}$ loses its stability at the critical concentration

$$(c_0)_m = \sum_{i=1}^{m} \frac{f_i - f_m}{f'_i}. \qquad (8)$$

In the limit of irreversible polynucleotide synthesis $(f'_k=0)$ the stationary state P_1 is stable within the entire range of total

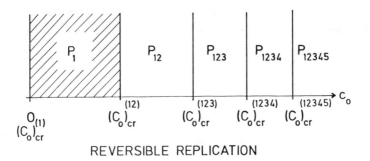

REVERSIBLE REPLICATION

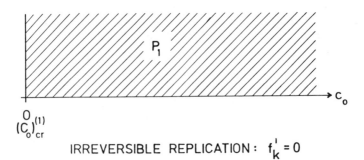

IRREVERSIBLE REPLICATION: $f'_k = 0$

Figure 6. Selection in the evolution reactor. The situation sketched here is analogous to the stability analysis in the CSTR shown in Figure 5. Here we have only one external parameter, the total concentration c_0. Note that the zero state has disappeared now. In the irreversible case $(f'_k = 0)$ selection occurs under all possible external conditions. This fact makes the evolution reactor particularly well suited for the study of selection and adaptation.

concentrations. Selection under the constraint of constant total concentration is shown graphically in Figure 6. The major effect of the dialysis reactor consists in an extension of the range of external parameters which lead to selection. Due to additional control devices it is impossible to dilute the templates out of the reactor and therefore the zero state P_0 which is stable for certain parameter ranges in the CSTR does not exist in the dialysis reactor. Apart from these obvious differences the conditions under which selection occurs are essentially the same in both reactors as an inspection of equations (6) and (8) shows. The irreversible case in the evolution reactor,

$$\dot{x}_k = x_k(f_k - \frac{1}{c_0}\Phi); \quad k=1,2,\ldots,n \qquad (7a)$$

360

with $\Phi(t) = c_0 \cdot \bar{E}(t) = \Sigma f_k x_k$, leads to selection for all values of the external parameter c_0. From equation (7a) we derive the following time dependence of the unspecific dilution flux $\Phi(t)$:

$$\frac{d\Phi}{dt} = \sum_{k=1}^{n} x_k \left(f_k - \frac{1}{c_0} \Phi\right)^2 \geq 0 \qquad (9)$$

The flux Φ, here the mean replication rate, is a non-decreasing function of time. It is optimized during the selection process. The rate constants f_k can be interpreted as global fitness factors and selection is understood as optimization of the mean fitness in the population considered.

It is useful to distinguish two different cases of the evolutionary selection process:

(i) Mutations are rare events. Then the population is almost always homogeneous. Only occasionally, just when the previously fittest type is replaced by a new fitter variant, we observe two different sequences simultaneously in the population.

(ii) Mutations occur frequently. The selection process cannot be considered independently of the production of mutants. Then populations are usually heterogeneous and we need a more sophisticated analysis which will be presented in the next section.

In case mutation is a rare event the evolutionary optimization process can be visualized by the scenario sketched in Figure 7. Stationary phases are interrupted by rather short transient periods during which a new fitter variant is selected. Since only advantageous mutants can replace previously selected types, the mean rate of replication represents a non-decreasing function of time. Increase in this quantity can be considered as the evolution criterion for this simple system. Adaptation to environmental conditions occurs stepwise through a succession of replacements of previous wildtypes by better adapted variants. Degradation of polynucleotides has not yet been included in the mechanism analysed here. Usually it plays no role since residence times in flow reactors are rather short. Introduction of individual degradation processes,

$$I_k \xrightarrow{k} (B); \quad k=1,2,\ldots,n \qquad (10)$$

causes no problem [9,10]: rates of replication need only be replaced by net rates of synthesis, $f_k \to f_k - d_k$, and all the expressions derived above apply to the extended mechanism as well (note that "net" refers now to the difference in the rates of synthesis and degradation).

24.3 Replication and Mutation

Mutation terms are introduced readily into the reaction mechanism of polynucleotide replication. In addition to reaction (1) the reaction network includes now all possible parallel reactions

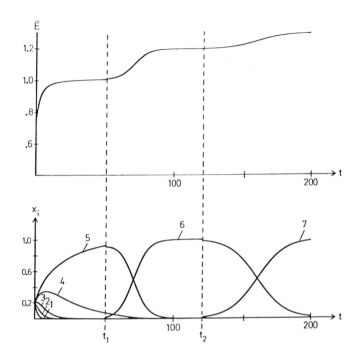

Figure 7. *Adaptation in the evolution reactor for the rare mutation case.* *The figure shows solution curves* $x_k(t)$ *of equation (7a) perturbed by occasional mutations.* *The individual numerical values are:* $f_1=0.3$, $f_2=0.65$, $f_3=0.82$, $f_4=0.96$, $f_5=1.01$, $f_6=1.2$ *and* $f_7=1.3$ $[t^{-1}]$. *The initial conditions are:* $x_1(0)=x_2(0)=\ldots=x_5(0)=0.2[M]$, $x_6(0)=x_7(0)=0$. *The mean (net) replication rate* $\bar{E}(0)$ *starts from an initial value,* $\bar{E}(0)=0.748$ *and increases steadily as the population becomes homogeneous:* $\bar{E} \to 1.01$ *and* $x_1, x_2, x_3, x_4 \to 0$ *as* $x_5 \to 1$ $[M]$. *At* $t=t_1=50$ $[t]$ *we introduce a fluctuation* $x_6=0 \to x_6=\delta$, *the appearance of* I_6, *which represents the rare formation of a favourable mutant.* *Now* I_5 *is replaced by* I_6 *as the mean (net) replication rate increases from 1.01 to 1.2.* *At time* $t=t_2=120$ $[t]$ *a still more efficient mutant,* I_7, *is formed and afterwards* I_6 *is replaced by the new type.* *Again* $\bar{E}(t)$ *shows a stepwise increase and approaches thereby the temporary optimum* $\bar{E}=1.3$.

leading to error copies:

$$A + I_j \xrightarrow{f_j Q_{kj}} I_k + I_j; \quad k, j = 1, 2, \ldots, n. \tag{11}$$

Instead of the n independent replication processes of equation

(1) we are dealing now with a network of n^2 reactions. We assumed irreversibility of the polymerization process which is in general agreement with the conditions *in vivo* and in test tube evolution experiments. The factors Q_{kj} describe the distribution of error-free copies and mutants among the polynucleotides synthesized on the template I_j. Every copy has to be either correct or erroneous and the elements in column of the nxn mutation matrix Q fulfil the conservation law

$$\sum_{k=1}^{n} Q_{kj} = 1. \qquad (12)$$

The replication-mutation system has been studied under the conditions of both flow reactors. In the evolution reactor the rate equations of the reaction mechanism (9) are of the simple form $(c_0 = \Sigma x_k)$

$$\dot{x}_k = (f_k \cdot Q_{kk} - \frac{1}{c_0} \Phi)x_k + \sum_{j \neq k} f_j \cdot Q_{kj} \cdot x_j; \quad k, j = 1, 2, \ldots, n. \qquad (13a)$$

It is useful to define a value matrix $W \div \{w_{kjk} = f_j \cdot Q_{kj}\}$. If degradation is to be considered explicitly, the diagonal elements of the value matrix are of the form: $w_{kk} = f_k \cdot Q_{kk} - d_k$. (In the following sections we indicate by "(net)" in parantheses that degradation may be taken into account simply by subtraction of the rate constants d_k.)

The differential equation (13a) may be written now in the form

$$\dot{x}_k = \sum_{j=1}^{n} w_{kj} \cdot x_j - \frac{x_k}{c_0} \Phi; \quad k = 1, 2, \ldots, n \qquad (13b)$$

with

$$\Phi = \sum_{j,k=1}^{n} w_{kj} \cdot x_j = \sum_{j,k1}^{n} f_j \cdot Q_{kj} \cdot x_j = \sum_{j=1}^{n} f_j \cdot x_j \qquad (13c)$$

as a consequence of Equation (12). Thus the mean (net) rate of polynucleotide synthesis is of the same general form as in the error-free system. The only nonlinear terms in Equation (13) are those involving the flux Φ. They can be removed by a general integral transformation [11,12].

Solutions of the kinetic equations (13) describing replication and mutation in the evolution reactor were derived, analysed and extensively discussed in the literature [11-15]. Here, we shall only repeat the most important results which distiguish the case of frequent mutation from the "rare mutant scenario" sketched in the previous section. (Essentially the same results were obtained also for the conditions in the CSTR [8]):

(i)   The continuos production of mutants leads to a distribution of sequences in the stationary state rather than to selection of a single fittest type.   This stationary distribution of sequences, called the "quasispecies", is unique in the sense that, independently of the particular initial conditions $[x_1(0), \ldots, x_n(0)]$, the solutions of the replication-mutation system (13) converge to it asymptotically.

(ii)   The stationary mutant distribution, the quasispecies is determined by the dominant (right-hand) eigenvector (ξ_m) of the value matrix W which corresponds to the largest eigenvalue (λ_m):

$$W \, \xi_m = \lambda_m \, \xi_m \tag{14}$$

with $\lambda_m = \max\{\lambda_k; k=1,2,\ldots,n\}$.   Frobenius theorem applies to the value matrix W and hence the largest eigenvalue is real positive and non-degenerate.   All components of the dominant eigenvector are strictly positive.   Apart from cases with kinetic degeneracy the quasispecies consists of a "master sequence" I_s and its most frequent mutants.   The master sequence is characterized as the most abundant sequence.   Commonly it has the largest selective value (w_{ss}).

(iii)   The concept of selection can be carried over to the case of frequent mutations if, instead of individual polynucleotide sequences, linear combinations of them as defined by the eigenvectors of the value matrix W are chosen as coordinate axes for the definition of concentrations (the orthogonal vectgors e_k span an Euclidean space to which the ordinary chemical concentrations refer):

$$x = (x_1, x_2, \ldots, x_n = \sum_{k=1}^{n} x_k \cdot e_k = \sum_{k=1}^{n} z_k \cdot \xi_k. \tag{15}$$

This transformation decouples the differential equation (13) and becomes formally identical to equation (7a)

$$\dot{z}_k = z_k (\lambda_k = \frac{1}{c_0} \, \Phi); \quad k=1,2,\ldots,n. \tag{7b}$$

Any particular choice of initial conditions $[x_1(0), \ldots, x_n(0)]$ corresponds to a defined linear combination of the eigenvectors ξ_k characterized by $[z_1(0), \ldots, z_n(0)]$.   The final state contains the dominant eigenvector exclusively: $[\bar{z}_m = c_0; \bar{z}_k = 0 \; \forall \; k \neq m]$ and hence, we can visualize the dynamics of equation (13) as competition and selection between the eigenvectgors of the value matrix W.

4.   The mean (net) rate of polynulceotide synthesis (or the

dilution flux Φ) converges asymptotically to the largest eigenvalue of the value matrix

$$\lim_{t \to \infty} \Phi = \lambda_m. \qquad (16)$$

The optimization principle (9), however, is no longer generally valid since

(i) the variables z_k need not be negative (as is always the case with the ordinary chemical concentrations x_k) and

(ii) the eigenvalues of the value matrix λ_k may be complex (Frobenius theorem states that the largest eigenvalue is real and positive. It makes no claim concerning other eigenvalues). Mean (net) rates of synthesis may increase or decrease monotonously, or even pass through an extremum in approaching λ_m [16].

Some important properties of the replication-mutation system (13) can be demonstrated best by means of appropriate models for mutation. One useful model [14] involves restriction to point mutations and uniform error rates (1-q) at all positions of the sequence. As further simplification binary sequences - these are sequences of two symbols $\{0,1\}$ instead of the naturally occurring polynucleotides with four symbols $\{G,A,C,U\}$ - are used. (This is not necessarily a restriction since doublets of binary digits can always be used to code for the four bases.) Then the elements of the value matrix are of the simple general form

$$w_{kj} = f_j \cdot q^{\nu-d(k,j)}(1-q)^{d(k,j)} = f_j \cdot q^{\nu}(\frac{1-q}{q})^{d(k,j)}. \qquad (17)$$

Herein, q represents the "single digit accuracy" of the replication process and hence 1-q is the frequency of errors at a particular position in the polynucleotide sequence. Accordingly, the range of physically meaningful values is restricted to $0 \leq q \leq 1$. We are only dealing with point mutations which leave the chain length unchanged and therefore all sequences are ν digits long. By d(k,j) we denote the "Hamming" distance of the two sequences I_j and I_k. This is the number of digits in which the two sequences differ.

Three limiting cases of replication accuracies are of particular interest [14]:

(i) The condition q=1 implies ultimate precision of replication and characterizes error-free, direct replication. Then, the stationary distribution contains the fittest type exclusively.

(ii) The condition q=0 represents also a case of ultimate precision. Every digit of the template is copied by incorporation of the complementary symbol into the newly synthesized strand. Hence we are dealing with error-free, complementary replication in this limit. In the stationary state only the two complementary sequences forming the fittest "plus-minus pair" are present.

(iii) The condition q=0.5 implies that correct and

complementary digits are incorporated into the new strand with equal probabilities. Consequently, there is no correlation in sequence between template and copy. This case has been called "random replication". Heredity and evolutionary optimization break down in this limit. The replication rates have no influence and thus the stationary distribution is uniform:

$$\bar{x}_1 = \bar{x}_2 = \ldots = \bar{x}_n = c_0/n.$$

The deterministic description by kinetic equations is not applicable to random replication. For all sequences which are sufficiently long to be of actual interest we are dealing with many more possible sequences than we have molecules or organisms in natural populations or in laboratory systems. Stationary concentrations would be much less than one molecule in the volume considered! Any appropriate description of random replication has to be based on a stochastic treatment of replication and mutation therefore (see e.g. [17]).

Systematic investigations of quasispecies as a function of chain length v and replication accuracy q yields three main results [14]:

(i) The value matrix of model systems defined by equation (17) has exclusively real eigenvalues (cf. Rumschitsky [18]).

(ii) For sufficiently long sequences ($v \geq 10$) plots of the quasispecies as a function of q (Figure 8) show three characteristic ranges which are separated by two rather narrow transition zones around two critical values q_{min} and q_{max}. In the range $0 < q < q_{min}$ we observe complementary replication with errors. Both ranges of organized replication are separated by the broad intermediate domain of random replication ($q_{min} < q < q_{max}$). The two critical values represent "error thresholds" beyond which polynucleotide sequences are unstable in successive replications. The values of the critical single digit accuracies depend on the relative replication rates of master sequences and their mutant distributions [9,10,14].

(iii) The two transition zones become exceedingly narrow with increasing chain length. This general result suggests the conjecture that we are dealing here with an analogue to a phase transition in the limit $v \to \infty$. Indeed, the replication-mutation system has much in common with (generalized) spin lattices [19,20]. The error thresholds correspond to the critical temperatures at which magnetic order-disorder transitions occur. The error threshold defines a maximum chain length, v_{max}, in case the single digit accuracy is constant and the chain length v is varied [9,10]. Such a situation is commonoly encountered in natural populations which use a more or less constant molecular machinery of replication. Examples are different prokaryotes, viruses and bacteria. Then the error thresholds determine the maximum length of RNA or DNA which can be transferred reliably over many generations and set limits to the genomes of these organisms. RNa viruses which were studied in great detail have

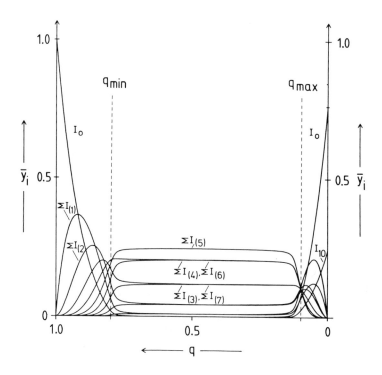

Figure 8. The quasispecies as a function of the single digit accuracy of replication (q) for a chain length $\nu=10$. We plot relative stationary concentrations of the master sequence, $\bar{y}_0$, the sum of the relative concentrations of all one error mutants, $\bar{y}_1$, of all two error mutants, $\bar{y}_2$, etc. We observe selection of the master sequence at $q=1$. At $q=0$ we have selection of a "master pair" consisting of I_0 and its complementary sequence, I_{1023} in our example. As rate constants we chose $f_0=10$ $[t^{-1}]$ for the master sequence and $f_k=1$ $[t^{-1}]$ $(k\neq 0)$ for all mutants. The entire domain of physically acceptable q-values is split into three regions: we observe direct replication in the range $1\geq q\geq q_{max}$, complementary replication in the range $0\leq q\leq q_{min}$ and finally random replication in the intermediate range $q_{min}\leq q\leq q_{max}$.

indeed lengths of genomes that lie very close to the thresholds.

24.4 Evolutionary Adaptation

Evolutionary adaptation can be visualized best as a process in an abstract "sequence space" [21]. The sequence space is a point space in which a single point is assigned to every binary sequence. In case of constant chain length ν the 2^ν points may be considered as the corners of a ν-dimensional cube. The sequences

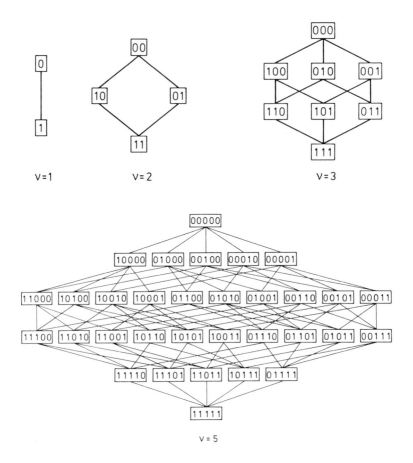

Figure 9. *The sequence space of binary sequences based on the two-letter alphabet (0,1). We present here the cases $\nu=1,2,3$ and 5. All pairs of sequences I_i and I_k with Hamming distance $d(i,k)=1$ are connected by straight lines. The graphs obtained are hypercubes of dimension ν.*

are arranged such that all nearest neighbours on the ν-cube $(I_{(k)}, I_{(k+1)})$ have Hamming distance $d(k,k+1)=1$. Now we draw lines connecting all pairs of sequences with Hamming distance $d=1$. Examples of low dimensional graphs obtained in this way are shown in Figure 9. In any realistic population we have many more possible sequences, or points, than actual molecules. Therefore only small fractions of the sequence space will be "occupied". A point is occupied if the corresponding sequence is actually present in the population under consideration. For the most part the space is empty.

Let us consider now a realistic, "quasistatitionary" distribution of sequences. By "quasistationary" we mean that the distribution is constant apart from minor fluctuations until a rare advantageous mutant appears and spreads in the population.

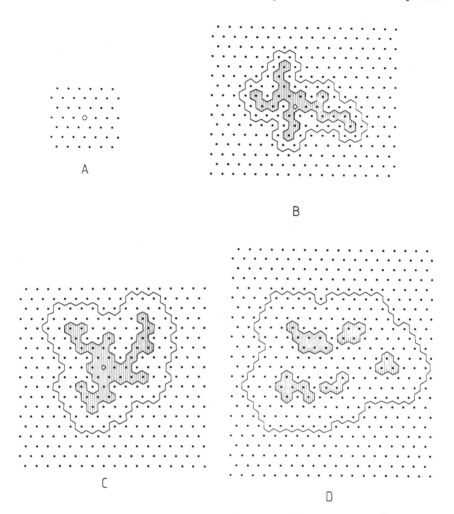

Figure 10. A schematic sketch of quasistationary distributions in sequence space. These distributions closely resemble quasispecies. The consist of a master sequence (o), a "core" of frequent mutants (hatched areas) and a periphery of rare mutants. We take "snapshots" of distributions and distinguish four different situations which are characteristic for different accuracies of replication:
A: Ultimate accuracy of replication. No mutants are formed and the distribution contains only the master sequence. B: Highly accurate replication. The distribution contaains a compact core surrounding the master sequence. it extends mainly along "ridges" of high fitness in sequence space. The core is surrounded by a rather narrow zone of rare mutants. C: Replication above error threshold. The core is still compact but it spreads out widely into sequence space. The range of rare mutants is much larger than in scenario B. D: Replication below error thresholds, characterized as random replication.

> There is no well defined master sequence. The core is split into several rapidly fluctuation zones of higher population numbers only. Hence, the area of rare mutants is enormously large.

Quasistationary distributions are like quasispecies. They are not scattered randomly in sequence space but they fill a well defined region since they consist mainly of a master sequence and some very frequent mutants. These sequences represent the "core" of the distribution since they are present in many copies each. In sequence space a quasistationary distribution of sequences looks like an "amoeba" with protrusions extending along "ridges" of high fitness (Figure 10). The core of the distribution is quasistationary and resembles closely the quasispecies formed by these sequences. This implies that all core sequences have been evaluated already and are present in concentrations determined by their (net) replication rates and mutation frequencies.

At the "periphery" of a quasistationary distribution in sequence space mutations are present in very few copies, eventually in a single copy only. These are the rare mutants which determine the course of evolution. Rare mutants appear and disappear stochastically. In case they are advantageous they may occasionally survive the stochastic phase and grow to larger numbers. Then, they destabilize the previously quasistationary distribution. During a transient selection period the old master sequence is replaced by the new, more efficient variant and a new quasispecies forms.

The general behaviour of such a replication-mutation ensemble of sequences has been studied recently by computer simulation [22]. Populations of a few thousand binary sequences of chain lengths about $\nu=100$ are replicated with constant single digit accuracy q. Secondary structures of the sequences are formed by a folding algorithm. The selective values of the individual sequences are determined by evaluation of the secondary structures according to common biophysical criteria. The accuracy of replication was found to be of fundamental importance for the migration of quasispecies in sequence space. We distinguish four different cases sketched in Figure 10:

(i) In the limiting case of error-free replication, q=1, the stationary population is homogeneous and therefore confined to a single point in sequence space. No evolutionary adaptation is possible.

(ii) Accurate replication corresponding to q-values very close to unity, $q=1-\delta$, leads to narrow quasispecies distributions. The few rare mutants which are formed at all under these conditions are very unlikely to be advantageous. The population is well localized in sequence space and adaptation is very slow.

(iii) Replication accuracies just above error threshold, $q=q_{min}+\delta$, allow to develop most widespread quasistationary distributions. Occurrence and fixation of advantageous mutants in the population reach optimum rates under these conditions. Quasistationary distributions are localized most of the time. During transient periods they migrate from regions of lower to regions of higher mean fitness. Adaptation is most efficient in systems which operate close to the maximum tolerable error frequency.

(iv)   No quasistationary distributions are formed if the accuracy of replication is beyond the error threshold of the system, $q < q_{min}$.   In this case the distribution cannot be localized in sequence space.   It "diffuses" away in the sense of a random walk.   Clearly, no evolution is possible under these conditions.

Condition (iii) appears to be optimum for the evolution of primitive systems like molecules in the test tube, bacteriophages, viruses or bacteria.   As pointed out in the previous section, experimental data on RNA bacterophages and viruses support this conjecture strongly.

The concept of "localization of mutant distributions in sequence space" as a model for direct or complementary replication with errors has been used also in a comprehensive study of quasispecies [15].

24.5 Second Order Autocatalysis

Second order autocatalysis requires at least two different catalytic activities.   RNA molecules, for example, may act as templates and catalysts for their own replication.   The general mechanism comprises a reaction network in which every type plays both parts, that means that every polynucleotide can act as template and as catalyst for all possible processes:

$$A + I_k + I_j \xrightarrow{f_j} 2\,I_k + I_j; \quad j,k=1,2,\ldots,n. \qquad (18)$$

This mechanism, when run under the conditions of the evolution reactor, leads to a set of coupled, essentially nonlinear, kinetic differential equations:

$$\dot{x}_k = x_k \left(\sum_{j=1}^{n} f_{kj} \cdot x_j - \frac{i}{c_0}\,\Phi \right); \quad k=1,2,\ldots,n \qquad (19)$$

with $\Phi = \sum_{k,j=1}^{n} f_{kj} \cdot x_j \cdot x_k$ and $\sum_{k=1}^{n} x_k = c_0$.

The rate constants f_{kj} are properly arranged in a square matrix $F=\{f_{kj}; k,j=1,\ldots,n\}$.   General properties of the coefficient matrix F can be used to classify special cases of (19).   It turned out to be useful to define a "concentration simplex" S_n which comprises the domains of physically meaningful values of concentration variables as shown in Figure 11:

$$S_n \doteq \{x \in R^n: \ x_k \geq 0, \ \sum_{k=1}^{n} x_k = c_0. \qquad (20)$$

The differential equation (19) is an example of a replicator equation [23].   One property of this class of equations is particularly important: the concentration simplex S_n is invariant under the flow of replicator equations.   No trajectories cross the boundaries of the simplex, since

$$x_k = 0 \rightarrow \frac{dx_k}{dt} = 0 \rightarrow \frac{d^2x_k}{dt^2} = 0 \rightarrow \ldots$$

The dynamics of equation (19) is extremely rich.   Depending on the choice of rate constants we find many different types of behaviour ranging from globally and asymptotically stable fixed points to multiple stationary states, oscillations and chaotic attractors.   Special cases of equation (19) are used in many different fields from nonlinear chemical kinetics and

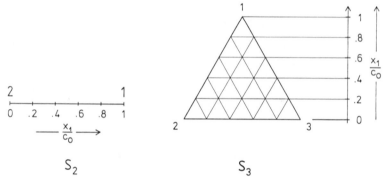

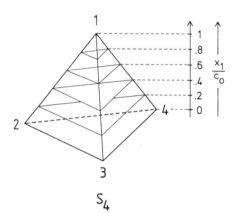

Figure 11.   Three examples of concentration simplices S_n (n=2,3 and 4).   The concentration simplex comprises the physically meaningful domains of concentration variables: $x_k \geq 0$ and $\Sigma x_k = c_0$.

morphogenesis to population genetics and game dynamics [23]. The n-dimensional differential equation (19) was shown to be equivalent to a generalized, multidimensional Lotka-Volterra equation of dimension n-1 [24] and so, there exists also a connection to the nonliner dynamics encountered in models of theoretical ecology.

Here we can only present a brief survey of the qualitative properties of some important special cases of equation (19) which

are classified according to general properties of the coefficient matrix F. We start with three particularly simple special cases:
(i)   the multidimensional Schlögl model [6],
(ii)   Fisher's selection equation of population genetics, and
(iii)   the case of cyclic symmetry.

In the first example, the Schlögl model, F is a diagonal matrix: $f_{kj}=f_k \cdot \delta_{kj}$ where δ_{kj} represent the Kronecker "delta" symbol ($\delta_{kj}=1$ if k=j and $\delta_{kj}=0$ if k≠j). This system is the simplest case of nonlinear, autocatalytic competition which is described by only n

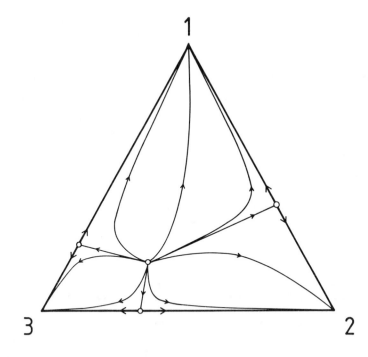

Figure 12.   The three basins of attraction of the three-dimensional Schlögl model according to equation (18a) with n=3 in the evolution reactor.   The simplex S_3 is split into three basins which have the corners 1, 2 and 3 as asymptotically stable point attracts.   The rate constant f_k determines the size of the basin of attraction.   Here we chose: $f_1=3$, $f_2=2$ and $f_3=1$ $[t^{-1}.M^{-1}]$.

reactions out of the n^2 different processes of the general mechanism:

$$A + 2 I_k \xrightarrow{f_k} 3 I_k; \quad k=1,2,\ldots,n. \tag{18a}$$

The kinetic equations of the multidimensional Schlögl model under the conditions of both flow reactors were investigated by qualitative analysis [8,10]. The system sustains multiple, stable stationary states. All equilibrium points in the interior of the concentration simplex S_n are unstable and hence the system leads to selection. The outcome of the selection process, however, is not unique. Depending on the initial conditions $[x_1(0),\ldots,x_n(0)]$ the concentration simplex is a local, asymptotically stable stationary state with its own basin of attraction (Figure 12). For example, the type I_k is selected in the basin of the state P_k, $[\bar{x}_k=c_0, \bar{x}_{j=0} \forall j \neq k]$. The larger the rate constant f_k, the larger is the basin of attraction of the corresponding stationary state P_k.

Fisher's selection equation describes the spreading of n genes, more precisly n "alleles" $(A_1, A_2,\ldots,A_n)$ at a single locus. The diploid genotypes are either homozygous (e.g. $A_k A_k$) or heterozygous (e.g. $A_k A_j$). The rate coefficients f_{kk} or f_{kj} are measures of the fitness factors of the phenotypes corresponding to the genotypes $A_k A_k$ or $A_k A_j$, respectively. Apparently, the fitness of $A_k A_j$ is identical to that of $A_j A_k$ – it does not matter which of the two alleles A_k or A_j sits on which arm of the chromosome. Consequently, the coefficient matrix F is always symmetric, $f_{jk}=f_{kj}$. The dynamics of the selection equation is more complicated than that of the Schlögl model. Stable stationary states may also occur in the interior of the concentration simplex S_n. Corners may, but need not, be stable equilibrium points. Fisher's selection equation has been studied extensively in the past (for further details see [25]).

Both special cases, the multidimensional Schlögl model and Fisher's selection equation share one important property: both differential equations can be transformed into generalized gradient systems and hence, we are dealing with dynamical processes which optimize some potential function $V(x_1,\ldots,x_n)$. Such gradient systems have several general properties. Particularly relevant for biophysical applications are:

(i) lack of oscillations in the variables $x_k(t)$ and

(ii) nonexistence of stable dissipative spatial structures under no flux boundary conditions [26].

Are there other special cases of equation (19) which can be interpreted as generalized gradient system? This question has been answered recently [27] for a very useful class of generalized

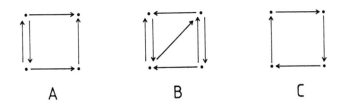

A   B   C

Figure 13. Three classes of directed graphs used to classify non-hyperbolic cases of the differential equation (19).
A: An example of a non-irreducible graph. Note that not every vertex can be reached from every vertex through a directed arc.
B: An example of an irreducible graph. Every vertex can be reached from every vertex through a directed arc.
C: A Hamiltonian arc. Note that neither A nor B contain Hamiltonian arcs. But all graphs containing a Hamiltonian arc are irreducible.

gradients based on a Riemannian metric which were introduced originally by Shahshahani [28]. Equation (19) is a Shahshahani gradient if, and only if, the coefficients fulfil the relation

$$f_{ij} + f_{jk} + f_{ki} = f_{ik} + f_{kj} + f_{ji}. \qquad (21)$$

It can be verified straight away that the multidimensional Schlögl model and Fisher's selection equation represent special cases of condition (21).

Cyclic symmetry of the differential equation (19) implies that the elements of the coefficient matrix F fulfil the condition

$$f_{k+1, j+1} = f_{kj} \text{ for all } j,k=1,2,\ldots,n; \quad k,j \bmod n. \qquad (22)$$

Then, the centre of the concentration simplex S_n, is always a stationary point, $\bar{x}_1 = \bar{x}_2 = \ldots = \bar{x}_n = c_0/n$. Clearly, systems with cyclic symmetry do not meet condition (21). Theorems on the occurrence of Hopf-bifurcations in this system were derived [29] and used to prove the existence of stable limit cycles in Lotka-Volterra equations with dimension $n \geq 3$ [24]. Further details can be found in [29].

In the more general cases of equation (19) it turned out to be useful to choose another property of the coefficient matrix F for classification: equation (19) is invariant to the addition of constants to the columns of F. We may remove this arbitrariness by reducing the diagonal elements to zero. The new matrix of coefficients

$$B \div (b_{kj} = f_{kj} - f_{jj}; k, j=1, \ldots, n), \qquad (23)$$

allows us to distinguish a special class from the completely general situation. The dynamical system (19) is nonhyperbolic if all coefficients of the matrix B (23) are nonnegative, in other words, if the original rate constants fulfil the condition

$$f_{ij}-f_{jj} \geq 0 \ (\forall i, j=1, \ldots, n).$$

The diagonal element of every column is smaller than or equal to the smallest off-diagonal element in the non-hyperbolic case. Then we can assign a directed graph to the differential equation (19): the vertices of the graph correspond to individual polynucleotide sequences I_k $9k=1,2,\ldots,n)$. If the element $b_{kj}>0$ we draw an edge $j \rightarrow k$. The graphs are now analyzed with respect to two properties (Figure 13):
  (i)   irreducibility and
  (ii) existence of a Hamiltonian arc.
  A directed graph is called irreducible if every vertex I_k can be reached from every vertex I_j through a directed arc. An arc is said to be Hamiltonian if it contains a directed circuit – this is an arc which returns to its starting point – which covers all vertices of the graph without self-intersection.

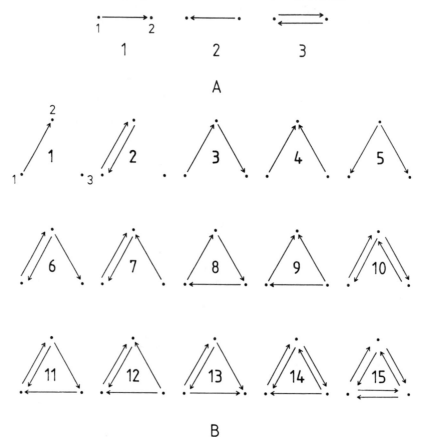

Figure 14 (a,b).

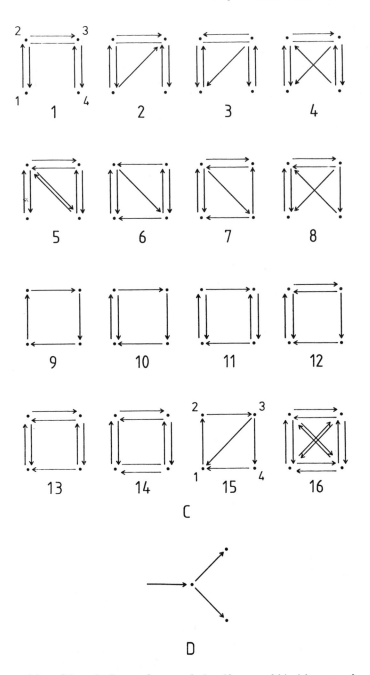

Figure 14. Directed graphs used in the qualitative analysis of non-hyperbolic cases of the differential equation (19).
A: *The case n=2. The graphs 1 and 2 are non-irreducible, 3 is irreducible.* B: *The case n=3. We show all 15 graphs which can be drawn up to a permutation of the vertices (1,2,3).* C: *The case n=4. We show only some selected graphs. 1-8 are*

irreducible graphs which do not contain a Hamiltonian arc. The corresponding differential equations lead to exclusion. The graphs 9-14 contain (a) Hamiltonian arc(s) but have no intersections. The corresponding differential equations are permanent. Graph 15 contains a Hamiltonian arc and a "pitchfork" element (D). It leads to exclusion for certain ranges of rate constants: $f_{13} > f_{43}$. *The graph 16 is an example of a permanent system with intersections. D: The "pitchfork" element.*

Two notions characterizing the long-time behaviour of equation (19) are important:
   (i). A differential equation leads to "exclusion" if at least

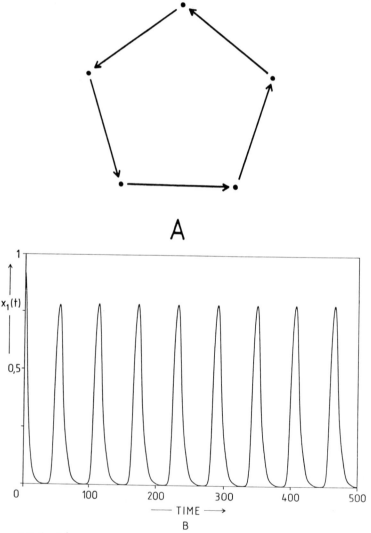

A

B

Figure 15 (a,b).

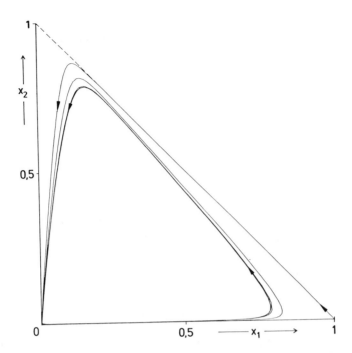

Figure 15. Solution curves and trajectories for the elementary hypercycle with n=5. The long time behaviour of the system is described by an asymptotically stable limit cycle. All rate constants were chosen equal: $f_1 = f_2 = \ldots f_5 = 1[t^{-1}.M^{-1}]$.

A: *The graph of the elementary hypercycle with n=5.*
B: *The solution curve $x_1(t)$. The four other curves $x_k(t)$ (k=2,...,5) are almost identical apart from a different initial phase.*
C: *The phase portrait of the elementary hypercycle with n=5. The trajectory shown starts near the corner $x_1=1$ and converges to the limit cycle.*

one type (e.g. I_ℓ) vanishes after sufficiently long time and hence the ω-limit of the orbits describing equation (19) are not disjoint from the boundary of the concentration simplex S_n.

(ii) "Permanence (or "cooperation"), on the contrary, means that no type vanishes in the limit of long times and hence, the ω-limits of (almost) all orbits lie in the interior of the concentration simplex S_n.

Complete qualitative analysis was performed for all nonhyperbolic cases of equation (19) up to dimension n≤4. Graphs of these dynamical systems are shown in Figure 14. For n=2 the situation is more or less trivial: 1 and 2 lead to exclusion, 3 is permanent.

There are 15 different graphs with n=3 whose properties are derived from two theorems:

(i)   If the non-hyperbolic system (19) with n=3 has a unique fixed point P_{123} in the interior of the concentration simplex S_3 then it is permanent and P_{123} is the ω-limit of every orbit in the interior of S_3.

(ii)   If the graph of the non-hyperbolic system (19) with n=3 contains no Hamiltonian arc, the system leads to exclusion.

Out of the 15 possible cases shown in Figure 14 we find permanence only in the systems 8, 11, 14 and 15.   According to these two theorems no stable limit cycles exist in nonhyperbolic systems with n=3.   This result implies also that there are no stable limit cycles in Lotka-Volterra systems of dimension n=2 [24].   Equation (19) with n=4 is more involved.   It is, nevertheless, possible to derive a condition for exclusion: If the graph of the non-hyperbolic equation (19) with n=4 is not irreducible then the system leads to exclusion.

There are 8 irreducible graphs for n=4 which do not contain a Hamiltonian arc (Figure 14).   Numerical integration of these systems indicates that we have always exclusion and supports the conjecture that the non-hyperbolic system has to have a graph with a Hamiltonian arc in order to be permanent.

The Hamiltonian arc, however, is a necessary but not sufficient criterion for permanence.   Systems belonging to the three graphs 9, 10 and 12 are permanent.   The remaining five systems (11, 13, 14, 15 and 16) are more difficult to analyse, although the corresponding graphs to contain Hamiltonian arcs: the long time behaviour may depend on the numerical values of the rate constants.   As an example we consider the graph 15:   it contains a "pitchfork" type element (Figure 14D) involving the vertices 3, 4 and 1.   Systems with such elements lead to exclusion for certain ranges of rate constants.   In our particular example we have

$$\lim_{t \to \infty} x_4(t) = 0 \text{ if } f_{13} > f_{43}.$$

In higher dimensions, n≥5, the analytic procedure proposed here becomes very clumsy because of the enormous number of graphs which lead to many different cases which have to be treated separately. We restrict our study therefore to one special case which is of particular importance because it leads always to permanence: the elementary hypercycle.

Elementary hypercycles represent a special class of non-hyperbolic systems.   They contain only one non-zero element in every row (or column) of matrix B: $b_{kj} = b_k \cdot \delta_{j,k-1}$ (k, j=1, ..., n; k, j mod n).   The corresponding graph consists only of a single Hamiltonian arc with n vertices (Figure 15).   Examples of hypercycles are some graphs in Figure 14:   A3, B8 and C9. Hypercycles are permanent according to a general theorem derived previously [30,31].   The properties of solutions of the hypercycle equation have been studied in detail [32,33].   For n≤4 the orbits converge to an asymptotically stable stationary state in the interior of the concentration simplex.   In higher dimension (n≥5) the systems approach an asymptotically stable limit cycle.   With increasing n the oscillations approach more and more the shape of rectangular concentration pulses (Figure 15).

Finally we present one example of a general hyperbolic system which is related to a three-dimensional Lotka-Volterra system [34]. The differential equation (19) with n=4 and the rate coefficients shown in Figure 16 leads to chaotic dynamics. The constant μ was chosen more or less arbitrarily as external parameter. The strange attractor of the dynamical system is reached along the direction of increasing μ-values through a Feigenbaum type cascade of bifurcations with period doubling.

24.6 Conclusions

The results presented in this contribution are summarized in Table 1. It is useful to distinguish four different classes of replication-mutation systems according to the two features:
(i)   (effective) first or higher order autocatalysis, and
(ii)   rare or frequent mutations.
The term "effective" points at the rate law which is actually in operation under the conditions applied. Systems with intrinsically higher order reaction mechanisms can be run under conditions where the first order law applies to the kinetic equations.   (Note that the inverse is usually not true since saturation phenomena always lead to a decrease of the effective order of a chemical process.)

First order autocatalysis with rare mutations represents the almost trivial case of pure selection dynamics. The type whose (net) fitness is largest is selected and the mean (net) fitness of the population is optimized during the selection process. The differential equations under the conditions of the CSTR or the evolution reactor are readily formulated as generalized gradient systems. The potential function for the system in the evolution reactor is linear. The stationary states can be characterized as "corner equilibria" (Figure 17). This notion is taken from theoretical economics and describes optimization in cases where the potential function does not assume a maximum value in the interior of the concentration simplex. A corner equilibrium need not be stable in higher dimenisional space. When a new, more efficient mutant is formed, the previous "quasistationary" state becomes unstable. Evolutionary adaptation is visualized as a succession of quasistationary states with stepwise increasing (net) replication efficiency.

First order autocatalysis with frequent mutations leads to a quasilinear differential equation and can be studied in great generality. One of the most important properties of these systems is global uniqueness of long time solutions. Independently of the particular initial conditions, these solutions converge always to the same asymptotically stable, stationary state. The stationary sequence distribution is characterized as quasispecies. Frequent mutations introduce restrictions into the optimization process. The mean fitness of the population does not always increase during selection of the quasispecies. Certain initial conditions lead to a decrease in, or even to oscillations of, the mean fitness. the frequency of replication errors is crucial for evolutionary optimization. The more mutants are formed, the faster is the adaptation process, but there is also a sharply defined error threshold beyond which evolution breaks down.

Second order autocatalysis has been studied in the rare mutation limit. No general approach comparable to that of quasilinear systems is available, but the analysis is facilitated largely by the general property of replicator equations:

$$x_k = 0 \rightarrow \frac{dx_k}{dt} = 0 \rightarrow \frac{d^2 x_k}{dt^2} = 0 \rightarrow \ldots$$

a

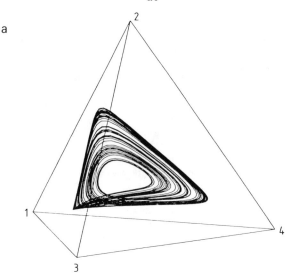

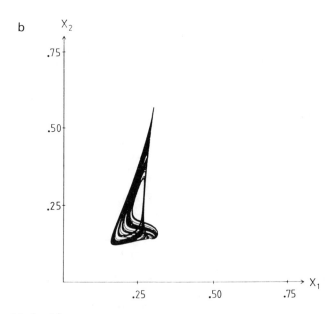

b

Figure 16 (a,b).

c
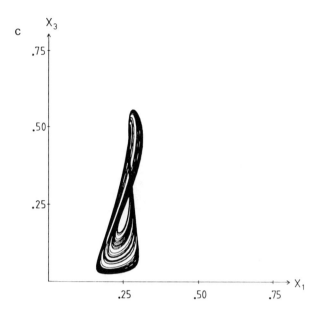

d
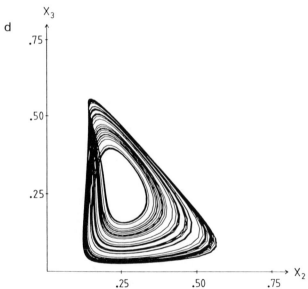

Figure 16.   Trajectories for an example of differential equation
(19) in the hyperbolic case.   The rate constants are given in
the matrix (units: $[t^{-1}.M^{-1}]$)

$F =$
0	0.5	-0.1	0.1
1.1	0	-0.6	0
-0.5	1.0	0	0
μ+0.2	0.2-μ	-0.2	0

with μ=1.43.   We show the phase portrait of the chaotic
attractor on the concentration simplex S_4 as well as some

projections on to two-dimensional coordinate planes (x_k, x_j).

The properties of special cases are summarized in Table 1. The existence or nonexistence of a potential function provides a very useful feature of classification. The systems for which potential functions are available, the multidimensional Schlögl model, Fisher's selection equation and some more general systems, apparently optimize mean fitnesses or populations, but different initial conditions may lead to different, only locally optimal solutions. Systems with cyclic symmetry, hypercycles and the more general cases do not sustain potentials since they do not meet the generalized symmetry condition (21).

The fourth possible combination of the two features, second order autocatalysis and frequent mutations can be modelled by the general reaction mechanism:

$$A + I_i + I_j \xrightarrow{k_j^{(i)}} I_k + I_i + I_j; \quad i,j,k=1,\ldots,n. \tag{24}$$

The polynucleotide I_k is produced as an error copy of I_i under catalysis of I_j. This reaction network gives rise to a very general kinetic equation

$$\dot{\phi} = \sum_k x_k f_k$$

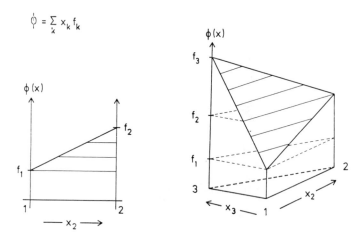

Figure 17. An example of a corner equilibrium. The linear potential function $\Phi(x_1,..,x_n)=\Sigma x_k.f_k$ *assumes its maximum value commonly at a corner of teh concentration simplex. We show an example in three dimensions* $(f_1=2,\ f_2=5$ *and* $f_3=9)$. *In the two-dimensional subspace* (x_1,x_2) *selection of* I_2 *occurs since the quasistable corner equilibrium lies at 2. This point is unstable in the dimension of the coordinate* x_3. *If* I_3 *is formed by a mutation process 2 becomes unstable and the system moves towards 3 which represents the new corner equilibrium.*

Table 1: Properties of replication-mutation systems

System	Mechanism of replication	Long time behaviour[1]		Existence of potential
First order autocatalysis				
Darwinian	$A+I_k \to 2I_k$	U	F Selection	YES[2]
Replication-mutation	$A+I_j \to I_k+I_j$	U	F Quasi-species	YES
Second order autocatalysis				
$A+I_j+I_k \xrightarrow{kj} 2_{jk}+I_j$				
Schlögl model	$f_{kj}=f_k \cdot \delta_{kj}$	M	F Selection	YES
Fisher's equ.	$f_{kj}=f_{jk}$	M	F	YES
Hypercycle equation	$f_{kj}=f_k \cdot \delta_{j,k-1}$		Permanence	NO
	n≤4	U	F	
	n≥5	U	C	
Cyclic symmetry	$f_{k+1,j+1}=f_{kj}$	U	F, as.[3], Permanence	NO
		X	F, us.[3]	NO
General case		X	F, C, S	NO

1. Abbreviations: U   Long time behaviour is unique; all initial conditions on the concentration simplex lead to the same long time dynamics.

    M   Multiple stable stationary states.

    X   Different possibilities depending on the choice of initial conditions and rate constants.

    F   Asymptotically stable fixed point(s).

    C   Asymptotically stable limit cycle.

    S   Strange attractor.

2. The potential function is linear. All optima of the potential function represent "corner equilibria".

3. Abbreviations: as. - The central fixed point is asymptotically stable.

    us. - The central fixed point is unstable.

$$\dot{x}_k = x_k (\sum_{j=1}^{n} f_{kj}^{(k)} x_j - \frac{1}{c_0} \Phi) + \sum_{j}^{n} \sum_{i \neq k}^{n} f_{kj}^{(i)} x_j x_i; \quad k=1,\ldots,n \quad (25)$$

which turns out to be exceedingly difficult to solve by the analytical techniques discussed here. It can be neither transformed into a linear problem, nor does it meet the general property of replicator equations. The investigation of general nonlinear replication-mutation networks has to rely therefore on numerical integration of special examples which we shall not discuss here.

So far we made the implicit assumption that the system can be described properly by means of differential equations. This implies populations which are so large that stochastic phenomena play no role. Although the use of kinetic equations is well justified in many cases there are examples where the deterministic description fails for principal reasons. One example has been discussed in section 24.3: if replication errors are so frequent that the system fails beyond the error threshold no realistic population is large enough to guarantee the existence of stationary mutant distributions. Instead, systems beyond error threshold drift randomly in sequence space.

A different but closely related problem arises when two types have identical or almost identical fitnesses. The evolution of populations with selectively neutral mutants has been analysed and discussed extensively in the rare mutation limit [35]. Frequent mutations make the case of selective neutrality much more involved. It matters then whether the two types in question are members of the same quasispecies or not. In the first case the two sequences are close relatives and their relative frequency is controlled by the balance of frequent mutations in the network. Random drift is thus restricted to distant sequences.

Acknowledgements

This work has been supported financially by the Stiftung Volkswagenwerk and the Fonds zur Förderung der wissenschaftlichen Forschung in österreich (Project no. 5286). Technical assistance in the preparation of the manuscript by Mr. J. König is gratefully acknowledged.

References

[1]   C.K. Biebricher, M. Eigen and W.C. Gardiner, Jr. (1983). *Biochemistry* 22, 2544.

[2]   C.K. Biebricher, M. Eigen and W.C. Gardiner, Jr. (1984). *Biochemistry* 23, 3186.

[3]   C.K. Biebricher, M. Eigen and W.C. Gardiner, Jr. (1985). *Biochemistry* 24, 6550.

[4]   B.L. Bass and T.R. Cech (1984). *Nature* 308, 920.

[5]   C. Guerrier-Takada, K. Gardiner, T. Marsh, N. Pace and S. Altmann (1983). *Cell* 35, 849.

[6]   F. Schlögl (1972). *Z. Phys.* 253, 147.

[7]   R. Lefever, G. Nicolis and I. Prigogine (1967). *J. Chem.*

Phys. **47**, 1045.

[8]   P. Schuster and K. Sigmund (1985). *Ber. Bunsenges. Phys. Chem.* **89**, 668.

[9]   M. Eigen (1971). *Naturwissenschaften* **58**, 465.

[10] M. Eigen and P. Schuster (1979). *The Hypercycle - a Principle of Natural Self-Organization,* Springer-Verlag, Berlin.

[11] C.J. Thompson and J.L. McBride (1974). *Math. Bioscience* **21**, 127.

[12] B.L. Jones, R.H. Enns and S.S. Rangnekar (1976). *Bull. Math. Biol.* **38**, 12.

[13] R. Feistel and W. Ebeling (1978). *Studia Biophysica* **71**, 139.

[14] J. Swetina and P. Schuster (1982). *Biop[hys. Chem.* **16**, 329.

[15] J.S. McCaskill (1984). *J. Chem. Phys.* **80**, 5194.

[16] P. Schuster and J. Swetina, *Stationary mutant distributions and evolutionary optimization.* To be publisheshed, 1987.

[17] L. Demetrius, P. Schuster and K. Sigmund (1985). *Bull. Math. Biol.* **47**, 239.

[18] D. Rumschitzky (1987). *J. Math. Biol.* **24**, 667.

[19] I. Leuthäusser (1986). *J. Chem. Phys.* **84**, 1884.

[20] L. Demetrius, *J. Chem. Phys.,* in press, 1987.

[21] M. Eigen (1985). *Ber. Bunsenges. Phys. Chem.* **89**, 658.

[22] W. Fontana and P. Schuster (1987). *Biophys. Chem.* **26**, 123.

[23] P. Schuster and K. Sigmund (1983). *J. theor. Biol.* **100**, 533.

[24] J. Hofbauer (1981). *Nonlinear Analysis* **5**, 1003.

[25] W.J. Ewens (1979). *Mathematical Population Genetics,* Springer-Verlag, Berlin.

[26] R.G. Casten and C.J. Holland (1978). *J. Diff. Equ.* **27**, 266.

[27] K. Sigmund (1985) in *Lotka-Voltera-Approach to Cooperation and Competition in'Dynamic Systems,* eds. W. Ebeling and M. Peschel, p. 63ff, Akademie-Verlag, Berlin.

[28] S. Shahshahani (1979). *Memoirs Am. Math. Soc.* **211**.

[29] J. Hofbauer, P. Schuster, K. Sigmund and R. Wolff (1980). *SIAM J. Appl. Math.* **38**, 282.

[30] P. Schuster, K. Sigmund and R. Wolff (1979). *J. Diff. Equ.* **32**, 357.

[31] J. Hofbauer, P. Schuster and K. Sigmund (1981). *J. Math. Biol.* **11**, 155.

[32] P.E. Phillipson and P. Schuster (1983). *J. Chem. Phys.* **79**, 3807.

[33] P.E. Phillipson, P. Schuster and F. Kemler (1984). *Bull. Math.Biol.* **46**, 339.

[34] A. Arneodo, P. Coullet and C. Tresser (1980). *Phys. Letters* **79A**, 259.

[35] M. Kimura (1983). *The Neutral Theory of Molecular Evolution,* Cambridge University Press, Cambridge, U.K.

NONLINEAR DYNAMICS IN THE WORLD ECONOMY: THE ECONOMIC LONG WAVE

Abstract

The economic crisis of the 1980s has revived interest in the economic long wave or Kondratiev cycle, an approximately 50 year cycle of prosperity and depression, with downturn periods in the 1830s, 1870s-90s, 1930s, and the 1980s.   Since 1975 the System Dynamics National Model has been the vehicle for the development of a dynamic, endogenous theory of the long wave.   This paper describes the integrated theory that has now emerged from extensive analysis of the full National Model and from simple models.   The theory of the long wave is summarized, and a simple model is discussed in detail.   Heuristic methods, simulation, and formal stability analysis are used to show that the long wave is a limit cycle strongly shaped by nonlinear processes.   The long wave arises from the interaction of two fundamental facets of modern industrial economies.   First, physical lags in the economy, information limitations, and bounded rationality in economic decision-making create the potential for oscillatory behaviour. For example, the physical lags in capital acquisition coupled with locally rational decision rules governing production and investment create fluctuations in capital investment.   Second, a wide range of self-reinforcing processes amplify the inherent oscillatory tendencies of the economy, leading to the long wave. These self-reinforcing processes involve macroeconomic linkages among capital investment, labour markets and workforce participation, real interest rates, inflation, and technological progress.   Laboratory experiments show that human subjects behave in accordance with the assumed decision rules of the model.

25.1 **Introduction**

The worldwide economic stagnation of the 1970s and '80s has challenged traditional economic theories and led to the abandonment of long cherished beliefs about the stability of the economic system.   The long-term nature of present difficulties and similarities to past periods of extended depression in the late 1800s and 1930s have revived interest in the economic long wave or Kondratiev cycle [1,2].   Numerous theories of the long wave have emerged in the past 10 years [3,4], including theories stressing innovation [5,6], employment and wages [7], resource scarcity

[8,9], and class struggle [10,11]. Since 1975 the System Dynamics National Model (NM) has provided an increasingly rich theory of the long wave [12-20]. Though the model focuses primarily on economic forces, the theory emerging from the NM is not monocausal: it relates capital investment; employment, wages, and workforce participation; inflation and interest rates; aggregate demand; monetary and fiscal policy; innovation and productivity; and even political values.

This paper surveys the theory of the long wave that has now emerged from the NM. Simple models and a variety of techniques, including heuristic, simulation, and formal stability analysis are used to show the long wave to be a limit cycle strongly shaped by nonlinear processes. Experimental tests show human managers

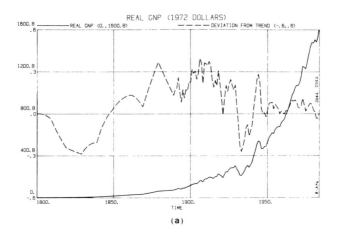

(a)

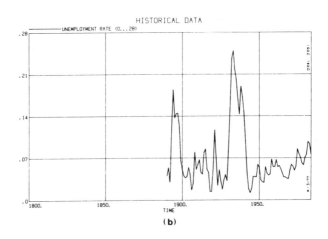

(b)

Figure 1. (a) *Real GNP in the United States, 1800-1984;* (b) *unemployment rate in the United States, 1890-1984.*

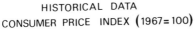

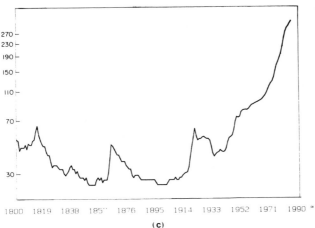

(c)

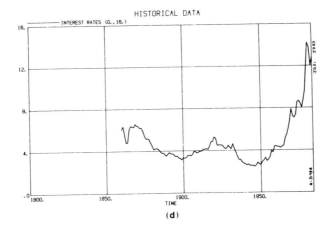

(d)

Figure 1. (c) consumer Price Index in the United States, 1800-1984; (d) interest rates in the United States, 1860-1984 (source [45]: 1860-1899, average yield on higher grade railroad bonds; 1900-1975, prime corporate bonds; 1975-1984, Moody's AAA bonds).

behave in accordance with the assumed decision rules of the model.

25.2 Multiple Modes of Economic Behaviour

Figure 1 a-d shows the behaviour of important economic variables in the United States from 1800 to 1984. The data exhibit several modes of behaviour. The behaviour of real GNP, for example, is dominated by the long-term growth of the economy, averaging 3.4 percent/year since 1800. GNP also exhibits the short-term business cycle, and there is a hint of longer-term fluctuations: output is lower than normal between 1830 and 1840, during the 1870s through 1890s, during the Great Depression, and from the 1970s to the present. These dates coincide with the timing of the long wave established by van Duijn [3].

The long wave is more apparent in the behaviour of unemployment, aggregate prices, and interest rates. Unemployment fluctuates strongly with the business cycle, but also shows major peaks during the 1890s and the 1930s. Unemployment rates in the 1980s are the highest since the Great Depression. Consumer prices

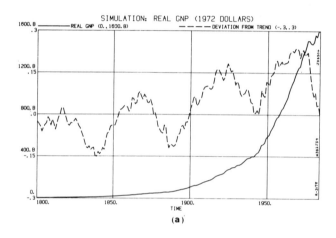

(a)

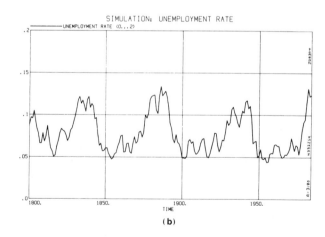

(b)

Figure 2.   (a) Simulated real GNP, 1800-1984;   (b) simulated unemployment rate, 1800-1984.

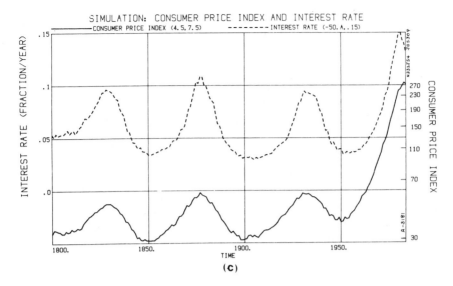

Figure 2(c) simulated interest rate and price level, 1800-1984.

likewise fluctuate over the business cycle but are dominated by the long wave, with peaks roughly coincident with the peaks of the long wave in real activity. An additional mode of behaviour develops after World War II as inflation has carried the price level to unprecedented levels, dominating the long-wave pattern in prices. (Note, however, that the "disinflation" of the 1980s is consistent with the deflationary forces of the long-wave downturn. The postwar inflation coincides with the expansion in the relative size of government from about 10 percent of GNP in the 1920s to about 35 percent in the 1980s.)

Interest rates likewise rise and fall with a roughly 50-year period, and are approximately in phase with the price level. Like prices, interest rates rose well above historic levels in the last decade as inflation reached double digit rates.

The data reflect the interaction of several distinct modes of behaviour, including long-run population growth and technological progress, the business cycle, the relative growth of government, post-war inflation, and the long wave. The interaction of the modes makes it difficult to establish the existence of the long wave through purely empirical means, especially since reliable numerical data are not available over a long enough period. (Difficulties in identification of long waves from empirical data are discussed in [21]. Anecdotal and other descriptive data (e.g [22]) are extremely useful and corroborate the timing of the long wave established through examination of the numerical data.)

Because the National Model represents behaviour at the microlevel of individuals and firms, it generates the multiple modes of economic behaviour that appear in the historical data. Compare the historical data against Figure 2a-c, which shows a simulation of the NM from 1800 to 1984. As shown in Figure 3, all the macroeconomic aggregates are generated endogenously, as are a host of variables at the sectoral level. The only exogenous variables are population, technological progress, and per capita government activity. In addition, a small amount of random noise has been added to production and ordering rates. The noise excites the business cycle and causes the behaviour in the time domain to be somewhat irregular.

Simulated unemployment, real GNP, interest rates, and prices all exhibit the long wave and business cycle. In addition, GNP exhibits the long-term growth of the economy, and prices show the postwar inflation due to the growth of government and the partial monetization of growing government deficits. Because historical data series are not used as inputs, the behaviour, and in particular the long wave, is the endogenous result of the interaction of the system components. It is not driven by the exogenous variables. Yet, the simulation captures the major patterns in the development of the economy over almost 200 years. (Population and technological progress, though exogenous, are assumed to grow at uniform fractional rates. Thus the long wave and its timing in the simulation are not due to exogenous

Exhibit 3: Major endogenous and exogenous variables in the National Model

Endogenous	Exogenous
GNP	Population
Consumption	Technological Progress
Investment	Authorized government
Saving	services per capita
Government Expenditure	Random noise in order rates
Tax rates	and production
Prices	
Wages	
Inflation rate	
Employment	
Unemployment	
Workforce participation	
Wealth	
Interest rates	
Money supply	
Private debt	
Public debt	
Banking system reserves	
Monetary policy	
(open market operations)	
Fiscal policy	
(transfer payments, government	
purchases, employment, deficit)	
Sectoral variables for the consumer	
goods and services sector and plant	
and equipment sector:	
Production	
Capacity	
Capital Stock	
Employment	
Investment	
Price	
Debt	
Dividends	
Return on investment	
Taxes	
Balance sheet	
Income statement	

Figure 3. Major endogenous and exogenous variables in the National Model.

variables.)

25.3 **Origin of the Long Wave**

The long wave is characterized by successive waves of overexpansion and decline of the economy, particularly the capital-producing sector. Overexpansion means an increase in the capacity to produce and in the production of plant, equipment, and goods relative to the amount needed to replace worn-out units and provide for growth over the long run. Overexpansion is undesirable because, eventually, production and employment must be cut back below discards to reduce the excess. The explanation for capital overexpansion and the long wave can be divided into two parts: first, the structure of individual firms contains inherently oscillatory processes. In isolation, these processes are stable, producing damped oscillations when excited. Second, however, a wide range of self-reinforcing or autocatalytic processes exist in the linkages between firms and among the production, financial, and household sectors of the economy. These positive feedbacks destabilize the inherently oscillatory structures within individual firms, lengthening the period and increasing the amplitude of oscillation. Through the process of entrainment the oscillations become coherent, leading to the long wave. Model analysis shows that the positive feedbacks cause a Hopf-bifurcation through which the equilibrium becomes unstable. The resulting divergent oscillations are bounded by a variety of nonlinearities, creating a limit cycle.

25.3.1 *Oscillatory Adjustment of Individual Firms to Shocks*

Consider a manufacturing firm in stationary equilibrium which experiences a sudden, unanticipated step increase in incoming orders. Assume for the moment constant returns to scale and that the firm is small relative to the labour, capital, materials, and other input markets, so that the prices of factor inputs can be considered constant. In the long run the firm will increase production commensurate with the change in orders. Labour force and capital stock will rise by the same amount.

But during the transient adjustment, production must exceed the new equilibrium level: delays in reacting to the change in orders and increasing production capacity imply the firm's inventory must fall. Backlog will rise. In addition, desired inventory levels will probably rise in proportion to orders. Therefore, production must rise above the equilibrium level long enough to rebuild inventory and work off the excess backlog. If production capacity rises above equilibrium in order to provide the necessary expansion of output, then the firm will experience excess capacity. The stability of a firm's response to disturbances can be classified according to which of three fundamental processes it relies on to alter production (Figure 4):

I   More intensive use of existing labour force and capital stock;

II   Increases in workforce and more intensive use of capital stock;

III   Increases in capital stock.

In the near term, Type I adjustments (overtime and longer workweeks) will be used. These adjustments can be scheduled and cancelled almost immediately. Type I adjustments thus involve a first-order negative feedback loop which implies a stable, nonoscillatory response to shocks. However, use of overtime and longer workweeks is limited by the cost of overtime, by the inability of workers to sustain long workweeks, and by the diminishing marginal efficiency of work effort resulting from more intensive use of the capital stock. Thus, while small amplitude disturbances can be handled by Type I adjustments, larger disturbances cause saturation of Type I response.

Expanding the workforce (Type II adjustment) can be accomplished only with a significant delay. The lag stems from the time required to create and fill vacancies, to train new employees, and for their productivity to rise to that of experienced workers. Thus Type II adjustments create a negative feedback loop with significant phase lag elements. Such feedback systems are oscillatory. The characteristic behaviour of models which portray Type II adjustments is damped oscillation with a periodicity of 3 to 7 years. These models generate many of the phase and amplitude characteristics of the short-term business cycle [23,24].

Like Type I adjustments, Type II responses are limited by diminishing returns as labour expands relative to existing capital stocks and by decreasing availability of workers. Thus nonlinear saturation effects constrain the magnitude of disturbances which can be handled by Type I and Type II adjustments alone.

Type III adjustment (expanding the firm's capital stock) involves even longer delays. The delays in planning for, ordering and constructing new plant and equipment are substantial, often

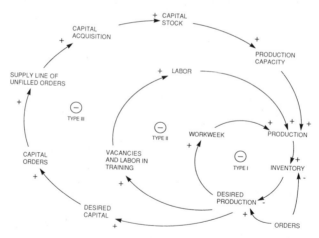

Figure 4. Three negative feedback loops which regulate production in individual firms. Type I adjustment: capacity utilization and workweek variations are nonoscillatory. Type II adjustments: variations in workforce produce 3-7 year business cycle fluctuations. Type III adjustments: variations in fixed capital investment produce 15-25 year fluctuations.

several years or more [25-27]. Type III adjustment thus creates the potential for oscillations of substantially longer period than the business cycle. Models of the interaction of capital investment with production generate damped oscillation with periods of 15-25 years and possessing many of the characteristics of the Kuznets or construction cycle [28-31].

The overall response of an individual firm to unanticipated shocks depends strongly on the nonlinear nature of the three different adjustment processes and size and duration of the disturbance. Small excursions can be dissipated through utilization effects alone. Larger disturbances will force much of the adjustment to occur through workforce expansion, leading to business cycle fluctuations. To the extent Type I and II adjustments fail to correct imbalances, the adjustment will be forced through the investment loops, producing 15-25 year fluctuations as well.

Yet simulations of the NM show the long wave is a 50-year fluctuation which does not die away. The long period, large amplitude, and persistent nature of the long wave arise from a wide range of self-reinforcing processes which operate in the economy as a whole. These positive feedback loops couple different firms to one another and to the household and financial sectors of the economy. The effect of these self-reinforcing processes is to amplify further the oscillatory tendencies of individual firms, lengthening the period and increasing the amplitude of the fluctuations produced by Type III adjustments. Analysis of the model isolates several distinct processes which contribute to the 50-year cycle of overexpansion and decline.

25.3.2 *Capital Self-ordering*

The capital producing sector (plant, equipment, and basic materials) differs from others due to the existence of "self-ordering". In order to expand capacity, producers of capital plant and equipment must order additional plant and equipment from each other. In the aggregate, the capital-producing sector acquires capital from itself, hence self-ordering. Though all sectors of the economy are linked to one another to some degree, self-ordering is strongest in the industries that produce capital plant and equipment, basic industries such as steel, and other heavy industry [32].

To illustrate, consider the economy in equilibrium. If the demand for consumer goods and services increases, the consumer-goods industry must expand its capacity and so places orders for new factories, equipment, vehicles, etc. (initiating Type III response). To supply the higher volume of orders, the capital-producing sector must also expand its capital stock and hence places orders for more buildings, machines, rolling stock, trucks, etc., causing the total demand for capital to rise still further in a self-reinforcing spiral of increasing orders, a greater need for expansion, and still more orders.

The strength of self-ordering depends chiefly on the capital intensity (capital/output ratio) of the capital-producing sector. It is easily shown [19,33] that the equilibrium multiplier effect created by self-ordering is given by:

$$1/(1-KCOR/KALC) \qquad (1)$$

where

KCOR = capital output ratio of the capital sector (years)
KALC = average lifetime of capital in the capital sector (years).

The multiplier effect can be derived by noting that in equilibrium (i) capital production equals the investment of the goods sector plus the investment of the capital sector: KPR=GINV+KINV; (ii) production is related to capital stock by the capital output ratio: KPR=KC/KCOR; (iii) the investment of the capital sector in equilibrium equals physical depreciation. In equilibrium, discards are given by the capital stock divided by the average life of capital: KINV=KC/KALC.

With KALC=20 years and KCOR=3 years (approximate values for the aggregate economy) the equilibrium multiplier effect is 1.18. In the long run, an increase in the demand for capital from the rest of the economy yields an additional 18% increase in total investment through self-ordering.

The long wave is an inherently disequilibrium phenomenon, however, and during the transient adjustment to the long run the strength of self-ordering is greater than in equilibrium. During disequilibrium a variety of additional positive feedback loops further augment the demand for capital. These include:

25.3.1.a *Amplification caused by inventory and backlog adjustments.* Rising orders deplete the inventories and swell the backlogs of capital-sector firms, leading to further pressure to expand and still more orders. During the downturn, low backlogs and involuntary inventory accumulation further depress demand, leading to still more excess inventory.

25.3.1.a *Amplification caused by rising lead times for capital.* During the long wave expansion, the demand for capital outstrips capacity. Capital producers find it takes longer than anticipated to acquire new capacity, causing capacity to lag further behind desired levels, creating still more pressure to order and further swelling the demand for capital.

25.3.1.c *Amplification caused by growth expectations.* Growing demand, rising backlogs, and long lead times during the long wave expansion all encourage expectations of additional growth in demand for capital. Expectations of future growth lead to additional investment, further swelling demand in a self-fulfilling prophecy. During the downturn, pessimism further undercuts investment.

Analysis of the NM shows the positive feedback loops created by self-ordering significantly reinforce the tendency of firms to amplify changes in demand. Once a capital expansion gets under way, the self-ordering loops amplify and sustain it until production catches up to orders, excess capacity is built up, and orders begin to fall.

At that point, the self-ordering loops reverse: a reduction in orders further reduces the demand for capital, leading to a contraction in the capital sector's output, rising inventories, falling backlogs and lead times, followed by declining employment, wages, aggregate demand, and production of goods and services. Capital production must remain below the level required for replacement and long-run growth until the excess capacity depreciates - a process that may take a decade or more due to the long lifetimes of plant and equipment. Once the capital stock is worn out, investment rises, triggering the next upswing.

To illustrate, consider the development of the U.S. economy after World War II. The capital stock of the economy was old and severely depleted after 15 years of depression and wartime needs. Demand for all types of capital equipment - roads, houses, schools, factories, machines - surged. A massive rebuilding began. In order to satisfy long-run demand, fill pent-up demand, and rebuild the capital and infrastructure, the capital-producing sector had to expand beyond the long-run needs of the economy. The overexpansion of the capital-producing sector was exacerbated by self-ordering: as the demand for consumer goods, services, and housing rose, manufacturers of capital plant and equipment had to expand their own capacity, further swelling the demand for structures, equipment, materials, transportation, and other infrastructure. By the late 1960s, the capital stock had been largely rebuilt, and investment began to slow to levels consistent with replacement and long-run growth. Excess capacity and unemployment began to show up in basic industries. Faced with excess capacity, investment in these industries was cut back, further reducing the need for capital and reinforcing the decline in investment as the economy moved through the 1970s and into the 1980s.

The capital self-ordering component of the long-wave theory predicts a growing margin of excess capacity, especially in heavy manufacturing industry, as the economy moves through the long-wave peak and into the downturn. Excess capacity is in fact one of the dominant symptoms of the malaise of the 1970s and 1980s, and has been amply documented elsewhere [20].

25.4 A Simple Model of Self-Ordering

In addition to analysis of the full national model, simple models have been developed to explore the dynamics of the long wave [19,34]. The simple model described below was specifically designed to test the role of capital self-ordering in the genesis of the long wave. The model shows how the investment and production policies of individual firms, though rational from the point of view of the individual actors, interact in the context of the whole system to produce the long wave.

The heart of the simple model is a representation of a typical capital producing firm (Figure 5). The model consists of two parts: first, the physical structure of capital accumulation, production, and orders; and second, the behavioural decision rules governing the rates of change in the various states. The physical structure of the model is quite simple.

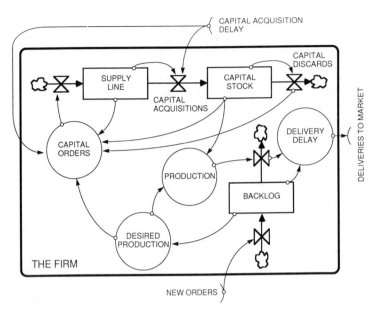

Figure 5.   Structure of simple capital self-ordering model. Boxes represent state variables, arrows and valves the rates of flow (time derivatives) of the states.   Clouds are sources and sinks of material or energy.   Dotted lines show information feedback from states.   Circles represent the decision rules of the model, which are typically nonlinear.   The model shown is therefore a third-order ordinary (nonlinear) differential equation.   It represents a typical capital-producing firm and responds to an exogenous demand for the firm's product.   When capital self-ordering is added, the firm's own capital orders add to the exogenous demand for the firm's product.

Orders arriving from customers accumulate in the backlog, which is depleted by production.   Production is determined by capacity and capacity utilization.   Utilization is a nonlinear function of the ratio of desired production to capacity.   Capacity is determined by the firm's capital stock and the capital/output ratio (assumed constant).   Capital stock is augmented by acquisitions and diminished by discards.   The average lifetime of capital is assumed to be constant and the discard process exponential.   Capital acquisition depends on the supply line (the backlog of unfilled orders for capital, including units under construction) and the average delay in acquiring capital.   The supply line is augmented as orders for capital are placed with suppliers, and diminished when construction is completed and the capital enters the firm's productive stock.

The key decision points in the system are the desired production decision and the capital investment decision.   Both decisions are formulated in accordance with the principles of bounded rationality [35,36].   They rely on information locally available to real firms and process that information in straightforward ways.   Desired production is formed from expected orders modified by a fraction of the discrepancy between the desired and actual backlog. The desired backlog is the backlog consistent with the expected demand and the normal construction time for capital. Expected demand is formed adaptively from past orders.

Three motivations for capital investment are assumed: first, to replace discards; second, to correct any discrepancy between the desired and actual capital stock; and third, to correct any discrepancy between the desired and actual supply line. Appropriate nonlinearities are included to ensure that capital orders are bounded between zero and a maximum fraction per year of the existing capital stock.

25.4.1 *Behaviour of the Simple Model*

To test the behavioural adequacy of the model, a series of partial model tests were conducted. Before exploring the dynamics of the full system, one must have confidence that the individual decision rules behave appropriately. Partial model testing considers whether a decision rule makes sense relative to the mental models and cognitive capabilities of the decision-makers, the information available to them, and the heuristics and decision aids used to process the information.

For example, the production scheduling decision was tested under the assumption that capacity imposed no constraint on the output. Likewise, the ability of the investment decision to respond to an exogenous increase in desired capacity was tested.

The partial model tests showed that the individual decision rules are locally rational. The response of the partial model to unanticipated shocks is smooth, stable, and appropriate. Next, the investment and production decisions were coupled together, testing the response of the firm as a whole to exogenous shocks.

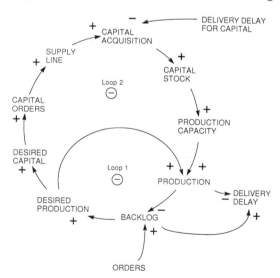

Figure 6. Causal structure of simple model, showing negative loops controlling production.

The result is a highly damped 20 year oscillation. The oscillation arises from the delay in increasing capacity, causing orders to accumulate in the backlog, and forcing capacity to expand above the long-run equilibrium level in order to both fill incoming orders and reduce the backlog to normal levels. In terms of the feedback structure, the model contains two negative feedback loops which can control the backlog (Figure 6). Loop 1 is a stable, first order looop which regulates backlogs through variations in capacity utilization. If an unanticipated increase in orders raises the backlog, desired production will rise, boosting capacity utilization and production, and reducing the backlog. Loop 1 thus represents Type I and II adjustments. The goal of loop 2 is the same as loop 1: to increase production so as to reduce the backlog to normal levels. However, loop 2 regulates production capacity, which is only acquired with a substantial lag, and corresponds to Type III adjustments. The behaviour, and in particular the stability, of the system depends upon which loop is dominant. The nonlinear capacity utilization function controls that dominance. Loop 1 can only operate if there is excess production capacity. If not, the burden of control is shifted to loop 2, a high-order, oscillatory loop.

Thus when the demand for capital is exogenous, the system, though oscillatory, is highly damped, at least for reasonable values of the parameters. However, as capital self-ordering is added, damping falls. Structurally, self-ordering creates a first-order positive feedback loop (loop 3 in Figure 7) which causes the capital-producing sector's own demand for capital to react to any change in the exogenous component of demand. As orders for capital are placed in an attempt to reduce the discrepancy between demand and capacity, self-ordering acts to

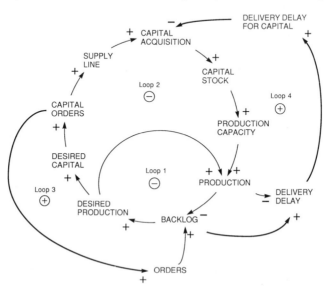

Figure 7. Causal structure of simple model, showing positive loops created by capital self-ordering.

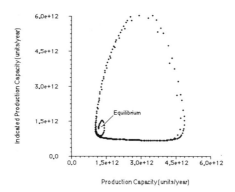

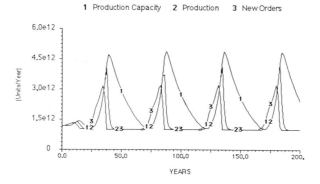

Figure 8. Behaviour of simple model, showing the limit cycle. Note that several cycles are required to reach full amplitude. Flow in the phase plot is clockwise.

increase the discrepancy by expanding desired production with each new order. The capital producing sector chases its own shadow. As self-ordering becomes stronger, the positive feedback loop gradually becomes dominant over the various first-order negative loops which dissipate disequilibrium pressures. When the capital/output ratio rises above a critical value, the model passes through a Hopf-bifurcation and becomes unstable around the equilibrium point. At this point, the behaviour shifts to a limit cycle with an approximately 50 year period (Figure 8). The locally unstable oscillation is bounded by various nonlinearities. Foremost among these is the capacity utilization function. Fundamentally, it is the fact that output can only rise as capacity grows that creates the disequilibrium, and the fact that output must fall as the backlog is depleted that limits it.

25.4.2 *Formal Stability Analysis of the Model*

Formal stability of the model confirms the heuristic analysis above and shows that the model is capable of generating a variety of additional modes, including various period-multiples and chaos

[34-37]. The formal analysis is based on a simplified version of the model. Simulations confirm that the simplification does not significantly alter the dynamics. Scaling the variables by the exogenous demand for capital from the goods sector allows the model to be expressed as a second-order nonlinear system:

$$x' = [f(z)/c - 1/\tau]x - 1 \qquad (2)$$

$$y' = [(g(z) - 1)/\tau]x - x' \qquad (3)$$

where

$$z = \frac{c}{x}\left(\frac{y}{p} + 1\right) \qquad (4)$$

$$x = \frac{\text{production capital}}{\text{desired production of goods}} = \frac{PC}{DPG} \text{ (units/units/year = years)}$$

$$y = \frac{\text{unfilled orders for capital}}{\text{desired production of goods}} = \frac{UOC}{DPG} \text{ (units/units/year = years)}$$

c = Capital/output ratio (years) = COR
τ = Average lifetime of capital (years) = ALC
p = Desired delivery delay for capital (years) = DDC

The variable z is the ratio of desired production to production capacity.

The nonlinear functions f(z) and g(z) portray the nonlinear capacity utilization and capital investment functions, respectively:

f(z)	g(z)
f(0)=0	g(0)=0
f(1)=1	g(1)=1
f(z)=a for z>2	g(z)=b for z>2
f'(1)=α 0$\leq\alpha\leq$1	g'(1)=β
f'(z)>0 for 0$\leq$z$\leq$2	g'(z)>0 for 0$\leq$z$\leq$2

These assumptions determine a unique equilibrium such that z=1 (production capacity and demand are equal). Consistent with eq. 1,

$$x_{eq} = c^*[1/(1-c/\tau)] \qquad (5)$$

The equilibrium supply line of unfilled orders depends only on the equilibrium capital discard rate and the capital acquisition delay:

$$y_{eq} = (x_{eq}/\tau)p \qquad (6)$$

Linearizing f(z) and g(z) about the equilibrium point z=1 and recognizing that the average life of capital must exceed the capital/output ratio (τ>c) results in local stability at the equilibrium point if

$$\beta - [\tau^*p]/c][1/\tau + \alpha/p - (1-\alpha)/c] < 0 \qquad (7)$$

Consistent with intuition, eq. 7 shows that more aggressive capital investment (larger β), corresponding to more aggressive use of Type III response, reduces stability. Likewise, for c>p, (the realistic case), less flexible capacity utilization (smaller α) reduces stability. Smaller α implies less reliance on stable type I adjustments; more on oscillatory Type III adjustments.

The effect of self-ordering is determined by the capital output ratio c. When c = 0 (no self-ordering), stability depends only on β, a parameter local to each individual firm's policies and management style. Equation 7 shows that higher capital/output ratios reduce stability, at least when capacity utilization is sufficiently flexible around equilibrium ($\alpha>0$), confirming the critical role of self-ordering in destabilizing the Type III adjustments of individual firms. For typical parameters, the system is stable for small c, producing damped oscillations with a period of about 20 years, consistent with type III adjustment processes. As c grows and the strength of self-ordering increases, damping rapidly falls. At some critical value of c the eigenvalues of the linearized system cross the imaginary axis and the system becomes locally unstable. The system goes through a Hopf-bifurcation and begins to exhibit a limit cycle. Further growth of c (to a point, at least) rapidly lengthens the period and boosts the amplitude of the cycle.

The effects of p and τ on stabilty are nonmonotonic and contingent on other parameters. Sensitivity analysis is presented in [19]. The results show the long wave is not caused by the "echo effect" that figures in some innovation theories of long waves. In these theories, an initial (and unexplained) coherent surge of innovations causes growth, later allegedly followed by a period of decline as the technologies mature, all supposedly at the same time, leading to the next pulse of replacement investment.

The existence and character of the limit cycle depends on the nonlinearities which constrain excursions of x and y well away from equilibrium. The model contains two crucial nonlinearities which shape the orbits of the system. First, capacity utilization is nonlinear, with utilization saturating at f(z)=a for large z. Second, capital investment is constrained to be nonnegative and saturates at a maximum value b when z is large.

The system can generate various period multiples and ultimately chaotic behaviour by disturbing it with a sinusoid with a period characteristic of the short-term business cycle (3-7 years). As the amplitude of the disturbance grows, the system becomes more chaotic (Figure 9). In the highly simplified model used here, the transition to chaos occurs in a region of parameter space well outside plausible values. Nevertheless, the system clearly contains the possibility of chaotic behaviour. Additional analysis is needed to determine whether more complete and realistic models can exhibit chaos with plausible parameters. The possibility of empirically distinguishing chaotic behaviour from

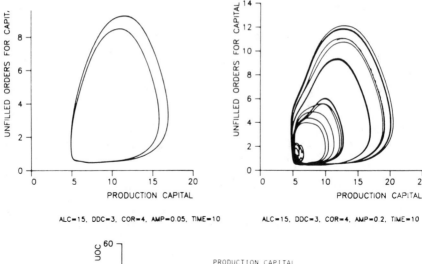

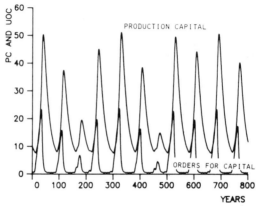

Figure 9. (a) With 5 percent sinusoidal modulation of incoming orders (DPG), the model exhibits period doubling. (b) With 20 percent sinusoidal modulation of incoming orders (DPG), the model exhibits chaotic behaviour (phase plot). (c) Time domain corresponding to conditions in (b).

more common damped oscillations disturbed by noise is limited by the short runs of data available and the existence of numerous other modes of behaviour in the economy. Yet each historic long wave, despite sharing many common features, has been unique in timing and severity, raising the possibility that the development of the world economy results more from inherently nonlinear processes than has been previously appreciated.

25.5 Experimental Tests of the Decision Rules

People are not like electrons: each is a unique individual. Further, the laws of human behaviour are much less stable than the laws of physics. It is therefore necessary to test the decision rules assumed in the model against the behaviour of real people. Since controlled experiments on the entire economy are not possible, a laboratory experiment was developed to test the decision rule governing capital investment, which the analysis above shows to be crucial in the capital self-ordering dynamic that leads to the long wave. To test the correspondence of the model to real behaviour, an interactive simulation 'game' was developed [38,39]. In the game, human subjects play the role of manager of the capital producing sector of the economy and are responsible for making investment decisions. The physical and institutional context in which the players make decisions is identical to the simulation model, and the player is given the full information set available in the original model. The only difference between the original model and the experiment is that investment decisions are specified in the latter by the player and in the former by the decision rule of the model.

A large sample of players has been tested, including undergraduate and graduate students at MIT; academicians in economics, operations research, and other technical fields from the U.S., Europe, and the Soviet Union; and business executives including several presidents and CEOs. Results show the behaviour generated by the human subjects to be strikingly similar to that of the simulation model. In both the model and the experiment,

1. The period of the cycle is approximately 50 years.
2. Output rises slowly due to the lags in acquiring capital, but falls precipitously, followed by a long depression while the excess capital depreciates. Capacity peaks after and higher than production.
3. Successive cycles occur despite the fact that there are no exogenous disturbances after an initial disequilibrating step increase in orders.
4. The amplitude of the cycle is much more sensitive to parameter variations than the period.

Figure 10 presents typical experimental results; Figure 11 compares the sample means for key aspects of the behaviour against the simulation model (Figure 8).

The experiment shows that real people behave in a fashion consistent with the assumptions of the model [46]. It is also interesting to note that while the human subjects tried to behave rationally, their behaviour is far from optimal. It is possible in the experiment to return the system to equilibrium rapidly and without generating any oscillation, yet only 8% of the subjects found equilibrium at all. The mean absolute deviation between supply and demand generated by the players was 31 times greater than the optimal level, clearly showing that people do not behave rationally or optimally even with perfect knowledge of the system structure and perfect information of system states. (The game can be played manually or on personal computers. Copies of the game and floppy disks for the IBM PC or Macintosh are available from the author at System Dynamics Group, E52-562, MIT, Cambridge, MA 02139 USA.)

25.6 **Additional Amplifying Processes in the Long Wave**

Self-ordering thus seems to be a sufficient cause of long waves. However, though it may be sufficient to generate the long wave, self-ordering is not the only process at work. Other positive feedback loops operate through a variety of macroeconomic linkages to further destabilize the economy. These include linkages

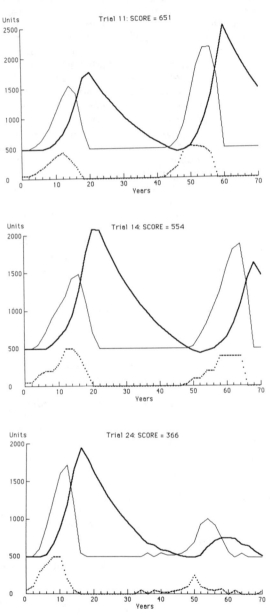

Figure 10 (see facing page for caption).

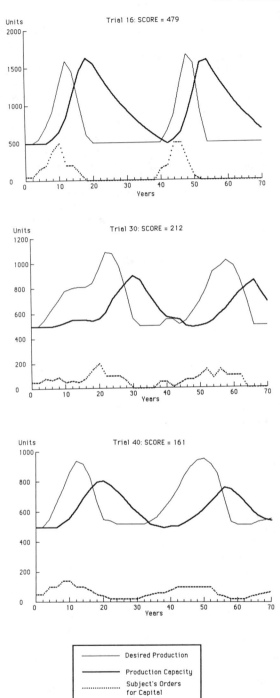

Figure 10. Typical experimental results. The experiment
duplicates the structure of the model except human players

substitute for the capital investment decision rule. *Players attempt to balance demand (Desired Production) and supply (Production Capacity) by ordering capital for their own use (New Orders from Capital Sector). The system begins in stationary equilibrium. The exogenous demand for capital from the consumer goods sector is the only external input to the system. It experiences a one-time step input of 10% in year 4. In 92% of the sample, players generate substantial recurrent fluctuations despite the constancy of the input after the initial step. Note that peak production capacity frequently exceeds the equilibrium production capacity of 560 units by a factor of 2 to 5.*

through the labour market and wages; prices, inflation, and interest rates; debt buildup and default; technological innovation; social innovation and organizational change; and political values [20,40-44].

Figure 11. Comparison of Model and Experimental Results. Source: [38].

	Model[a]	Experiment (N=49)	
		Mean	Std. Deviation
Period of cycle (years)	49	46	13
Peak value, Production Capacity (units/period)	2300[b]	2200	3900
Peak value, Capital Order Rate (units/period)	750[b]	630	930
Minimum value, Capital Order Rate (units/period)	0	4	11
Minimum Fraction of Demand Satisfied (%)	38	48	14

a   Computed from simulation shown in Figure 8.
b   Computed by scaling simulation to the equilibrium order rate in the experiment.

25.7 Conclusion

The economist interested in long-term economic change is hampered by inadequate data, inability to conduct controlled experiments, and host of other problems which constrain physical scientists to a much smaller degree. The research reported here therefore attacks the problem of long waves from a variety of perspectives. These include a detailed, large scale simulation model of economic evolution, simple simulation models, formal mathematical analysis of model properties, empirical studies, and laboratory experiments to examine human decision-making behaviour. Results show the long wave to be a complex phenomenon which influences a wide range of economic and social factors. The long wave is the result of the interaction of the physical structure of the economy and the decision-making of individuals and firms. It is fundamentally a disequilibrium phenomenon and essentially nonlinear. It is generated endogenously, and does not depend on random shocks such as gold discoveries or wars to account for its persistence or for turning points.

The long wave is controversial, and it is fair to ask why one should bother to analyze such a phenomenon. There is little appreciation among the public, policy-makers, or academics of the world economy as a dynamic, evolving system, a system which endogenously produces multiple modes of behaviour. In public policy, this has meant most policies are reactive in nature: they are responses to particular events. Similarly, most scholarly analysis views all fluctuations as policy mistakes or as part of the business cycle. Such a short-term perspective has led to the adoption of policies which are ineffective or counterproductive. Policies suited to the short duration of the business cycle are failing to solve the more persisent problems generated by the downturn of the long wave. At the least, then, analysis of the dynamic processes which create long-term economic change can illuminate the search for appropriate policies. By suggesting historical analogies, providing a consistent, unified explanation for current difficulties, and offering a means to test policies and anticipate side-effects, the stresses of the economic transition may be understood and overcome.

Acknowledgements

The contributions of my colleagues Jay Forrester, Alan Graham, David Kreutzer, Erik Mosekilde and Peter Senge are gratefully acknowledged. This work was supported by the Sponsors of the System Dynamics National Model Project. I am solely responsible for any errors.

References

[1]   N. Kondratiev (1935). *Review of Economic Statistics* **17**, 105.
[2]   N. Kondratiev (1984). *The Long Wave Cycle*, tr. G. Daniels, Richardson and Snyder, distributed by E.P. Dutton, New York.
[3]   J.J. Van Duijn (1983). *The Long Wave in Economic Life,* Allen and Unwin, London.
[4]   C. Freeman (1983). *The Long Wave and The World Economy,* Butterworths, Boston.
[5]   J. Schumpeter (1939). *Business Cycles,* McGraw-Hill, New York.
[6]   G. Mensch (1979. *Stalemate in Technology,* Ballinger,

Cambridge MA.

[7] C. Freeman, J. Clark and L. Soete (1982). *Unemployment and Technical Innovation: A Study of Long Waves and Economic Development*, Greenwood Press, Westport, Connecticut.

[8] W. Rostow (1975). *Journal of Economic History* **35**, 719.

[9] W. Rostow (1978). *The World Economy: History and Prospect*, University of Texas Press, Austin.

[10] E. Mandel (1980). *Long Waves of Capitalist Development*, Cambridge University Press, Cambridge.

[11] E. Mandel (1981). *Futures* **13**, 332.

[12] J.W. Forrester (1981). *Futures* **13**, 323.

[13] J.W. Forrester (1979) in *Economic Issues of the Eighties*, eds. N. Kamrany and R. Day, Johns Hopkins University Press, Baltimore.

[14] J.W. Forrester (1977). *De Economist* **125(4)**, 525.

[15] J.W. Forrester (1976). *Futures* **8**, 195.

[16] A. Graham and P. Senge (1980). *Technological Forecasting and Social Change* **17**, 283.

[17] P. Senge (1982). *The Economic Long Wave: A Survey of Evidence*, System Dynamics Group, MIT, Cambridge MA 02139.

[18] J. Sterman (1985). *Futures* **17**, 104.

[19] J. Sterman (1985). *Journal of Economic Behaviour and Organization* **6(2)**, 17.

[20] J. Sterman (1986). *System Dynamics Review* **2(2)**, 87.

[21] J.W. Forrester, A. Graham, P. Senge and J. Sterman (1983). *An Integrated Approach to the Economic Long Wave*, System Dynamics Group, MIT, Cambridge MA 02139.

[22] S. Rezneck (1968). *Business Depression and Financial Panic: Essays in American Business and Economic History*, Greenwood Publishing Co., New York.

[23] L. Metzler (1941). *Review of Economic Statistics* **23**, 113.

[24] N. Mass (1975). *Economic Cycles: An Analysis of Underlying Causes*, The MIT Press, Cambridge MA.

[25] P. Senge (1978). *Ph.D. Dissertation*, Sloan School of Management, MIT.

[26] D. Jorgenson (1963). *American Economic Review* **53**, 247.

[27] T. Mayer (1960). *Journal of Business* **33**, 127.

[28] G. Low (1980) in *Elements of the System Dynamics Method*, ed. J. Randers, The MIT Press, Cambridge MA, 76.

[29] N. Forrester, *Ph.D. Dissertation*, MIT, 1982.

[30] S. Kuznets (1930). *Secular Movements in Production and Prices*, Houghton Mifflin, New York.

[31] B. Hickman (1963). *American Economic Review: Papers and Proceedings* **53**, 490.

[32] J. Sterman (1982). *Amplification and Self-Ordering: Causes of Capital Overexpansion in the Economic Long Wave*, System Dynamics Group, MIT, Cambridge MA 02139.

[33] R. Frisch (1965) in *Readings in business Cycles*, eds. R. Gordon and L. Klein, Richard D. Irwin, Homewood, Illinois.

[34] S. Rasmussen, E. Mosekilde and J. Sterman (1985). *System Dynamics Review* **1**, 92.

[35] H.A. Simon (1979). *American Economic Review* **69**, 493.

[36] J. Morecroft (1983). *Omega* **11(2)**, 131.

[37] M. Szymkat and E. Mosekilde (1986). *Global Bifurcation*

Analysis of an Economic Long Wave Model, presented at the 2nd European Simsulation Congress, Antwerp, Belgium, September.

[38]   J. Sterman (1987). *Management Science* **33(12)**, 1572.

[39]   J. Sterman and D. Meadows (1985). *Simulation and Games* **16(2)**, 174.

[40]   E. Mosekilde and S. Rasmussen (1986). *European Journal of Operational Research* **25**, 27.

[41]   D. Dickson (1983). *Science* **219**, 933, 25 February.

[42]   J. Sterman (1983). *Science* **219**, 1276, 18 March.

[43]   A. Kleinknecht (1984). *Cambridge Journal of Economics* 8, 251.

[44]   N. Rosenberg and C. Frischtak (1983). *American Economic Association Papers and Proceedings* **73(2)**, 146.

[45]   S. Homer (1977). *A History of Interest Rates,* 2nd ed., Rutgers University Press, New Brunswick, NJ.

[46]   J. Sterman, *Organisational Behaviour and Human Decision Processes,* forthcoming, 1988.

COMPUTER MODELLING OF THERMODYNAMIC PROPERTIES OF VERY DILUTE, CHARGED HARD SPHERES IN A DIELECTRIC CONTINUUM

26.1 Introduction

The simplest model possible of an electrolyte solution is to assume the ions to be charged hard spheres immersed in a dielectric continuum. The hard spheres are assumed to have the same dielectric constant as the continuuum. In spite of the artificial nature of such a model, it is generally believed to be useful at least for very dilute, electrolyte solutions, where the thermodynamic properties are dominated by the electrostatic long range interactions. After an intelligent, but quite unsuccessful attempt by Milner [1,2] to calculate the electrostatic mean energy for a collection of point particles starting from rigorous statistical mechanics, the well known analytic results for charged hard spheres were obtained by Debye and Hückel in 1923 [3]. The Debye-Hückel law is used in electrochemistry for the proper extrapolation to infinite dilution of electrode potentials or for calculating the limiting thermodynamic properties of dilute electrolyte solutions. However, the price paid for obtaining analytical solutions was high. A number of further drastic and unrealistic assumptions were made, which have ever since cast doubt about the validity of the Debye-Hückel (DH) law. For example, the peculiar mixture of microscopic and macroscopic concepts (combination of Poisson's equation of electrostatic and the Boltzmann distribution of ions around a central ion) is hard to accept. Also, the linearisation of the Poisson-Boltzmann equation leads to absurd negative local concentrations of coions to the central ion, and the nonlinear Poisson-Boltzmann equation has been shown to be inconsistent.

Today, it is generally believed from experimental and theoretical evidence, that the so-called **limiting law** of Debye and Hückel (DHLL) is valid at exteme dilutions. However, it has hitherto not been known whether or not the DH-law correctly describes the **first deviation** from DHLL at somewhat higher concentrations. This question is of crucial importance for electrochemical standardization, since the validity of DHLL is normally restricted to such low concentrations, that it is very difficult to measure anything with confidence (small difference between large experimental quantities). Therefore, one is normally forced to work with slightly higher concentrations, where deviations from DHLL are expected. Theoretically, there exists now a number of more satisfactory approaches starting from modern statistical mechanical approximation techniques [4,5,6]. They all agree about the DHLL, whereas the DH-law is more suspicious. Nevertheless, the first expansion in concentration is identical with the first expansion of the DH-law for some of those theories [7].

Recently [8], we have begun to study the DH-system at extreme dilutions by Monte Carlo techniques [9-14] in order to obtain rigorous results for the first deviation. At the same time, we have an excellent opportunity to check the convergence properties of Monte Carlo simulations in systems with far reaching electrostatic forces, by comparison with the DHLL-limit.

In the present paper we review these results, and present some new ones at even more extreme dilution and for unequal sizes of the ions (not taken into account in the classical DH-theory). Furthermore, we commence an investigation of the range of validity of a special integral equation approximation, which may be seen as the first step in an expansion in the concentration of the electrolyte.

26.2 Monte Carlo Results

We use the Metropolis method [9] using periodic boundary conditions and the energy is calculated by the minimum image method. The same method was applied to the primitive electrolyte model by Card, Valleau and Rasaiah [11,12], but those authors did not study the properties down to the DHLL, because of lack of computing power. Therefore, they missed the region where the primitive model might have some correlation with reality.

The problem is that the so-called Debye-Hückel shielding length L(DH) increases inversely proportional to the square root of the concentration. Therefore, at low concentrations one has to take at least interactions between 500 particles into account in order to be able to extrapolate to the proper thermodynamic limit of an "infinitely large system". Thus, there are 500 "nearest neighbours" with nonlinear electrostatic interactions and with all possible configurations in the real space. Furthermore, it shows up to be necessary with around 1 million Metropolis configurations in each Monte Carlo simulation in order to obtain precise values for the excess energy E(ex), the excess heat capacity C_v(ex), the osmotic coefficient ϕ and the radial distribution functions g++(r) and g+-(r). We have also calculated the excess electrostatic Helmholtz free energy ΔF(ex) in the same Markov chain by means of the Salsburg-Chestnut method [15-17]. Although we have shown [8] that this method is in principle exact for an infinite Metropolis Markov chain - quite contrary to what is often claimed [10,14] - the values seem to have converged completely for the simulations with many particles even after 1 million configurations. The Helmholtz free energies seem to differ systematically by 5-10% after extrapolation to an infinite number of particles. Some results are given in Table 1. Notice, that for the **restricted, primitive model** (RPM), there are only two dimensionless

416

Table 1. Extrapolated Monte Carlo values for the thermodynamic properties of the restricted primitive model [8]. Values obtained from DHLL and DH are also given (in some cases DHX-values) N = number of particles B = 1.546.

$\rho^* =$	$6.4 \ 10^{-5}$	$1.25 \ 10^{-4}$	$2.5 \ 10^{-4}$
-E(ex)*100/NkT:			
MC	2.684±.007	3.728±.008	5.166±.016
DHLL	2.726	3.809	5.387
DH	2.633	3.630	5.036
DHX	2.661	3.681	5.125
-ΔF(wx)*100/NkT:			
MC			
(Salsburg-Chesnut)	1.917(±.018)	2.651(±.024)	3.774(±.042)
DHLL	1.817	2.540	3.591
DH	1.770	2.449	3.416
C_V(ex)*100/Nk;			
MC	1.348±.015	1.827±.015	2.593±.020
DHLL	1.363	1.905	2.694
DH	1.272	1.730	2.354
$(1-\phi)$*100:			
MC	0.8655±.0037	1.1868±.00037	1.6067±.0083
DHLL	0.9087	1.270	1.796
DH	0.849	1.155	1.570
DHX	0.8556	1.1669	1.5911

Each MC-value is based on simulations with N = 32,44,64,100,216 and 512 ions with about 1 million Metropolis configurations for each N. The sampled averages were plotted against 1/N and fitted to the most significant polynomium at the 90% degree of significance (2nd or 1st order). From these, the extrapolated valutes at 1/N = 0 and their uncertainties were found. Uncertainties were also estimated from running averages over blocks of 50,000 configurations with the same results.

parameters: the dimensionless concentration $\rho^* \equiv a^3 \rho$ (a = common diameter of the ions, = density of ions) and the **Bjerrum parameter** $B \equiv z^2 e_0^2/(4\pi\epsilon kTa)$ (z = number of elementary charges e_0 on each ion, ε the absolute dielectric permittivity, k Boltzmann's constant, T = absolute temperature).

From Table 1 it is evident - apart from the Salsburg-Chesnut values which deviate to the "wrong" side from DHLL - that the MC-values for all three concentrations are **significantly different** from the DHLL as well as the DH-values. Incidentally, they are situated in between these two theories. The MC-values are quite close to the DHX-values, which are obtained taking the energy integral over radial distribution functions which are obtained using effective pair potentials $w_{ij}(r)$ given as the shielded electrostatic interactions of the DH-theory ($w_{ij}(r) \equiv -kT\ln g_{ij}(r)$). Thus, DHX is a hybrid model with a quite weak theoretical justification, but it is known to work quite well also at higher concentrations. The DHX-theory also often yields reasonable estimates of the radial distribution functions (see later). Nonetheless, the MC-values of E(ex) and of $1-\phi$ are still significantly different from the DHX-values.

26.3 **Towards the Ultimate Limit**

As expected, the values of E(ex)/NkT, C_v(ex)/Nk and $1-\phi$ creep nearer and nearer to DHLL, when the concentration is lowered.

Table 2. Preliminary MC-results for B=1.546 and $\rho^* = 2 \ 10^{-5}$.

No. of ions	Configs.	$\dfrac{-E(ex)*100}{NkT}$	$\dfrac{-\Delta F(ex)*100}{NkT}$	$\dfrac{C_v(ex)*100}{Nk}$	g+-(a)	g++(a)
100	350,000	1.747	1.398	0.5900	4.0	0.26
					±.5	±.08
100	470,000	1.741	1.394	0.5960	3.1	0.32
					±.3	.07
100	565,000	1.738	1.398	0.5961	3.5	0.32
					±.4	±.06
216	285,000	1.637	1.279	0.6377	3.1	0.17
					±.4	±.06
216	425,000	1.620	1.248	0.6417	3.0	0.20
					±.4	±.05
216	825,000	1.625	1.243	0.6801	5.2	0.26
					±.5	±.05
512	210,000	1.564	1.155	0.7320	3.4	0.34
					±.4	±.10
512	400,000	1.554	1.122	0.7383	2.9	0.36
					±.4	±.07
512	540,000	1.554	1.124	0.7296	3.4	0.27
					±.4	±.06
1000	170,000	1.612	1.240	0.7329	6.3	0.52
					±.7	±.06
1000	210,000	1.597	1.246	0.7075	5.6	0.53
					±.6	±.05
1000	250,000	1.603	1.229	0.7225	5.8	0.52
					±.6	±.05

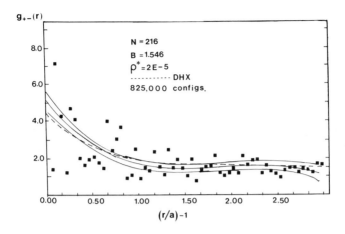

Figure 1.   Radial distribution function g+-(r) sampled during a MC-simulation with N = 216 and 825,000 configurations.   From the highly   scattered   points   (sampled   in   spherical   shells   of thickness 0.05 contact radii) the most significant polynomium at the 95% level is extracted (standard error belt of smoothed values shown), which is practically indistiguishable from g+-(r) from the DHX-theory B = 1.546, $\rho* = 2.10^{-5}$.

However, E(ex)/NkT is still about 1.5% lower than the DHLL-value and $1-\phi$ is about 5% lower. It should be mentioned that B=1.546 in water at $25^{o}C$ for a uni-univalent electrolyte corresponds to a diameter a = 4.62 Å, and the three dimensionless concentrations in Table 1 correspond to molar concentrations equal to $1.078 \ 10^{-3}$, $2.106 \ 10^{-3}$ and $4.210 \ 10^{-3}$ mol/dm^3, respectively.   Thus, we have still not furnished the ultimate proof of the limiting law by MC-calculations.   Therefore,   we   are   presently   performing MC-calculations for a dimensionless concentration as low as 2.10^{-5} corresponding to a molar concentration equal to $3.369 \ 10^{-4}$.   The preliminary results are given in Table 2.

   From Table 2 it appears that E(ex) and C_{v}(ex) require at least half a million configurations to converge to a precision of 3 digits. The higher N, the slower convergence. ΔF(ex) has not yet converged, especially for N = 512 and 1000. We have also listed the contact values g+-(a) and g++(a) found by fitting the sampled radial distribution functions in the interval [a,4a] to the most significant least square polynomium.   An example is shown in Figure 1.   The scatter is considerable (also for the runs with 1 million configurations of which Table 1 is an extract) due to the rare encounters in very dilute systems.   The contact values are necessary   in   order   to   calculate   the   osmotic   coefficients. Fortunately,   however, the contribution of the contact term to the osmotic coefficient is a smaller and smaller fraction of 1/3 of E(ex)   when   the   concentration   diminishes,   so   the   osmotic coefficients   may   be   determined   with   confidence.   For   higher

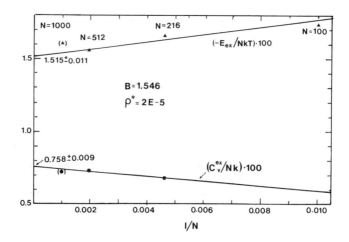

Figure 2. Extrapolation to an "infinite thermodynamic system"
$-E(ex)/NkT$ *and* $C_V(ex)/Nk$ *vs.* $1/N$. *The values for N = 1000 have*

not quite converged after 250,000 configurations.

concentrations, the radial distribution functions are easily found
with great precision. The part of the radial distributions
functions close to the ions is not very much affected by the
number of ions in the simulations. The values of $g+-(a)$ and
$g++(a)$ seem not to be incompatible with the DHX-values
$\exp(\pm B/[1+\kappa a]) = 4.55$ and 0.220, where κ is the inverse DH
shielding length. At the present concentration and B, the value
of κa is $1.971.10^{-2}$. For N = 512 and 1000, the edge length of the
central simulation box is 5.81 and 7.26 times longer than L(DH).
This might seem to be enough to be an "infinite system", but
Figure 2 shows that it is still necessary with an extrapolation to
$1/N = 0$. On Figure 3 we have plotted the extrapolated values of
$-E(ex)/NkT$ against κa for the four concentrations considered
above. It is seen that the excess energy of the lowest
concentration still differs 0.5% from the DHLL and that the DH-law
drastically exaggerates the first deviation from the DHLL. The
real law is between DHLL and DH no matter how dilute the system
is, and it is not far from DHX.

26.4 Different Sizes of the Ions

We have also performed a few calculations on systems with
different cationic and anionic radii. Figure 4 shows the radial
distribution functions for N = 100, $\rho^* = 5.10^{-4}$, B = B(+-) = 1.546
after 855,000 configurations. The contact distance a(++) is 3
times the distance a(--). The DH-contact distance a(+-) is
shortly denoted a. The solid curves are calculated from a simple
generalization of DHX.

$$g+-(s) = \exp[B^*\exp(-\kappa a^*s)/[(1+\kappa a)(s+1)]] \quad s \geq 0 \qquad (1)$$

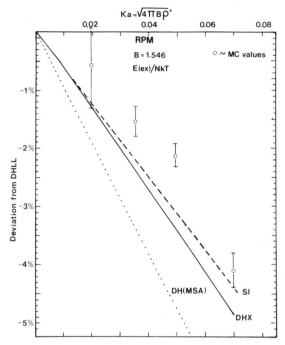

Figure 3. The deviation of the MC-calculations of E(ex)/NkT from DHLL as compared to the DH (and MSA) law as a function of κa (square root of concentration). DHX and SI results are also shown. (SI = simple integral equation, see the last section).

$$g_{++}(s) = \exp[-B^*\exp(-\kappa a^*s)/[(1+1.5\kappa)(s+1.5)]] \quad s \geq 0 \quad (2)$$

$$g_{--}(s) = \exp[-B^*\exp(-\kappa a^*s)/[(1+0.5\kappa a)(s+0.5)]] \quad s \geq 0 \quad (3)$$

$$s = \begin{cases} (r/a) - 1 & (+-) \\ (r/a) - 1.5 & (++) \\ (r/a) - 0.5 & (--) \end{cases} \quad (4)$$

The correspondence between the MC-calculations and the generalized DHX-formulae is striking. The MG-samplings (sampled in s-intervals of 0.05) seem to distribute themselves neatly Gaussian around the DHX-curves. If this holds true for higher N and for some other examples too, then we might have solved the classical electrolyte problem to a very high degree of approximation by very simple formulae (all the mechanical properties of general electrolyte mixtures like energy, osmotic pressure and compressibility can then be found by simple numerical integrations).

The excess energy is seemingly not altered very much, when we vary the ratio of the radii a(++)/a(--) away from unity and maintain a(+-) and B(+-). For a similar simulation as in Figure 4, but with N = 512, we obtain after 270,000 configurations an

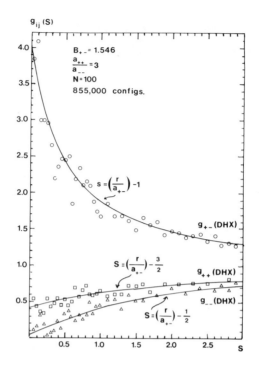

Figure 4. The radial distribution functions g+-(s)(0), g++(s)(□) and g--(s)(△) for an electrolyte system with cationic diameter three times the anionic diameter. B(+-) = 1.546. ρ* = 5.10⁻⁴. N = 100 and 855,000 configurations. Solid curves are calculated from the generalized DHX-theory.

excess energy $-E(ex)/NkT = 7.253 \ 10^{-2}$ as compared to $7.180 \ 10^{-2}$ for equal radii and the same number of configurations. (Probably $E(ex)$ has almost converged after 270,000 configurations, at the moderate concentration considered, since the value after 1 million configurations and equal radii is found to be $7.164 \ 10^{-2}$). Thus, the main influence of uneven radii will be on the contact terms (corrected excluded volume) contributing to the osmotic coefficient in moderately dilute systems.

26.5 A Simple Integral Equation Solution (SI)

A convenient point of departure for many modern statistical mechanical theories of fluids is the Ornstein-Zernike (OZ) equation:

$$h_{ab}(r_{12}) = c_{ab}(r_{12}) + \sum_s \rho_s \int c_{as}(r_{13}) h_{sb}(r_{32}) dr_3 \quad (5)$$

where $h(r) \equiv g(r) - 1$ and $c_{ab}(r)$ is the **direct correlation function** between species a and b (e.g. + and -), which for a wide class of fluid Hamiltonians and for almost all densities (ρ) and temperatures (the compressibility should be finite) aproaches the dimensionless pair potential between a and b, when r_{12} tends to infinity. Replacing simply c_{ab} by the negative, electrostatic potential $-\phi_{ab}/kT$ for all values of r, a simple integral equation is obtained, the solution of which yields the classical DHLL-result, where the extension of the ions is not accounted for. In the mean spherical approximation theory (MSA), one further forces $g_{ab}(r)$ to be zero for $r < a_{ab}$, the contact distance between species a and b. This leads to analytical results for the resticted primitive model (RPM) with only one contact distance and to simple iterative schemes for a pocket calculator for ions of different radii. The MSA yields far better correspondence with MC-data at higher densities than the DH-theory, but at the densities considered here, the DH and MSA results are almost identical and the g++ and g-- become negative at high dilution, so MSA is not a real improvement in comparison to DH.

One may use a less drastic approximation than in MSA, however. It is possible to write a **formal** density expansion of the direct correlation function

$$c(r) = f(r) + \rho \int \triangle\ dr_3 + (\rho^2/2)\iint [2\ \square + \bowtie +$$

$$4\ \boxtimes + \mathparagraph\mathparagraph + \bowtie + \boxtimes\ ]\ dr_3\ dr_4 = f(r) + \rho C_1 + \ldots. \tag{6}$$

The electric + hard sphere pair potential is called u_{ab} and Mayer's f-function is given by

$$f_{ab}(r) = \exp(-u_{ab}/kT) - 1. \tag{7}$$

The graphs in eq. (6) represent products of similar f-functions as indicated by the connections. From this point of view, the simplest possible integal equation (SI) is obtained by the choice $c_{ab}(r) = f_{ab}(r)$. It is seen that MSA requires an additional linearization of $\exp(-\phi_{ab}/kT)$. The present approximation has been discussed previously from a general, theoretical point of view by Rushbrooke and Scoins [18], but as far as we know, it has never before been used in practical calculations for Coulomb systems. Clearly this approximation is better to describe correlations between ions close to each other in dilute systems, than MSA (or DHLL), since $|\phi_{ab}/kT| \gg 1$ close to the ions. Indeed, MSA and DHLL may be characterized as **low correlation theories** rather than as **low concentration theories**.

3D-Fourier transformation (^) of the OZ-equation for the RPM gives rise to the following relations (using the spherical symmetry)

$$\hat{h}_I(k) = \hat{c}_I(k)/[1 - \rho^*\hat{c}_I(k)] \qquad (I = A,B) \tag{8}$$

$$\hat{\gamma}_I(k) = \rho^*\hat{c}_I^2(k)/[1 - \rho^*\hat{c}_I(k)] \tag{9}$$

$$\gamma_I(k) \equiv h_I(k) - c_I(k); \quad \hat{\gamma}_I(k) \equiv \hat{h}_I(k) - \hat{c}_I(k) \tag{10}$$

$$h_A(r) = [h{+}{+}(r) + h{+}{-}(r)]/2; \quad h_B(r) \equiv [h{+}{+}(r) - h{+}{-}(r)]/2. \tag{11}$$

For the SI-approximation we have ($t = r/a$):

$$c_A(t) = \begin{cases} \cosh[B/t]-1 & t \geq 1 \\ -1 & t < 1 \end{cases} \tag{12}$$

$$c_B(t) = \begin{cases} -\sinh[B/t] & t \geq 1 \\ 0 & t < 1. \end{cases} \tag{13}$$

The Fourier transforms of $c_I(t)$ are easily found, and from eq. (8), the Fourier transforms $h_I(t)$ are calculated directly. The excess energy is then given by

$$E(ex)/NkT = (\rho^*/4\pi^2)\int_0^\infty \hat{U}(k)\hat{h}_B(k)k^2\, dk \tag{14}$$

with normalized ++ potential given as

$$U(t) \equiv \begin{cases} 0 & t < 0 \\ B/t & t \geq 0 \end{cases} \tag{15}$$

Since $\hat{U}(k) = 4\pi B\cos(k)/k^2$, we obtain from eq. (14):

$$E(ex)/NkT = (B\rho^*/\pi)\int_0^\infty \hat{h}_B(k)\cos(k)dk. \tag{16}$$

If one wants to find both radial distribution functions (i.e. h++ and h+-), it is advantageous to use eqs. (9), since $\gamma_I(t)$ is continuous in the point $t = 1$, so the back-transformation of $\hat{\gamma}_I(t)$ in the vicinity of $t = 1$ may be performed without problems.

In the concentration range considered in this paper, the SI-radial distribution functions yield somewhat higher values than the DHX-values, a difference which increases with increasing density and with decreasing value of t. A comparison of some values is given in Table 3. The reason is easy to understand. Since we have not forced g++(t) and g--(t) to be zero for t <1, the values of the local concentrations at contact are somewhat exaggerated, since the ions to a certain extent interpenetrate each other (contrary to the assumptions in the model). We have assumed that DHX is close to the real distribution functions (cf. the MC simulations). This immediately suggests an improved

Table 3.   Comparison between SI and DHX radial distribution
            functions at three small concentrations. The argument
            is t = r/a.

	SI		DHX	
	g+-(t)	g++(t)	g+-(t)	g++(t)
$\rho^*=6.4\ 10^{-5}$				
t = 1.0	4.64	0.267	4.45	0.225
t = 1.2	3.57	0.330	3.44	0.291
t = 1.6	2.58	0.434	2.49	0.401
t = 2.0	2.11	0.515	2.06	0.486
t = 3.0	1.623	0.650	1.590	0.629
t = 5.0	1.313	0.785	1.296	0.772
$\rho^*=1.25\ 10^{-4}$				
t = 1.0	4.62	0.289	4.36	0.229
t = 1.2	3.55	0.351	3.37	0.296
t = 1.6	2.56	0.455	2.44	0.409
t = 2.0	2.095	0.536	2.016	0.496
t = 3.0	1.604	0.670	1.560	0.641
t = 5.0	1.295	0.803	1.274	0.785
$\rho^*=2.5\ 10^{-4}$				
t = 1.0	4.59	0.320	4.24	0.236
t = 1.2	3.53	0.382	3.28	0.305
t = 1.6	2.53	0.486	2.38	0.421
t = 2.0	2.069	0.566	1.962	0.510
t = 3.0	1.580	0.699	1.520	0.658
t = 5.0	1.273	0.829	1.244	0.804
t = 6.0	1.208	0.864	1.185	0.844

version of SI, i.e. one should force g++ and g+- to be 0 for t < 0
by an appropriate choice of c(t) for t < 0 (like in the MSA
model).   This model is under investigation at present.   Likewise
are higher order aproximations investigated, taking into account
higher order terms in eq. (6) after suitable renormalization
("chain graphs", "water melon graphs").

The correspondence between E(ex) values calculated from SI is
quite acceptable at low densities (see Fig. 3), since the
thermodynamic properties are dominated by the "tail" of the
distribution functions in very dilute systems.   Preliminary
calculations show, however, that models with "forced" internal
g-functions or models taking account of the chain graphs in the
C_i-terms in eq. (6), is indeed improving the values at higher
densities.   Publications of these results will be forthcoming.

Acknowledgements

We are indebted to the Technological Council of Denmark (Teknologistyrelsen) for financial support to the performance of the computer simulations. Jørgen Birger Jensen (Fysisk-Kemisk Institut), Bjørn Malmgren-Hansen (Fysisk-Kemisk Institut) and Hans Boye Nielsen (Radiometer A/S) are thanked for useful discussions and suggestions.

REFERENCES

[1]   S.R. Milner (1912). *Phil. Mag.* **23**, 551.
[2]   S.R. Milner (1913). *Phil. Mag.* **25**, 742.
[3]   P. Debye and E. Huckel (1923). *Phys. Zeitschr.* **24**, 185.
[4]   J.E. Mayer (1950). *J. Chem. Phys.* **18**, 1426.
[5]   P.M.V. Resibois (1968). *Electrolyte Theory,* Harper & Row, New York, Evanston, London.
[6]   H. Falkenhagen und W. Ebeling (1971). *Theorie der Elektrolyte,* S. Hirzel Verlag, Leipzig.
[7]   T.S. Sørensen, P. Sloth and M. Schrøder (1984). *Acta Chem. Scand.* **A38**, 735.
[8]   P. Sloth,, T.S. Sørensen and J.B. Jensen (1987). *Faraday Transactions 2,* **83**, 881.
[9]   N. Metropolis, A.W. Rosenbluth, M.M. Rosenbluth, A.H. Teller and E. Teller (1953). *J. Chem. Phys.* **21**, 1087.
[10]   W.W. Wood (1968) in *Physics of Simple Liquids,* eds H.N.V. Temperley, J.S. Rowlinson and G.S. Rushbrooke, North-Holland Publ. Co., Amsterdam.
[11]   D.N. Card and J.P. Valleau (1970). *J. Chem. Phys.* **52**, 6232.
[12]   J.C. Rasaiah, D.N. Card and J.P. Valleau (1972). *J. Chem. Phys.* **56**, 248.
[13]   J.P. Valleau and S.G. Whittington (1977) in *Statistical Mechanics. Part A: Equilibrium Techniques,* Ch. 4, ed. B.J. Berne, Plenum Press, New York and London.
[14]   J.P. Valleau and G.M. Torrie (1977) in *Statistical Mechanics. Part A: Equilibrium Techniques,* Ch. 5, ed. B.J. Berne, Plenum Press, New York and London.
[15]   Z.W. Salsburg, J.D. Jacobson, W. Fickett and W.W. Wood, *J.* (1959). *Chem. Phys.* **30**, 65.
[16]   D.A. Chesnut and Z.W. Salsburg (1963). *J. Chem. Phys.* **38**, 2861.
[17]   D.A. Chesnut (1963). *J. Chem. Phys.* **39**, 2081.
[18]   G.S. Rushbrooke and H.I. Scoins (1953). *Proc. Royal Soc.* **A216**, 203.

NOISE AND CHAOS IN DRIVEN JOSEPHSON JUNCTIONS

27.1 Introduction

Despite its long history, the study of the complex responses of driven nonlinear systems is presently in a period of great activity, in large part because the advent of powerful computing techniques has enabled computer experiments and simulations to occupy a bridge position between real experiments on physical systems and purely analytical investigations. Much of the philosophical fascination of the subject stems from the realization that noisy, chaotic motions can result from the purely determinstic response of simple nonlinear systems to an ideal, noiseless drive. Motivated by this consideration, many simulations have been made, and thresholds of subtle instabilites computed to 6-figure precision. Real physical systems, however, are always subject to some level of extrinsic noise from their coupling to the external heat bath of effectively infinitely many degrees of freedom, if not also to larger and less ideal disturbing influences.

It is the purpose of this paper to describe experimental measurements and simulations of the response of Josephson junctions to a combined dc and ac drive current. This system provides an excellent example of a nonlinear system whose properties are very well understood in fundamental terms, and in which the effects of natural noise on the complex nonlinear response are easily seen, but are not so large as to dominate the problem.

27.2 The Pendulum And The Josephson Junction

It is well known that the equations governing the motion of a driven pendulum and a driven Josephson junction are mathematically the same, if the latter is represented by the so-called RCSJ (resistively and capacitively shunted junction) model [1]. However, the physically realistic regimes for the two systems differ widely. To illustrate the consequences of these differences in scale, we summarize the two examples in parallel.

In the absence of damping and drive, the pendulum energy is

$$E = -MgL \cos \theta + 1/2 \ ML^2 \dot{\theta}^2 \tag{1}$$

from which elementary arguments yield a (small-amplitude) resonance frequency

$$\omega_o = (g/L)^{1/2} \sim 10 \ s^{-1} \tag{2}$$

for a 10 cm pendulum arm.

The corresponding energy expression for the Josephson junction is

$$E = -(\hbar I_c/2e)\cos \phi + 1/2 \ C(\hbar\dot\phi/2e)^2 \tag{3}$$

from which the same argument yields the so-called plasma resonance frequency

$$\omega_p = (2eI_c/\hbar C)^{1/2} \sim 10^{12} \ s^{-1} \tag{4}$$

for typical junction parameters. Here I_c is the critical current of the junction, and C is its capacitance. The voltage V across the junction is given by the Josephson frequency relation $2eV = \hbar\dot\phi$, while the current through it is given by the Josephson current-phase relation $I = I_c\sin\phi$.

The difference by a factor of 10^{11} in characteristic frequency implies that the Josephson junction can explore enormous numbers of drive cycles per second, an aid in observing infrequent events. But the very high frequency makes it unfeasible to follow the motion in time; one can only measure time averages such as the voltage, which corresponds to ϕ averaged over very many cycles of the driven motion. If this is plotted vs. the slowly-swept bias current, one obtains the so-called I-V characteristic of the junction. To get insight into the details of the motion, one must resort to simulations, either analog or digital, and presume that a simulation has caught the essence of the true physical motion if the simulated I-V curve reproduces the "fingerprint" of the experimental one, i.e. some unique feature on the I-V curve associated with a narrow range of parameters. As with real fingerprints, one assumes that complex patterns do not match by accident, but one cannot prove a unique correspondence.

27.3 How Important is Noise?

As soon as one adds any damping to the system, the same forces which can damp the motion by coupling energy out to the environment can also couple energy in. This coupling is modelled by a fluctuating force, related by the fluctuation-dissipation theorem to the damping, whose effect is to maintain 1/2 kT of energy per degree of freedom in equilibrium. Applying this principle to the pendulum, (1) implies a thermal noise amplitude of vibration

$$\delta\theta_{rms} = (kT/MgL)^{1/2} \sim 10^{-10} \tag{5}$$

428

at room temperature for M = 1 kg and L = 10 cm. This suggests that thermal noise should not appreciably affect the driven motion unless it is computed or observed with such enormous precision as to pick up one part in 10^{10}.

Applying the same approach to the Josephson junction potential energy in (3), we find

$$\delta\phi_{rms} = [\frac{kT}{\hbar I_c/2e}]^{1/2} \tag{6}$$

If we use standard theoretical relations [2] to relate I_c to the normal-state resistance of the junction R_N and the transition temperature of the superconductor T_c, this can be transformed to

$$\delta\phi_{rms} = [\frac{T}{T_c}]^{1/2} \ [\frac{R_N}{1.38\hbar/e^2}]^{1/2} \ \sim 0.01 \tag{6a}$$

Here $\hbar/e^2$ = 4109 ohms is the quantum unit of resistance, and we have assumed the representative values $T/T_c \sim 1/2$ and R_N = 1 ohm.

Since this noise-induced fluctuation is of order 1%, it can be expected to have very significant effects in washing out subtle structures found in noise-free simulations, without changing the overall pattern too markedly. It should be noted, however, that if $R_N \geq \hbar/e^2$, then $\delta\phi \geq 1$, and phase fluctuations become very important in high resistance junctions.

27.4 Noise in Simulations vs. Real Experiments

When one desires to build noise into simulations in a quantitative way so as to test a model more closely against experimental data, one faces an ambiguity implicit in the irregular nature of noise. Because infrequent extreme noise excursions can play a key role in triggering a switch from motion about one attractor to motion about another, presumably the simulation depends more on the peak than on the rms value of noise, and for random signals the relation between peak and rms value depends on the duration of the sample. If one runs the system through N_a drive cycles at a given bias condition, there are three possibilities: (a) no switch is triggered, (b) a few switches are triggered, or (c) many switches are triggered. Since (b) and (c) will lead to exploration of the same motions and lead to similar average values, the critical distinction is between (a) and (b). If the simulation ran through the same N_a as the experiment, it would be clear that the noise amplitude in the simulation should bear the same ratio to the depth of the potential well as in the experiment, i.e. one should have the same value of the parameter

$$\gamma = \hbar I_c/ekT \cong 2/(\delta\phi_{rms})^2 \tag{7}$$

in the simulation as in the experiment, where T describes the mean square noise current level i_n^2 = 4kTB/R in bandwidth B and we have used (6) to obtain the second form. But if N_a differs so radically as between 10^{12} in the experiment and 10^4 in the simulation, as is typical, the noise temperature in the simulation must be increased to have the same likelihood of reaching condition (b) with the fewer available attempts. For Gaussian noise in a 2-dimensional $(\phi, \dot{\phi})$ phase space, the probability that an excursion equalling or exceeding a critical amplitude $\Delta\phi$, sufficient to switch to another attractor, is reached in N_a attempts is roughly

$$N_a \, e^{-\gamma(\Delta\phi)^2}/4.$$

Setting this probability equal to unity to obtain a **threshold condition**, we have

$$\gamma(\Delta\phi)^2/4 = \ln N_a \qquad (8)$$

Since $\Delta\phi$ is fixed by the attractors of the system, it follows that to maintain the same threshold condition, one should scale γ with $\ln N_a$, and hence the effective temperature T^* should scale inversely with $\ln N_a$ to reproduce correctly the effect of these infrequent noise peaks. For the typical numbers quoted above, we should use

$$T^*_{sim} = \frac{\ln 10^{12}}{\ln 10^4} \times T_{exp} = 3 \, T_{exp}. \qquad (8a)$$

It is clear that this procedure is not rigorous, since the increased T^* increases the mean-square noise above the correct value, in order to compensate for the radically fewer opportunities in a short simulation to sample the very high noise peaks. It should provide an improved simulation, however, in those cases where the most crucial effect of noise is occasionally to switch the system from one attractor to another, since the typical smaller rms excursions tend to average out even if they are magnified by this procedure. In fact the data below support this procedure, which is based on a similar one introduced by Danchi et al. [3] in the simulation of premature switching out of the zero-voltage state of a dc-biased junction.

27.5 Experimental Observations and Simulations

The experiments reported here were performed on two nominally identical Nb-aSi-Nb junctions, with plasma frequencies near 400 GHz, and with Nb thin-film antennas forming an integral part of the devices [4,5]. Their I-V curves were measured under various levels of irradiation from a gas laser, optically pumped by a commercial CO_2 laser, and operating at $f_L = \omega_L/2\pi = 419$ GHz. The stability of the laser output was sufficiently good (a few percent) that successive sweeps of an I-V curve were essentially the same. The laser output level was monitored in arbitrary units by a pyroelectric detector, and converted to normalized ac drive current i_{ac} using a single scaling factor chosen to give the best fit to data taken over a wide range of power levels.

The experimental I-V curves are compared with the output of an analog simulator of the Magerlein [6] design, modified to include a piece-wise linear approximation to the nonlinear quasiparticle resistance R, switching from the leakage resistance R_1 below the gap voltage to the normal resistance R_N above the gap. The simulator finds the time evolution of the junction phase as governed by the RCSJ equation:

$$\frac{d^2\phi}{dt^2} + \frac{1}{\beta_c^{1/2}}\frac{d\phi}{dt} + \sin\phi = i_{dc} + i_{ac}\sin\left[\frac{\omega_L t}{\omega_p}\right] + i_n(t) \qquad (9)$$

where $\beta_c = (\omega_p RC)^2$, t is time in units of ω_p^{-1}, and i is current in units of I_c. The plotted output is $\langle\dot\phi\rangle$ ($\propto V$) vs. i_{dc}.

The near coincidence of the laser frequency with the natural frequency of the junctions not only gives efficient coupling to the nonlinear circuit, but also allows exploration of the sensitive dependence of the I-V characteristics on the exact frequency ratio $\omega_L/\omega_p \sim 1$. In fact, the distinct I-V "fingerprints" of the two junctions enabled us to determine [5] rather accurately the 10% difference in their plasma frequencies by comparison with simulations, despite the fact that data could be taken at only one frequency. As described below, this determination was confirmed by using a magnetic field to depress I_c of the junction with the higher ω_p to the same value as that of the other junction, at which point the "fingerprints" became very similar. Thus, the junctions have well-characterized parameters, and the I-V curves are well fitted by simulations over a wide range of bias parameters.

Without going into the details, let us now examine several representative sets of I-V curves, identifying the varying types of both phase-locked and chaotic motions underlying the dc I-V curves, and demonstrating the importance of adding the correct amount of noise in the simulations to obtain good agreement with the experimental curves.

Figure 1 shows a family of I-V curves of junction #1 taken at a succession of drive levels from the laser operating at 419 GHz. In contrast to normal, non-chaotic situations [3], one notices an immense amount of low-frequency noise on the traces, and also the presence of a prominent subharmonic step at n = 2/3 and the absence of the usual n = 1 step. The inset shows a noiseless

Driven Josephson junctions

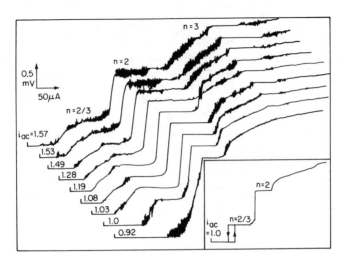

Figure 1.   I-V curves of Junction 1 irradiated with 419 GHz
radiation. i_{ac} is normalized to i_c.   The curve in the box is
the simulated I-V curve for i_{ac}=1.0 with no added noise.   The
hysteretic behaviour is found if continuity in ϕ and $\dot{\phi}$ is
maintained as i_{dc} is increased;   random phase space initial
conditions can produce solutions on either step.

simulation of the I-V curve for i_{ac} = 1.0, with a fitted value of
ω_L/ω_p = 1.07 for the junction plasma frequency.   The simulation
correctly predicts the step sequence 0, 2/3, 2, but, being
noiseless, shows hysteretic switching between the 0 and 2/3 step
instead of the experimental continuous noisy I-V curve there. A
similar family of experimental I-V curves for junction #2 is shown
in Fig. 2, together with simulations.   For this junction the
"fingerprint" step sequence is primarily 0, 1/2, 2, with only a
vestige of n = 2/3.   The analog simulation without noise (Fig.
2(b)) shows a bit of the 2/3 step , followed by an extended n =
1/2 step, using a fitted plasma frequency such that ω_L/ω_p = 0.97.

In the simulation with noise (Fig. 2(c)) one has an I-V curve very
similar to the experimental one, in that the 2/3 step is washed
out, as is the hysteretic switching between n = 1/2 and 2.   The
appropriate amount of noise added to accomplish this was found to
correspond   to   T*   ≈   20K,   whereas   the   theoretical
quasiparticle-tunnelling low-frequency noise current density at
voltage V and temperature T defines an effective T_{eff} by [3]

$$S_I = \frac{4kT_{eff}}{R} \frac{2eV}{R} \coth \frac{eV}{2kT} \qquad (10)$$

432

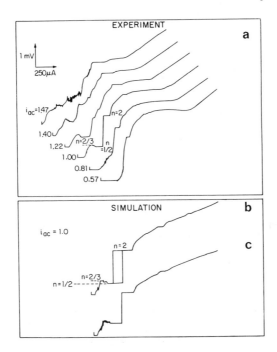

Figure 2. (a) I-V curves of Junction 2 irradiated with 419 GHz radiation. The distinctive feature of the I-V curve with $i_{ac}=1.0$ is the appearance of the 1/2 step with a little bump on the left edge. (b) Simulated I-V curve with $\omega_L/\omega_p=0.97$ and $i_{ac}=1.0$. (c) The same as (b) except adding a 20K white Gaussian noise. Note the strong resemblance to the experimental curve, $i_{ac}=1.0$, in Fig. 2(a).

which would give $T_{eff} \approx$ 5K on the n=1/2 step and ~10K in the n=2 step. (We use the low-frequency limit (10), rather than its frequency-dependent generalization, for several reasons: (1) we have shown experimentally that the hysteresis is strongly reduced only by noise at frequencies $\lesssim \omega_p/2$, where the low-frequency limit is a good approximation for the values of V and T appropriate to our experiments; (2) the technical simplicity of modelling white noise; and (3) there is some question about the appropriateness of including frequency-dependent zero-point photon noise in this context.) Presumably the argument leading to (8a) explains why a T^* three-times greater than these values of T_{eff} gives a much better fit to the data.

Comparison of Figs. 1 and 2 shows that the two nominally identical junctions have quite distinct fingerprints in terms of the sequence of subharmonic steps that are observed. The simulations indicate that these differences stem from the slightly different plasma frequencies of the two junctions, such that ω_L/ω_p

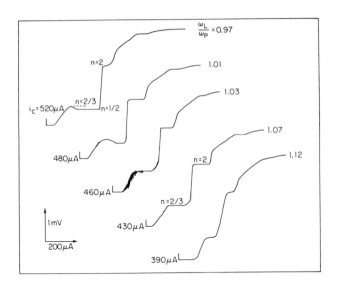

Figure 3. Experimental I-V curves of Junction 2 irradiated at 419 GHz. All the I-V curves are taken at about the same level of laser power corresponding to $i_{ac} \approx 1.0$. From the top to the bottom, the critical current is magnetically suppressed successively; thus the ratio ω_L/ω_p is increased successively. Note close similarity between the curve at $\omega_L/\omega_p = 1.07$ and that in Fig. 1 for $i_{ac} = 1.0$.

= 1.07 for junction 1 and 0.97 for junction 2. We were able to test this interpretation experimentally [5] by using a magnetic field to lower I_c and hence ω_p of junction 2 to bring it into line with junction 1. The resulting I-V curves at constant laser power level $i_{ac} \approx 1.0$ are shown in Fig. 3. As expected, the prominent step at n = 1/2 is displaced by a prominent step at n = 2/3 as the field is increased. In fact, at this point the entire I-V curve of junction 2 is transformed to closely resemble that of junction 1. This experiment and the simulations show how very sensitive the family of I-V curves is to the ratio of frequencies ω_L/ω_p, which allows this parameter to be determined with an accuracy of a few percent. By comparison, the dependence on β_c is found [5] to be relatively weak, variations of up to 50% being required to give a qualitative change. Thus the comparison of experiment and simulation gives an excellent measure of ω_p, without requiring that either the calibration of the laser-driven i_{ac} or the damping parameter β_c be precisely known.

In Figure 4, we show another interesting comparison of experiment and simulation. In this case, the laser power is the maximum available, ~14 times higher than in the previous figure, so that i_{ac} = 3.7. Here the noiseless simulation gives a step sequence 0,2,1,2,3,4, whereas the experimental data show only a rounded, negative-resistance behaviour in place of the first two non-zero steps. Addition of 40K of noise to the simulation, as shown in Fig. 4(c) yields at least a qualitative fit to the experimentally observed I-V curve. The need to add so much noise was initially surprising, but it is largely explained by the fact that at the higher dc voltage of the n=2 step, the shot noise is greater than the thermal noise, and (10) gives $T_{eff} \approx$ 10K. Combined with the factor of 3 from (8a), this gives an estimate of 30K, compared to the 40K empirically required, which is probably within the uncertainties of our approximations. The fact that (8) and (10) can account for the need to use such different values of T* in the simulations of Figs. 2 and 4 provides evidence that the agreement is significant and not merely fortuitous. Moreover, they confirm the need to include appropriate levels of noise in simulations of real Josephson junctions if any sort of quantitative comparison is intended.

The comparisons between experimental data and simulations in Figs. 1-4 illustate the high quality of the description of real Josephson junctions given by the RCSJ-model equation (9), particularly when thermal and shot noise (10) are included, adjusted for the effect of a factor of 10^8 in effective sampling time as described by (8). This gives us confidence that our

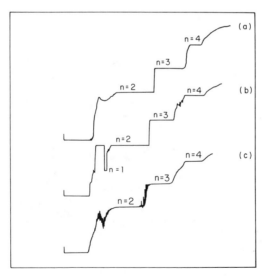

Figure 4.   (a) Experimental I-V curves of Junction 2 with a laser power 13.9 times higher than that of one with i_{ac}=1.0 in the Fig. 2(a).   (b) Simulated I-V curve with ω_L/ω_p=0.97 and i_{ac}=3.73.   (c) The same as (b), except adding 40K white Gaussian noise.

simulation procedure is sufficiently reliable to allow its use in interpreting the nature of the chaotic motions underlying the parts of the I-V curves between the phase-locked harmonic and subharmonic steps.

As an example, consider the data in Figure 1, particularly the trace at i_{ac} = 1. A prominent feature in the experimental trace is the noisy "ramp" between the 0 and 2/3 step. The noiseless simulation shows this as a bistable region, with solutions at both n=0 and n=2/3. Addition of noise in an analog simulation rounds this to a curve (not shown) resembling the experimental one. However a deeper insight is obtained by going to the higher precision of a digital simulation [7]. This reveals that, for any combination of dc and ac bias in this region, the noiseless simulation converges to a **periodic** motion of period 1 (zero step) or 3 (2/3 step) depending on the precise initial condition for $(\phi, \dot{\phi})$. Without noise, there is no chaos. In Figure 5 we display the basins of attraction for these two motions, for the particular bias conditions i_{dc} = 0.15 and i_{ac} = 1. That is, an initial point in the white region will converge to the 2/3 step periodic solution; any point in the black region will converge to the zeroth step. We note that despite the fine-grained 100x100 mesh, the basins are subdivided and interleaved on the finest scale that

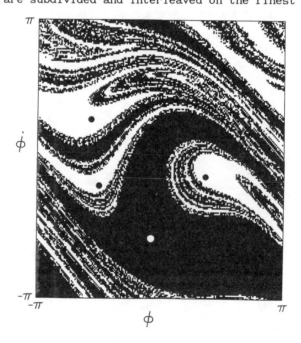

Figure 5. Basins of attraction for i_{ac}=1 and i_{dc}=0.15. The white region is the basin of the 2/3 step solution, corresponding to the three-point Poincaré section indicated by the centres of the black circles. The black region is the basin of the zeroth step solution, with one-point Poincaré section (white circle). No noise is included.

is probed.   If the mesh size is reduced by a further factor of 50,
one still finds subdivision of black and white regions at the
resolution limit of the mesh.   This behaviour is characteristic of
a **fractal** structure, and in fact the basin boundary in this figure
has a dimension of approximately 1.75, approaching the limiting
value of 2.0, at which the boundary would fill the entire plane.
As has been pointed out by Grebogi et al. [8], periodic attractors
separated by a fractal basin boundary should show a sensitivity to
initial conditions which diverges as the dimension approaches 2.
In a previous paper [7], we have reported a correlation between
the computed fractal dimension of the basin boundaries and the
measured noise amplitude as a function of i_{dc}, which reflects this

consideration.   In that work we also studied the effect on the
Poincaré sections associated with these periodic attractors when
noise was added.   The results are shown in Fig. 6(a,b) for bias
levels $i_{dc} = 0.18$ and $i_{ac} = 1$.   For small amounts of noise added
(1K), the 3-fold attractor, which in the absence of noise would be
represented by 3 dots after the initial transient had decayed, is
slightly blurred, but in the time sample considered appears to
retain a simple phase-locked structure.   However, when 50K of
noise is added (Fig. 6b) the Poincaré section spreads over a large
region in phase space, including substantial weight in the
vicinity of the n = 0, one-fold attractor.   This indicates that
the noise is sufficient to switch the motion frequently between

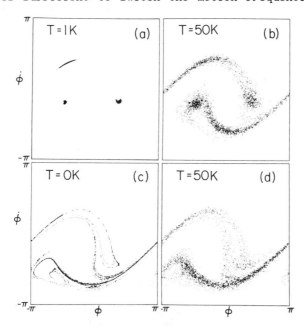

*Figure 6.   Effect of added Johnson noise on the Poincaré
sections for the system: (a,b) in a fractal basin boundary
regime at $i_{dc}=0.18$ and $i_{ac}=1$, and (c,d) in an "intrinsic"
chaotic regime, at $i_{dc}=0.15$ and $i_{ac}=1.5$.*

the vicinities of two periodic attractors.

It is illuminating to compare this behaviour with the effect of noise on the motion governed by an aperiodic strange attractor such as that shown in Fig. 6c. Even in this noiseless simulation, the Poincaré section samples widely distributed parts of phase space, and the addition of 50K of noise only blurs this a bit further. In fact, in the presence of noise, the two Poincaré sections shown in Figs. 6b,d are quite similar, as is the level of low-frequency noise observed in the real experiment at the corresponding bias points.

27.6 Conclusion

The work reported here shows that the experimental response of real Josephson junctions to ac drive at a frequency near their plasma frequencies can be very well simulated using the resistively and capacitively shunted junction model. However, to obtain this good simulation, one must include the gap structure in the quasiparticle damping resistance, and also noise including shot as well as thermal components. Moreover, in the usual case in which the simulation runs for many fewer cycles than the real experiment, the rms noise level (T*) must be increased by a factor of order 3 better to simulate the effect of noise peaks. In the presence of realistic amounts of noise, the Poincaré section of the actual motion is found to be qualitatively similar, whether the noiseless simulation reveals a chaotic strange attractor or two periodic attractors with basins of attraction separated by a boundary of fractal dimension >1.

Acknowledgements

The author is pleased to acknowledge the contributions of his students Qing Hu, who did the analog simulations and most of the experimental work, and Marco Iansiti, who did most of the digital simulations, and those of his faculty colleagues J.U. Free, R.M. Westervelt, and C.J. Lobb in many helpful discussions and suggestions.

This research was supported in part by the ONR contract N00014-83-K-0383 and by the Joint Services Electronics Program under contract N00014-84-K-0465.

References

[1]   B.A. Huberman, J.P. Crutchfield and N.H. Packard (1980). *Appl. Phys. Lett.* **37**, 750.

[2]   M. Tinkham (1975). *Introduction to Superconductivity,* p. 194, Krieger Publishing Co.

[3]   W.C. Danchi, J.B. Hansen, M. Octavio, F. Habbal and M. Tinkham (1984). *Phys. Rev.* **B30**, 2503.

[4]   Qing Hu, J.U. Free, M. Iansiti, O. Liengme and M. Tinkham (1985). *IEEE Trans.* **MAG-21**, 590.

[5]   Qing Hu, C.J. Lobb and M. Tinkham (1987). *Phys. Rev.* **B35**, 1687.

[6]   J.H. Magerlein (1978). *Rev. Sci. Inst.* **49**, 486.

[7]   M. Iansiti, Qing Hu, R.M. Westervelt and M. Tinkham (1985). *Phys. Rev. Lett.* **55**, 746.

[8]   C. Grebogi, S.W. McDonald, E. Ott and J.A. Yorke (1983). *Phys. Lett.* **99A**, 415;   C. Grebogi, E. Ott and J.A. Yorke (1983).   *Physica* **7D**, 181.

COLLECTIVE-COORDINATE METHODS FOR SOLITON DYNAMICS

Abstract

An introduction is given to useful collective coordinate methods for studying the dynamics of solitons (kinks) in **non**-integrable systems.   Several examples involving the nonlinear Klein-Gordon class of systems (e.g. sine-Gordon, ϕ^4, double sine-Gordon, etc.) are discussed, including the influence of perturbations on single-kink   dynamics,   the   interaction   of   two   kinks   or kink-antikink pairs, etc.   In all cases, emphasis is placed on using physical intuition to guide the choice of collective coordinates.   While most of the examples involve classical systems, the use of collective coordinates in studying the **quantum** dynamics of solitons is also discussed.

28.1   Introduction

The general problem of the dynamics of systems having a large number of coupled degrees of freedom has a plethora of interesting and important contexts and applications, of which only a few examples can be discussed in these proceedings.   Thus, it is particularly important to emphasize underlying themes which have broad applicabiilty and utility.   One of these has to do with characterizing complex systems by first identifying the important natural "modes" of behaviour and then studying the dynamical evolution of the system by focussing attention on a (hopefully) small subset of quantities which specify the role of these modes in the system as a whole.   A familiar example is the "normal mode" representation of the harmonic vibration of a large molecule or solid, in which one specifies the amplitude of each important mode and   thus   shifts   attention   from   the   individual   atoms   to   the **collective** behaviour of the interacting system.   If the motion is particularly simple with just one mode "excited" then one can describe   the   motion   of   the   entire   system   by   specifying   the dynamics of just one quantity, namely the amplitude of that particular mode.

   While these comments are painfully boring to most of us, it is nevertheless   helpful   to   recall   this   simple   lesson   from undergraduate study since the motivation for identifying natural modes carries over to the nonlinear systems of interest here. Indeed, since the behaviour of nonlinear systems is far more rich and complex than that of their linear counterparts, it is often the case that, short of direct numerical simulations using large computers, the only hope we have of "taming" the dynamics of such systems lies in the study of **nonlinear** "natural modes". While, of course, not all nonlinear systems actually have identifiable nonlinear modes, there are a few classes of systems which do (so-called "integrable" systems; see articles by Kruskal and Bullough, Pilling and Timonen, in this volume) and for these the exact behaviour can be described solely in terms of a "complete set" of such modes. From a fundamental perspective, there is nothing more to say about such exactly integrable systems - in principle, everything is known (or could be known) about them.

The systems of concern here are those which do *not* have this nice property, but at least have some simple identifiable kink "modes" or excitations and perhaps some two-kink solutions (either exact or approximate). In order to narrow the scope, we shall restrict our attention to the so-called nonlinear Klein-Gordon (NKG) class [1] of kink-bearing systems in one space and one time dimension. These are briefly introduced in Section 28.2 below. As examples we shall treat the dynamics of single kinks in perturbed (i.e. real!) systems (Section 28.3), kink-antikink collisions (Section 28.4), and kink-kink "molecules" (Section 28.5). In Section 28.6 we briefly discuss the quantum dynamics of single kinks, including the tunnelling of kinks through barriers. Finally in Section 28.7 we summarize and indicate some future directions for the development and implementation of collective-coordinate methods for kink dynamics.

28.2 Nonlinear Klein-Gordon Models

All of the examples treated in this article are described by Lagrangians of the nonlinear Klein-Gordon (NKG) form for a field $\Phi(x,t)$ in (1+1) dimensions:

$$L = \int_{-\infty}^{\infty} dx \ \{\frac{1}{2} \Phi_t^2 - \frac{1}{2} \Phi_x^2 - U(\Phi) + \lambda v(x,t)\Phi\}, \qquad (1)$$

where $v(x,t)$ is a localized perturbation (λ is a small parameter) which will be of interest in Section 28.3, but is absent in Sections 28.4 and 28.5. The nonlinear local "potential" function $U(\Phi)$ has at least two degenerate minima so that it supports single-kink solutions of the unperturbed ($\lambda = 0$) equation of motion derived from (1). Examples are the sine-Gordon [1,2] ($U = 1 - \cos \Phi$), Φ-four [1,3] ($U = [\Phi^2 - 1]^2/4$), and double sine-Gordon [4] ($U = 4 \ \text{sech}^2 R[1 + \cos(\Phi/2) + \sinh^2 R\{1 - \cos \Phi\}/4]$) potentials. These will be discussed in sections 28.3, 28.4 and 28.5, respectively.

The equation of motion following from the Lagrangian (1) has the NKG form:

$$\Phi_{tt} - \Phi_{xx} + U'(\Phi) - \lambda v(x,t) = 0. \qquad (2)$$

If we denote a static, classical kink solution of the unperturbed equation ($\lambda = 0$) by $\phi_c(x)$, then a kink moving with velocity $v(v<1)$ can be obtained from $\phi_c(x)$ by a "Lorentz boost" [1], and thus there exist an infinite number of related solutions which differ simply in the value of the kink velocity. Even for the static kink there exists an infinite number of degenerate solutions $\{\phi_c(x-X)\}$ which differ only in the location of the kink centre (X) along the x-axis. This is, of course, simply a consequence of the translational symmetry of the (unperturbed) Hamiltonian H (obtained from L above) and X is in fact a parameter which characterizes the elements of the continuous symmetry group consisting of all possible translations which commute with H. The promotion of this parameter X to the role of a dynamical variable identifies a natural collective coordinate which can be used to simplify the treatment of the dynamics of the perturbed system, as we shall see in the next section. The (continuous symmetry) collective coordinate method is also extremely useful in developing the quantum mechanical description [5-10] of the nonlinear system starting with the classical kink solution (see Section 28.6). This approach dates back to Bogolubov's 1950 treatment [11] of the polaron problem and in fact is a quantum version of the classical 1937 Bogolubov-Krylov method [12] in the theory of nonlinear oscillations.

The existence of the translational degree of freedom for the kink makes itself apparent in another way when one examines the spectrum of small oscillations about the unperturbed kink. If one writes the field as $\Phi(x,t) = \phi_c(x) + \psi(x,t)$, where ψ is assumed to be small, then the equation governing ψ has the form [1]:

$$\psi_{tt} - \psi_{xx} + U''(\phi_c(x))\psi = 0. \tag{3}$$

Assuming a solution with harmonic time dependence, $\psi(x,t) = f(x)\exp(i\omega t)$, one arrives at an ordinary differential equation for the spatial function, $f(x)$, which has a pseudo-Schrödinger form,

$$\frac{d^2 f}{dx^2} + U''(\phi_c(x))f = \omega^2 f, \tag{4}$$

where it can be seen that $U''(\phi_c(x))$ plays the role of a "potential function". It can be shown in general that this potential always has a "bound state" with eigenvalue $\omega^2 = 0$ and eigenfunction $f_0(x)$ proportional to the spatial derivative of the kink profile, $\phi_c'(x)$. There may be additional bound states with $\omega^2 > 0$ which correspond to localized shape oscillations of the kink. In addition to these bound states there are continuum states with ω^2 lying above the maximum of $U''(\phi_c(x))$ and these correspond to perturbed (by the kink) versions of the linear solutions of (2) (with $\lambda = 0$) in the absence of the kink, commonly referred to as "phonons" [1]. What is the physical significance of the zero-frequency mode? The answer is quite simple and can be appreciated by adding $\psi_0(x) = \alpha f_0(x) = \alpha\phi_c'(x)$ to $\phi_c(x)$, with α a small parameter, to obtain the full field $\Phi(x)$ corresponding to the kink plus the small zero-frequency "oscillation":

$$\Phi(x) = \phi_c(x) + \alpha\phi_c'(x) \overset{\sim}{=} \phi_c(x + \alpha). \tag{5}$$

We see that the result (5) is simply a kink profile whose centre has been shifted by a small amount $(-\alpha)$. Thus, this mode is referred to as the "translation mode" [1-8,13] of the kink and its existence "restores" the continuous translational symmetry broken by the introduction of the kink in the system (this is guaranteed according to Goldstone's theorem [14]), i.e. the translational symmetry is now manifested in the family of degenerate kink solutions which can be centred anywhere on the x-axis.

Note that Eq. (5) is only valid for **small** α and this limits the usefulness of the small oscillation corresponding to the translation mode for describing shifts in the kink position due to perturbations [13]. Indeed, certain perturbations lead to unbounded growth of α with time and hence secularities occur in the perturbation theory [13]. While these can often be removed by a "bootstrap" technique in which the unperturbed kink is continually redefined to absorb the successive small changes in its position, a more natural approach is to make a change of variables to a new set which includes the kink position as a collective coordinate that can be as large or small as necessary in response to the perturbation at hand. This advantage is the primary motivation for the perturbation method described in the next section.

28.3 Single-Kink Dynamics in the presence of Localized Perturbations

In this section we describe how the dynamics of single kinks in the presence of perturbations can be treated using a collective coordinate method developed by Gervais et al. [5,6] and Tomboulis [7] for the unperturbed kink quantization problem and extended by Horovitz, Flesch and Trullinger [2,15] to treat classical kink dynamics with perturbations present. The method is applied to the specific example [2] of a sine-Gordon kink "collision" with a localized perturbation.

The equation of motion for the full field $\Phi(x,t)$ is given by (2), where $v(x,t)$ represents the localized perturbation and λ is a small dimensionless parameter giving the strength of the perturbation. Other types of perturbations can be treated with slight modifications of the method [15]. What we seek is an equation of motion for the position of the kink centre, X. To this end we make a canonical transformation [2,5-7,15] from the original field variables $\Phi(x,t)$ and its conjugate momentum $\Pi_0(x,t)$ to new variables $X(t)$, its conjugate, $p(t)$, $\psi(x,t)$ and its conjugate $\pi(x,t)$:

$$\Phi(x,t) = \phi_c(x-X) + \psi(x-X,t) + \chi_0(x,t), \tag{6a}$$

$$\Pi_0(x,t) = \pi(x-X,t) - \frac{\phi_c'(x-X)}{M_0 + \xi} \{p + \int_{-\infty}^{\infty} dx\ \pi(x,t)\psi'(x,t)\}$$

$$- \dot{\chi}_0(x,t), \tag{6b}$$

where the quantities M_0 and ξ are defined by

$$M_0 \equiv \int_{-\infty}^{\infty} dx\ [\phi_c'(x)]^2 \ , \ \xi \equiv \int_{-\infty}^{\infty} dx\ \phi_c'(x)\psi'(x,t). \tag{7}$$

In our units M_0 is the dimensionless classical mass of the kink. As we shall see, the quantity ξ can be interpreted as a (time-dependent) renormalization of the kink mass.

The first term on the right hand side of (6a) represents a kink centred at $X(t)$. The ψ field accounts for the interaction of the kink with the perturbation. It is localized about the kink centre and has appreciable magnitude only when the kink is in the region of the spatially localized perturbation (e.g. an impurity). In addition, $\psi(x,t)$ must account for any extended phonons radiated as a consequence of the collision of the kink with the perturbation. However, it has been shown [2,15] that, for static perturbations $v(x)$, such phonons do not appear until at least third order in the perturbation series in λ.

The final contribution to the field is the "background" response, χ_0, of the system in the absence of the kink. It is advantageous to account explicitly for this background response because, unless the perturbation is switched on and off in some fashion, this response is present long before and after the kink interacts with the perturbation. Thus, by accounting for χ_0, the field ψ only has to account for the changes which occur when the kink interacts with the perturbation. The background response satisfies (2),

$$\chi_{0_u} - \chi_{0_{xx}} + U'(\chi_0) - \lambda\ v(x,t) = 0, \tag{8}$$

but with boundary conditions forcing $\chi_0(x,t)$ to be localized to the vicinity of the perturbation $v(x,t)$. If one requires χ_0 to first order in λ only, then the linearized version of (8) may be employed where, with suitable normalization of U, we have $U'(\chi_0) = \chi_0$. In this case what is required is the particular solution of the linearized equation which is localized about $v(x,t)$. In principle, however, one can solve the full equation (8) by numerical methods if need be.

The transformation

$$\{\Phi(x,t),\Pi_0(x,t)\} \rightarrow \{X(t),p(t),\psi(x,t),\pi(x,t)\}, \qquad (9)$$

does not conserve the number of degrees of freedom [5-7], and hence constraints must be imposed on the new variables. This is done [5-7] by forcing the continuous field ψ and its conjugate π to be orthogonal to the translation mode, $\phi_c'(x)$:

$$\int_{-\infty}^{\infty} dx \; \phi_c'(x)\psi(x,t) = 0, \qquad \int_{-\infty}^{\infty} dx \; \phi_c'(x)\pi(x,t) = 0. \qquad (10)$$

Using the Dirac formalism for constrained systems [16], the transformation (9) described by (6) can be shown to be canonical [15]. With such a transformation one may derive the equations of motion for the new variables using the standard rules of Hamilton-Jacobi theory. The details of this procedure are given in [15] and here we simply state the resulting equation of motion for the variable of primary interest, namely the collective coordinate, $X(t)$, specifying the kink position:

$$M_0\ddot{X} = (1+\xi/M_0)^{-1} \{\int_{-\infty}^{\infty} dx \; \phi_c'(x-X)[U'(\Phi(x,t))-U'(\chi_0(x,t))]$$

$$+ (1+\dot{X}^2)\int_{-\infty}^{\infty} dx \; \psi'(x,t)\phi_c''(x)$$

$$- 2\dot{X} \int_{-\infty}^{\infty} dx \; \phi_c'(x)[\pi'(x,t) - \dot{\chi}_0(x+X,t)]\}. \qquad (11)$$

The first term involving U' on the right hand side of (11) contains the full field $\Phi(x,t)$ given by (6a) and no assumption has yet been made regarding the smallness of $\chi_0(x,t)$. Note that the right hand side of (11) does not contain the perturbation function, $v(x,t)$, explicitly. All information regarding $v(x,t)$ is now contained in the background response function $\chi_0(x,t)$ (as a consequence of having used (8) in obtaining (11)) and in the response of the ψ field to the perturbation [2,15]. The fact that the explicit appearance of $v(x,t)$ can be eliminated from (11) is simply due to having already accounted for the "screening" of the perturbation. Note also that the right-hand side of (11) has what appear to be dissipative terms. However, since the system has no coupling to external degrees of freedom such as a heat bath, these "dissipative" terms can only represent a transfer of energy between the internal degrees of freedom. In the present example, the energy transfer is from the kink motion to the "phonon field" $\psi(x,t)$. A perturbation expansion shows [2,15] that to lowest order, no energy transfer occurs, and in second order, the energy given to the phonon field during the "collision" is ultimately given back to the translational motion of the kink.

To first order in λ, the equation of motion (11) for the kink centre takes a very simple form:

$$M_0\ddot{X} = - \frac{\partial V(X,t)}{\partial X} \quad , \quad V(X,t) \equiv \int_{-\infty}^{\infty} dx \; \chi_0(x+X,t)[\phi_c''(x) - \phi_c(x)].$$

(12)

Thus, to first order the kink behaves as a Newtonian particle of mass M_0 moving in the effective potential $V(X,t)$. If one proceeds to second order [2,15] there appear several additional forces on the kink as well as a renormalization of its mass, but it nevertheless still behaves as a Newtonian particle. A Green function method [15,17] has been developed to determine the first order ψ field which is required to evaluate the effective second-order force on the kink. We refer the reader to [2], [15] and [17] for the details of this procedure and here we content ourselves with illustrating the first order dynamical response of the kink with a specific example [2,15].

We consider a sine-Gordon kink initially travelling to the right which encounters the time-independent perturbation

$$v(x) = e^{-(x-x_0)^2} - e^{-(x+x_0)^2} ,$$

(13)

which is similar to the "impurity" potential studied some time ago by Fogel et al. [13]. In integrating the equation of motion, the following initial conditions and parameters were used:

$$X(t=0) = -20.0, \; \dot{X}(t=0) = 0.30, \; \lambda = 0.04, x_0 = 5.0,$$

(14)

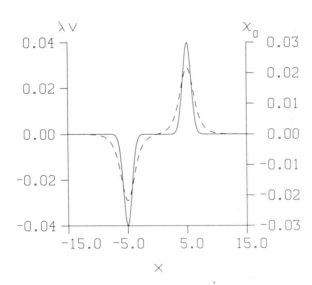

Figure 1. The perturbation $\lambda v(x)$ (see Eq. (13) in the text) and the linear response $\chi_0(x)$ (dashed curve) it generates (see Eq. (8)). (Taken from Ref. [2].)

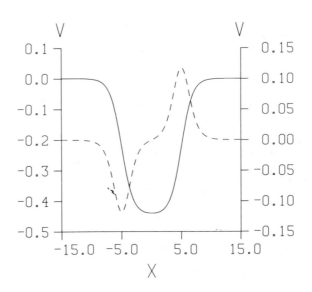

Figure 2. The effective potential V(X) (see Eq. (12)) and its derivative (dashed curve). (Taken from Ref. [2].)

Plots of v(x) along with the linear response χ_0 it generates are shown in Figure 1 (from Ref. [2]). The background response χ_0 is clearly localized near the perturbation. Figure 2 (from Ref. [2]) shows the effective potential V(X) experienced by the kink "centre-of-mass" in first order and since it is "attractive" the Newtonian kink should increase its velocity upon entering the perturbation region and then return to its original velocity upon leaving the region. The first-order motion for the kink position X is shown in Figure 3 (from Ref. [2]) and confirms this expectation. Several other types of perturbations in NKG systems are discussed in [15].

28.4 Kink-Antikink Collisions

In this section we briefly describe the successful application by Campbell et al. [3] of an approximate collective coordinate analysis to explain the appearance and nature of narrow resonance structure in the collisions of kinks in the ϕ^4 theory with their corresponding antikinks. A similar analysis has been applied to the resonance structure of kink-antikink collisions in a modified sine-Gordon model [18] and more recently to the double sine-Gordon model [19].

For the potential $U(\Phi) = (\Phi^2 - 1)^2/4$, the equation of motion (2) in the absence of external perturbations has the following static kink and antikink solutions:

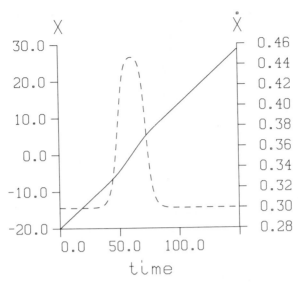

Figure 3. The first-order kink position and velocity (dashed curve) as a function of time during collision with the perturbation. (Taken from Ref. [2].)

$$\Phi_{K(\bar{K})} = \pm \tanh \left(\frac{x-x_0}{\sqrt{2}} \right),$$ (15)

where the + sign denotes the kink and the − sign denotes the antikink. The parameter x_0 specifies the centre of the kink (antikink) profile.

If a kink and an antikink are launched towards each other with an initial large separation, the topology of the double-well ϕ^4 potential prohibits them from passing completely through each other. The two basic final outcomes observed in numerical simulations [3] of such a collision are either that the kink and antikink "bounce" and hence reverse their direction (this is not perfectly elastic due to some "phonon" emission) or they collapse into a localized, oscillatory, quasi-bound state (in which the kinks are trapped by their mutual attraction) which ultimately decays by radiating "phonons". For the "bound" state to form, the "collision time" for the kinks must be long enough for them to shed their excess kinetic energy. Thus, one expects [3] trapping only at low relative velocity while at high velocity one expects (inelastic) reflection.

Detailed numerical simulations of kink-antikink collisions by Campbell et al. [3] are summarized in Figure 4 (taken from Ref. [3]), where the average final velocity [3], v_f, of the kinks is plotted versus their initial velocity, v_i. For low initial velocities ($v_i < 0.193$) the final velocity is always zero implying that the trapped state is formed and there are no kinks in the

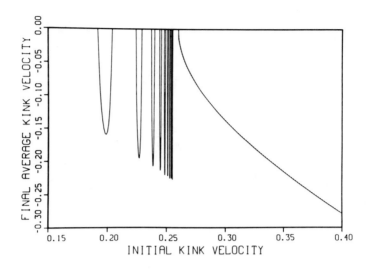

Figure 4. The final average kink velocity [Ref. 3] after a kink-antikink collision as a function of the initial kink velocity. A final velocity of zero implies mutual capture into a bound state. (Taken from Ref. [3]).

final asymptotic state. For high velocities ($v_i > v_c = 0.2598$) the kinks were found always to reflect. In between these two extremes is a sequence of regions called "reflection windows" [3] in which trapping and reflection alternate. To explain this remarkable phenomenon, Campbell et al. [3] first noted that the reflection windows involved the kinks bouncing not once, but twice, before escaping to infinity; after the first bounce the kinks recede to finite separation and then return to collide again. Between the two collisions some of the kinetic energy must be "stored" temporarily until it is released back to the translational motion of the kinks during the second bounce. Campbell et al. [3] proposed that the internal shape-oscillation mode of the ϕ^4 kink is the temporary repository of this energy. This is an additional "bound state" oscillation of the type considered in Section 28.2 above. The frequency of the shape mode is given by an eigenvalue of (4), namely $\omega_s = (3/2)^{1/2}$, and the corresponding eigenfunction is

$$f_s(x) \propto \tanh \frac{x-x_0}{\sqrt{2}} \operatorname{sech} \frac{x-x_0}{\sqrt{2}}. \tag{16}$$

The fact that this mode is localized about the kink (and a corresponding one is localized about the antikink) is the physical motivation for the hypothesis that it provides the collective degree of freedom in which to store the translational energy of the trapped kink and antikink in a two-bounce window. That such modes be candidates is suggested by the fact that for most of the time between bounces the kinks are sufficiently separated that the storage mode is likely to be found in the spectrum of localized oscillations about an isolated kink. The ϕ^4 kink has just one such mode as described above. Interestingly, other models [18,19] which lack an internal shape mode do not exhibit this reflection window phenomenon.

The scenario [3] is then that the first reflection sets up shape oscillations which carry more energy than the kinetic energy of the initially separated kink and antikink (the kinks gain kinetic energy as they approach since their interaction potential is attractive), and thus the kink and antikink are bound after the first collision. If the second collision occurs at just the right instant when the phase of the shape mode oscillation is appropriate, then the shape mode is extinguished and enough kinetic energy is restored that the kink and antikink are unbound. The condition for this timing to be just right has the form [3]

$$\omega_s T = \delta + 2\pi n, \tag{17}$$

where T is the time between the first and second bounce, δ is an offset phase, and n is an integer. Campbell et al. [3] developed a quantitative theory based on this scenario by treating the kink-antikink collisions using a phenomenological potential function $V(x_0)$ for their interaction which depends on their separation. This potential is obtained by integrating the energy density for the **Ansatz** function Φ_A appropriate to a static kink-antikink pair at large separation $2x_0$:

$$\Phi_A \equiv \Phi_K(x - x_0) + \Phi_{\bar{K}}(x + x_0) + 1. \tag{18}$$

Allowing x_0 to become time-dependent and integrating the energy density appropriate to the time-dependent **Ansatz** they obtain an effective Hamiltonian for $x_0(t)$. A detailed analysis [3] of the collisions based on this Hamiltonian gives good agreement with the numerical simulations [3]. Thus, by a judicious and physical choice of a collective coordinate (x_0 in this case) they were able to obtain semi-quantitative agreement with exact simulations, and more importantly, verify the simple physical hypothesis for the **cause** of the reflection windows. A more general **Ansatz** [3] which includes the width of the kinks as another dynamical variable is currently under investigation [20] with the objective of describing the energy transfer into and out of the shape mode in a more quantitative fashion.

28.5 Kink-Kink "Molecules"

In this section we briefly describe the recent work by Willis et al. [4], in which the internal oscillation of the "4π kink" in the double sine-Gordon model is viewed as a nonlinear vibration of two bound "2π kinks" against each other in a sort of kink-kink "molecule" [4]. The separation of the two "subkinks" is treated as what may be called [3] a **parametric** collective coordinate in contrast to the symmetry-related collective coordinate (single-kink position) discussed in Section 28.3, i.e. the parameter specifying the separation of the kinks is promoted to a dynamical variable.

The shape of the double sine-Gordon potential $U(\Phi)$ (see Section 28.2) is such that midway between two successive absolute minima (say, at -2π and 2π) there lies an elevated minimum (see Fig. 5) for $R > 1.25$. As the 4π-kink profile evolves from -2π to 2π, it "pauses" in the intermediate minimum at 0 so that the total profile looks the same as two 2π-kinks joined together. The degree of separation of the two subkinks in the static 4π-kink depends on the parameter R (see Fig. 5) in the potential U given above in Section 28.2. Indeed the analytic form of the full kink can be written as [4,19]

$$\phi(x, R, X) = \phi_{SG}[(x - X) + R] - \phi_{SG}[R - (x - X)], \quad (19)$$

where $\phi_{SG}(x) = 4 \tan^{-1} \exp(x)$ is the 2π sine-Gordon kink profile and X measures the centre of the 4π-kink.

Willis et al. [4] construct a canonical transformation, similar to that in [5-7], based on two collective coordinates, $X(t)$ and $R(t)$, and a "radiation" field similar to ψ in Section 28.3 above,

$$\Phi(x, t) = \phi(x, X(t), R(t)) + \psi(x, t), \quad (20)$$

together with the appropriate conjugate momenta. They require two

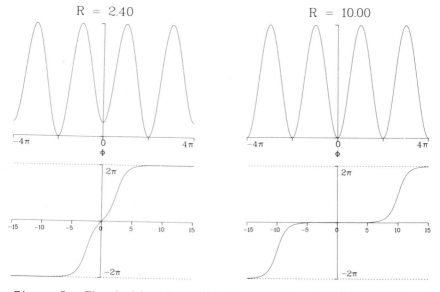

Figure 5. The double sine-Gordon potential (top plots) and kink solutions (bottom plots) for two values of R. (Taken from Ref. [19].)

additional constraints due to the extra variable R(t) and use
Dirac's method to obtain the Hamiltonian in terms of the new
variables and consequently the equations of motion which govern
them.   As a special case, they consider the vibrational motion of
the kink-kink molecule (with X = ψ = 0) and find an effective
potential well for the vibrational "collective" coordinate R(t).
The frequency of small oscillations agrees well with that obtained
in [19] and, more importantly, the general formalism sets the
stage for investigations of the influence of the "radiation" field
on the vibration of the molecule and also of the influence of
external perturbations.

28.6 Quantum Dynamics of Kinks

The problem of how to quantize the classical kink solutions
discussed above is well-documented in Refs. [5]-[10], where the
primary interest in kinks is as models of baryons.   One of the
features of the quantum theory which deserves attention is the
role of the self-coupling constant [5-8], g, in the nonlinear term
in the Lagrangian.   For example, the sine-Gordon potential is
written as U(Φ) = g^{-2}(1-cos[gΦ]).   Whereas this constant can be
scaled away in the classical limit (indeed, we already did so in
Section 28.2), this cannot be done in the quantum version of the
theory, and in fact, one must be careful to identify the proper
starting Lagrangian for the physical system of interest.   Usually,
one must appeal to a measured observable in order to fix the value
of g (with due regard for required renormalizations of g [5-8]).
In the "heavy kink" regime where the kink mass is large compared
to the "phonon mass" (i.e. when the kink energy is large compared
to the lowest-energy phonon) then the coupling constant is small
and a perturbation expansion [5-8] in powers of g proves to be
convenient.   The kink mass is of order g^{-2} and this is the
leading term in the total energy.   In order to account for kink
kinetic energy, it is necessary [5-8] to carry the expansion at
least to order g^2.   Since the phonon field $\psi(x,t)$ is of order $g^{1/2}$
[5-8], this means that one must include effects coming from ψ^4
terms as well as ψ^2 in order to be consistent.   This is an
important point and has to be kept in mind when considering the
quantum dynamics of kinks;   it is not consistent to treat the kink
motion without also including "anharmonicity" in the phonons.

The (renormalized) kink can be viewed [8] as a quantum particle
with due regard to the phonon influence and this means that
intuition regarding ordinary quantum dynamics of particles is
helpful in the problem of quantum kink dynamics as well.   We
should pause here to remark on why one should be interested in the
quantum motion of kinks in the first place.   The motivation lies
in the several contexts in condensed matter [1] where the kink is
essentially a one-dimensional object confined to a single chain of
atoms or molecules (or small group of chains) so that its energy
(and mass) are intrinsic quantities rather than extrinsic (such as
in the case of domain walls [1] where the energy scales with the
area of the wall).   In such systems, the kink mass may be small

enough to require consideration of the quantum aspect of its motion, i.e. when the de Broglie wavelength becomes comparable to the length scales of interest. One example is the conducting polymer polyacetylene [21] where the soliton mass is estimated to be only a few times the electron mass at most.

Several types of questions arise when considering the quantum dynamics of kinks in condensed matter contexts, where we must go beyond the description [5-8] of free-particle motion and include, for example, the influence of impurities and other defects which can cause kink scattering (or trapping in bound states). The quantum collective coordinate method based on the transformation given in Section 28.3 above is particularly useful in treating these types of problems since it forms a good meld with intuition regarding quantum dynamics. An effective potential similar to that in (12) appears in the lowest order Schrödinger equation treatment [22] and also in a more rigorous treatment [23] (through fourth order in the phonon field) employing a functional integral approach for the S matrix. This latter approach is basically a generalization of the work by Gervais et al. [5,6] to include the influence of an external perturbation of the type considered in Section 28.3 above. Thus, there are two small parameters, g and λ, which appear in a double perturbation expansion. The functional integral formalism provides a simple way to account for the phonon influence on, for example, the transition probability for a kink to tunnel through an impurity barrier. Since it is the kink which is of main interest, one can integrate out the phonon field (keeping fourth order terms via a cumulant expansion) to obtain a theory involving an effective action [5,6,23] which is a functional of the collective coordinate X alone. This action resembles that for a particle (the kink) (with phonon-induced renormalizations) in the presence of an effective potential barrier $V_{eff}(X)$. The WKB approximation can then be used, for example, to obtain the tunnelling probability for the kink. This and other problems involving kink dynamics are currently under investigation.

28.7 Summary and Discussion

In this brief introduction we have discussed some of the ways in which collective coordinate methods provide useful pictures for studying soliton dynamics, not only for perturbed systems but also for such problems as kink-antikink collisions and kink-kink "molecular" vibrations. These examples by no means exhaust the possibilities and indeed there are several other uses which we did not have space to discuss but which are nevertheless quite important. Of these we mention quantum statistical mechanics of kink-bearing systems [24,25], soliton modes in polymers [21,26-28], and polaron transport [29]. All of these areas will continue to reap benefits from the collective coordinate approach and one can expect to see advances in the treatment of several other problems as well. Among these is soliton Brownian motion in the presence of thermal noise [30] which is currently being studied via a Langevin approach [31] and a general Fokker-Planck approach [32] utilizing the canonical transformation described in

Section 28.3. Another general area of kink phenomena where the collective coordinate approach is expected to be quite fruitful is in the description of the interaction of external probes with kinks and in the identification of kink signatures in various signals. Yet another area deals with the role of kinks in the chaotic behaviour of driven systems (see articles by Bishop, McLaughlin and Overman in this volume). There is certainly a great deal left to do in exploiting this natural approach to the simplication of many-degree-of-freedom systems.

Acknowledgements

It is a pleasure to thank the several of my colleagues who have contributed to my (partial) understanding of collective coordinate methods. Among these I am especially indebted to B. Horovitz, J. Zmuidzinas, R. Flesch and D. Campbell. I also wish to acknowledge the USC Faculty Research and Innovation Fund and the U.S. Department of Energy for partial support of the work described in Section 28.3.

References

[1]   For a review of nonlinear Klein-Gordon kinks in condensed matter, see A.R. Bishop, J.A. Krumhansl and S.E. Trullinger (1980). *Physica* **1D**, 1.

[2]   R.J. Flesch, S.E. Trullinger and B. Horovitz (1987) in *Nonlinearity in Condensed Matter,* eds. A.R. Bishop, D.K. Campbell, P. Kumar and S.E. Trullinger, Springer Series in Solid State Sciences, Vol.69, Springer-Verlag, Berlin.

[3]   D.K. Campbell, J.F. Schonfeld and C.A. Wingate (1983). *Physica* **9D**, 1.

[4]   C.R. Willis, M. El-Batanouny and P. Sodano (1987) in *Nonlinearity in Condensed Matter,* eds. A.R. Bishop, D.K. Campbell, P. Kumar and S.E. Trullinger, Springer Series in Solid State Sciences, Vol. 69, Springer-Verlag, Berlin; C.R. Willis, M. El-Batanouny, S. Burdick, R. Boesch and P. Sodano (1987). *Phys. Rev.* **B35**, 3496.

[5]   J.L. Gervais, A. Jevicki and B. Sakita (1975). *Phys. Rev.* **D12**, 1038.

[6]   J.L. Gervais and A. Jevicki (1976). *Nucl. Phys.* **B110**, 93.

[7]   E. Tomboulis (1975). *Phys. Rev.* **D12**, 1678.

[8]   R. Jackiw (1977). *Rev. Mod. Phys.* **49**, 681.

[9]   O.A. Khrustalev, A.V. Razumov and A. Yu. Taranov (1980). *Nucl. Phys.* **B172**, 44.

[10]   K.A. Sveshnikov (1983). *Teor. Mat. Fiz.* **55**, 361.

[11]   N.N. Bogolubov (1950). *Ukr. Mat. Zh.* **2**, 3; (1970). *Selected Works,* pp. 499-520, Vol. 2 (in Russian), Naukova Dumka, Kiev.

[12]   N.N. Bogolubov and N.M. Krylov (1937). *Introduction to Nonlinear Mechanics* (in Russian), p. 328, Izd. Akad. Nauk. UkrSSR, Kiev.

[13]   M.B. Fogel, S.E. Trullinger, A.R. Bishop and J.A. Krumhansl (1976); *Phys. Rev. Lett.* **36**, 1411; (1977). *Phys. Rev.* **B15**, 1578. 1977.

[14]   See, for example, G.S. Guralnik, C.R. Hagen and T.W.B. Kibble (1968) in *Advances in Particle Physics,* eds. R.L. Cool and R.E. Marshak, Vol. 2, Wiley, New York.

[15]   B. Horovitz, R.J. Flesch and S.E. Trullinger, to be published.

[16]   P.A.M. Dirac (1964). *Lectures on Quantum Mechanics,* Academic, New York.

[17]   R.J. Flesch and S.E. Trullinger (1987). *J. Math. Phys.* **28**, 1619.

[18]   D.K. Campbell and M. Peyrard (1983). *Physica* **9D**, 33; (1986). *Physica* **18D**, 47.

[19]   D.K. Campbell, M. Peyrard and P. Sodano (1986). *Physica* **19D**, 165.

[20]   D.K. Campbell and R.J. Flesch, work in progress; see also S. Jeyadev and J.R. Schreiffer (1984). *Synth. Met.* **9**, 451.

[21]   See, for example, S. Kivelson (1986) in *Solitons,* eds. S.E. Trullinger, V.E. Kzaharov and V.L. Pokrovsky, *Modern Problems in Condensed Matter Sciences Series,* Vol. 17, Ch. 6, North-Holland, Amsterdam.

[22]   S.E. Trullinger (1979). *Sol. St. Commun.* **29**, 27.

[23]   J.S. Zmuidzinas, B. Horovitz and S.E. Trullinger, unpublished.

[24]   T. Miyashita and K. Maki (1983). *Phys. Rev.* **B28**, 6733; (1985) **B31**, 1836, and references therein.

[25]   F.G. Mertens and H. Büttner (1986) in *Solitons,* eds S.E. Trullinger, V.E. Zakharov and V.L. Pokrovsky, *Modern Problems in Condensed Matter Sciences Series,* Vol. 17, Ch. 15, North-Holland, Amsterdam.

[26]   M.J. Rice and E.J. Mele (1980). *Sol. St. Commun.* **35**, 487; see also M.J. Rice (1983). *Phys. Rev.* B28, 3587.

[27]   M.J. Rice, S. Jeyadev and S.R. Phillpot (1987). *Phys. Rev.* **B34**, 5583.

[28]   A.R. Bishop, D.K. Campbell, P.S. Lomdahl, B. Horovitz and S.R. Phillpot (1984). *Synth. Met.* **9**, 223.

[29]   T. Holstein (1981). *Mol. Cryst. & Liq. Cryst.* **77**, 235; H.-B. Schüttler and T. Holstein (1983). *Phys. Rev. Lett.* **51**, 2337.

[30]   S.E. Trullinger, M.D. Miller, R.A. Guyer, A.R. Bishop, F. Palmer and J.A. Krumhansl (1978). *Phys. Rev. Lett.* **40**, 206, 1063(E); D.J. Kaup (1984). *Phys. Rev.* **B29**, 1072.

[31]   B. Horovitz, unpublished.

[32]   R.J. Flesch and S.E. Trullinger (1987). *Bull. Am. Phys. Soc.* **32**, 673, and to be published.

SPIRAL WAVES IN THE BELOUSOV-ZHABOTINSKII REACTION

29.1 Introduction

The Belousov-Zhabotinskii reaction involves the oxidation of certain carboxylic acids by bromate ions in the presence of a suitable transition metal ion catalyst.   In the early 1950s B.P. Belousov was studying this reaction as an analog of the oxidative decarboxylation of organic acids in living cells when he discovered that the reaction oscillates back and forth between oxidized and reduced states for many cycles.   Despite the easy reproducibility of these oscillations and the careful mechanistic studies carried out by Belousov, he was never able to have his work published in a refereed journal because at that time editors and reviewers were convinced that sustained oscillations never occurred in chemical systems [1].

In the early 1960s A.M. Zhabotinskii was given Belousov's recipe by his thesis advisor, S.E. Schnol'.   Picking up where Belousov left off, Zhabotinskii delved into the mechanism and generalizability of the reaction.   His first publications on this subject appeared in 1964 [2], and news of the reaction reached the West in 1968 when Zhabotinskii reported his work at an international conference in Prague [3].   At this time chemists in Europe and America readily received news of a simple homogeneous chemical oscillator because the theoretical studies of Prigogine, Lefever and Nicolis [4] had begun to break down the barriers of suspicion that surrounded the phenomenon of chemical oscillations.

It was soon discovered by Zhabotinskii, Winfree and others that, in addition to sustained oscillations in well-stirred reaction mixtures, the BZ reaction (as it was becoming known) would also support spatial waves of oxidation which propagate through thin unstirred layers of reagent [5,6].   A thin layer need not be spontaneously oscillatory to support oxidation waves.   Indeed, the non-oscillatory, excitable mixture of Winfree [6] is particularly convenient for studying chemical waves.   When carefully prepared, the medium will remain (for a long time) uniformly in a reduced state, but if perturbed sufficiently, a single circular wave of oxidation will propagate away from the point of perturbation until it collides with the boundary of the dish and disappears.   If less care is taken in preparing the mixture, expanding concentric waves are emitted periodically from pacemakers distributed randomly in space.   The pacemakers are apparently small inhomogeneities (dust particles?) that push the reagent locally into an oscillatory mode. If an expanding circular wave is broken, the loose ends will wind around into a pair of counter-rotating spiral waves [6,7].

From even a cursory glance at a collection of target patterns (expanding concentric waves) in a Petri dish, it is immediately apparent that they come with a variety of spatial wavelengths. Since all the waves move at roughly the same speed (approx. 5mm/min), the variation in wavelength reflects a variation in temporal period of oscillation at the pacemaker. On the other hand, all spiral waves in a certain dish rotate with the same frequency and present to the viewer spirals of identical pitch.

A fundamental qualitative question of pattern formation in the BZ reaction is why target patterns come in many frequencies but spirals in only one. A truly satisfactory theory should also account quantitatively for the observed unique period of rotation of spiral waves in terms of the known mechanism of the BZ reaction [8,9].

29.2 Excitability and Travelling Waves

A quantitative theory of spiral waves in two-dimensional excitable media can be developed by singular perturbation methods. Pioneering work on this problem was carried out by Mikhailov and Krinskii [10] and by Fife [11]. Here we summarise the theory developed by Keener [12] and applied specifically to the BZ reaction by Keener and Tyson [13].

First, consider the behaviour of the BZ reaction in zero spatial dimension, i.e. in a well-stirred beaker. Under non-oscillatory, excitable conditions, there exists a locally stable, time-independent solution to the kinetic equations. In this resting state the transition metal ion is reduced (red in case of Fe^{2+}) and the key reaction intermediate, bromous acid ($HBrO_2$), is present in very low concentration. Small perturbations from the resting state are damped out, but a sufficiently large boost in bromous acid concentration triggers the autocatalytic production of $HBrO_2$. High levels of bromous acid cause the transition metal ion to be oxidized (red → blue) and subsequently bromous acid is destroyed by rapid disproportionation. At low bromous acid concentrations the transition metal ion recovers its reduced state (blue → red) by oxidizing various organic acids in the mixture. This sequence of events is typical of excitable kinetics.

If we string together excitable elements in a one-dimensional continuum, we can observe waves of excitation propagating along the spatial dimension at a speed characteristic of the medium. These waves of constant shape and velocity have been known for many years [14]; indeed, they were first described over eighty years ago by Luther for oxidation-reduction reactions [15].

In the BZ reaction, waves are driven through space by the autocatalytic production of bromous acid in the wavefront and the diffusion of $HBrO_2$ to regions of low bromous acid concentration ahead of the wave front, pushing these regions across the excitation threshold and triggering the autocatalytic reaction there. Behind the wave front the transition metal ion is oxidized, bromous acid is removed by disproportionation, and finally the transition metal ion is converted slowly back to the reduced state (the "recovery" phase).

As the medium recovers excitability, it may be challenged to propagate another oxidation wave. If sufficient time elapses between the first and second waves to allow nearly full reduction of the transition metal ion, then the second wave will travel just as fast as the first wave. But if the time between waves is short and the medium is not fully recovered, then the second wave will travel more slowly than the first because the medium is less excitable. If we imagine a periodic train of waves travelling along a one-dimensional continuum, it is clear that wave speed will depend upon the temporal spacing of the waves and that (generally speaking) wave speed will decrease as temporal period decreases. The dependence of wave speed on period, $c = \sigma(T)$, is called the dispersion relation of the medium.

29.3 Travelling Waves in Two Dimensions

In two spatial dimensions we can imagine a train of periodic plane waves travelling through the medium, and for reasons of symmetry such waves must satisfy the dispersion relation just described for a one-dimensional medium. Not so obviously, though, expanding target patterns and rotating spiral waves must also satisfy the dispersion relation, $c = \sigma(T)$, when we interpret c as the normal velocity of waves far from the centre of the pattern. This constraint on periodic patterns follows because, far from the centre, the waves have negligible curvature and move through space like periodic plane waves.

Apparently target patterns need **only** satisfy the dispersion relation. Depending upon the precise chemical properties of the inhomogeneity at the centre of a pattern, targets can arise with any temporal period (above a certain minimum value), and the dispersion relation then determines the asymptotic speed of propagation. The wavelength of the target pattern sufficiently far from the centre is $\lambda = cT = T\sigma(T)$. In this way we produce a one-parameter family of target patterns.

For spiral patterns to be uniquely determined, they must be constrained by a second relationship between c and T. The second relation is implied by geometric constraints on the propagation of waves in two spatial dimensions. Consider an arbitrary wave front in two dimensions. By transforming to a coordinate system travelling with the wave front, Keener [12] has shown that the normal velocity N of the wavefront is equal to the velocity c of plane waves decremented (or incremented) by an amount proportional to the curvature K of the front:

$$N = c - DK \qquad (1)$$

where D is the diffusion coefficient of the "autocatalytic" species.

Now, consider a rigidly rotating spiral wave, the parametric equations for which are

$$x = r \cos[\theta(r) - \omega t]$$

$$y = r \sin[\theta(r) - \omega t]$$

Here $\theta(r)$ determines the shape of the spiral (at fixed t) and ω is the angular frequency of rotation ($\omega = 2\pi/T$). Our problem is to determine both $\theta(r)$ and ω. Since N depends on $\theta'(r)$ and ω, and K depends on $\theta'(r)$ and $\theta''(r)$, equation (1) is really an ordinary differential equation for the unknown function $\theta(r)$ in terms of two parameters c and ω. Applying end conditions (at r = 0 and r → ∞, say) to this ODE, we obtain a typical eigenvalue problem which determines a unique c for each value of $\omega = 2\pi/T$. A rough-and-ready approximation to this "curvature" constraint is

$$c = \sqrt{6\pi D/T}. \tag{2}$$

Notice that the curvature relation depends on the diffusion coefficient of the "autocatalytic" species but is **independent** (to a first approximation) of details of the kinetic mechanism of excitability.

Spiral waves, we contend, lie at the intersection of the dispersion relation, $c = \sigma(T)$, and the curvature relation (2). To test this idea, we must know the dispersion curve for the BZ reaction. Dockery, Keener and Tyson [16] have calculated the dispersion relation for the Oregonator model [9] of the BZ reaction. All the parameters in this model are known from experiments independent of measurements on propagating waves [17]. Nonetheless, with no free parameters, the dispersion relation implied by the Oregonator agrees remarkably well with Winfree's observations of dispersion in the BZ reaction (Fig. 1). Furthermore, the predicted location of spiral waves at the intersection of the dispersion relation and the curvature relation, is very close to the observed location of spiral waves in the c-T plane.

29.4 Spiral Waves in Other Contexts

The theory outlined here should apply to spiral waves in other media that show excitability similar to the BZ reaction. For instance, a close analogy exists between chemical waves in the BZ reaction and waves of biological activity during the aggregation phase of the life cycle of the cellular slime mould, *Dictyostelium discoideum* [18]. Starving cells signal each other and move toward the centre of an aggregation domain to form a multicellular slug. This activity is organized either in expanding concentric rings (with variable period) or in rotating spirals (with unique period). The travelling waves are driven by the production, diffusion, and degradation of cyclic AMP in a manner completely analogous to the role of bromous acid in the BZ reaction. Since cyclic AMP and bromous acid have approximately the same diffusion coefficient, spiral waves in both contexts should satisfy the same

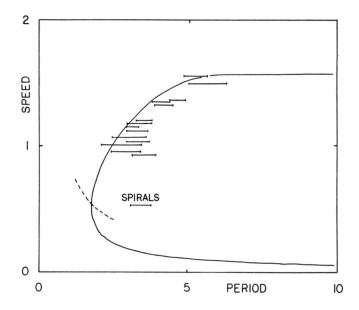

Figure 1.   Dispersion relation (solid line) and curvature relation (dashed line) (from Ref. [13], used by permission). Speed and period are given in dimensionless terms [13].   The dispersion relation was calculated by Dockery et al. [16] for Oregonator kinetics.   The curvature relation is given approximately by Eq. (2); for corrections to (2) see Ref. [13]. The unmarked bars show wave speeds as a function of time interval between waves, for plane waves propagating inside a capillary tube filled with excitable BZ reagent (A.T. Winfree, unpublished observations).   The bar marked "SPIRALS" shows the asymptotic wave speed and rotation period of spiral waves in this reagent.   According to theory, plane waves should be distributed along the dispersion curve, and spiral waves should be found only at the intersection of the dispersion relation and the curvature relation.

curvature constraint (2) with $D = 10^{-5} cm^2 s^{-1}$.   As Figure 2 shows, this theoretical expectation is roughly confirmed by experimental observations.

Heart tissue is an excitable medium that supports waves of neuromuscular activity [19].   The normal ventricular contraction, which originates at the Purkinje fibres and spreads wavelike across the ventricle, is analogous to an expanding concentric wave in the BZ reaction.   Some abnormal beats, such as flutter and fibrillation, are thought by many to arise from one or more high frequency spiral waves rotating in the heart tissue.

Neural tissue, such as the retina or regions of the cerebral cortex, can also be considered as a two-dimensional array of spatially coupled, excitable elements [20].   Waves of neural activity in such tissues have been measured by many investigators, and there is even evidence for rotating spiral waves in preparations of rabbit brain and chicken retina.

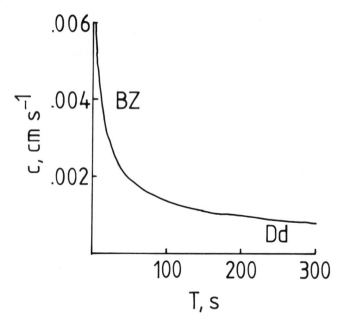

Figure 2.   The curvature constraint (2) for $D = 10^{-5} cm^2 s^{-1}$.   The approximate locations of spiral waves in the BZ reaction and in Dictyostelium aggregation are indicated by the symbols "BZ" and "Dd" respectively.

Finally, spiral galaxies are undoubtedly the grandest and loveliest examples of rotating spiral waves [21].   Though there are important differences between spiral galaxies and spiral chemical waves (in particular, the differential rotation of the material composing spiral galaxies has no counterpart in a fluid motion of the liquid chemical medium), there are striking similarities in terms of excitation and recovery.   At the leading edge of an arm of a spiral galaxy, new stars are being born by gravitational collapse of cool molecular clouds.   Behind the leading edge stars are ageing, dying and returning their material to dust clouds.   At first the material in these clouds is too hot for new stars to nucleate, but after a cooling-down (recovery) phase the dust clouds regain their excitability and another wavefront of star formation can pass.

Acknowledgement

This work was supported in part by the National Science Foundation (Grant DMS 85-18367).

References

[1]   A.T. Winfree (1984).   *J. Chem. Educ.* **61**, 661.
[2]   A.M. Zhabotinskii (1964).   *Proc. Acad. Sci. USSR (Phys. Chem. Sect.)* **157**, 701;   (1964).   *Biophysics* **9**, 329.
[3]   B. Chance, E.K. Pye, A.K. Ghosh and B. Hess (1973).

Biological and Biochemical Oscillators, Academic Press, New York.

[4]   I. Prigogine and R. Lefever (1968). *J. Chem. Phys.* **48**, 1695; R. Lefever (1969). *J. Chem. Phys.* 49, 4977; G. Nicolis (1971). *Adv. Chem. Phys.* **19**, 209.

[5]   A. N. Zaikin and A. M. Zhabotinskii (1973). *Nature* **225**, 535; (1973). *J. Theor. Biol.* **40**, 45.

[6]   A. T. Winfree (1972). *Science* **175**, 634.

[7]   A. T. Winfree (1974). *Scientific American* **230**, 82, June.

[8]   R. J. Field, E. Körös and R. M. Noyes (1972). *J. Am. Chem. Soc.* **94**, 8649.

[9]   R. J. Field and R. M. Noyes (1974). *J. Chem. Phys.* **60**, 1877.

[10]   A. S. Mikhailov and V. I. Krinskii (1983). *Physica* **9D**, 346.

[11]   P. C. Fife (1984) in *Non-Equilibrium Dynamics in Chemical Systems,* eds. C. Vidal and A. Pacault, p. 76, Springer-Verlag, Berlin.

[12]   J. P. Keener (1986). *SIAM J. Appl. Math.* **46**. 1039.

[13]   J. P. Keener and J. J. Tyson (1986). *Physica* **21D**, 307.

[14]   R. A. Fisher (1937). *Ann. Eugen.* **7**, 355; A. Kolmogorov, I. Petrovsky and N. Piscounoff (1937). *Bull. Univ. Moscow, Ser. Int. Sect. A.* **1**, 1; D. G. Aronson and H. F. Weinberger (1975). *Lect. Notes Math.* **446**, 5.

[15]   R. Luther (1906). *Z. Electrochemie* **12**, 596.

[16]   J. D. Dockery, J. P. Keener and J. J. Tyson, *Physica D* in press.

[17]   J. J. Tyson (1985) in *Oscillations and Travelling Waves in Chemical Systems,* eds. R. J. Field and M. Burger, p. 93, Wiley, New York; R. J. Field and H. D. Försterling (1986). *J. Phys. Chem.* **90**, 5400.

[18]   A. J. Durston (1973). *J. Theor. Biol.* **42**, 483; F. Alcantara and M. Monk (1974). *J. Gen. Microbiol.* **85**, 321; J. D. Gross, M. J. Peacy and D. J. Trevan (1976). *J. Cell. Sci.* **22**, 645; K. J. Tomchik and P. N. Devreotes (1981). *Science* **212**, 443.

[19]   G. R. Mines (1914). *Trans. R. Soc. Can.* **4**, 43; I. S. Balakhovskii (1965). *Biophysics* **10**, 1175; M. A. Allessie, F. I. M. Bonke and F. J. G. Schopman (1973). *Circ. Res.* **33**, 54; S. Dillon and M. Morad (1981). *Science* **214**, 453.

[20]   M. Shibata and J. Bures (1973). *J. Neurophysiol.* **35**, 381; H. Petsche, O. Prohaska, P. Rappelsberger, R. Vollmer and A. Kaiser (1974). *Epilepsia* **15**, 439; N. A. Goroleva and J. Bures (1983). *J. Neurbiol.* **14**, 353.

[21]   A. Toomre (1977). *Ann. Rev. Astron. Astrophys.* **15**, 437; J. V. Feitzinger, J. Spicker and J. A. Stüwe (1985). *Umschau* **12**, 750; L. S. Schulman and P. E. Seiden (1986). *Science* **233**, 425.

COMPLEMENTARY MODELLING, VISUALIZATION AND NUMERICAL DIAGNOSTICS

Continuum nonlinear dynamical systems in more than one space dimension are almost all mathematically intractable. For example, the generic equations of fluid dynamics, the incompressible Euler equations in two dimensions,

$$d_t\omega \equiv \omega_t - \psi_y\omega_x + \psi_x\omega_y = 0, \quad \Delta\psi = -\omega, \quad (x,y)\in IR^2,$$

is an example with *no* time-dependent analytical solutions if the initial condition is two or more localized vorticity distributions, $\omega(x,y,0)$.

A major goal of the study of nonlinear dynamical systems is an analytical understanding that will allow accurate predictions of state variables or functions of state variables over moderate-to-long times. To accomplish this objective, it is necessary to use more than one model to represent complementary aspects of a dynamical process. The results of simulations with these models must be visualized and quantitative feature extraction must be performed using numerical diagnostic algorithms. This permits the judicious investigator to "separate" complex interactions among coherent structures into simpler and hopefully analytically tractable parts. This synergetic approach was used in the discovery of the soliton [1,2].

The essential problems in inviscid, incompressible two-dimensional hydrodynamics were not posed until computer simulations showed: **axisymmetrization** of noncircular distributions of vorticity; **merger** of like-signed vorticity regions; **binding** of opposite-signed vorticity regions; and **entrainment** in a host vortex of small regions of irrotational fluid or opposite-signed vorticity. In all these processes one also observes **gradient intensification** and **filamentation** of vorticity due to relative transport of vorticity. Three key questions in evolving inviscid or nearly-inviscid two-dimensional flows are:

1.   What are the mechanisms by which smooth vorticity distributions aggregate or "condense" into near-circular regions of vorticity (that is the mechanisms of axisymmetrization, gradient intensification and filamentation)?

2.   What are the mechanisms by which two or more like-signed regions of vorticity merge or oppposite-signed regions of vorticity bind or entrain?

3.   How do the filaments (small scale and/or low-lying vorticity regions) which arise in axisymmetrization, merger and binding affect the long time evolution of the large-scale structures?

Axisymmetrization, merger and binding may be considered fundamental physical space interactions in two-dimensional turbulent flow [3,4].   In spectral jargon, merger corresponds to an "upward" energy cascade and filamentation accounts for the "downward" enstrophy cascade.   The filamentation of vorticity during axisymmetrization, merger and binding is best understood by considering the local corotating or cotranslating streamfunctions.

In a recent review, Melander, Overman and Zabusky [5] discussed computational and mathematical vortex dynamics emphasizing two-dimensional aspects.   Here we focus briefly on complementary models and numerical diagnostics in this field.   We recall that in 1888 Bassett noted, "If therefore more than one vortex exists in the fluid the effect of any one of the vortices upon the others will be to produce a motion of translation combined with a deformation of their cross sections.   The mathematical

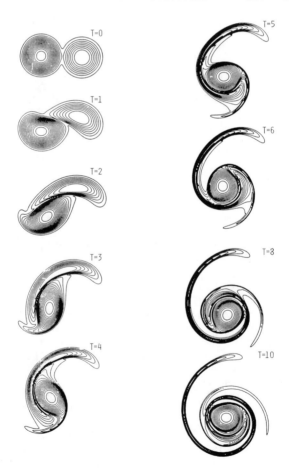

Figure 1(a).

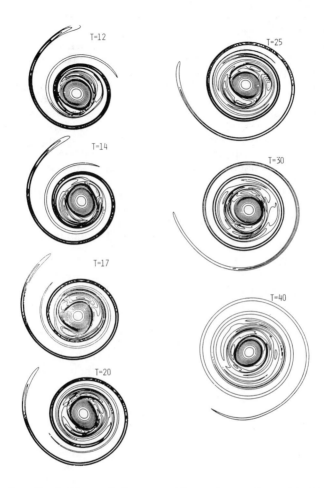

Figure 1. Evolution of two smooth regions of vorticity obtained with a pseudospectral code (with ν_4 hyperviscosity) on a 256^2 mesh. The left vortex (which is victorious) has max $\omega_2 = 10$, and the right vortex has max $\omega_1 = 5$, such that $\Gamma_2/\Gamma_1 = 2.0$ and $A_2/A_1 = 1.0$, which corresponds to the filled cirlce in Figure 2. Although the initial state is resolved, the convection of vorticity leads to steep vortex gradients and fine vortex filaments.

difficulties of solving this problem (evolution of interacting localized vortex regions in two dimensions) when the initial distribution of the vortices and the initial forms of their cross sections are given are very great; and it seems impossible in the present state of analysis to do more than obtain approximate solutions in certain special cases" [6]. Beginning in the early 1970's finite-difference, spectral, pseudospectral and vortex-in-cell models and algorithms became very popular. The first three must include a dissipative mechanism to resolve small-scale structures. However, if gradient-scale lengths form which are smaller than five lattice intervals then significant truncation and aliasing errors may arise. Thus, none of these schemes are capable of a general study of the Euler equations or even of "very high" Reynolds number flows at "moderate" times, despite the increasing availability of supercomputer resources. However, aspects of these flows can be better studied with the contour dynamical algorithms [7,8,9,10] and the moment model [11].

Contour dynamics, a generalization of the "waterbag" model [12] is a free boundary-integral evolutionary method that is ideally suited for incompressible, inviscid or nearly-inviscid two-dimensional flows. The countours are the boundaries of **constant density** that are the sources of the flow; e.g. of constant vorticity regions in the homogeneous Euler equations and of constant mass density regions in the stratified Euler equations. Generally, the velocities of the contours are obtained from integrals (homogeneous Euler) or integral equations (stratified Euler, etc.) on the contours. Thus, in CD the evolution of plane curves describes the nonlocal and nonlinear dynamics of two-dimensional fluid and plasma systems. It is a natural technique for flows in unbounded media because the Green's function has a simple form. At preintermediate [13] and especially long [14] times, one must utilize topological-change or "surgery" algorithms which interconnect and clip contours. In effect, these introduce a smallest scale into the problem. Dritschel has automated this process with a robust algorithm which he calls "contour surgery" [14].

The moment model is derived by assuming that the vortex regions which create the flow are well-separated. A moment representation for each region is introduced and truncated after second moments (elliptical representation) and thus one obtains two additional degrees of freedom for each centroid (since local circulation is conserved). Although an asymptotic description, the model works well for closely interacting vortices and resolves the initial stages of vortex merger. In particular, the model becomes integrable when applied to the symmetric merger of two identical vortex regions, and yields explicit necessary-and-sufficient conditions for merger [15]. Although the moment model is derived under the assumption of **uniform** vorticity, the conditions for merger are in agreement with our high-resolution spectral simulations containing initially **smooth** circular vortex distributions. This indicates a certain degree of universality of the merger conditions.

Recently, we examined the problem of **asymmetric** merger, a more realistic problem, and thereby one richer in prameters. One may ask: which vortex "core" will be the **victor** in a merger? [16,17] Figure 1 shows vorticity contours during the evolution of two initial circular smooth regions of vorticity, obtained with a pseudospectral code. (See references [4] and [15] for details on initial conditions and methods.) Here the left vortex region dominates for it shears out the right region. In the process of doing this: vorticity gradients are enhanced (in fact, gradients are limited by numerical resolution); an intense and a weak

filament are beginning to form at t = 4;   a pocket of irrotational
fluid is formed between 10 < t < 12;   a nearly three-fold inner
state is evident at t = 17;   and at t = 40 one sees a near
axisymmetric core surrounded by a near-spiral of vorticity.   The
complexity of the processes involved are best visualized by colour
cinemas, a sample of which is given in the colour plates.   Plate
shows three successive zooms on the vorticity at t = 17.   In the
upper right is the total field and in the upper left a 4x
magnification.   The near three-fold symmetric core is evident.   If
we zoom in on the region (in the square) containing the
irrotational pocket, we obtain the background figure on the plate.
Plate shows the vorticity transport term **u**. $\nabla\omega$ at t = 4 and t =
17.   The latter shows a zoom around the central region and the
near 3-fold symmetric structure (red regions).

A cinematic view enhances the discovery and appreciation of new
closely-coupled dynamical processes.   This "easy" recognition of
new phenomena is usually not available when viewing still colour
plates.   Also, choosing the most "informative" colour map is
presently a trial-and-error process and pleasing results are

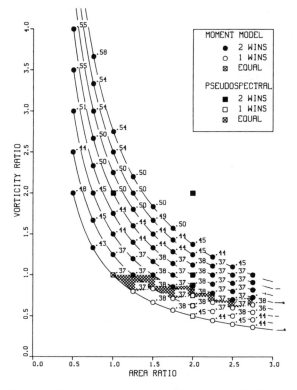

*Figure 2.   Asymmetric vortex merger diagram.   Each point on this
figure is obtained from ten simulations of the moment model
where the parameters are fixed and the intercentroid distance
varies.   If the initial state corresponds to points above (or
below) the cross-hatched line, then the No. 2 (or No. 1) vortex
dominates.*

obtained by those with an artistic sensibility. We must advocate procedures for archiving these cinematic results in an accessible and inexpensive manner.

It would be terribly expensive to answer the question of the vortex victor with a pseudospectral code or even with a contour dynamical code. However, the moment model can give excellent insights into the victory process as shown in Figure 2. (Each point on Figure 2 was determined by ten moment model simulations with eight ordinary differential equations.) Here the vorticity ratio $\gamma = (\omega_2/\omega_1)$ is plotted against the area ratio $(\alpha = A_2/A_1)$ where the No. 1 vortex is to the right of the No. 2 vortex. Each curve corresponds to $\Gamma_2/\Gamma_1 = \gamma\alpha = $ constant. (Symmetric merger corresponds to $\Gamma_2/\Gamma_1 = 1.0$.) The cross-hatched band shows the transitional region below which No. 1 dominates and above which No. 2 dominates. All pseudospectral simulations performed gave "victory" results in agreement with the moment model.

In all these problems, a better analytical understanding awaits more simulations with information-laden numerical diagnostics obtained in an interactive visualization and numerical diagnostic environment [18,19]. In communicating new research results, the computational mathematician is obliged to provide quantitative and cogent diagnostics so that the reader can assess the validity, generality and accuracy of his claims. Rigorous standards in this area of validation through multi-model simulations and asymptotical analysis are beginning to be formulated.

Acknowledgements

All the results on axisymmetrization, symmetric and asymmetric merger, and the moment model were made in close collaboration with Dr. Mogens V. Melander. The latter's creative computational and mathematical work is responsible for much of our present state of understanding. This work was supported by the Office of Naval Research and the U.S. Army Research office. The coloured plates were produced in our Interactive Graphics and Numerical Diagnostics Laboratory on equipment provided by SUN Microsystems, Inc. and the National Science Foundation.

References

[1]   N.J. Zabusky (1981). *J. Comput. Phys.* **43**, 195.
[2]   N.J. Zabusky (1984). *Phys. Today* July 36–45.
[3]   J.C. McWilliams (1984). *J. Fluid Mech.* **146**, 21.
[4]   M.V. Melander, J..C. McWilliams and N.J. Zabusky (1987). *J. Fluid Mech.* **178**, 137.
[5]   M.V. Melander, E.A. Overman and N.J. Zabusky (1987). *Applied Num. Math.* **3**, 59.
[6]   A.B. Basset. *A Treatise on Hydrodynamics, II,* p. 42, Dover.
[7]   N.J. Zabusky, M.H. Hughes and K.V. Roberts (1979). *J. Comput. Phys.* **30**, 96.
[8]   G.S. Deem and N.J. Zabusky (1978) in *Solitons in Action,* eds. K. Lonngren and A. Scott, p. 277, Academic Press, New York.

[9]   N.J. Zabusky (1981).   *Ann. New York Acad. Sci.* **373**, 160.

[10]   H.M. Wu, E.A. Overman, II and N.J. Zabusky (1984).   *J. Comput. Phys.* **53**, 42.

[11]   M.V. Melander, N.J. Zabusky and A.S. Styczek (1986).   *J. Fluid Mech.* **167**, 95.

[12]   H. Berk and K.V. Roberts (1970).   *Methods of Computational Physics* **9**, 87.

[13]   E.A. Overman and N.J. Zabusky (1982).   *Phys. Fluids* **25**, 1297.

[14]   D.G. Dritschel (1988).   *J. Comput. Phys., in press.*

[15]   M.V. Melander, N.J. Zabusky and J.C. McWilliams (1987).   *J. Fluid Mech.*, in press.

[16]   M.V. Melander, N.J. Zabusky and J.C. McWilliams (1987). *Phys. Fluids* **30**, 2610.

[17]   C. Seren, M.V. Melander and N.J. Zabusky (1987).   *Phys. Fluids* **30**, 2604.

[18]   N.J. Zabusky (1986).   *Physica* **18D**, 15.

[19]   K.-H. Winkler et al. (1987).   *Phys. Today*, 28-39, October.

AN APPLICATION OF SHILNIKOV'S THEOREM TO LINEAR SYSTEMS WITH PIECEWISE LINEAR FEEDBACK

Complicated homoclinic trajectories leading to different types of Shilnikov bifurcations may exist in a very simple system.

In many recent papers hopes for understanding the onset of chaotic oscillations by means of Shilnikov's theorem have been expressed.

31.1 Shilnikov's Theorem [3], [6], [7], [9], [10]

$$\dot{x}(t) = g(x(t), \mathcal{H}), \qquad (1)$$

where g: R^3 x R → R^3, g (0, $\mathcal{H}$) = 0 for every $\mathcal{H}$ from some neighbourhood of 0 and g is such that the Hartman-Grobman linearization [11] holds for (1) at 0. Suppose also that the Jacobi matrix at 0, i.e. $[\frac{\partial g}{\partial x}(x, \mathcal{H})]_{x=0}$ has for $\mathcal{H}$ = 0 the eigenvalues γ, $\alpha \pm i\beta$, where γ, α, β, $\in$ R, $\beta > 0$, α and γ of opposite signs. Finally, assume that for $\mathcal{H}$ = 0 system (1) has a homoclinic orbit (HO) with respect to 0∈R^3.

(i)   If $|\frac{\gamma}{\alpha}| > 1$, then in a neighbourhood of HO the Poincaré return map for (1) with $\mathcal{H}$ = 0 has a countable set of Smale's horseshoes. For all sufficiently small $|\mathcal{H}|$ these horseshoes still exist since they are structurally stable.

(ii)   If $|\frac{\gamma}{\alpha}| < 1$, then in an appropriately chosen neighbourhood of HO there is no periodic orbit of (1) with $\mathcal{H}$ = 0, but for sufficiently small $|\mathcal{H}|$, $\mathcal{H} \neq 0$ an exactly one periodic orbit is generated from the disappearing HO, stable if $\gamma > 0$ and unstable if $\gamma < 0$.

The properties of the Smale's horseshoe suggest that its existence is close, but not equivalent, to the existence of a strange attractor (compare properties of the Smale's horseshoe discussed below with definition of the strange attractor).

The existence of strange attractor is strictly connected with the existence of transversal HO but it need not be the HO described in (i). In fact, in [8] the generation of a Smale's horseshoe from the HO with respect to a periodic orbit is proved. This periodic orbit may be, for example, the one described in (ii) for $|\mathcal{H}|$ large enough (after its possible destabilization). It seems, however, that the practical importance of Shilnikov's theorem relies on the possibility of better arrangement of physical and/or computer experiments, towards the exhibition of strange attractors, by contraction of "suspected" ranges of the bifurcation parameter $\mathcal{H}$.

31.2   In this paper we shall apply Shilnikov's theorem to the following linear system with piecewise-linear element in the feedback loop,

$$\begin{bmatrix} \dot{x} \\ \dot{y} \\ \dot{z} \end{bmatrix} = \begin{bmatrix} 0 & 1 & 0 \\ 0 & 0 & 1 \\ -1 & -B & -A \end{bmatrix} \begin{bmatrix} x \\ y \\ z \end{bmatrix} + f(x) . \begin{bmatrix} 0 \\ 0 \\ 1 \end{bmatrix} , \quad A>0, \ AB>1, \tag{2}$$

$$f(x) = \{ \begin{matrix} -m_0 x & x \leq x_0 \\ m(\lambda)x+n(\lambda), & x \geq x_0 \end{matrix} \}, \quad m_0, x_0 \text{ are fixed constants, } x_0$$

> 0, $m(\lambda) = \lambda^3 + A\lambda^2 + B\lambda + 1$, $n/\lambda) = -m_0 x_0 - x_0 m(\lambda)$, λ is a bifurcation parameter. System (2) describes a class of 3rd order oscillators (see below for biological and electrical examples). Another motivation for the study of the system (2) follows from the fact that if f is an arbitrary locally-Lipschitz function, satisfying the sector conditions: $1 - AB < \dfrac{f(x)}{x} < 1 \ \forall \ x \neq 0$, $f(0)$ = 0, then (2) has the globally asymptotically stable equilibrium point at $0 \in R^3$. Shilnikov's theorem enables us to find an example of a piecewise-linear function f which violates the sector conditions and gives rise to a system with periodic orbits and/or very complicated trajectories. Two classes of systems (2) will be considered:

Class I
$-m_0 < -AB+1$, $\lambda \in \{z \in R: z>0, 3z^2 + 2Az + (4B-A^2)>0\}$, $\tag{3}$

Class II
$-m_0 > 1$, $\lambda < -A$, $A^2 - 4B < 0$. $\tag{4}$

For both classes (2) has two equilibrium points $0, N = (\dfrac{n}{1-m}, 0, 0) \in R^3$. The system linearized at 0 has eigenvalues γ, $\alpha \pm i\beta$, where

$$\alpha = -0.5 \ (A + \gamma), \ \beta = 0.5 \ \sqrt{3\gamma^2 + 2A\gamma + (4B - A^2)}. \tag{5}$$

The same formulas hold for the linearization at N with λ instead of γ.

$|\dfrac{\gamma}{\alpha}| > 1$ for Class I and those systems of Class II for which $AB+2A^3+1<-m_0$. $\tag{6}$

$|\frac{\gamma}{\alpha}| < 1$, $\gamma > 0$ for those systems of Class II for which
$AB + 2A^3 + 1 > -m_0$. (7)

By Shilnikov's theorem, for the values of λ for which (2) has HO with respect to 0, Smale's horseshoes are generated in the case (6), while in the case (7) stable periodic orbits arise from vanishing HOs. The translation of the origin to N enables us to apply the Shilnikov's theorem once again and now we have

$|\frac{\lambda}{\alpha}| > 1$ for those systems of Class I for which $\lambda > A$ and for Class II, (8)

$|\frac{\lambda}{\alpha}| < 1$, $\lambda > 0$ for those systems of Class I for which $\lambda \in (0, A)$.
(9)

Therefore, if the HO with respect to N exists for λ then by Shilnikov's theorem horseshoes are generated if (8) holds while a stable periodic orbit is created from HO if (9) is satisfied. This way the verification of the assumptions of Shilnikov's theorem is reduced to the question of existence of HOs.

31.3 A necessary condition for HO with respect to 0 exist is the existence of a pair (T, λ), $T < 0$ for Class I ($T > 0$ for Class II), λ is such that:

$$x(T; Q_0) = x_0, \quad \psi(x(T; Q_0), y(T; Q_0), z(T; Q_0)) = 0, \quad (10)$$

where $\psi(x, y, z) = (\gamma^2 + A\gamma + B)x + (\gamma + A)y + z$ and $(x(t; Q_0),$ $y(t; Q_0), z(t; Q_0)$ is the solution to (2) with the initial condition $Q_0 = (x_0, \gamma x_0, \gamma^2 x_0)$, Q_0 is an eigenvector corresponding to γ which ends at the plane $x = x_0$, while $\psi = 0$ is the complementary linear manifold. Adding the requirement that the above solution tends to 0 as $t \to -\infty$ (for Class I) or $t \to +\infty$ (for Class II) in the plane $\psi = 0$ one obtains a necessary and sufficient condition for the existence of HO. Similarly, the necessary condition for the existence of HO with respect to N is the existence of a pair (T, λ) satisfying (10) with T of opposite sign (t in the case of additional requirements for necessary and sufficient conditions), Q_0 replaced by $Q_N = (x_0, x_0(1+m_0)\lambda^{\frac{1}{1-m}}, x_0(1+m_0)\lambda^2\frac{1}{1-m})$ and ψ redefined as $\psi(x, y, z) = (\lambda^2 + A\lambda + B)x + (\lambda + A)y + z + \frac{n}{\lambda}$. Let us introduce the function F: $R^4 \to R$, $F(t, c_1, c_2, c_3) = w_1 e^{\alpha t} \cos\beta t + w_2 e^{\alpha t} \sin\beta t + w_3 e^{\gamma t}$, where w_1, w_2, w_3 are the functions of (c_1, c_2, c_3)

defined as follows $w_3 = \dfrac{c_1\gamma^2 + c_2\gamma + c_3}{(\gamma - \alpha)^2 + \beta^2}$, $w_1 = c_1 - w_3$, $w_2 = \dfrac{1}{\beta}[c_2$

$+ (\alpha+\gamma)c_1 + (\alpha-\gamma)w_3]$, with α, β given by (5). For $\mu = 1+m_0$, the vector

$$(F(t,x(0),Ax(0) + y(0),Bx(0) + Ay(0) + z(0)), F(t,y(0),z(0) +$$
$$Ay(0), -\mu x(0)), F(t,z(0), -By(0) - \mu x(0)))^T \qquad (11)$$

determines the solution of (2) in the halfspace $x \leq x_0$, starting from the initial point $(x(0), y(0), z(0))^T$, $x(0) \leq x_0$ for such t for which its first component is less or equal to x_0. Replacing $x(0)$ by $x(0) - \dfrac{1}{1-m}$, $x(0) \geq x_0$; γ by λ, taking $\mu = 1-m$ and adding

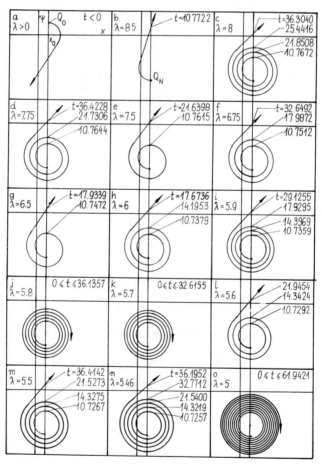

Figure 1.

the term $\dfrac{n}{1-m}$ to the first component of the vector (11) one obtains a formula for the solution of (2) in the halfspace $x \geq x_0$, valid for such t for which the first component is greater or equal to x_0. The Newton procedure can be used to find $\overline{T}$ such that $x(\overline{T}, x(0), y(0), z(0)) = x_0$. This will be needed for gluing together the segments of trajectories. To localize the solutions of (10) we fix a sequence of values of λ and observe the trajectories, startng from Q_0 with appropriate direction of time, in the plane (x, ψ). If for two different values λ_1, λ_2 of λ there exists a time interval $[t_1, t_2]$ such that for λ_1 the trajectory crosses the straight line $x=x_0$ with the opposite sign of ψ which it does the same for λ_2 then by continuity arguments, a solution to (10) exists in the rectangle $[t_1, t_2] \times [\lambda_1, \lambda_2]$. By the simple HO with respect to 0 we mean HO for which (10) holds and additionally $x(t; Q_0) \geq x_0 \ \forall \ t \in [0, T]$. If the pair (T, λ) corresponding to the simple HO is localized then the exact solution to (10) can be determined by careful application (since this system has countably many solutions) of the iterative scheme

$$T_{j+1} = T_j - \frac{x(T_j; Q_0) - x_0}{y(T_j; Q_0)} \ , \ \lambda_{j+1} =$$

$$- \frac{n(\lambda_j)}{\psi(x(T_j; Q_0), y(T_j; Q_0), z(T_j; Q_0)) - \dfrac{n(\lambda_j)}{\lambda_j}} \tag{12}$$

$j = 0, 1, 2, \ldots, (x(.; Q_0), y(.; Q_0), z(.; Q_0))^T$ is a solution to (2) in the halfspace $x \geq x_0$, starting from Q_0.

The similar procedure can be applied to the HO with respect to N. In this case Q_0 should be replaced by Q_N and for simple HO we have $x(t; Q_N) \leq x_0 \ \forall t \in [0, T]$.

31.4 Example 1: O.E. Rössler's system [1,2]. The substitutions $x = x_3 - \dfrac{67}{188}$, $y = x_2 - x_3$, $z = x_1 - 2x_2 + x_3$ transform the standard Rössler system with bifurcation parameter r to the form (2) with A=B=3, $m_0 = 8.4$, $x_0 = \dfrac{95}{1316}$, $m(\lambda) = (\lambda+1)^3 = 8.4r$. In a wide range of allowed (by (3)) values of λ the trajectories have the general shape shown in Figure 1a. Hence we conclude that there is no NO with respect to 0. Figure 1b-1o show the trajectories starting from Q_N,

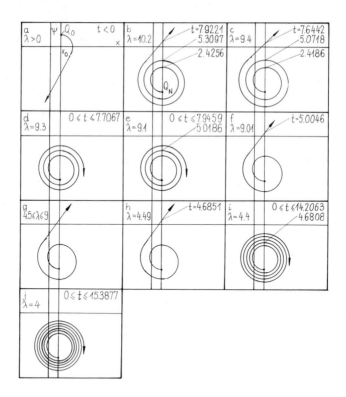

Figure 2.

observed in (x, ψ)-plane (the scale and precise shape are neglected as unessential) for different fixed λ. This allows us to deduce that there are HO in the intervals, (8,8.5) (the simple HO), (7.5,7.75), (6.5,6.75), (5.9,6), (5.5,5.6). The exact data for simple HO are found to be: T = 10.77216478, λ = 8.498305443 (r = 102.0138429). All estimated values of λ correspond to the generation of Smale's horseshoe (by (8)). Note that in [1,2] chaotic oscillations were found by computer simulations for λ = 4.4243074 (r = 19).

Example 2. A transformed model of asymmetric RC generator of chaos, considered in [4] has the form (2) with A=5, B=6, m_0 = $\frac{16913}{512}$, $x_0 = \frac{1}{5}$. The results of calculations are presented in Figure 2. Again there is no HO with respect to 0, while in the intervals (10.2,10.3) (simple HO with T=2.426254614, λ = 10.287103039), (9.3,9.4), (9.01,9.1), (4.4,4.49), HOs with respect to N have been found. By (8), (9) the last value corresponds to the stable periodic orbit (its existence has been confirmed by computer simulation) and other values to Smale's horseshoes.

Acknowledgement

Supported by the Polish Ministry of Science and Higher Education under the contract PR.I.02.ASO.2.2.1986.

References

[1] O.E. Rössler (1981). *Res. Notes Math.* 46, 50; (1984). Z. *Naturforsch* **38a**, 342.

[2] C. Sparrow (1981). *J. Math. Anal. Appl.* **83**, 275; (1984). *J. Stat. Phys.* **35**, 645.

[3] A. Arneodo et al. (1981). *Comm. Math. Phys.* **79**, 573; (1982). *J. Stat. Phys.* **27**, 171.

[4] Ogorzalek, Grabowski (1985). *Proc. 8th Nat. Conf. Electronics, Poznan, Poland* (in Polish), 367–371.

[5] L. Chua et al. (1985). *Res. Mem. UCB/ERL, M*, Berkeley Univ. California, **85**, 102.

[6] J. Guckenheimer and P. Holmes (1986). *Nonlinear Oscillations, Dynamical Systems and Bifurcations of Vector Fields,* Springer.

[7] L.P. Shilnikov (1965). *Doklady ANSSR* **160**, 558; (1963). *Mat. Sbornik* (in Russian), **61(103)**, 443.

[8] L.P. Shilnikov (1968). *Mat. Sbornik* **77(119)**, 461.

[9] P. Gaspard et al. (1983). *Phys. Lett.* **97A**, 1; (1983). J. *Stat. Phys.* **31**, 449; (1984). **35**, 697.

[10] C. Tresser (1984). *Ann. Inst. Poincaré* **40**, 441; (1984). *J. Physique* **45**, 837.

[11] Ph. Hartman (1964). *Ordinary Differential Equations,* J. Wiley, N.Y.

THE SELF-ORGANIZATION HYPOTHESIS FOR 2D NAVIER-STOKES EQUATIONS

32.1 Introduction

In this short contribution we will sketch a rigorous proof of the "self-organization hyopthesis" for the 2D Navier-Stokes equations (see [1] for a detailed analysis).

In the physical literature this hypothesis is roughly described by saying that the long-time behaviour of the system must be related to the solution of a certain minimization problem. If the system admits two functionals that are conserved when dissipation is neglected, the minimization problem is that of minimizing the functional that decays fastest in the presence of dissipation on level sets of the slower decaying one.

For the problem under consideration, the energy E and enstrophy W serve as such functionals and W is minimized at prescribed E. Arguments that are usually used to support this hypothesis arise from considering non-linear mode interactions (which show an inverse and a normal cascade for the spectral energy and the enstrophy density respectively), together with the selective dissipation due to the viscosity (see [2,3,4] and the references therein).

In this paper we describe how a simple analysis of the deterministic equations (instead of the spectral densities) will show that for a large class of initial data this minimization problem indeed determines the spatial structure of the asymptotic motion.

After the completion of this work it appeared that FOIAS and SAUT [5] have considered this asymptotic behaviour in great detail and also for more general situations. The presentation below is simpler and follows closely the idea behind the self-organization hypothesis. In fact, this line of reasoning has proved to be useful in other cases too, and a description for more general classes of dissipative Poisson systems will be published elsewhere (see also [8]).

32.2 Preliminaries

The Navier-Stokes equations for a flow in a plane with uniform density $\rho = 1$ can be described with the Eulerian velocity $\underline{v}$ and the component of the vorticity ω perpendicular to the plane, $\omega =$

2D Navier-Stokes equation

rot $\underline{v}$ = $(0,0,\omega)$, like

$$\nabla.\underline{v} = 0 \text{ and } \omega_t + \underline{v}.\nabla\omega = \nu\Delta\omega,$$

where $\nu > 0$ is the viscosity coefficient. In order to study 2D flow structures on a period grid, let R = $[0,\pi]$ x $[0,\pi]$ and take as boundary conditions $\underline{v}.\underline{n}$ = 0 on ∂R. An appropriate continuation of a solution on R will then provide a smooth solution on a periodic grid with period 2π in both directions. With the boundary condition ω = 0 on ∂R, there is a one-to-one correspondence between $\underline{v}$ and ω when related as above, and we can consider all the functionals as they appear below as functionals of ω.

For the Euler equations, i.e. when ν = 0, the energy E and the enstrophy W are constants of the motion and given by

$$E = \int \frac{1}{2} \underline{v}^2 \text{ and } W = \int \frac{1}{2} \omega^2$$

(integrations over R). Let us first consider the minimization problem referred to in the self-organization hypothesis. The solutions are, apart from a scaling factor, independent of the prescribed value of E, and they are the same as those of the minimization problem for the functional Q which is the Rayleigh quotient of W and E:

$$Q = W/E.$$

For further appreciation it is important to note that Q depends only on the so-called **spatial structure** of a function ω, and not on its amplitude:

$$Q(\omega) = Q(\hat{\omega}), \text{ where } \hat{\omega} = \frac{\omega}{||\omega||}$$

(with $||\ ||$ the L_2-norm on R).

The critical points of Q (on functions ω satisfying ω = 0 on ∂R) are precisely the eigenfunctions of the Laplace operator on R.

For $\underline{k} \in \mathbf{N}$ x $\mathbf{N}$ these eigenvalues and eigenfunctions are

$$\lambda_{\underline{k}} = |\underline{k}|^2 = k_1^2 + k_2^2, \quad \Omega_{\underline{k}} = \sin k_1 \, x. \sin k_2 y.$$

Denote the increasing set of eigenvalues by μ_k, k$\in$N and let $\hat{\Omega}$ be the normalized eigenfunction $\Omega_{(1,1)}$.

Then the variational characterization of the lowest eigenvalue shows that

$$\mu_1 = Q(\hat{\Omega}) = \min \{Q)\omega)|\omega = 0 \text{ on } \partial R\}.$$

Now it is to be observed that with every $\Omega_{\underline{k}}$, which is an exact, time-independent solution of the Euler equations, the functions

$$\Omega_{\underline{k}} \exp\left(-\nu\lambda_{\underline{k}}\, t\right)$$

are exact solutions of the Navier-Stokes equations, which will be called planar Taylor vortices (after Taylor [6]).

32.3 Viscous Evolution of the Functionals E and W

In the presence of dissipation due to viscosity, $\nu > 0$, the functionals E and W are no longer constant on a solution; their time derivatives are given by

$$\dot{E} = -\nu \int \omega^2 \text{ and } \dot{W} = -\nu \int (\nabla\omega)^2.$$

If we introduce the **dissipation rate quotient** as the functional

$$\Lambda = \dot{W}/\dot{E},$$

the time derivative of Q can be expressed like

$$\dot{Q} = -2\nu Q[\Lambda-Q].$$

Using the Cauchy-Schwartz inequality it is easily shown that $\Lambda(\omega) - Q(\omega)$ is sign definite (positive) and zero only if $\hat{\omega} = \hat{\Omega}_{\underline{k}}$ for some $\underline{k}$. Since $Q(\omega) \geq \mu_1 > 0$ for all ω, we obtain the following fundamental result:

Proposition 1: For any solution $\omega(t)$ it holds that $\dot{Q}(\omega(t)) \leq 0$, and $\dot{Q}(\omega(t)) = 0$ iff $\omega(t)$ is some planar Taylor vortex.

32.4 Self-organization

As a consequence of this result we find that for any initial value ω_o the value $Q(\omega(t))$ at the corresponding solution decreases monotonically to some definite value, to be denoted by $q(\omega_o)$, which must necessarily be μ_k for some $k \in \mathbf{N}$. These quantisized limit values determine manifolds $I_k := \{\omega_o | q(\omega_o) = \mu_k\}$ that are invariant for the flow. In particular, if $Q(\omega_o) < \mu_2$ then $\omega_o \in I_1$, and for $\omega_o \in I_1$ we have $Q(\omega(t)) \to \mu_1$, from which it follows that $\hat{\omega}(t) \to \hat{\Omega}$. Using a more refined analysis of the difference $\Lambda(\omega) - Q(\omega)$ near $\hat{\Omega}$, we can prove:

Proposition 2: For the solution $\omega(t)$ with initial data $\omega_o \in I_1$, the convergence of the spatial structure to $\hat{\Omega}$:

$$\hat{\omega}(t) = \frac{\omega(t)}{||\omega(t)||} \to \hat{\Omega} \text{ as } t \to \infty,$$

can be more precisely described as follows:

$$\omega(t) = \alpha(t) \; \hat{\Omega} + \xi(t),$$

where $\alpha(t)$ is a scalar function satisfying $c_1 \exp(-\nu\mu_1 t) \le |\alpha(t)| \le c_2 \exp(-\nu\mu_1 t)$ for some constants c_1 and c_2, and ξ is orthogonal to $\hat{\Omega}$ (in L_2-sense) and decreases exponentially faster than $\alpha(t)$:

$$||\xi(t)\}|| = |\alpha(t)| . 0(e^{-\delta t}), \text{ with } \delta = \nu(1-\mu_1/\mu_2)(\mu_2-\mu_1).$$

For initial data $\omega_0 \in I_k$, $k \ne 1$, the limiting behaviour can also be studied. However, these limiting structures will be unstable since, as a consequence of proposition 1, any perturbation ξ with $Q(\xi) < \mu_k$ at a time T sufficiently large will imply $Q(\omega(T) + \xi) < \mu_k$, and thus $q(\omega(T) + \xi) < \mu_k = q(\omega(T))$.

Hence, practically speaking, in the presence of random perturubations all solutions will eventually have the asymptotic behaviour as described in proposition 2. This result specifies and justifies the self-organization hypothesis mentioned before.

32.5 Remarks

(1) Any solution $\omega(t)$ defines a curve $t \to (E(\omega(t)), W(\omega(t))$ in a 2-dimensional E-W diagram. The functional Q determines the angle of a point with the E-axis, and Λ is the tangent to the curve. The self-organization phenomenon is reflected in the fact that this solution curve is tangent to the $Q = \mu_1$-line at the origin.

(2) In a spectral setting, Q and Λ turn out to be the mean squared wave number with weight function the spectral energy and enstrophy density respectively. The results described here show that both these means tend to the smallest wavenumber $\underline{k} = (1,1)$.

(3) The asymptotic results above, and in [5], do not explain the appearance of small scaled vortices as has been observed in recent computer simulations by McWilliams [7].

References

[1]   E. van Groesen (1988). *Physica* **148A**, 312.
[2]   A. Hasagawa (1985). *Adv. Physics* **34**, 1.
[3]   R.H. Kraichnan and D. Montgomery (1980). *Rep. Prog. Phys.* **43**, 547.
[4]   Z. Yoshida, T. Uchida and N. Inoue (1984). *Phys. Fluids* **227**, 1785.
[5]   C. Foias and J.C. Saut (1984). *Ind. Univ. Math. J.* **33**, 459.
[6]   G.T. Taylor (1923). *Philos. Mag.* **46**, 671.
[7]   J.C. McWilliams (1984). *J.F.M.* **146**, 21.
[8]   E. van Groesen (1988). A deterministic approach towards self-organisation in continuous media. *Cont. Math.*, to appear.

33   *H.M. Isomäki, J. von Boehm and R. Räty*

CHAOS AND THE STRUCTURE OF THE BASIN BOUNDARIES OF A DISSIPATIVE OSCILLATOR

Abstract

We study the chaotic motion of a damped sinusoidally-driven impact oscillator. We find the Feigenbaum bifurcations and chaos with the calculated universality numbers δ = 4.672, θ = 0.179 and ρ = 0.879 agreeing closely with the theoretical ones for one-dimensional maps: 4.699, 0.187 and 0.892, respectively. The winding number q (the number of rebounds within the unit of time) forms an incomplete devil's staircase in the region 3/4 > q > 3/5. The three coexisting attractors with q = 1/2, 3/5 and 7/10 have the basins of the initial conditions exhibiting fractal boundaries with the Cantor set structure. The calculated fractal dimension is $\approx$ 1.65.

We study the chaotic motion [1] of the damped sinusoidally-driven impact oscillator [2,3]

$$\ddot{x}(t) + 0.4\ \dot{x}(t) + x(t) = \cos(\omega t). \qquad (1)$$

When $x(t_i)$ = 0 an impact occurs, $\dot{x}(t_i^+)$ = $-\dot{x}(t_i^-)$, causing the nonlinearity in the motion. ω is the system parameter. The number of rebounds within the unit of time is the winding number q which decreases with increasing ω. We consider the motion only for ∞ > q $\geq$ 1/2 because it is expected that the motions are qualitatively similar for 1 > q > 1/2 and 1/n > q > 1/(n+1) n = 2,3... [2]. For ∞ > q $\geq$ 1, q forms a harmless staircase. With q = 1 we find the Feigenbaum bifurcations and chaos with the calculated universality numbers δ = 4.672, θ = 0.179 and ρ = 0.879 agreeing closely with the theoretical ones for one-dimensional maps: 4.699 [4], 0.187 [5] and 0.892 [6], respectively. For 1 $\geq$ q $\geq$ 3/4 the motions are mainly chaotic and irregular. For 3/4 > q > 3/5 we find an incomplete devil's staircase [2]. The dynamics in this region is very complicated because the attractors of the devil's staircase coexist with attractors having q = 7/10 and 1/2 [2]. Each attractor has a set of initial conditions or a basin consisting of points finally approaching the particular attractor. The motion depends strongly on the structure of the basin boundaries, which may be smooth or fractal [7-10].
    In Figure 1 we present the calculated basins at ω = 3.2 where

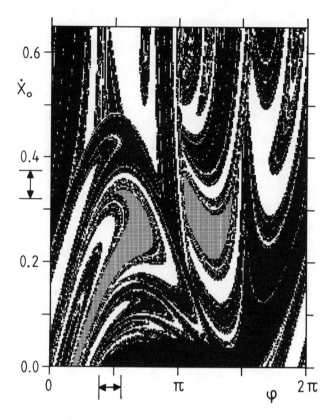

Figure 1. $(\varphi, \dot{x}_o)$*-basins at* ω = 3.2. φ *and* φ + 2π *are equivalent. The grey, black and white regions denote basins of the attractors with q = 1/2, 3/5 and 7/10, respectively.*

the attractors with q = 1/2, 3/5 and 7/10 coexist. Due to the piecewise linearity of the motion the most convenient initial conditions are the phase of the driving force φ = ωt_i and the velocity of the particle $\dot{x}_o$ = $\dot{x}(t_i^+)$ both given at the impact x = 0. In all Figures 1-3 the grid in the $(\varphi, \dot{x}_o)$-plane is 200x260 and the grey, black and white regions denote the basins of the attractors with q = 1/2, 3/5 and 7/10, respectively. Besides large islands the basins exhibit complicated fine structure. All the 1+3+7 = 11 stable fixed points are clearly inside the respective islands in Figure 1. To explore the fine structure more closely we have magnified nested subregions by the factors of 12^m, m = 1,2,...,5 starting from the rectangle denoted by arrows in Figure 1. The 12 and 12^5 = 248 832 times magnifications are shown in Figures 2 and 3, respectively. These magnifications and all the intermediate ones exhibit similar parallel striped structure. We conjecture that this would continue similarly in further magnifications. We stress that the graininess in Figures

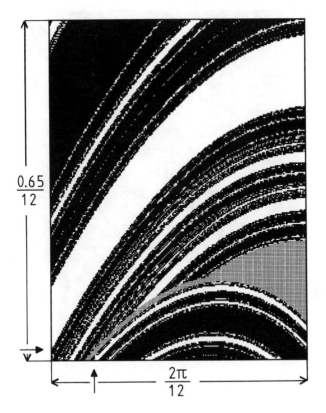

Figure 2. Magnification by a factor of 12 of the rectangle ($2\pi/12$ x $0.65/12$) denoted by the arrows in Figure 1, $1.225 \le \varphi \le 1.749$, $0.32 \le x_o \le 0.3742$.

1-3 is due to the finite grid in the calculation and is not a basin property. The (inifinitely) narrow disconnected stripes form transversally a Cantor set due to the existence of a Smale horseshoe associated with the homoclinic structure developed by the crossing of the stable and unstable manifolds of a fixed saddle point [7,8,10]. The large islands in the basins do not belong to the horseshoe and thus do not reveal any fine structure. The horseshoe is not an attractor - although it can cause long chaotic transients in the motion - and therefore the fractal basin boundary is found despite the fact that the three attractors in our problem are periodic.

The fractal dimension D characterizes the fractal boundary and is calculated as follows. D is defined for a set by the equation [1]

$$nr^D = 1, \ r \to 0, \qquad (2)$$

where n is the relative number of identical pieces of area covering the set and r is the relative linear scale. If the number of the set-covering pieces of area A in scale r is N then N

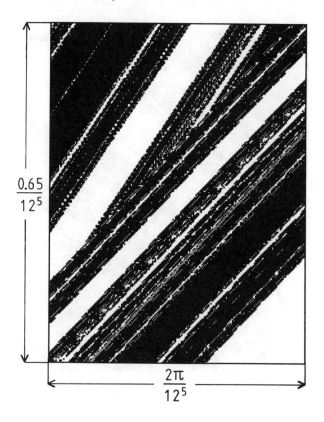

$\frac{0.65}{12^5}$

$\frac{2\pi}{12^5}$

Figure 3. Magnification by a factor of 12^5 of the rectangle $(2\pi/12^5 x0.65/12^5)$ denoted by the arrows in Figure 2, 1.313 907 6 $\leq \varphi \leq 1.313$ 932 9, 0.321 854 34 $\leq \dot{x}_o \leq 0.321$ 856 95.

$= nN_o$ and $A = r^2 A_o$ where N_o and A_o are the respective starting values. The total area of the cover in scale r is

$$NA = nr^2 N_o A_o. \tag{3}$$

To get D we first calculate the uncertainty fraction f [8] that is asymptotically (r → 0) proportional to NA in Eq. (3). f denotes the relative area of the uncertain phase plane points which consist of the initial conditions $(\varphi, \dot{x}_o)$ converging to an attractor other than that found with perturbed initial conditions (e.g., $(\varphi \pm \Delta\varphi, \dot{x}_o)$). The uncertain phase plane area is asymptotically (r → 0) proportional to NA (Eq. 3), i.e. f ~ NA, f_o ~ $N_o A_o$. Hence

$$\frac{f}{f_o} \sim nr^2. \tag{4}$$

D is calculated using the uncertainty exponent α which is from Eqs. (2) and (4)

$$\alpha = \lim_{r\,0} \frac{\ell nf/f_o}{\ell nr} = 2 - D. \tag{5}$$

f is calculated by perturbing φ by $\pm r\Delta\varphi$, where $\Delta\varphi$ is 0.05 times the horizontal width of the phase plane subregion (see, e.g., Figures 1-3) and r = 1,0.9,...,0.1. All the 200x260 = 52,000 grid points are considered as initial conditions.

The calculated fractal dimensions for the magnifications 12^m, m = 0,1,...,5 are D = 1.67, 1.65, 1.57, 1.68, 1.60, 1.70, respectively. This close agreement indicates that D is independent of the selected subregion although the global forms of the subregions may differ considerably (see,, e.g. Figures 1-3). Thus the basin boundary exhibits statistical self-similarity with the fractal dimension D $\approx$ 1.65.

Finally it is interesting to compare the dynamics of a realistic physical system studied here and the behaviour of the widely studied complex analytic maps [1,9]. These maps have two-dimensional basins in the complex plane. The basin boundaries - the Julia sets [1] - may exhibit fractal structure [9] that, however, fully differs from the parallel striped one found here. Moreover, the impact oscillator has chaotic attractors implying the existence of a positive Lyapunov exponent whereas the complex analytic maps, due to the Cauchy-Riemann conditions, cannot exhibit chaotic attractors [7,9].

In conclusion, the damped sinusoidally-driven impact oscillator exhibits the Feigenbaum bifurcations leading to chaos with the calculated universality numbers which agree closely with the corresponding theoretical numbers for one-dimensional maps. Beyond the chaotic region the winding number forms an incomplete devil's staircase. In the region where the devil's staircase exists we find three coexisting attractors having a fractal basin boundary of the Cantor set type. The calculated fractal dimension is $\approx$ 1.65.

Acknowledgements

The authors would like to thank Mrs. Tuula Aalto for typing the manuscript.

References

[1]   H.G. Schuster (1984). *Deterministic Chaos,* Physik Verlag, Weinheim,

[2]   H.M. Isomäki, J. von Boehm and R. Räty (1985) in *Festkörperprobleme (Advances in Solid State Physics),* XXV, 83; (1985). *Phys. Lett.* **107A**, 343.

[3]   H.M. Isomäki, J. von Boehm and R. Räty (1988). *Phys.*

Lett. **126A**, 484.

[4] M.J. Feigenbaum (1978). *J. Stat. Phys.* **19**, 25.

[5] E.N. Lorenz (1980). *Ann. N.Y. Acad. Sci.* **357**, 282.

[6] J.A. Ketoja and J. Kurkijärvi (1986). *Phys. Rev.* **A33**, 2846.

[7] C. Grebogi, E. Ott and J.A. Yorke (1983). *Phys. Rev. Lett.* **50**, 935; (1983). *Phys. Rev. Lett.* **51**, 942.

[8] C. Grebogi, S.W. McDonald, E. Ott and J.A. Yorke (1983). *Phys. Lett.* **99A**, 415.

[9] S.W. McDonald, C. Grebogi, E. Ott and J.A. Yorke (1985). *Phys. Lett.* **107A**, 51.

[10] F.C. Moon and G.-X. Li (1985). *Phys. Rev. Lett.* **55**, 1439.

TECHNIQUE TO TRACE BIFURCATION POINTS OF PERIODIC SOLUTIONS

We have developed a computational tool to follow curves of bifurcation points for periodic solutions in 2-parameter systems of ordinary differential equations (ODEs) and iterated maps (IMs). We shall describe the elements used to formulate the algorithms and the basic steps in the algorithms. We give two applications of the tool.   A full account of the algorithm is under preparation. An adaptation of the tool to the NAG-library will be undertaken.

A general formulation of the task stated above is this: Let S be the 2-dimensional surface of zero points of a map $Q: H \times R^2 \to H$, and let $G: S \to R$ be a real function defined on the surface. Let B be the curve on S of zero points of G. We shall be interested in tracing the curve B. First we shall specify the map Q, next the real function G and last how to trace the zero points of G. We shall formulate the map Q in two cases.

We first examine the IM

$$x'(t) = x(t+1) = f(x(t);c),$$

$$t = 0,1,2,\ldots, \quad x \in R^n, \quad c \in R^2, \quad f \text{ smooth},$$

for which we consider periodic solutions of period T (T is an integer). We let x_p denote the periodic solution for a fixed set of parameters c, then by definition of a periodic solution

$$x_p(T) = f^T(x_p(0);c) = x_p(0),$$

where f^T means f composed with itself T times. We can then, at least in a neighbourhood of x_p, define the map

$$Q: R^n \times R^2 \to R^n, \quad Q(u;c) = f^T(u;c) - u.$$

A zero point of Q, where we take $H = R^n$, is a T-periodic solution of the IM. With this formulation, we can determine stable and unstable periodic solutions with equal ease. All zero points of Q make up the surface S.

We next examine the autonomous ODE

$$\dot{x}(t) = f(x;c)$$

$$0 \leq t, \ x \ \varepsilon \ R^n; \quad c \ \varepsilon \ R^2, \ f \text{ smooth.}$$

Let $x_p(t)$ be a periodic solution with period T. Then $x_p(T) = x_p(0)$. We let H be the n-1 dimensional hypersurface in R^n containing $x_p(0)$ and transversal to the trajectory through $x_p(0)$. We introduce a coordinate system in H. Let the n-1 vector u(0) be any point in H expressed in the coordinate system of H. We let the n-vector x_u be the same point expressed in the coordinates of R^n. Let $\phi(t)$ be the trajectory starting at x_u at t = 0. (If $x_u = x_p$ then $\phi(t) = x_p(t)$.) This trajectory will, at least if x_u is in a neighbourhood of $x_p(0)$, strike H at time τ, where τ is near to T, in the point $x_u(\tau)$. Let u(τ) be the coordinates of $x_u(\tau)$ expressed in the coordinate system of H. The map Q is then taken to be

$$Q: HxR^2 \to H, \ Q(u(0);c) = u(\tau)-u(0).$$

The surface S of zero points of Q corresponds to periodic solutions of the ODEs.

We now go on to define the map G. We note that the Poincaré map P is defined by P = Q+I. The stability of a periodic solution is then found examining the eigenvalues λ_i of DP, where DP can be evaluated in any point of S. If one of the eigenvalues, λ_j say, has modulus 1, then the point of S is a bifurcation point for a periodic solution. We define

$$G = |\lambda_j|-1,$$

and then G = 0 in bifurcation points. In general the zero points of the function G will make up a curve, which we denote B. In order to define G properly, we can introduce an ordering of the eigenvalues, such that

$$\lambda_k < \lambda_{k+1} \text{ if } |\lambda_k| < |\lambda_{k+1}|.$$

Then G can be defined using the jth eigenvalue as above.

We shall now give the basic steps in the algorithm used to trace the curve B on the surface S. For convenience we begin taking A as a point on B. The tangent plane of the surface S at the point A is spanned by two vectors t_p and t_L. We shall take t_p to be a unit-vector parallel to the tangent of the curve B and t_L to be a unit-vector orthogonal to t_p (P = parallel, L = lateral).

To obtain another point on B we take a step from A in the t_p direction. We reach a point, which is not necessarily on S, so

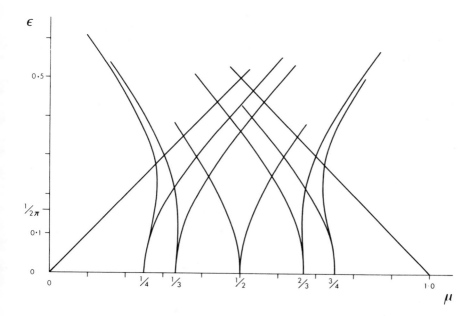

Figure 1. Curves of saddle node bifurcation points. The straight lines emanating from μ = 0 and μ = 1 can be worked out analytically. The wedges (= tongues) have been computed. The period of the solutions within the tongues is (from left or right) 4, 3, 2, 3, 4 respectively.

we use Newton's method to move orthogonal to the plane spanned by t_P and t_L back to the surface S. In that point on S we reach, we evaluate G. We then take first an increment in the positive direction of t_L and next in the negative direction of t_L. In both cases, having taken a step in the direction of t_L we reach points which are not necessarily on S, so we use Newton's method to move orthogonal to the plane spanned by t_P and t_L back to the surface S, whereafter G is evaluated. If the stepsize along t_P and the increments along t_L are suitable G will not have the same sign in all three points. We now use the bisection method on the smallest interval where G changes sign to find the zero point of G.

This formulation caters for saddle node, period doubling, and torus bifurcations without further adjustments.

As our first example we examine the IM

$$x(t) = x(t+1) = x(t) + μ + ε \cos(2πx(t)) \qquad (\text{mod } 1)$$

$$t = 0,1,2,\ldots, \; x\varepsilon R/Z, \; 0 < μ < 1, \; 0 < ε$$

cf. [2] p. 303. We have found the Arnold tongues for low periods, see Fig. 1. First we obtained a periodic solution. We followed it to a saddle node bifurcation. Then we traced the curve of

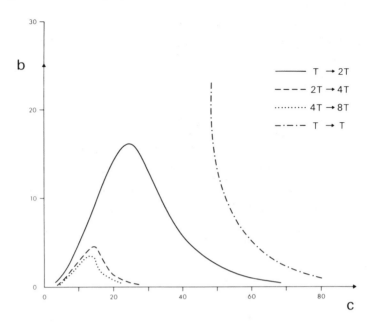

Figure 2. The Rössler system. The dune shaped curves are curves of period doubling bifurcations (T → 2T, 2T → 4T, and 4T → 8T). The cliff to the right is a curve of saddle node bifurcations (T → T).

saddle node bifurcation points. This map is an orientation preserving circle map for $0 < \varepsilon < \dfrac{1}{2\pi}$. More bifurcation points have been reported in [1], and a recent review is provided in [5].

As our second example we examine Rössler's system, which is an autonomous system of ODE [3]

$$\dot{x} = \quad - y - \quad z$$

$$\dot{y} = \quad x + ay$$

$$\dot{z} = b \quad - (c-x)z$$

where

$$0 \le t, \ (x,y,z) \ \varepsilon \ R^3, \ (a,b,c) \ \varepsilon \ R^3.$$

We shall consider $0 < z$, and $a = \dfrac{1}{5}$, $0 < b$, $0 < c$. Originally Rössler found a period doubling cascade increasing c keeping $a = b = \dfrac{1}{5}$. The three first bifurcations have been followed, see Fig. 2. We have found an inverted period doubling sequence and a saddlenode bifurcation for higher values of c. The region in which chaos can take place is confined to the small area "underneath" the curve of the third period doubling. The ODEs

494

have been solved using LSODA [4]. It turns out that H can be
taken as a the coordinate halfplane y = 0, x < 0, z > 0.

References

[1]   L. Glass and R. Perez (1982). *Phys. Rev. Lett.* **48**, 1772.
[2]   J. Guckenheimer and P. Holmes (1983). *Nonlinear oscillations, dynamical systems, and bifurcations of vector fields*, Springer Verlag.
[3]   A. J. Lichtenberg and M. A. Lieberman (1983). *Regular and Stochastic Motion*, Springer Verlag.
[4]   L. Petzold (1983). *SIAM J. Sci. Stat. Comput.* **4**, 136.
[5]   J. M. T. Thompson and H. B. Stewart (1986). *Nonlinear dynamics and chaos*, John Wiley & Sons.

QUALITATIVE RESEARCH ON SOLITONS OF NONLINEAR SCHRODINGER EQUATION WITH EXTERNAL FIELD

35.1 Introduction

In his book "Nonlinear Wave Mechanics - a causal interpretation" [1] Louis de Broglie proposed using the nonlinear wave equations to give a clear and causal interpretation to the wave-particle dualism - the fundamental myth in microphysics. According to his idea, the elementary particle is represented as a small region of large amplitude of a nonlinear wave, i.e. the soliton in modern terms.

Following his idea, we study the nonlinear Schrödinger equation with external field

$$i\frac{\partial\psi}{\partial t} + \frac{\partial^2\psi}{\partial\kappa^2} + f(|\psi|^2)\psi = U(x,t)\psi. \tag{1}$$

We write the complex-valued solution $\psi(x,t)$ frequently in the form

$$\psi(x,t) = \varphi(x,t)e^{i\theta(x,t)}.$$

If a solution $\psi(x,t)$ vanishes rapidly as $x \to \infty$, we can see

$$\int_{-\infty}^{\infty}\psi^*\psi dx = \int_{-\infty}^{\infty}\varphi^2 dx = \text{const}, \tag{2}$$

$$\frac{d}{dt}\langle x\rangle \equiv \frac{d}{dt}\frac{\int_{-\infty}^{\infty}\psi^*x\psi dx}{\int_{-\infty}^{\infty}\psi^*\psi dx} = \frac{2\int_{-\infty}^{\infty}\varphi^2\frac{\partial\theta}{\partial x}dx}{\int_{-\infty}^{\infty}\varphi^2 dx}, \tag{3}$$

$$\frac{d^2}{dt^2}\langle x\rangle = \frac{-2\int_{-\infty}^{\infty}\psi^*(\frac{\partial U}{\partial x})\psi dx}{\int_{-\infty}^{\infty}\psi^*\psi dx} \equiv -2\langle\frac{\partial U}{\partial x}\rangle.. \tag{4}$$

Equation (4) corresponds to the Ehrenfest theorem in quantum mechanics.

A solution $\psi = (x,t) = \phi e^{i\theta}$ of (1) is called the weak soliton solution, if the amplitude function $\phi(x,t)$ satisfies the following conditions:

(i) $\quad 0 < \int_{-\infty}^{\infty} \varphi_s^2(x,t)dx < +\infty$

(ii) There exists a function $s(t)$, $\dot{s} = \dfrac{ds}{dt} \neq 0$, such that

$$\phi(x,t) = \phi(x - s(t)).$$

Here "weak" means that we have not taken the stability into account in the definition. We can easily see that, according to (4), the soliton moves according to the second Newton's law of motion.

Since we cannot expect to obtain the exact solution of (1) in general cases, the qualitative research and numerical test are the main tools in the study of this equation.

35.2 Conditions of the Existence of Soliton Solution

The weakened defition of soliton has already put a strong constraint on the form of (1). In fact, we have proved the following [2]:

Theorem 1. The equation (1) has a soliton solution, if and only if the following two conditions are both satisfied:

(i) There exists such a function $s(t)$, $\dot{s} \neq 0$, that $U(x,t)$ can be rewritten as

$$U(x,t) = V(x - s(t)) - \frac{1}{2}\ddot{s}x + h(t) = V(\xi) - \frac{1}{2}\ddot{s}x + h(t), \quad (5)$$

where $\xi = x - s(t)$, $\ddot{s} = \dfrac{d^2s}{dt^2}$ and $h(t)$ is an arbitrary definite function.

(ii) The following nonlinear Sturm-Liouville problem

$$\begin{cases} \dfrac{d^2\varphi}{d\xi^2} - V(\xi)\varphi + f(\varphi^2) + \lambda\varphi = 0 \\ 0 < \int_{-\infty}^{\infty}\varphi^2 d\xi < +\infty \end{cases} \quad (6)$$

has an eigenvalue λ with corresponding eigenfunction $\phi(\xi)$.

When (i) and (ii) are both satisfied, the soliton solution of (1) has the following form:

$$\begin{cases} \psi_\lambda(x,t) = \varphi_\lambda(x - s(t))e^{i\theta_\lambda(x,t)} \\ \theta_\lambda(x,t) = \frac{1}{2}\dot{s}x - \lambda t - \int_o^t [h(t') + \frac{1}{4}(\dot{s}(t'))^2]dt' + \theta_o \end{cases} \tag{7}$$

where λ and $\phi\lambda(\xi)$ are the eigenvalues and the eigenfunction of (6).

35.3 Nonlinear Sturm-Liouville Problem

It is easy to see that λ and $\phi(x)$ in the bound stationary state solution $\psi(x,t) = \phi(x)e^{-i\lambda t}$ of another NLS equation

$$i\frac{\partial\psi}{\partial t} + \frac{\partial^2\psi}{\partial x^2} + f(|\psi|^2)\psi = V(x)\psi \tag{8}$$

satisfy the same nonlinear Sturm-Liouville problem (6). Using the Schauder fixed point theorem, we have proved the following two theorems for this problem [3,4]:

Theorem 2. If the real function $f(x)$ in (6) is continuous on $[0,+\infty)$, and if the continuous real function $V(x)$ goes to $-\infty$ as $x \to \infty$, then for any positive number M and any non-negative integer n the nonlinear Sturm-Liouville problem has an eigenvalue $\lambda(M)$ and its corresponding eigenfunction $\phi(x,M)$ which has exactly n zeros, such that

$$\begin{cases} ||\varphi_n(x,M)|| \equiv \sup\{|\varphi_n(x,t)| | x \in (-\infty,\infty)\} \\ \mu_n - \bar{f}(M) \le \lambda_n(M) \le \mu_n - \underline{f}(M) \end{cases} \tag{9}$$

where μ_n is just the (n+1)-th eigenvalue of the corresponding linear Sturm-Liouville problem

$$\begin{cases} \frac{d^2\varphi}{dt^2} - V(x)\varphi + \mu\varphi = 0 \\ < \int_{-\infty}^\infty \varphi^2 dx < +\infty, \end{cases} \tag{10}$$

and

$$\begin{cases} \bar{f}(M) \equiv \sup\{f(p^2) | \rho\in[0,M]\} \\ \underline{f}(M) \equiv \inf\{f(\rho^2) | \rho\in[0,M]\}. \end{cases} \tag{11}$$

Theorem 3. If the real function $f(x)$ is continuous on $[0,+\infty)$ and $f(0) = 0$, if the continuous real function $V(x)$ satisfies

$$\lim_{x \to \infty} \inf V(x) = 0$$

and if the number $n(V)$ of the discrete eigenvalues of the relevant linear Sturm-Liouville problem (10) is not equal to zero, then for any non-negative integer $m \le n(V)$, there exists a positive number ϵ_m, such that, for any positive numvber $M \in (0, \epsilon_m)$, the nonlinear problem (6) has an eigenvalue $\lambda_m(M)$ and its corresponding eigenfuntion $\phi_m(x, M)$, which has exactly m zeros and satisfies the conditions (9) (note: we should use m instead of n).

35.4 Harmonic Oscillator

Using the above results, we can rigorously prove that the NLS equation of the harmonic oscillator

$$i\frac{\partial\psi}{\partial t} + \frac{\partial^2\psi}{\partial x^2} + 2|\psi|^2\psi = Kx^2\psi, \quad K > 0 \tag{12}$$

has the soliton solution which moves as an oscillator with a frequency proportional to $k^{1/2}$ [2]. This fact was noticed by Kaup and Newell [5], but their method (the singular perturbation) can only be used for small k.

35.5 Stability of Soliton

We have proved the following [2]:

Theorem 4. If the conditions in theorem 1 are satisfied, then at any given time t_o, among all the solutions of (1) which satisfy the following constraints

$$\int_{-\infty}^{\infty}\psi^*\psi dx = M > 0$$

$$\langle x\rangle|_{t=t_o} \equiv \frac{\int_{-\infty}^{\infty}\varphi^2(x,t)x dx}{M}\bigg|_{t=t_o} = x(t_o) \tag{13}$$

$$\frac{d}{dt}\langle x\rangle|_{t=t_o} = \frac{2\int_{-\infty}^{\infty}\varphi^2\frac{\partial\theta}{\partial x}dx}{M}\bigg|_{t=t_o} = \dot{s}(t_o),$$

only the soliton solution can render the energy the extremum (the minimum or the saddle point), where the energy is defined by

$$E(\psi) = \int_{-\infty}^{\infty}[|\frac{\partial\psi}{\partial x}|^2 + U(x,t)|\psi|^2 - F(|\psi|^2)]dx \tag{14}$$

and $F(x) = \int_{0}^{x}f(x')dx'$.

This theorem shows that the weak soliton is stable or metastable in the sense of dynamics.

By introducing a generalized measure, we can more rigorously prove the nonlinear stability of the solition solution of the NLS equation [6]

$$i\frac{\partial \psi}{\partial t} + \frac{\partial^2 \psi}{\partial x^2} + 2|\psi|^2 \psi = 0. \tag{15}$$

It is interesting that the numerical test shows that the solitons of (12) can also re-emerge after a collision with each other [7].

References

[1]   L. de Broglie (1960).   *Nonlinear wave mechanics,* Elsevier, New York.

[2]   K. Y. Guan (1983).   *KEXUE TONGBAO, Special Issue,* **89**.

[3]   K. Y. Guan (1984).   *Acta Mathematicae Applicatae Sinica* **1(1)**, 8.

[4]   K. Y. Guan (1985).   *Acta Mathematica Sinicva* **26(3)**, k 341.

[5]   D. J. Kaup and A. C. Newell (1978).   *Proc. Roy. Soc. Lond.* **A361**, 413.

[6]   K. Y. Guan and W. N. Everitt, *Generalized Measure and Stability of Nonlinear Schrödinger Equation,* to appear.

[7]   H. Y. Wang (1985).   *Journal of Computational Physics* **2(1)**, 124.

MELNIKOV THEORY APPLIED TO A VARIABLE LENGTH PENDULUM

Abstract

The Melnikov function for prediction of Smale-horseshoe chaos is applied to a driven damped pendulum with variable length. Depending on the parameters, it is shown that this dynamical system undertakes heteroclinic bifurcations which are the source of the unstable chaotic motion. Furthermore, the Melnikov method is applied to study subharmonic bifurcations and the theoretical predictions are confirmed by direct numerical calculations.

36.1   Introduction

In this letter we investigate the heteroclinic and subharmonic bifurcations of a pendulum with variable length using the so-called Melnikov method [1-2].   This method is a global perturbation theory which is able to predict analytically the occurrence of homoclinic (heteroclinic) and subharmonic bifurcations when some parameters of the system are varied. Furthermore, it gives sufficient conditions for the occurrence of transversal intersection between the stable and unstable branches of the homoclinic (heteroclinic) orbits.   The ability of the subharmonic Melnikov function to predict the occurrence of subharmonic solutions in parameter space is tested against numerical simulations.

36.2   The Physical Model

We consider a simple damped pendulum with unit mass and variable length in the gravity field, driven by an oscillating external torque.   It can be shown that the equation of motion which governs this system is [3]

$$\frac{d}{d\tau} (\ell^2 \frac{dx}{d\tau}) + \ell g \sin(x) = -r\frac{dx}{d\tau} + r_1 \sin(\omega_1\tau). \tag{1}$$

Here $x(\tau)$ is the angular displacement from the vertical at time τ, r is the damping constant, r_1 is the amplitude of the oscillating external torque with frequency ω_1 and g denotes the gravitational constant.   The length $\ell(\tau)$ of the pendulum is assumed to vary with time τ according to

$$\ell(\tau) = a + b \sin(\omega\tau), \quad a \gg b > 0, \quad \omega > 0. \tag{2}$$

Introducing $H = b/a \ll 1$, $\omega_0^2 = g/a$, and redefining time τ as $t = \omega_0\tau$ we obtain to order $O(H)$

$$[1 + 2H \sin(\frac{\omega}{\omega_0}t)]\ddot{x} + 2 \frac{\omega}{\omega_0} H \cos(\frac{\omega}{\omega_0}t)\dot{x} +$$

$$[1 + H \sin(\frac{\omega}{\omega_0}t)] \sin(x) = - \frac{r}{\omega_0^2 a^2} \dot{x} + \frac{r_1}{\omega_0^2 a^2} \sin(\frac{\omega_1}{\omega_0}t). \tag{3}$$

To indicate the assumed order of magnitude of the constants entering Eq. (3) we set $\dfrac{r}{\omega_0^2 a^2} = \varepsilon\beta$, $\dfrac{r_1}{\omega_0^2 a^2} = \varepsilon\rho_1$, $\dfrac{\omega_1}{\omega_0} = \bar{\omega}_1$, $\dfrac{\omega}{\omega_0} = \bar{\omega}$, and $\dfrac{\omega}{\omega_0}H = \varepsilon\bar{\omega}\bar{H}$, where $\varepsilon \ll 1$. Assuming $H \ll \dfrac{\omega}{\omega_0}H \ll 1$, $H \ll \varepsilon\beta$, and $H \ll \varepsilon\rho_1$ Eq. (3) can be approximated by

$$\ddot{x} + \sin(x) = \varepsilon[-\beta\dot{x} + \rho_1 \sin(\bar{\omega}_1 t) - 2\bar{\omega}\bar{H} \cos(\bar{\omega}t)\dot{x}].$$

36.3 The Heteroclinic Melnikov Function

Eq. (4) can be written as a system of two first-order ordinary differential equations which reads after dropping the bars

$$\begin{cases} \dot{x} = y \\ \dot{y} = -\sin(x) + \varepsilon[-\beta\dot{x} + \rho_1\sin(\omega_1 t) - 2\omega H \cos(\omega t)\dot{x}]. \end{cases} \tag{5}$$

When $\varepsilon = 0$ system (5) becomes the classical simple pendulum equations which are known to be completely integrable. In order to caclulate the heteroclinic Melnikov function [4-5] it is necessary to know the heteroclinic solution of the unperturbed system. This solution is given by

$$\begin{cases} \bar{x}(t) = \pm 2 \tan^{-1}(\sinh(t)) \\ \bar{y}(t) = \pm 2 \operatorname{sech}(t) \end{cases} \tag{6}$$

and it separates the oscillating solutions from the rotating ones. The Melnikov function for system (5) is [4]

$$M^{\pm}(t_0) = \int_{-\infty}^{\infty} \bar{y}(t) \{-\beta\bar{y}(t) + \rho_1 \sin[\omega_1(t + t_0)] +$$

$$- 2\omega H \cos[\omega(t + t_0)]\bar{y}(t)\} dt. \tag{7}$$

Substituting Eqs. (6) into Eq. (7) and making the integral we obtain the following expression for the Melnikov function

$$M^{\pm}(t_0) = -8\beta \pm 2\pi\rho_1 \text{sech}(\frac{\pi}{2}\omega_1) \sin(\omega_1 t_0) +$$

$$- 8\pi\omega^2 H \text{ cosech}(\frac{\pi}{2}\omega) \cos(\omega t_0). \tag{8}$$

If $\sin(\omega_1 t_0) = \pm 1$ and $\cos(\omega t_0) = -1$, corresponding to the ratios $\frac{\omega_1}{\omega} = \frac{1+4s}{2+4t}$ and $\frac{\omega_1}{\omega} = \frac{3+4s}{2+4t}$, respectively, s and t are integers, the Melnikov function will have infinitely many zeros when

$$\frac{H}{\beta} \geq \frac{\sinh(\frac{\pi}{2}\omega)}{\omega^2} [\frac{1}{\pi} - \frac{1}{4} \frac{\rho_1}{\beta} \text{ sech}(\frac{\pi}{2}\omega_1)]. \tag{9}$$

Note that the conditions on the ratios of ω_1 and ω imply that these frequencies are commensurable. The equality in the formula (9) characterizes the onset of the heteroclinic bifurcations with an approximation of $O(1)$ for sufficiently small ε. The strict inequality in formula (9) characterizes the existence of transverse heteroclinic intersection points between the local stable and unstable branches of the heteroclinic orbits and as a result the system (5) will exhibit Smale horseshoe chaos [4]. Furthermore, it is seen from inequality (9) that adding an external driving torque with amplitude ρ_1 leads to chaotic behaviour in a larger region of the parameter space $(\omega, H/\beta)$.

36.4 The Subharmonic Melnikov Function

In this section we shall test predictions obtained from the subharmonic Melnikov theory by direct numerical simulations. In order to do so, we need the analytical expressions of the periodic solutions of the unperturbed pendulum, i.e. Eqs. (5) for $\varepsilon = 0$. There are two qualitatively different periodic solutions, the oscillating ones and the rotating ones. In the following we shall only consider the oscillating solutions which read [2]:

$$\{ \begin{array}{l} x_{os}(t,k) = \pm 2 \sin^{-1}[k \text{ sn } (t,k)] \\ y_{os}(t,k) = \pm 2k \text{ cn } (t,k). \end{array} \tag{10}$$

In Eqs. (10) sn and cn are Jacobi elliptic functions and $k < 1$ is the elliptic modulus [6]. If the two driving frequencies ω and ω_1 are commensurable, the period of the perturbation in the system (5) is $T = pT_1 = qT_2$. Here $T_1 = 2\pi/\omega_1$ and $T_2 = 2\pi/\omega$ are the periods of the two terms $\sin(\omega_1 t)$ and $\cos(\omega t)$, respectively, and

ω_1/ω = p/q with p and q relatively prime.   The period of the oscillating unperturbed solutions (Eqs. (10)) is equal to 4 K(k) where K(k) denotes the complete elliptic integral of the first kind [6].   In the resonance case where 4 K(k) = (m/n)T with m and n relatively prime, subharmonic solutions may exist for system (5).   A necessary condition for the occurrence of subharmonic solutions can be obtained from the subharmonic Melnikov function which is defined as [4]:

$$M_{os}^{m/n}(t_0) = \int_0^{mT} y_{os}(t,k)\{-\beta y_{os}(t,k) + \rho_1 \sin[\omega_1(t+t_0)] + \quad (11)$$

$$-2\omega H \cos[\omega(t + t_0)]y_{os}(t,k)\}dt.$$

Insertion of the solution (10) into the expression (11) and making the integral gives for n = 1, odd mp and even mq

$$M_{os}^{m/1}(t_0) = -16\beta[E(k) - k'^2 K(k)] \pm 4\pi\rho_1 \text{sech}(\omega_1 K'(k))\sin(\omega_1 t_0)$$

$$+ -16\pi\omega^2 H\text{cosech}(\omega K'(k))\cos(\omega t_0). \qquad (12)$$

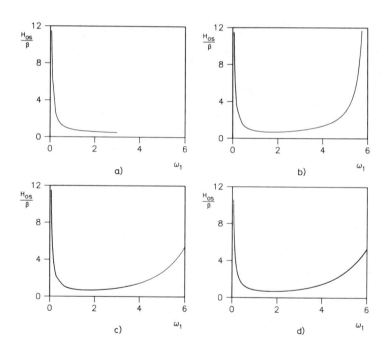

Figure 1.   *Subharmonic curves in the* $(\omega_1, H/\beta)$ *parameter plane for the oscillating case, (a,b,c), as obtained from formula (13) using (mp,mq)* = *(3.2), (6,4) and (9,6), respectively. Heteroclinic bifurcation curve, (d), obtained from formula (9).* $\rho_1/\beta = 0.2.$

In Eq. (12) k' is the complementary modulus defined by $k'^2 = 1 - k^2$, $K'(k) = K(k')$, and $E(k)$ is the complete elliptic integral of the second kind [6]. Following the argument in the heteroclinic case, we can find a necessary condition for the occurrence of subharmonics of period mT corresponding to the ratios $\frac{\omega_1}{\omega} = \frac{1+4s}{2+4t}$ and $\frac{\omega_1}{\omega} = \frac{3 + 4s}{2 + 4t}$, where s and t are integers. This condition reads

$$\frac{H}{\beta} \geq \frac{\sinh(\omega K'(k))}{\omega^2} \left[\frac{1}{\pi}[E(k) - k'^2 K(k)] - \frac{1}{4}\frac{\rho_1}{\beta} \operatorname{sech}(\omega_1 K'(k))\right].$$

$$= H_{os}/\beta \tag{13}$$

The bifurcation curves in the $(\omega_1, H/\beta)$ parameter plane which separate regions with and without occurrence of subharmonics are given by the equality in formula (13). Three subharmonic bifurcation curves are shown in Fig. 1 a-c for $\rho_1/\beta = 0.2$ and for $(mp, mq) = (3,2)$, $(6,4)$ and $(9,6)$, respectively. Figure 1d depicts the corresponding heteroclinic bifurcation curve as obtained from formula (9). It can be shown that the subharmonic curves converge to the heteroclinic bifurcation curve when $m \to \infty$ [2,4]. It is interesting to note that the presence of an external driving torque stabilizes the subharmonic solutions in the sense that H_{os} decreases when ρ_1 increases.

The prediction for subharmonic bifurcation by the Melnikov theory (formula (13)) was tested by a numerical simulation. For parameter values $2 \sin^{-1}(k) = 1.6$, $\beta = 0.1$, $\rho_1 = 0.1$, $\omega_1 = \pi mp/2K(k)$, and $\omega = \pi mq/2K(k)$ with $mp = 1$ and $mq = 2$, formula (13) yields the critical value $H_{os} = 0.0147$. Figure 2 shows numerical solutions to Eqs. (5) for the same values of k, β, ρ_1, ω_1 and ω. For $H = 0.0144$ we get the stable subharmonic solution shown in Fig. 2a while a decrease in H to $H = 0.0143$ makes the subharmonic solution unstable as seen in Fig. 2b. The prediction by Melnikov's theory thus only deviates by 3%.

36.5 Conclusion

We have shown that a pendulum with variable length exhibits Smale horseshoe chaos for certain ranges of parameter values using the Melnikov theory. The pendulum is driven by two sinusoidal oscillating terms with commensurable frequencies. One term comes from an external torque and the other one comes from the varying length. The presence of the external oscillating torque leads to a larger area of the parameter plane where Smale horseshoe chaos may be present ($\rho_1 \neq 0$ in formula (9)). If only system (5) is considered we see from formula (9) (Fig. 1d) that when ω_1 goes to zero we do not get Smale horseshoe chaos. This is in agreement with the well-known classical phenomenon that the action variable

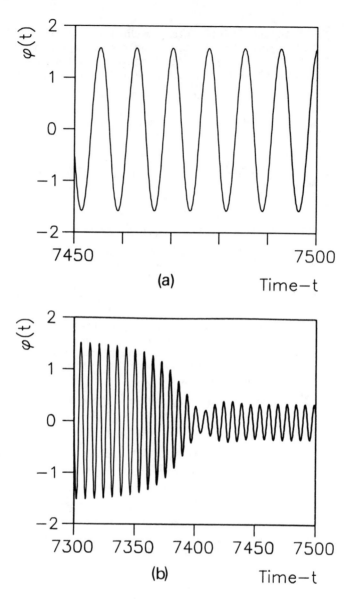

*Figure 2.   Numerical solution of Eqs. (5) for 2 $sin^{-1}(k)$ = 1.6,
β = 0.1, ρ_1 = 0.1, ω_1 = $\pi mp/2K(k)$, and ω = $\pi mq/2K(k)$ with mp = 1
and mq = 2.   (a) H = 0.0144.   (b) H = 0.0143.*

is an adiabatic invariant for this system when its length is
varied sufficiently slowly.

Furthermore, by using the subharmonic Melnikov function we have
obtained the bifurcation diagrams for the occurrence of
subharmonic orbits. By adding an external oscillating torque, the
critical value H_{os} decreases implying a stabilization of the

subharmonic solutions. Finally, the subharmonic analysis has been tested against direct numerical simulations and good agreement has been demonstrated.

Acknowledgements

The financial support of the Consiglio Nazionale della Ricerche, Roma, Italy to two of the authors (M.B. and V.M.) and of the European Research Office of the United States Army (through contract No. DAJA-45-85-C-0042) is acknowledged.

References

[1]   V.K. Melnikov (1963). *Trans. of Moscow Math. Soc.* **12**, 1.
[2]   M. Bartuccelli, P.L. Christiansen, V. Muto, M.P. Soerensen and N.F. Pedersen, to be published.
[3]   T. Levi-Civita and U. Amaldi (1927). *Lezioni di Meccanica Razionale,* Zanichelli, Bologna.
[4]   J. Guckenheimer and P. Holmes (1983). *Appl. Math. Sciences* *42,* Springer-Verlag, Berlin.
[5]   B.D. Greenspan and P.J. Holmes (1983) in *Nonlinear Dynamics and Turbulence,* 172, eds. G. Barenblatt, G. Iooss and D.D. Joseph, Pitmann, London.
[6]   P.F. Byrd and M.D. Friedman (1954). *Handbook of Elliptic Integrals for Engineers and Physicists,* Springer-Verlag, Berlin.

HOPF BIFURCATION IN FUNCTIONAL-DIFFERENTIAL SYSTEMS - COMPUTATIONAL ASPECTS

37.1 Introduction

The analysis of the nonlinear functional-differential equations arises in the study of models involving various kinds of delays. They occur in population dynamics, physiological models or in the description of some physical effects. In the present paper we will concentrate on computational aspects which one has to take into consideration in practical application of the Hopf bifurcation theorem for RFDE (retarded functional-differential equations). We will limit ourselves to the investigation of equations of the form

$$\dot{x}(t) = L(\alpha)x_t + f(\alpha, x_t) \tag{1}$$

where x_t denotes an element of the Banach space $C=C^o([-r,0],\mathbb{R}^n)$ (with the supremum norm) associated with the unknown function x by means of the relation $x_t(\theta)=x(t+\theta)$ for $\theta\in[-r,0]$, $L(\alpha)$ denotes the family of continuous linear operators from $\mathbb{C}$ to $\mathbb{R}^n$ depending on a parameter α in some Banach space (usually $\mathbb{R}$). The Riesz theorem implies that $L(\alpha)$ can be represented as Stieltjes integral of the form

$$L(\alpha)\phi = \int_{-r}^{0} [d\eta_\alpha(\theta)]\,\phi(\theta) \tag{2}$$

where η_α is a matrix function with bounded variation. In what follows we assume that (2) has no singular part, that is

$$L(\alpha)\phi = \sum_{k=0}^{\infty} A_k(\alpha)\,\phi(-r_k(\alpha)) + \int_{-r}^{0} A(\alpha,\phi)\,\phi(\theta)\,d\theta \tag{3}$$

where $r_o(\alpha)=0$, $r_k(\alpha)\in(0,r]$ for $k>0$, $A_k(\alpha)$ and $A(\alpha,\theta)$ are real nxn matrices. We assume also that $\phi \to f(\alpha,\phi)$ is a C^1 mapping from $\mathbb{C}$ into $\mathbb{R}^n$ satisfying

$$f(\alpha,0) = 0, \quad \frac{\partial f}{\partial \phi}(\alpha,0) = 0 \qquad (4)$$

for α belonging to some neighbourhood of $\alpha=\alpha_o$. As in the standard setting for the Hopf bifurcation theorem for RFDE, cf. [1], we assume about the differentiability with respect to α that for any continuously differentiable ϕ in $\mathbb{C}$ $L(\alpha)\phi$ and $f(\alpha,f)$ are continuously differentiable in α in some neighbourhood of $\alpha=\alpha_o$.

That means that these mappings do not have to be differentiable in α in product space topology with ϕ in $\mathbb{C}$. This implies that we admit the delays as bifurcation parameters.

37.2 Linear Stability Analysis

The linear part of (1)

$$\dot{x}(t) = L(\alpha) \; x_t \qquad (5)$$

has the characteristic matrix

$$\Delta(\alpha,\lambda) = \lambda \; I - L(\alpha)(e^{\lambda \cdot} I) \qquad (6)$$

whose determinant is an entire function of λ. In general we assume that the characteristic equation

$$\det \Delta(\alpha,\lambda) = 0 \qquad (7)$$

is satisfied for $\alpha=\alpha_o$ and $\lambda = \pm\lambda_o = \pm i$. Let λ_o be the simple root of $\det \Delta(\alpha_o,\lambda)$ and λ_o and $-\lambda_o$ be the only two roots of this function lying on the imaginary axis. It follows then from the implicit function theorem that there exists a C^1 function $\lambda(\alpha)$ defined in the neighbourhood of $\alpha=\alpha_o$ such that

$$\det \Delta(\alpha,\lambda(\alpha)) = 0 \text{ and } \lambda(\alpha_o) = \lambda_o. \qquad (8)$$

We will assume that for $\alpha<\alpha_o$ all other roots of (7) lie in the left half of the complex plane. There are different methods of checking the validity of the above assumption, methods based on the principle of argument, e.g. the method proposed in [4], the so-called τ-decomposition method originally introduced by Neimark in [3], or methods developed recently by many authors using algebraic elimination in the case of commensurate delays, cf. [2]. The characteristic equation may be then written in the form

$$\det \Delta(\alpha,\lambda) = \sum_{k=0}^{m} \sum_{j=0}^{n_k} q_{kj}(\alpha) \; \lambda^j \; e^{k\tau(\alpha) \lambda} \qquad (9)$$

where $q_{kj} \in \mathbb{R}$, $\tau(\alpha) > 0$, $q_{kn_k} \neq 0$, $n_m \geq n_k$ for $k=0,1,2,\ldots,m-1$.

For systems of this kind we propose the following numerical procedure for checking the above spectral assumptions.

We start with finding the intervals containing all α's satisfying

$$\sum_{\substack{k=0 \\ n_k=n_m}}^{m} q_{kn_k}(\alpha)\, z^k = 0 \Rightarrow |z| < 1 \quad \text{(stability of the} \quad\quad (10)$$
$$\text{"difference part"),}$$

$$\sum_{k=0}^{m} q_{ko}(\alpha) \neq 0 \quad\quad\quad \text{(no root at origin),} \quad\quad (11)$$

and

$$\forall\, \mu \in \mathbb{R}\ \exists\, k \in \{0,1,2,\ldots,m\}: \quad \sum_{j=0}^{n_k} q_{kj}(\alpha)\, (i\mu)^j \neq 0 \quad\quad (12)$$

(no common polynomial factor with a root on the imaginary axis). All the above conditions can be checked with standard polynomial techniques (e.g. Jury's test, Stodola's criterion).

In the second step we compute N_o, i.e. the number of the right half plane roots of $\sum_{k=0}^{m} \sum_{j=0}^{n_k} q_{kj}(\alpha)\, \lambda^j$ (polynomial in λ) using for example the standard Routh test.

Finally we look for those possible values α_o which are the boundary points of the intervals containing all α's satisfying both conditions (10), (11), (12) and the condition

$$N_o + 2 \sum_p \varepsilon_p \text{ Entier } \left[\frac{\beta_p(\alpha) - \tau(\alpha)\, \mu_p(\alpha)}{2\pi} \right] = 0 \quad\quad (13)$$

where the summation runs over all solutions of the following system of two polynomial equations in μ and ν ($\rho = 1$)

$$\begin{bmatrix} \nu & -1 \\ 1 & \nu \end{bmatrix}^m \begin{bmatrix} \text{Re} \sum_{k=0}^{m} \sum_{j=0}^{n_k} q_{kj}(\alpha)\, (i\mu)^j\, (\rho\frac{i-\nu}{i+\nu})^k \\ \text{Im} \sum_{k=0}^{m} \sum_{j=0}^{n_k} q_{kj}(\alpha)\, (i\mu)^j\, (\rho\frac{i-\nu}{i+\nu})k \end{bmatrix} = \begin{bmatrix} 0 \\ 0 \end{bmatrix}, \quad (14)$$

$$\beta_p = \begin{cases} 2 \arctan \nu_p, & \text{for } \nu_p \geq 0 \\ 2\pi - 2 \arctan \nu_p, & \text{for } \nu_p < 0 \end{cases} \quad\quad (15)$$

$$\varepsilon_p = \left\{ \begin{array}{l} 1, \text{ when } \dfrac{\partial \rho}{\partial \mu} > 0 \\[2ex] -1, \text{ when } \dfrac{\partial \rho}{\partial \mu} < 0 \end{array} \right. \tag{16}$$

(in the case of the derivative equal to 0 one should take the sign of the first nonzero derivative if its order is odd, otherwise the point is irrelevant). The most convenient method to solve the system (14) is to use a recursive procedure for computing its resultant in μ. Having done that we can immediately find corresponding solutions in ν using subresultants. Then the derivative in (16) can be evaluated by implicit differentiation. The last thing left is the identification of the term in (13) for which the argument of the Entier function has an integer value for $\alpha = \alpha_o$. Let us assume that it is unique (otherwise we have more than one purely imaginary root of (9)) and that the corresponding eigenvalue $i\mu_p$ is simple. We take this eigenvalue as λ_o with suitable rescaling and parametrization reversal if needed.

37.3 Transversality Condition

The next assumption guarantees the uniqueness of the periodic orbit appearing when α passes the critical value α_o. We assume that $\operatorname{Re}\lambda'(\alpha_o) \neq 0$ (' will denote here differentiation with respect to α). We will discuss here computational methods for checking this condition. Using the theory presented in [1] one can associate with the family of linear equations (5) and the sets of corresponding eigenvalues $\{\lambda(\alpha), -\lambda(\alpha)\}$ decompositions of the space C in the form $C = P_\alpha \oplus Q_\alpha$ where P_α is a two-dimensional subspace consisting of real parts of functions being linear combinations of ϕ_α and $\bar{\phi}_\alpha$ where $\phi_\alpha = e^{\lambda(\alpha) \cdot} b(\alpha)$ and $b(\alpha)$ is a complex n-vector satisfying $\Delta(\alpha, \lambda(\alpha)) b(\alpha) = 0$. Let us define also functions $\psi(\alpha)$ and $\bar{\psi}(\alpha)$ from the space formally adjoint to C. Let $\psi(\alpha) = a(\alpha) e^{-\lambda(\alpha) \cdot}$ where $a(\alpha)$ is a complex row vector of dimension n satisfying $a(\alpha)\, \Delta(\alpha, \lambda(\alpha)) = 0$. We impose on ϕ_α and ψ_α the normalization condition

$$\psi_\alpha(0)\, \phi_\alpha(0) - \int_{-r}^{0} \int_0^\theta \psi_\alpha(\xi-\theta)\, [d\eta_\alpha(\theta)]\, \phi_\alpha(\xi)\, d\zeta = 1 \tag{17}$$

Then it can be shown for families with C^1 dependence on α that

$$\operatorname{Re} \lambda'(\alpha) = \operatorname{Re} [a(\alpha)\, L'(\alpha)\, (e^{\lambda(\alpha) \cdot} b(\alpha))]. \tag{18}$$

Unhappily the result [1, chapter 7, lemma 3.9] stating that

$$\begin{bmatrix} \mathrm{Re}\lambda'(\alpha) & \mathrm{Im}\lambda'(\alpha) \\ -\mathrm{Im}\lambda'(\alpha) & \mathrm{Re}\lambda'(\alpha) \end{bmatrix} = \begin{bmatrix} \underline{a}(\alpha) \\ \underline{a}(\alpha) \end{bmatrix} L'(\alpha)\{[b(\alpha),\bar{b}(\alpha)]\exp\begin{bmatrix} \mathrm{Re}\lambda(\alpha) & \mathrm{Im}\lambda(\alpha) \\ -\mathrm{Im}\lambda(\alpha) & \mathrm{Re}\lambda(\alpha) \end{bmatrix}\}$$

does not hold true, cf. [5]. However, it can be proved that the trace of the matrix standing on the right side of the above formula is really equal to 2 Re $\lambda'(\alpha)$, which can be used for practical computational purposes.

37.4 Conclusions

We have shown how certain algebraic techniques can be used to obtain efficient procedures for checking the assumptions of the Hopf bifurcation theorem for RFDEs. The numerical process proposed for checking the spectral assumption involves only polynomial algorithms. The whole "transcendental nature" of the discussed equations is hidden in the relation (15). The transversality condition is checked using the method avoiding complex numbers computations.

Acknowledgement

This work was supported by the Polish Ministy of Higher Education and Technology under the contract PR.I. ASO 2.1/1986.

References

[1] J.K. Hale (1977). *Theory of functional-differential equations,* Springer, New York.
[2] D. Hertz, E.I. Jury and E. Zeheb (1984). *J. Frankl. Inst.* **318**, 143.
[3] Yu. I. Neimark (1949). *Prikl. Mat. Mekh.* **13**, 349.
[4] G. Stépán (1979). *Coll. Math. Soc. J. Bolyai* **30**, 971.
[5] M. Szymkat 91986). *Hopf bifurcation theorem and its generalization,* Inst. Math. Jagiell. Univ., Krakow.

38 *M.P. Soerensen, T. Schneider, A. Politi, E. Tosatti and*
    M. Zannetti

NUMERICAL RESULTS ON THE RELATIONSHIP BETWEEN CLASSICAL DIFFUSION AND QUANTUM SYSTEMS

Classical diffusion in spatially-continuous random media is investigated numerically by using the mapping to a disordered quantum system. Calculation of the characteristic function allows estimating the integrated density of states and the inverse localization length.

In previous articles [1-3], we established the relationship between a diffusion problem of a classical particle in a random environment [4] and an associated quantum system, the latter describing the motion of a quantum particle in a random potential, so that localization might occur [5-7]. In doing so, we use the connection between the Langevin equation, describing classical diffusion, and the Fokker-Planck equation. The latter is then reduced to an imaginary-time Schrödinger equation, defining the Hamiltonian of the associated quantum system.

The localization properties of the eigenstates are most conveniently studied by calculating the characteristic function of the quantum problem [8]. Its imaginary part yields the integrated density of states while the inverse of the real part is the localization length. Two different models, denoted by A and B, are studied. Model A corresponds to a trapped classical particle, while model B exhibits sublinear long-time diffusive behaviour [9] of the Sinai type. Finally, we consider the standard Anderson model to illustrate the differences with respect to quantum models generated by mapping from an underlying classical diffusion problem.

To map the diffusion of a classical particle in a random medium on to an associated Schrödinger equation with a random potential, we proceed as follows [1,3]: diffusion is described by the Langevin equation

$$\dot{x} = -\frac{\partial W}{\partial x} + \eta(t), \qquad (1)$$

where x denotes the position of the particle diffusing in the random potential W(x), and $\eta(t)$ is Gaussian white noise with $<\eta(t)> = 0$ and $<\eta(t)\eta(t')> = \sigma\delta(t - t')$. The associated Fokker-Planck equation can be transformed into the imaginary-time Schrödinger equation [1,3]

517

$$-\frac{\partial \psi}{\partial t} = \mathcal{H}\psi, \quad \mathcal{H} = -\frac{\sigma}{2}\frac{\partial^2}{\partial x^2} + V(x). \tag{2}$$

The random quantum potential $V(x)$ and the classical potential $W(x)$ are related by the Riccati equation

$$V(x) = \frac{1}{2\sigma}\left(\frac{\partial W}{\partial x}\right)^2 - \frac{1}{2}\frac{\partial^2 W}{\partial x^2}. \tag{3}$$

The eigenfunctions and eigenvalues of $\mathcal{H}$ are given by $\mathcal{H}\phi_n = \lambda_n\phi_n$ and, by construction, the ground state is $\phi_0 \sim \exp(-W/\sigma)$ with zero energy, i.e. $\lambda_0 = 0$.

Expanding the transition probability of the Langevin process in terms of the eigenfunctions ϕ_n of $\mathcal{H}$, it can be shown that the time-dependent mean-square displacement satisfies [1,3]

$$\lim_{t\to\infty} <<x^2(t)>> = <<\phi_0|x^2|\phi_0>>, \tag{4}$$

$<<...>>$ denoting averages with respect to the Gaussian noise, and to the probability measure of the random drift potential W. Eq. (4) implies that in the quantum analog, diffusive motion leads to a ground state with inifinite variance, which in turn excludes exponential localization. On the other hand, an exponentially localized ground state corresponds to a classical particle which is trapped.

To characterize localization, we introduce the characteristic function [8]

$$\Gamma(\omega) = \int_0^\infty \ln(\Omega + z)\rho(z)dz. \tag{5}$$

The inverse exponential localization length $\gamma'(\omega)$ and the integrated density of states $N(\omega) = \gamma''(\omega)$ then follow from [1,3]

$$\Gamma(-\omega + i0^+) = \gamma'(\omega) + i\pi\gamma''(\omega). \tag{6}$$

To illustrate the above mapping, we performed numerical simulations on two specific models referred to as A and B [9]. Model A is defined by the classical potential

$$W(x) = \sum_{\ell=1}^{N} q_\ell[\theta(x - (\ell - \tfrac{1}{2})a) + \theta(-x-(\ell - \tfrac{1}{2})a)], \tag{7}$$

where

$$\theta(y) = \int_{-\infty}^{y} \frac{1}{\sqrt{\pi\varepsilon}} \exp(-\frac{z^2}{\varepsilon})dz. \tag{8}$$

The amplitudes q_ℓ are independent random variables distributed according to

$$P(q_\ell) = \begin{cases} \dfrac{1}{2\Delta} & \text{for } q_0 - \Delta \leq q_\ell \leq q_0 + \Delta \\ 0 & \text{otherwise.} \end{cases} \qquad (9)$$

In the numerical simulations, the following parameter values were used: $\varepsilon = 0.01$, $\Delta = 0.2$, $q_0 = 1$, $a = 2$, $N = 300$, and $\sigma = 2$. The resulting potential $W(x)$ is symmetric around $x = 0$, and increases monotonically with $|x|$, leading to a ground-state wave function $\phi_0 \sim \exp(-W/\sigma)$ with an exponentially decaying envelope [1]. Hence, the mean square displacement of the quantum particle is finite and the diffusing classical particle gets trapped. The associated quantum system has been solved numerically using Gill's fourth-order Runge-Kutta method with space-step $\Delta x = 0.005$ and boundary condition $d/dx(\ln\phi(x = 0)) = 0$. Figure 1a shows the numerical results for the inverse localization length $\gamma'(\omega)$, and the integrated density of states $N(\omega)$ as obtained from averaging over five independent replicas. $N(\omega)$ was obtained in terms of node counting [6]. The figure also reveals a gap ($\Delta\omega \sim 0.05$) between the ground and excited states. For the excited states, the inverse localization length is about one order of magnitude smaller than for the ground state and, as ω increases, the degree of localization decreases further.

Model B is an example where sublinear diffusion, as studied by Sinai [9], is expected to occur. In this model, the drift potential is defined by

$$W(x) = \sum_{\ell=1}^{N} q_\ell\, \theta(x - (\ell - \tfrac{1}{2})a) \qquad (10)$$

describing a one-dimensional random walk. The random, independent amplitudes q_ℓ are distributed according to

$$P(q_\ell) = \begin{cases} \dfrac{1}{2\Delta} & \text{for } -\Delta \leq q_\ell \leq \Delta \\ 0 & \text{otherwise.} \end{cases} \qquad (11)$$

The function $\theta(y)$ is given by eq. (8). In the numerical simulations, the following parameter values were used: $\varepsilon = 0.01$, $\Delta = 2$, $a = 1$, $N = 1200$, and $\sigma = 2$. Sinai proved that in the long-time limit the mean-square displacement evolves as $\langle\langle x^2(t)\rangle\rangle \sim \ln^4 t$. Accordingly, the associated quantum system possesses an extended ground state. In Fig. 1b, we show numerical results of $\gamma'(\omega)$ and $N(\omega)$ for model B, using the same space-step and boundary condition as for model A and averaging over 10 replicas. The real part of the characteristic function rapidly approaches zero when $\omega \to 0$, signalling the extended ground state.

To illustrate the differences with respect to a standard continuous Anderson model [5], we next consider the Schrödinger equation with potential

$$V(x) = \sum_{\ell=1}^{N} q_\ell \theta(x - (\ell - \tfrac{1}{2})a), \qquad (12)$$

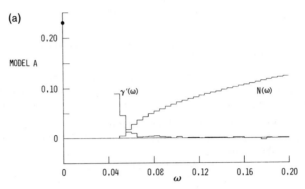

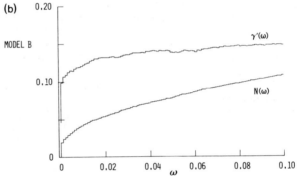

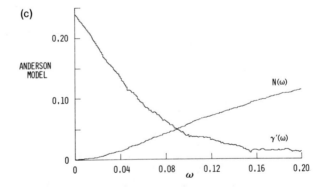

Figure 1. Numerical results for the integrated density of states $N(\omega)$ and the inverse localization length $\gamma'(\omega)$ [eq. (5)] for the continuum models. (a) Model A [eqs. (7)-(9)] for $\varepsilon = 0.01$, $\Delta = 0.2$, $q_0 = 1$, $a = 2$, $N = 300$, and $\sigma = 2$. (b) Model B[eqs. (10)-(11)] for $\varepsilon = 0.01$, $\Delta = 2$, $a = 1$, $N = 1200$, and $\sigma = 2$. (c) The Anderson model [eq. (12)] for $\varepsilon = 0.01$, $\Delta = 0.2$, $a = 1$, $N = 1200$, and $\sigma = 2$.

where $\theta(y)$ is given by eq. (8), and independent amplitudes q_ℓ are distributed according to eq. (11) with parameters $\varepsilon = 0.01$, $\Delta = 0.2$, $a = 1$, $N = 1200$, and $\sigma = 2$. The Schrödinger equation was solved using the above Runge–Kutta algorithm and with the same space-step and boundary condition as for model A. In Fig. 1c, we again show the inverse localization length $\gamma'(\omega)$ together with the integrated density of states $N(\omega)$ averaged over five independent replicas. The finite value of $\gamma'(\omega)$ for low frequencies reveals localization of the corresponding eigenstates and, in particular, of the ground state. For higher values of ω, the inverse localization length $\gamma'(\omega)$ decreases.

In summary, we have numerically studied the connection between diffusion in random media and the motion of a quantum particle in a random potential. In terms of the characteristic function, the integrated density of states and the inverse localization length have been numerically obtained for two different models. The ground-state wave functions are exactly known, yielding exponential localization in model A and an extended ground state ini model B. This difference becomes particularly transparent by again emphasizing the diffusive behaviour of the classical analogs. In model A, the particle becomes trapped, while diffusion occurs in model B. Accordingly, the mean-square displacement diverges in the latter case, excluding exponential localization of the ground-state wave function by construction.

Acknowledgement

It is our pleasure to acknowledge the financial support from the IBM Zurich Research Laboratory, IBM Denmark, and IBM Italy.

References

[1]   T. Schneider, M.P. Soerensen, E. Tosatti and M. Zannetti (1986). *Europhys. Lett.* **2**, 167.

[2]   T. Schneider, M.P. Soerensen, A. Politi and M. Zannetti (1986). *Phys. Rev. Lett.* **56**, 2341.

[3]   T. Schneider, in these Proceedings.

[4]   S. Alexander, J. Bernasconi, W.R. Schneider and R. Orbach (1981). *Rev. Mod. Phys.* **53**, 175.

[5]   P.W. Anderson (1958). *Phys. Rev.* **109**, 1492.

[6]   B.J. Halperin (1967). *Adv. Chem. Phys.* **13**, 123.

[7]   H. Kunz and B. Souillard (1980). *Commun. Math. Phys.* **78**, 201.

[8]   T.M. Nieuwenhuizen (1982). *Physica* **113A**, 173.

[9]   Ya. Sinai (1982) in: *Proc. Berlin Conference on Mathematical Problems in Theoretical Physics*, p. 12, eds. R. Schrader, R. Seiler and D.A. Uhlenbrock, Springer-Verlag, Berlin.

39 *S. Wojciechowski*

CONSTRUCTION OF INTEGRABLE RICCATI SYSTEMS BY THE USE OF LOW DIMENSIONAL LIE ALGEBRAS

Abstract

A method of constructing large multiparameter family of integrable dynamical systems of few degrees of freedom with quadratic nonlinearities is presented. It is based on the use of low dimensional Lie algebras and their Casimir operators. Some examples are given.

Riccati systems of equations (R-E)

$$\dot{e}_k = \gamma_k + \sum_{r=1}^{n} \beta_k^r e_r + \sum_{r=1}^{n} \sum_{s=1}^{n} \alpha_k^{rs} e_r e_s \qquad (1)$$

naturally arise as first nonlinear terms of Taylor expansion of a general vector field (γ_k, β_k^r, α_k^{rs} are here parameters). They also appear in many applications like chemical rate equations and ecological systems. There is not, as yet, a general method of studying a given system of R-E and, in particular, determining whether it has integrals of motion. Here we approach this problem differently by constructing R-E with *a priori* given integrals.

The purpose of this paper is to outline a method which leads to the construction of a large number of multiparameter families of integrable equations (1). This may be useful in the classification of all equations of this type.

The method is based on the use of real low dimensional Lie algebras and their invariants known as Casimir operators. We will make use of complete classification of all indecomposable Lie algebras of dimension ≤5 [1] and of the nilpotent algebras of dimension six [2]. Their invariants have been found and have been listed in [3] together with the commutation relations for the algebras.

For a given Lie algebra

$$[E_i, E_j] = \sum_{k=1}^{n} c_{ij}^k E_k \qquad (2)$$

we can define, on a space of dynamical variables $e_1, \ldots, e_n$, a

Lie-Poisson bracket (LPb) [4]:

$$\{e_i, e_j\} = \sum_{k=1}^{n} c_{ij}^{i} e_k \tag{3}$$

which is also bilinear, skew-symmetric and satisfies the Jacobi identity. This definition of the LPb extends to arbitrary functions f(e) by setting

$$\{e_k, f(e)\} = \sum_{r=1}^{n} \{e_k, e_r\} \frac{\partial f}{\partial e_r}.$$

The Casimir operators $C_k(E_1, \ldots, E_n)$, k=1,...,n on a Lie algebra become the Casimir functions $C_k(e_1, \ldots, e_n)$ which have zero LPb (commute) with all dynamical variables.

If on a space of dynamical variables $e_1, \ldots, e_n$ we define Hamilton's dynamical equations by the formula

$$\dot{e}_k = \{e_k, H(e_1, \ldots, e_n)\} \qquad k=1, \ldots, n \tag{4}$$

then the Casimir functions are integrals of motion which are functionally independent of the hamiltonian $H(e_1, \ldots, e_n)$. If the hamiltonian were a function H=f(C(e)) of the invariant then the equations of motion would be trivial.

With the foregoing in mind we examine now the list of low dimensional Lie algebras and their Casimir operators given in [3]. There is only one real algebra $A_{1,1}$ of dimension one; it has one Casimir operator which is its single element. There is one real algebra $A_{2,1}$ of dimension two. The commutation rule is $[E_1, E_2] = E_2$, it is solvable and has no Casimir operator. The hamiltonian equations (4) have an integral of motion - the hamiltonian itself - and are obviously integrable: trajectories are labelled by the value of the hamiltonian H.

The first nontrivial examples are in the dimension three. There are 9 real three-dimensional Lie algebras, two of which depend on parameters and hence constitute continua of algebras. Each algebra has one invariant C(e) which together with the hamiltonian H(e) gives two integrals of motion. Each integral defines a foliation of $\mathbb{R}^3$ into two dimensional manifolds and trajectories are the intersections of leaves. They are labelled by the values of H and C.

A dynamical system for which trajectories are determined by the values of integrals of motion is called **completely degenerate.** All three-dimensional dynamical systems defined above are, therefore, completely degenerate. If the manifold of constant value of one of the integrals C or H is compact then the motion is strictly periodic. Solving equations of motion reduces then to finding the proper time parametrization of trajectories.

A well known example of such a system is the Euler integrable case of motion of rigid body around centre of mass. Equations of motion are generated by the hamiltonian $H = \alpha_1 e_1^2 + \alpha_2 e_2^2 + \alpha_3 e_3^2$ and by the LPb of the su(2) algebra: $\{e_1, e_2\} = e_3$, $\{e_2, e_3\} = e_1$, $\{e_3, e_1\} = e_2$. The Casimir function $C = e_1^2 + e_2^2 + e_3^2$ is physically interpreted as total angular momentum of the rigid body.

As another example we take the Lie algebra $A_{3,7}^a$

$$\{e_1, e_3\} = ae_1 - e_2, \quad \{e_2, e_3\} = e_1 + ae_2 \quad (a>0)$$

which depends on the parameter a and has an invariant

$$C = (e_1^2 + e_2^2)[(e_1 + ie_2)/(e_1 - ie_2)]^{ia} =$$

$$(e_1^2 + e_2^2)\exp(-2a \ \mathrm{arctg}(e_2/e_1)); \quad i = \sqrt{-1}.$$

The most general hamiltonian which under the LPb generates R-E is of the second order

$$H = \alpha_1 e_1^2 + \alpha_2 e_2^2 + \alpha_3 e_3^2 + \beta_1 e_2 e_3 + \beta_2 e_3 e_1 + \beta_3 e_1 e_2 + \gamma_1 e_1 + \gamma_2 e_2 + \gamma_3 e_3$$

and the dynamical equations read

$$\dot{e}_1 = \{e_1, H\}$$

$$= a\beta_2 e_1^2 - a\beta_1 e_2^2 - 2\alpha_3 e_2 e_3 + 2a\alpha_3 e_3 e_1 + (a\beta_1 - \beta_2)e_1 e_2 + \gamma_3 ae_1 - \gamma_3 e$$

$$\dot{e}_2 = \{e_2, H\}$$

$$= \beta_2 e_1^2 + a\beta_1 e_2^2 + 2a\alpha_3 e_2 e_3 + 2\alpha_3 e_3 e_1 + (\beta_1 + a\beta_2)e_1 e_2 + \gamma_3 e_1 + a\gamma_3 e_2$$

$$\dot{e}_3 = \{e_3, H\}$$

$$= -(2a\alpha_1 + \beta_3)e_1^2 - (2a\alpha_2 - \beta_3)e_2^2 - (a\beta_1 - \beta_2)e_2 e_3 - (\beta_1 + a\beta_2)e_3 e_2$$

$$+ (2\alpha_1 - 2\alpha_2 - 2a\beta_3)e_1 e_2 - (a\gamma_1 + \gamma_2)e_1 - (a\gamma_2 - \gamma_1)e_2$$

These equations depend on 10 parameters (!) and are not only integrable but even completely degenerate having two independent integrals C and H. Integrable R-E generated by other 3-dimensional Lie algebras can be written down in the same way.

There are 12 Lie algebras of dimension four, 5 of which depend on parameters. Either of these algebras have two Casimir invariants C_1, C_2 and the Hamiltonian equations following from the general hamiltonian of the second order are completely degenerate. The remaining 4 have no invariant, i.e. the LPb is nondegenerate.

We need an additional integral of motion to render the Hamiltonian equations integrable in the sense of the Liouville-Arnold theorem [5]. Such an integral may only exist for particular values of the parameters in the hamiltonian and have to be determined by other methods.

Five dimensional Lie algebras can be used to generate integrable Riccati equations in a completely analogous way. There are 40 such algebras, 15 of which have three Casimir invariants C_1, C_2, C_3 which, together with the hamiltonian H, render the dynamical system completely integrable. The remaining 25 algebras have only one invariant, C_1 say, and we need an additional integral of motion to have an integrable hamiltonian H.

Six dimensional Lie algebras are classified only in nilpotent case, and these can be employed in a similar way.

So far we have considered generation of integrable R-E by the direct use of LPb. Another way is to perform reduction of a more dimensional integrable R-E. Reduction of R-E leads to another R-E only if an invariant relation between the variables is linear. For instance if one of the equations is of the form $\dot{e}_k = e_k f(e)$ then the condition $e_k = 0$ is compatible with the equations. Sometimes it may be even a linear integral of motion which defines global foliation in the space of dynamical variables. An example is provided by the algebra $A_{4,1}$ defined by the LPb relations $\{e_2, e_4\} = e_1$, $\{e_3 e_4\} = e_2$. Casimir invariants are $C_1 = e_1$, $C_2 = e_2^2 - 2e_1 e_3$ and therefore the equation on e_1 is trivial: $\dot{e}_1 = 0$. Thus the 4-dimensional algebra $A_{4,1}$ gives rise to effectively 3-dimensional integrable R-E involving constant terms and, therefore, not possible to generate directly as $\dot{e}_k = \{e_k, H(e)\}$ from 3-dimensional Lie algebra.

We have discussed here R-E which follow from a quadratic hamiltonian. The method applies, however, to any hamiltonian even with coefficients depending on time. Then the Casimir invariants are always integrals but the time-dependent hamiltonian is not.

Restriction to Lie algebras is not necessary and has only this advantage that Lie algebras and their Casimir operators are best known. More general algebras can be used for generation of dynamical systems. If an algebra is skew then the hamiltonian still remains to be an integral. Otherwise algebras possessing Casimir operators are of interest. Such a generalized approach seems to be very promising in the case of n=3 and gives a chance of characterizing all integrable R-E - at least in the homogeneous case when H(e) contains no linear terms. Research on these questions is in progress.

Acknowledgement

This research was partially supported by NFR contract F-FU 8677-100,KTO: 518677100-1.

References

[1]   G. M. Mubarakzyanov (1963).   *Izv. Vysshikh. Uchebn. Zavedenii Mat.* **1(32)**, 114; (1963).   **3(34)**, 99; (1963).   **4(35)**, 104.
[2]   V. V. Morozov (1958).   *Izv. Vysshikh. Uchebn. Zavedenii Mat.* **4(5)**, 161.
[3]   J. Patera, R. T. Sharp, P. Winternitz and H. Zasenhaus, *J.* (1976).   *Math. Phys.* **17**, 986.
[4]   A. Weinstein (1983).   *J. Diff. Geometry* **18**, 523.
[5]   V. I. Arnold (1978).   *Mathematical methods of classical mechanics,*   Springer, Berlin.

FINITE-TIME THERMODYNAMICS

40.1 Motivation

From its infancy, over 150 years ago, thermodynamics has provided limits on work or heat exchanged during real processes. The first problem treated in a systematic way was how much work a steam engine can produce from the burning of one ton of coal. With true scientific generalization Sadit Carnot concluded that *any* engine taking in heat from a hot reservoir at temperature T_H has to deposit some of that heat in a cold reservoir (e.g. the surroundings), whose temperature we call T_L; the largest fraction of the heat which can be converted into work is

$$\eta_C = 1 - T_L/T_H, \qquad (1)$$

traditionally known as the Carnot efficiency. Later related criteria of performance emerged, such as the effectiveness ε, the ratio of the actual work supplied by a work source to the change in availability of the system.

All these criteria have long been common currency for thermodynamic studies in physics, chemistry, and engineering. They all share one characteristic: the ideal to which any real process is compared is a **reversible process**. Stated in a different way, **traditional thermodynamics is a theory about equilibrium states and about limits on process variables for transformations from one equilibrium state to another.** Nowhere does time enter the formulation, so these limits must be the lossless, reversible processes which proceed infinitely slowly and thus take an infinite length of time to complete. Are these reversible limits close enough to real performances to be useful comparisons and criteria? For some processes the answer is yes – large steam turbines, and the industrial fusion of hydrogen and nitrogen to form ammonia are two examples – but in most cases more realistic limits to the performance of real processes are needed. Some processes, such as the acoustic heat pump and the motion of microorganisms by use of flagella, even exist which have no reversible limits, they rely on irreversibilities for their operation. Is it possible then to find general bounds for the costs of operating those processes in finite time, and can one use such bounds to find better criteria of merit for real processes?

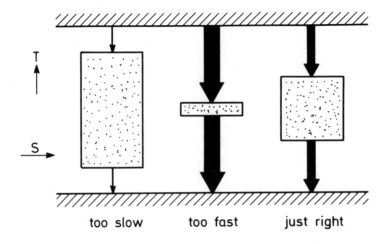

T

S

too slow   too fast   just right

Figure 1. In this pseudo temperature-entropy plot for a heat engine operating between a hot and a cold reservoir, the area of each rectangle is proportional to the amount of work produced in each cycle. For slow operation, losses associated with heat transfer to and from the engine are small, but there are only a few cycles per second. Rapid operation achieves many cycles per second, but each one delivers little work. Maximum power is obtained at a compromise between rate and losses, and is therefore necessarily accompanied by a loss of efficiency compared with reversible operation.

40.2   Finite-time Thermodynamics

Finite-time thermodynamics does just that. It is the **extension of traditional thermodynamics to deal with processes which have explicit time or rate dependences.** These constraints, of course, imply a certain amount of loss, or entropy production, which is at the heart of the questions posed above. Figure 1 shows in an intuitive way why losses are unavoidable.

The spectrum of methods developed so far in finite-time thermodynamics include the construction of time-dependent thermodynamic potentials, optimal control theory to calculate the optimal time path, and the use of a thermodynamic length to bound dissipation in a process. These methods will be elaborated on below.

The processes for which, for example, the generalized potentials and time paths are to be constructed are not supposed to be extremely detailed descriptions of a specific real system, including all interconnections and losses – that would be duplicating the highly accurate, but far from general, simulations of the practical engineer. Rather, finite-time thermodynamic systems and processes are idealized models which contain what is typical of a certain class of real systems. The construction of such **generic models** to represent broad classes of processes is central to finite-time thermodynamics. Each generic model should contain all the important qualities of the type of real systems studied, but not the individual details which would only obscure the physical content and make calculations very difficult or impossible. We are already used to such models in traditional thermodynamics: the Carnot engine is, for example, the highly idealized reversible representation of all heat engines.

The first "improvement" of the Carnot engine is the endoreversible engine shown in Fig. 2 (endoreversible means that all irreversibilities are located in the coupling of the system (engine) to its surroundings, there are no internal irreversibilities). It is a Carnot engine, signified by the triangle, with the simple constraint that it be linked to its reservoirs through **finite** heat conductances. The maximum efficiency of this engine is of course $\eta_C = 1 - T_L/T_H$, obtained at zero rate so that losses across the resistors vanish, but when the system operates to produce **maximum power**, the efficiency of the engine is only [1]

$$\eta_w = 1 - \sqrt{T_L/T_H}, \qquad (2)$$

surprisingly independent of the magnitude of the conductances. This efficiency shows much closer agreement [1] with the empirically determined efficiencies of power plants than does η_C.

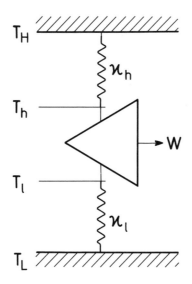

Figure 2. An endoreversible engine has all its losses associated with its coupling to the environment, there are no internal irreversibilities. This is illustrated here as resistances in the flows of heat to and from the working device indicated by a triangle. These unavoidable resistances cause the engine proper to work across a smaller temperature interval, $[T_h; T_l]$, than that between the reservoirs, $[T_H; T_L]$, one which depends on the rate of operation.

40.3 **Generalized Potentials**

In traditional thermodynamics the decrease in a thermodynamic potential $\mathcal{P}$ from state i to state f is equal to the amount of work that is produced when a reversible process carries out the transition under the given constraints, and hence is the upper bound for the amount of work produced by any other process with the same constraints,

$$W \leq W_{rev} = \mathcal{P}_i - \mathcal{P}_f. \tag{3}$$

The constraints are usually the constancy of some state variables like pressure, volume, temperature, entropy, chemical potential, particle number, etc. However, it is possible to introduce almost any constraints, including constraints on time or rate, and use the same type of Legendre transformation which led to Eq. (3) to derive generalized thermodynamic potentials with the same bounding quality on the work recovered during a constrained process [2].

40.4 **Thermodynamic Length**

In an attempt to develop a more transparent way of calculating all the well known partial derivatives in traditional thermodynamics Weinhold identified [3] the quantities $\partial^2 U/\partial X_i \partial X_j$, where U is the internal energy, and X_i and X_j are extensive variables, as a metric on the abstract space of equilibrium states of a system. He only used the metric at single points to calculate scalar products, but if, instead, one calculates lengths between equilibrium states a wide range of dynamic information becomes available. The length L_U itself turns out [4] to be the change in some molecular velocities, depending, of course, on the constraints of the process (isobaric, isochoric, etc.).
Even more important is the bound [5]

$$-\Delta A \geq L_U^2 \, \varepsilon/\tau, \tag{4}$$

where $-\Delta A$ is the loss of availability during a process, ε is the internal relaxation time of the system, and τ is the duration of the process. For comparison, traditional therodynamics only states that $-\Delta A \geq 0$, so the bound Eq. (4) is a considerable improvement and actually displays the cost of haste. A similar bound on the entropy production during a process,

$$\Delta S^u \geq L_S^2 \, \varepsilon/\tau, \tag{5}$$

where the length L_S is calculated with the entropy metric $\partial^2 S/\partial X_i \partial X_j$, has been tested for both microscopic and macroscopic systems [6]. The two metrics (in the same coordinates!) are related [7] by the usual factor of $-T$.
For complicated systems the internal relaxation time ε may be

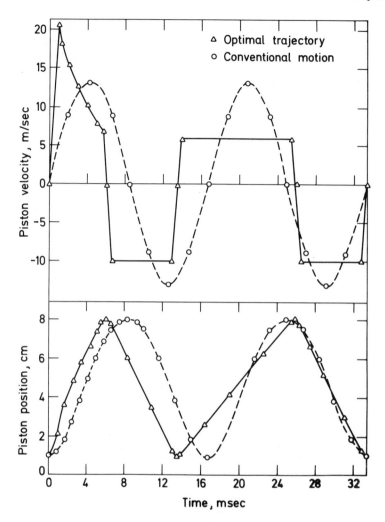

Figure 3. The time path of the Otto cycle has been optimized for maximum work output with all parameters fixed at values representative of a real car engine. By extracting work rapidly before the hot combustion gases cool too much on the cylinder wall, and by reducing friction in the remaining three strokes through constant piston speed, the power output can be increased by 8 to 15%.

difficult to obtain. In those cases the new method of simulated annealing has proven very successful, and we are currently developing methods for calculating ε dynamically.

40.5 Optimal Path

In some cases knowledge of the maximal work that can be extracted during a process, e.g. calculated by one of the methods mentioned above, may not be sufficient. One needs to know **how** this maximum work can be extracted, i.e. the time path of the thermodynamic variables. In reversible thermodynamics this is either irrelevant (any reversible path will do) or obvious (the path is given by the reversible constraint, but without time consideration, of course). In finite-time thermodynamics a specific time path is required, and one must often resort to optimal control theory to obtain it.

The optimal path for the endoreversible engine of Fig. 1 yielding maximum power at the efficiency in Eq. (2) has been calculated by Rubin [8] and is, not surprisingly, quite Carnot-like.

A more realistic example is the optimization [9] of the Otto cycle (ordinary gasoline engine), where a variation of the time path, keeping the speed and fuel consumption fixed, results in an increase of engine efficiency of up to 15%. Figure 3 compares the standard and optimal time paths.

More extensive reviews of finite-time thermodynamics may be found in Ref. [10].

Acknowledgements

I am indebted to all my collaborators, especially R. Stephen Berry and Peter Salamon.

References

[1] F.L. Curzon and B. Ahlborn (1975). *Am. J. Phys.* **43**, 22.

[2] P. Salamon, B. Andresen and R.S. Berry (1977). *Phys. Rev.* **A15**, 2094.

[3] F. Weinhold (1975). *J. Chem. Phys.* **63**, 2479, 2484, 2488, 2496.

[4] P. Salamon, B. Andresen, P.D. Gait and R.S. Berry (1980). *J. Chem. Phys.* **73**, 1001; (1980). **75**, 5407E.

[5] P. Salamon and R.S. Berry (9893). *Phys. Rev. Lett.* **51**, 1127.

[6] P. Salamon, J.D. Nulton and R.S. Berry (1985). *J. Chem. Phys.* **82**, 2433; T. Feldmann, B. Andresen, A. Qi and P. Salamon (1985). *J. Chem. Phys.* **83**, 5849.

[7] P. Salamon, J.D. Nulton and E. Ihrig (1984). *J. Chem. Phys.* **80**, 436.

[8] M.H. Rubin (1979). *Phys. Rev.* **A19**, 1272, 1277.

[9] M. Mozurkewich and R.S. Berry (1981). *Proc. Natl. Acad. Sci. U.S.A.* **78**, 1986; (1982). *J. Appl. Phys.* **53**, 34.

[10] B. Andresen, P. Salamon and R.S. Berry (1984). *Physics Today* **37(9)**, 62; (1984). B. Andresen, R.S. Berry, M.J. Ondrechen and P. Salamon, *Acc. Chem. Res.* **17**, 266.

,41   H.C. Fogedby, K. Osano and H.J. Jensen

QUANTUM CORRECTIONS TO THE SPECIFIC HEAT OF THE EASY-PLANE FERROMAGNETIC CHAIN

41.1 Introduction

One-dimensional nonlinear soliton bearing condensed matter systems have received much attention in recent years. In this context the properties of the magnetic salts $CsNiF_3$ and $(C_6H_{11}NH_3)CuBr_3$ (CHAB) are of particular interest. In the long wavelength limit these salts are described by the one-dimensional spin model [1,2]

$$H = \int dx [\frac{1}{2} j(\frac{d\bar{S}}{dx})^2 + a(S^z)^2 - hs^x].$$

(1)

This model is obtained from the discrete spin-1 and spin-1/2 Hamiltonians for $CsNiF_3$ and CHAB. The model parameters for $CsNiF_3$ and CHAB are [1,3]

$$j/k = 23.6K, \quad a/k = 9K, \quad h/k = .16H \text{ K (H in kG)} \qquad (2)$$

and [2]

$$j/k = 110.K, \quad a/k = 5.5k, \quad h/k = .13H \text{ K (H in kG)} \qquad (3)$$

for a **classical** spin field the model in Eq. (1) is **soliton bearing** [1]. Most recently, specific heat data obtained for $CsNiF_3$ [4] and CHAB [2] have been compared with the theoretical predictions for the **classical** sine-Gordon model [1], which is taken as an approximation for Eq. (1) in the limit of large anisotropy [1], and the observed peak features have been attributed to the soliton excitations of that model [1].

The anisotropy parameters for $CsNiF_3$ and CHAB are, however, not large enough to justify the use of the sine-Gordon model [5] and a variety of investigations have examined the importance of out-of-plane fluctuations [5,6]. The discrepancy between the experimental data and the classical predictions, however, remains large and it has become clear [6] that quantum effects must be included in order to approach quantitative agreement.

41.2 Classical Calculation of the Specific Heat

We take as our starting point the steepest descent analysis [5] of the classical partition function $Z = \int \Pi d^3 S \exp(-H/kT)$. In the low temperature limit the leading contributions to Z arise from *static* linear spin wave configurations and thermally excited *static* nonlinear soliton configurations. We find the classical specific heat per spin, $c = -Td^2(F/N)/dT^2$, $F = -kT\log Z$ (N is the number of spins):

$$c = k + k(8\pi)^{-1/2} g(\beta)(E_0/jS^2)(E_0/kT)^{5/2} \exp(-E_0/kT) \qquad (4a)$$

$$g(\beta) = [1 + (1+\beta)^{1/2}][2 + (1+\beta)^{1/2}]/[\beta-3]^{1/2}\beta^{1/2}. \qquad (4b)$$

Here $E_0 = 8S(jhS)^{1/2}$ is the soliton energy [1] and $\alpha = a/j$ the reduced anisotropy parameter. We also define the derived strength parameter $\beta = 2aS/h$. For large anisotropy $\beta \gg 1$, $g(\beta) = 1+3\beta^{1/2}$, the out-of-plane fluctuations are suppressed and we obtain the classical sine-Gordon result for c, independent of [1].

41.3 Quantum Calculations of the Specific Heat

In the quantum mechanical case the aim is to evaluate the **quantum** partition function $Z = \text{Tr}[\exp(-H/kT)]$. This is, however, a much more difficult task. Here we limit ourselves to a **semi-classical** evaluation of Z valid for **large** S. In the ideal gas approximation [1,5] which holds for $E_0 \ll kT$ it is justified to treat the soliton as a *classical* excitation and only quantize the linear vibration spectrum about the soliton. To leading order in ℏ the Hamiltonian H is represented by a set of quantum oscillators

$$H = E_0 + \frac{1}{2} \sum_n [|p_n|^2 + \omega_n^2 |q_n|^2] \qquad (5)$$

where $[p_n, q_n] = -i\hbar\delta_{nm}$.

The spectrum has the form [7] $\{0, \omega, \omega_k\}$. The zero frequency mode corresponds to the **classical** translational motion of a soliton with mass [8] (ψ' is the trigamma function)

$$M = \frac{64(S\hbar)^2}{E_0(\beta-3)} [\frac{3}{4} + \frac{1}{8} \frac{\beta}{(1+\beta)^{1/2}} \psi'(\frac{1}{2} + \frac{1}{2}(1+\beta)^{1/2})]. \qquad (6)$$

The discrete finite frequency solution is a localized vibration mode of the soliton. The energy of the mode lies below the spin wave band

$$\hbar\omega_k = jS[(k^2+\kappa^2+2\alpha)(k^2+\kappa^2)]^{1/2} \qquad (7)$$

and is to leading order in β^{-1} given by [7]

$$\hbar\omega = \hbar\Delta(1 - \frac{8}{9}\beta^{-2}),\qquad(8)$$

where $\hbar\Delta = \hbar\omega_{k=0}$ is the spin wave gap and $\kappa = (h/jS)^{1/2}$ is the inverse correlation length. The spin waves are phase shifted by the soliton giving rise to a reduction of the density of states, $\rho_k = N/2\pi + \Delta_k$. To leading order in β^{-1} [7]

$$\Delta_k = \frac{1}{2\pi}\frac{d}{dk}(\delta_0(k) + \delta_1(k))\qquad(9a)$$

$$\delta_0(k) = 2\tan^{-1}(\kappa/k)\qquad(9b)$$

$$\delta_1(k) = 2\,\frac{k^2 + \frac{1}{3}\kappa^2}{k^2 + \alpha + \kappa^2}\frac{\kappa}{k}.\qquad(9c)$$

The momentum of the soliton is $p = M^{1/2}p_0$. Introducing furthermore the number operators n and n_k for the localized mode and the spin waves, and the density of states ρ_k we can express H in Eq. (5) in the form

$$H = \Delta_{sw} + E + p^2/2M + \hbar\omega n + N\int\frac{dk}{2\pi}\hbar\omega_k n_k + \int dk\Delta_k\hbar\omega_k n_k\qquad(10)$$

Here $\Delta_{sw} = N\int(dk/2\pi)\hbar\omega_k\gamma_k$ is the spin wave contribution to the ground state energy. In the present context Δ_{sw} does not contribute to the specific heat c and we set $\Delta_{sw}=0$. More importantly, the localized mode and the phase shifted spin waves give rise to a correction to the soliton gap,

$$E = E_0 + \hbar\omega\gamma + \int dk\Delta_k\hbar\omega_k\gamma_k\qquad(11)$$

The localized mode and spin wave zero point energies $\hbar\omega\gamma$ and $\hbar\omega_k\gamma_k$ are **undetermined** parameters. Since n and n_k are constants of motion, the Hamiltonian H in Eq. (10) is in diagonal form and it is straightforward to evaluate the quantum partition function Z. In accordance with the ideal gas phenomenology [1], we treat the solitons as a Boltzmann gas of particles with gap energy E and mass M. On the other hand, the local vibration mode associated with the soliton and the spin wave gas are treated according to Bose statistics. For the **quantum** specific heat per spin we finally obtain $c=c_{sw}+c_{sol}$, where the spin wave and soliton parts are given by

537

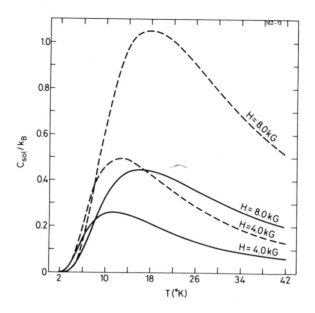

Figure 1. *Soliton contribution c_{sol} to the specific heat as a function of temperature for two different fields. The dashed curves show the classical results; the full-drawn curves the quantum results.*

$$c_{sw} = k \int \frac{dp}{2\pi} \left[\frac{\hbar\omega_p/2kT}{\sinh(\hbar\omega_p/2kT)}\right]^2 \qquad (12)$$

$$c_{sol} = k \frac{2}{2\pi\hbar} (2\pi MkT)^{1/2} [1-\exp(-\hbar\omega/kT)]^{-1} (E/kT)^2$$

$$\exp(-E/kT)\exp[-\int dp \Delta_p \log(1-\exp(-\hbar\omega_p/kT))] \qquad (13)$$

41.4 Discussion and Conclusion

The expressions for c_{sw} and c_{sol} hold in the semiclassical-ideal gas-long wavelength-large anisotropy limit.

In a recent paper Mikeksa and Frahm [9] have also discussed the specific heat of the easy-plane ferromagnet within a semiclassical quantization scheme. Comparing our data with their specific heat plots, we find some disagreement which we believe is due to a different treatment of phase shift effects.

Choosing $\gamma_k = \gamma = 1/2$ for the spin waves and the localized mode, and parameters for $CsNiF_3$ we have in Fig. 1 depicted the soliton contribution to the classical and quantum expressions for the specific heat.

In Fig. 2 we have shown $\Delta c = c_T(H)-c_T(0)$ using parameters for

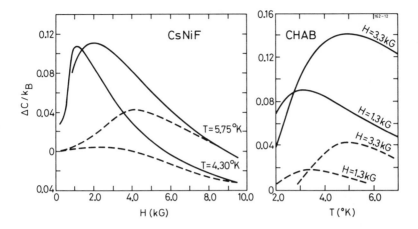

Figure 2. Specific heat differences $\Delta c = c_T(H) - c_T(0)$ as a function of magnetic field for two different temperatures. The dashed curves are the experimental results; the full-drawn curves the theoretical quantum results. Specific-heat differences $\Delta c = c_H(T) - c_{H=0}(T)$ as a function of temperature for two different fields. The dashed curves are the experimental results; the full-drawn curves the theoretical quantum results.

$CsNiF_3$ at two temperatures and $\Delta c = c_H(T) - c_{H=0}(T)$ in the case of CHAB for two fields.

The figures clearly show the considerable reduction in c_{sol} caused by the inclusion of quantum effect. The remaining substantial quantitative discrepancy is, of course, to be expected given the approximations in the present scheme. In that context it is of particular interest that **quantum** Monte Carlo calculations for the easy-plane discrete spin-1 model are now emerging [10], providing a theoretical calibration.

References

[1]   H. J. Mikeska (1978). *J. Phys.* **C11**, L29.
[2]   K. Kopinga, A. M. C. Tinus and W. J. M. de Jonge (1984). *Phys. Rev.* **B29**, 2868.
[3]   J. K. Kjems and M. Steiner (1978). *Phys. Rev. Lett.* **41**, 1137.
[4]   A. P. Ramirez and W. P. Wolf (1982). *Phys. Rev. Lett.* **49**, 227; (1985). *Phys. Rev.* **B32**, 1639.
[5]   H. C. Fogedby, P. Hedegard and A. Svane (1985). *Physica* **132B**, 17.
[6]   O. G. Mouritsen, H. J. Jensen and H. C. Fogedby (1984). *Phys. Rev.* **B30**, 498.
[7]   H. J. Mikeska (1982). *Phys. Rev.* **B26**, 5213.
[8]   H. J. Mikeska and K. Osano (1983). *Z. Physik* **B52**, 111.
[9]   H. J. Mikeska and H. Frahm (1986). *J. Phys.* **C19**, 3203.
[10] G. M. Wysin and A. R. Bishop (1986). *Phys. Rev.* **B34**, 3377.

BLOW-UP IN NONLINEAR SCHRODINGER EQUATIONS

The Nonlinear Schrödinger Equation (NSE)

$$iu_t + \nabla^2 u + f(|u|^2)u = 0 \qquad (1)$$

is one of the basic evolution models for nonlinear waves in many branches of physics. With a cubic nonlinearity $f(|u|^2) = q|u|^2$ it is a generic equation describing the slowly varying envelope of a wave-train in nonlinear dispersive systems and $q = \partial\omega/\partial|u|^2$ is related to the nonlinear frequency shift. In some important application $u(\mathbf{x},t)$ is an envelope only in time. In such cases the NSE can adequately model very spiky oscillatory structures, such as Langmuir solitons in plasmas and Davydov solitons on an α-helix protein.

In this article we give a brief review of some basic properties of this equation, with emphasis on the blow-up problem. A more detailed review is found in [1].

The solution $u(\mathbf{x},t)$ is said to blow up if some suitably defined norm goes to infinity in finite time. The appropriate norm is obtained by formulating the Cauchy-problem in the Sobolev space W^1, where the norm is defined as $\|u\|_{W^1} = (\|\nabla u\|_2^2 + \|u\|_2^2)^{1/2}$. Here $\|u\|_2$ is the usual L_2-norm. An unbounded Sobolev-norm implies that $\max|u|$ is unbounded, which conforms with our intuitive understanding of "blow-up" of a solution.

The NSE exhibits the following conserved integrals:

"mass": $N = \int|u|^2 d^D x,$

"momentum": $P = \frac{1}{2}\int i(u^*\nabla u - u\nabla u^*)d^D x,$ and $\qquad (2)$

"energy": $H = \int(|\nabla u|^2 - F(|u|^2))d^D x,$

where $F(|u|^2) \equiv \int_0^{|u|^2} f(s)ds$. We also note the possibility of solitary wave solutions $u(\mathbf{x},t) = R_n(\mathbf{x};\lambda)\exp(i\lambda t)$, where $R_n(\mathbf{x};\lambda)$ are bound-state solutions to the eigenvalue equation:

$$\nabla^2 R_n - \lambda R_n + f(R_n^2)R_n = 0. \tag{3}$$

Of main interest is the ground-state solution R_0, which is positive, radially symmetric and monotonically decreasing with increasing $r \equiv |\mathbf{x}|$. For a power nonlinearity, $f(|u|^2) = |u|^{2\sigma}$, the ground state exists for $\sigma < \infty$ if $D \leq 2$, and for $\sigma < 2/(D-2)$ if $D > 2$, where D denotes the number of spatial dimensions. Stability of the ground state is ensured for $\sigma < 2/D$.

From Eqs. (1) and (2) one can derive a "virial theorem":

$$I(t) = 4Ht^2 + Bt + C + 4\int_0^t dt' \int_0^{t'} dt'' A(t'') \tag{4}$$

for "the moment of intertia" $I = \int |\mathbf{x} - \langle \mathbf{x} \rangle|^2 |u|^2 d^D x$. Here, the quantities A, B and C are given as:

$$A = (D+2)\int F(|u|^2)d^D x - D\int f(|u|^2)|u|^2 d^D x$$

$$B = \dot{I}(t=0) = 4\,\mathrm{Im}\int \mathbf{x}(u_0^* \nabla u_0)d^D x \tag{5}$$

$$C = I(t=0) = \int |\mathbf{x}|^2 |u_0|^2 d^D x$$

where $u_0(\mathbf{x}) = u(\mathbf{x}, t=0)$. If $A \leq 0$ and one of the following conditions is satisfied:

$$H < 0, \text{ or } H = 0 \text{ and } B < 0, \text{ or } H > 0 \text{ and } B \leq -4\sqrt{HC} \tag{6}$$

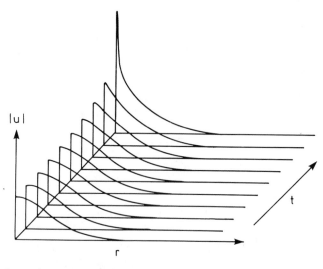

Figure 1. An example of blow-up in the 2-D cubic Schrödinger equation (CSE). The CSE is solved in radial symmetry. Note that a narrow spike blows up at the origin before a total collapse has time to develop.

it follows from Eq. (4) that $I \to 0$ in a finite time. Since this implies that the total "mass" is concentrated at one point at a time $t_c < \infty$, this phenomenon is termed **total collapse**. It is easy to show that total collapse implies blow-up at a time $t_o \leq t_c$. Typically, blow-up occurs before the collapse has become total, i.e. $t_o < t_c$. This behaviour is illustrated in Fig. 1. Note that the criteria for total collapse are sufficient, but not necessary, for blow-up.

There are also some sufficient criteria for non-blow-up, i.e. finite Sobolev-norm for all times. For the NSE global existence of the solution to the Cauchy-problem is equivalent with non-blow-up, and for the case of power nonlinearities $f(|u|^2) = q|u|^{2\sigma}$ some quite general existence theorems have been established. Some of the criteria for global existence and blow-up are summarized in Table 1.

Table 1. Global existence (GE) or blow-up (BU) of solutions to the power nonlinearity NSE. The criterion for blow-up in the case $\sigma > \frac{2}{D}$ is only necessary. Sufficient criteria are obtained by including the requirements (6).

σ:	$< \frac{2}{D}$	$= \frac{2}{D}$	$> \frac{2}{D}$	
$\|u_o\|_2$:	$< \infty$	$< \|R_o\|_2$	$\geq \|R_o\|_2$	$\leq \infty$
$q > 0$	GE	GE	BU	BU
$q < 0$	GE	GE	GE	GE

For the power nonlinearity the following similarity solution can be derived from the simple scale invariance $t' = \lambda^2 t$, $x' = \lambda x$, $u' = \lambda^{1/\sigma} u$:

$$u(x,t) = [t_o - t]^{-\frac{1}{2\sigma}} v(\eta) \exp[i \nu \ln(t_o - t)] \ , \quad \eta \equiv \frac{x}{(t_o - t)^{1/2}} \quad (7)$$

However, the function $v(\eta)$ is **non-localized**, i.e. $\|v(\eta)\|_2 = \infty$. Still, numerical solutions indicate that Eq. (7) is a local attractor for the collapsing spike if $\sigma > 2/D$, and in a certain stage of spike development if $\sigma = 2/D$. On the other hand, for $\sigma = 2/D$ there are **localized** solutions:

$$u(x,y) = [\alpha(t_o-t)]^{-D/2}R_o(\eta;\lambda)\exp i[\frac{-\frac{1}{4}r^2+\alpha^{-2}\lambda}{t_o-t} + \phi_o],$$

$$\eta \equiv \frac{x}{\alpha(t_o-t)} \tag{8}$$

which are associated with the conformal invariance $t' = \beta-1/\alpha^2 t$, $x' = x/\alpha t$, $u' = (\alpha t)^{D/2}u \exp(-ix^2/4t)$. These solutions are **unstable**; certain perturbations alter the temporal collapse behaviour, other perturbations prevent blow-up altogether. In the spike region certain terms may become asymptotically small compared to other terms, such that certain generalizations to the conformal transformation above also leave the NSE invariant in this region. The appropriate transformation has the form;

$$u(\mathbf{x},t) = g^{D/2}v(\eta,\tau)\exp[-\frac{i}{4}\eta^2\frac{g'}{g}]$$

$$\tau = \tau(t), \quad \eta = g(\tau)\mathbf{x}, \quad \frac{d\tau}{dt} = g^2(\tau) \tag{9}$$

which yields a transformed Schrödinger equation:

$$iv_\tau + \nabla_\eta^2 v - V(\eta,\tau)v = 0 \tag{10}$$

with a time-dependent potential $V(\eta,\tau) = -|v|^{2\sigma}-a(\tau)\eta^2/4$, where $a(\tau) \equiv g''/g$. We assume that $g(\tau)$ is chosen such that $v(0,\tau)$ relaxes to a constant as $\tau \to \infty$. It is then a reasonable conjecture that the inner solution relaxes to the ground state $R_o(\eta;\lambda)\exp(i\lambda\tau)$ and matches on to an outer radiative solution $Cg^{-D/2}\exp(i\eta^2 g'/4g)$ in the vicinity of the outer turning point. In the original variables this radiative field corresponds to an outer quasistationary field of magnitude C. Carrying through this matching procedure, we find that $a(\tau) = \tau^{-1}[C_o+C_1\tau^{-1/2}+C_2\tau^{-1}+...]$ as $\tau \to \infty$, and by expanding $v(\eta,\tau)$ in powers of $\tau^{-1/2}$ we obtain to the leading order in the inner region:

$$u(\mathbf{x},t) = (\frac{\lambda}{4C_o}\frac{|\ln(t_o-t)|}{t_o-t})^{D/4}[R_o(\xi)$$

$$+ \frac{\mu\sqrt{C_o}}{\lambda|\ln(t_o-t)|}(\frac{D}{2}R_o(\xi)+\xi\frac{dR_o}{d\xi})]_{\xi=\sqrt{\lambda}gr} \tag{11}$$

$$\times \exp i[\frac{\lambda}{16C_o}|\ln(t_o-t)|^2 + \frac{\mu}{4\sqrt{C_o}}|\ln(t_o-t)|]$$

Since the methods leading to Eq. (11) are essentially heuristic, the nature of the blow-up singularity remains a controversial question, which may only be settled by more complete analytical developments supported by detailed numerical studies. However, recent numerical results [2] on the blow-up in the 2-D cubic Schrödinger equation indicate that an asymptotic regime is not reached for peak amplitudes up to 10^4 times the initial amplitude. But it seems that the rescaled profile of the spike approaches the ground state solution, while the temporal development of the spike maximum seems not to approach the development given in Eq. (11).

References

[1]   J. Juul Rasmussen and K. Rypdal (1986). *Physica Scripta* **33**, 481;   K. Rypdal and J. Jull Rasmussen (1986). *Physica Scripta* **33**, 498.

[2]   D.W. McLaughlin, G.C. Papanicolaou, C. Sulem and P. Sulem (1986). *Phys. Rev.* **A34**, 1200.

THE SOLITON LASER: A COMPUTATIONAL TWO-CAVITY MODEL

43.1 Introduction

In 1980, L.F. Molleauer, R.H. Stolen and J.P. Gordon [1] observed picosecond pulse narrowing of solitons on optical fibres experimentally. This observation was used in the soliton laser proposed and designed by Mollenauer and Stolen [2] in 1984. It was demonstrated experimentally that this device could produce stable pulses of width 210 fsec and later [3] width down to 100 fsec. As illustrated in Fig. 1, the soliton laser is a double-cavity system consisting of a synchronously pumped, mode-locked colour-centre laser [4] coupled to an optical single-mode fibre acting as a control cavity. The present work presents results obtained in Ref. [5].

43.2 The Colour-Centre Laser

The laser cavity shown in Fig. 1 formed by the mirrors M_o and M_1 consists of a lasing medium (a F-centre crystal) and a pair of birefringent tuner plates [6]. The laser is mode-locked by synchronous pumping through the mirror M_1 from a Nd:YAG laser. The pumping frequency, ω_M, tuned to match the cavity round trip time, T_c, is locking regions of the large laser bandwidth on to different polarization directions. The birefringent tuner plates operate as a bandpass filter restricting the bandwidth to one region with a single polarization direction. This allows us to use the scalar version of the Maxwell-Bloch equations [7] as the model equations for the laser cavity. The bandwidth of the colour-centre laser (CCL) is very large compared with the signal bandwidth [2], the polarization is forced to follow the electric field (the rate equation approximation). The relaxation of the inversion population to its equilibrium value is described by the longitudinal relaxation constant, $\gamma_{\shortparallel}$.

$$E_z + \dot{E} = (\alpha D - \kappa)E \qquad \text{(1a)}$$

$$\dot{D} = \gamma_{\shortparallel} - \gamma_{\shortparallel}(1 + |E|^2/I_s)D. \qquad \text{(1b)}$$

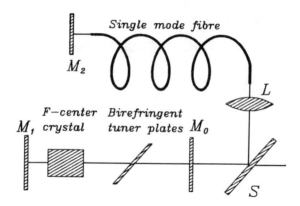

Figure 1.   Schematic diagram of the soliton laser.

The gain α and the saturation I_s will be used as model
parameters. As seen from Eq. (1b) the system will saturate at a
certain field intensity through a speed-up of the relaxation of
the inversion to its equilibrium value.

The synchronized pumped CCL is modelled as an actively
mode-locked system by introducing a modulated cavity decay rate

$$\kappa = \kappa_o + A_M[1 - \cos(\omega_M(z - t))]. \qquad (2)$$

Here $\omega_M = 2\pi/L_c$, where L_c is the normalized unfolded cavity
length.

The action of the birefringent tuner plates is taken to be that
of a bandpass filter, with a centre frequency equal to the centre
frequency of the CCL, ω_o, and a bandwidth $\Omega_p \ll \gamma_\perp$.   In the
frequency domain the transfer function of the plates can then be
written as

$$B(\omega) = \frac{1 - q_o}{1 + \left(\dfrac{\omega-\omega_o}{\Omega_p}\right)^2} + q_o. \qquad (3)$$

The parameter q_o, the minimum of the transfer function, depends
on the material and the number of plates.   By expanding the
electric field in cavity modes and approximating the field
intensity with its spatial average, $I(t) = \langle|E(z,t)|^2\rangle$, we obtain
a system of coupled ordinary differential equations for the
normalized cavity modes and inversion

$$\dot{E}_n = (\alpha D - \kappa_o) E_n + \beta_n(E_{n+1} - 2E_n + E_{n-1}) \qquad (4a)$$

$$\dot{D} = \gamma_\shortparallel - \gamma_\shortparallel (1 + I(t)/I_s)D. \tag{4b}$$

The cavity modes are advanced in time according to Eqs. (4a) and (4b) by using a forward finite difference approximation to the time derivatives.

43.3 The Optical Fibre

The appropriate evolution equation for optical pulse propagation within the fibre is the perturbed nonlinear Schrödinger equation [8] (PNLS):

$$i E_z + \frac{1}{2} E_{tt} + E|E|^2 + i\gamma E = 0. \tag{5}$$

Analytical solutions to the NLS are available to us through the original work by Zakharov and Shabat [9], who showed that NLS possesses soliton solutions. In particular, Satsuma and Yajima [10] have investigated initial value problems for the NLS, using the inverse scattering transform. Initial pulses of the form

$$E(0,t) = E_o \, \text{sech}(t/T_o), \tag{6}$$

produce N discrete eigenvalues in the scattering problem, N being an integer satisfying $|N - E_o T_o| < \frac{1}{2}$. The solution consists of the N-soliton and radiation [10]. The radiation is "peeled off" the pulse during the propagation on the fibre. It is important to note that the N-solitons ($N \geq 2$) are deeply modulated pulses which are periodic in z except for a constant phase shift, returning to

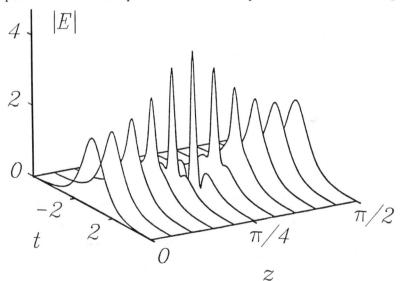

Figure 2.   Propagation of the 2-soliton on a fibre of length equal to one soliton period. $|E| = |E(z,t)|$ is shown.

the same profile given by Eq. (6) for distances equal to multiples of the soliton period [10]

$$z_o = \frac{\pi}{2} T_o^2. \tag{7}$$

Because z_o scales with T_o^2, a broader pulse requires a longer distance in the fibre to complete one period, and to a given fibre length there corresponds a pulse of definite width which goes through one period between the fibre ends. In other words, if the periodicity length of the pulse can be shortened then the pulse is temporally narrowed. In Fig. 2, the evolution of the 2-soliton, $E(0,t) = 2 \, \mathrm{sech}(t)$, through one period is shown. The overall phase shift is $\pi/4$ for the 2-soliton.

If the pulse differs slightly from the exact 2-soliton the "period" is also changed. The term "period" refers to the soliton part of the solution. With perturbed initial condition the solution to NLS is not strictly periodic. Eq. (5) is solved numerically by a split-step Fourier method with γ being a loss term localized at the boundaries [11].

43.4 The Double Cavity System

As shown in Fig. 1 the soliton laser is a double cavity system, consisting of a CCL cavity bounded by the mirrors M_1 (reflectivity $R_1 \sim 100\%$) and M_o (reflectivity $R_o < 100\%$), and a fibre cavity bounded by M_o and the movable mirror M_2 (reflectivity $R_2 \sim 100\%$). The lens L focusses the pulse launch into the fibre (transmission coefficient T_L). The output from the cavity system is through the beam splitter S (reflectivity $R_S < 100\%$). The pulses propagating into the laser and out of the laser are denoted E_{TL} and E_{FL} respectively. Similarly, pulses into the fibre and out of the fibre are denoted E_{TF} and E_{FF} respectively. The coupling between the two cavities is described by the equations

$$E_{TL} = R_o E_{FL} + (1 - R_o^2)^{1/2} R_S e^{i\phi} E_{FF} \tag{8a}$$

$$E_{TF} = ((1 - R_o^2)^{1/2} R_S E_{FL} + R_S^2 R_o e^{i\phi} E_{FF}) T_L. \tag{8b}$$

Here $e^{i\phi}$ is an overall phase factor modelling small dynamical adjustments by the mirror M_2 in order to ensure that the pulses E_{FL} and E_{FF} are superimposed in phase. The stability of the device turns out to depend on this phase adjustment. In Fig. 3 the results from a simulation with a fairly long fibre ~150 m are shown. The CCL cavity emits stable pulses of temporal full width at half maximum (FWHM) of 7.5 psec. When the fibre cavity is connected, initially these broad pulses perform only a small

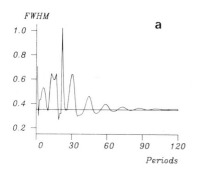

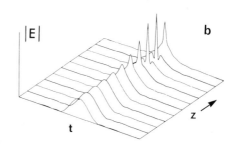

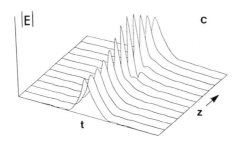

Figure 3. Simulation with normalized fibre length $\ell = 0.032$. $R_o = 83\%$, $R_S = 65\%$, $T_L = 100\%$, $\kappa_o = 0.2454$, $\gamma_{\shortparallel} = 0.01473$, $\Omega_c = 1.473$, $A_M = 2$, $q_o = 0.5$, $\Omega_p = 2.454$, $\alpha = 0.4909$, and $I_s = 300$. (a) The normalized width (FWHM) of the pulses at output from the soliton laser versus cavity round trip times (periods). The pulse width is compressed by the fibre to the theoretical width (indicated by horizontal line). FWHM = 0.35 after 120 periods. (b) First propagation period of the pulse along the fibre. (c) Last propagation period (after 120 periods) of the pulse along the fibre, showing the typical 2-soliton oscillation.

fraction of a 2-soliton oscillation on the fibre, returning with a much smaller temporal width (see Fig. 3b). In Fig. 3a the system is seen to stabilize after 100 cavity round trips (periods). Then the system emits stable pulses with a temporal width of 3.5 sec.

In Fig. 3c the propagation of the pulse along the fibre in the last period of the simulation is shown indicating that the stable operation of the soliton laser is a configuration where the 2-soliton performs a slightly longer period of oscillation than the actual fibre length. This period of oscillation is shifted, giving rise to a pulse at exit from the fibre of width slightly smaller than the width of the input pulse. The interaction with the pulse from the CCL cavity then compensates for this discrepancy.

43.5 **Discussion**

A genuine two-cavity numerical model of Mollenauer-Stolen's soliton laser is developed. The dynamical stabilization obtained experimentally by Mitschke and Mollenauer [3] is mimicked by an overall phase factor in the coupling equations (7). The introduction of absorbing boundaries in the equation for the fibre enables us to reduce the pulse spacings, without introducing interaction between adjacent pulses, in comparison with the experimental device. In this way a substantial reduction in the number of modes needed for the numerical calculations is achieved. In the experimental work [3] a pulse compression factor of 75 was demonstrated. We believe that this level may also be reached in our computational model, if 5-10 times as many cavity modes are included. The corresponding increase of the temporal resolution in the fibre must also be introduced.

Acknowledgements

The authors are grateful to L.F. Mollenauer for sending preprints of his papers. J.N. Elgin is thanked for his continued interest in this work. The financial support from the Danish Council for Scientific and Industrial Research is acknowledged.

References

[1] L.F. Mollenauer, R.H. Stolen and J.P. Gordon (1980). *Phys. Rev. Lett.* **45**, 1095.

[2] L.F. Mollenauer (1985). *Phil. Roy. Trans. Lond.* **A315**, 333.

[3] F.M. Mitschke and L.F. Mollenauer (1986). *IEEE J. Quantum Electron.* **QE-22**, 2242.

[4] L.F. Mollenauer, N.D. Vieira and L. Szeto (1982). *Opt. Lett.* **7**, 414.

[5] P. Berg, F. If, P.L. Christiansen and O. Skovgaard, accepted by *Phys. Rev. A*; F. If, P.L. Christiansen, J.N. Elgin, J.D. Gibbon and O. Skovgaard (1986). *Opt. Comm.* **57**, 350; F. If, P.L. Christiansen, J.N. Elgin, J.D. Gibbon and O. Skovgaard, *Physica* 23D, 362 (in press).

[6] I.J. Hodgkinson and J.I. Vukusic (1978). *Opt. Comm.* **24**, 133; I.J. Hodgkinson and J.I. Vukusic (1978). *Appl. Opt.* **17**, 1944.

[7] H. Haken (1985). *Light Vol. II: Laser Dynamics,* North-Holland, New York, 98.

[8] A. Hasegawa and F. Tappert (1973). *Appl. Phys. Lett.* **23**, 142.

[9] V.E. Zakharov and A.B. Shabat (1972). *Sov. Phys. JETP* **34**, 62.

[10] J. Satsuma and N. Yajima (1974). *Prog. Theor. Phys. Suppl.* **55**, 284.

[11] F. If, P. Berg, P.L. Christiansen and O. Skovgaard, accepted by *J. Comp. Phys.*

44   *J.N. Elgin and J.B. Molina Garza*

ON THE TRAVELLING WAVE SOLUTIONS OF THE MAXWELL-BLOCH EQUATIONS

Abstract

A possible class of travelling wave solutions to the Maxwell-Bloch equations is presented and discussed. These are given by periodic solutions of the Lorenz equations; in particular, solutions for negative values of the parameter σ give stable travelling wave solutions with phase velocity less than the speed of light. The bifurcation pattern of periodic solutions to the Lorenz equations with negative σ is discussed in detail.

It is well known that the amplitude of a homogeneously broadened single-mode laser can become unstable only in the bad cavity limit where the linewidth κ of the singly excited resonant cavity mode exceeds the sum of the relaxation rates of the polarization ($\gamma_\perp$) and population ($\gamma_\shortparallel$) [1]. This criterion is removed for multimode laser as first demonstrated by Risken and Nummedal [2] for the case when the initially excited single cavity mode is exactly resonant with the gain medium, and by several other authors [3] for the case when an additional detuning parameter is introduced into the system. For the latter case there are two types of instability regions, now generally known as regions of phase instability and amplitude instability [4]. The region of amplitude instability can be considered a simple extension of that first studied by Risken and Nummedal. The region of phase instability occurs immediately above threshold in a manner similar to that observed in inhomogeneously broadened laser systems. The stability properties of the initially excited single mode have been extensively reviewed in the references cited [2-3] and it is not our intention to discuss these further here. Rather, we consider one possible class of final state solutions to the laser equations, namely travelling waves simply related to the set of periodic orbits of a low dimensional dynamical system: the Lorenz equations [5].

In appropriately normalized units, the laser equations for a resonant ring cavity system are [2];

$$c \frac{\partial E}{\partial x} + \frac{\partial E}{\partial t} = \kappa(P - E)$$

$$\frac{\partial P}{\partial t} = \gamma_{\perp} ES - \gamma_{\perp} P \qquad (1)$$

$$\frac{\partial S}{\partial t} = \frac{\gamma_{\shortparallel}}{2} \lambda (E^{*}P + EP^{*}) - \gamma_{\shortparallel}(S - \lambda - 1).$$

Here E and P are the (complex) electric field and polarization slowly varying envelopes respectively, S is the (real) population inversion, λ is a pump parameter ($\lambda=0$ at threshold, and is greater than zero above threshold). The cavity mirrors impose the periodicity conditions

$$F(x+L, \ t) = F(x, t) \qquad (2)$$

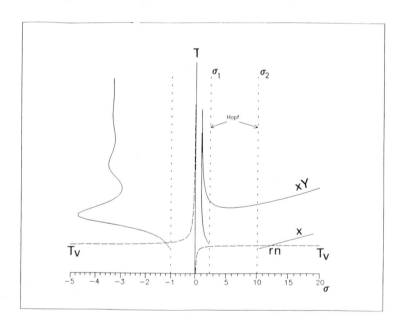

Figure 1. Period of three simple orbits in the Lorenz equations plotted against σ (for b=.5 and r=16). Here we plot the period of the X orbit. XY symmetric orbit and the symmetric principal periodic orbit for negative σ. T_v (equation 5) is also shown in this figure for both positive and negative σ. There is a Hopf bifurcation of the origin at $\sigma = -1$, and two Hopf bifurcations of the fixed points C_1 and C_2 at σ_1, σ_2. The first homoclinic orbit of the origin is indicated by a broken line, there is a second one at $\sigma \simeq 44$. A heteroclinic orbit connects C_1 and C_2 at $\sigma = -3.3$ (notice the different horizontal scales for negative and positive σ).

where F is either E, P or S.   Equations (1) have a constant amplitude solution, whose stability properties have been discussed elsewhere [2-4].   This solution corresponds to continuous laser operation in a single mode. We consider travelling wave solutions of the form

$$E(x,\ t) = E(t-x/v)$$
$$P(x,\ t) = P(t-x/v)$$
$$S(x,\ t) = S(t-x/v)$$

in which case equations (1) are replaced by

$$\frac{dE}{d\tau} = \kappa'(P - E)$$

$$\frac{dP}{d\tau} = \gamma_\perp ES - \gamma_\perp P \qquad\qquad (1)'$$

$$\frac{dS}{d\tau} = -\frac{\gamma_{\shortparallel}}{2}\ \lambda(E^*P + EP^*) - \gamma_{\shortparallel}(S - \lambda - 1)$$

where $\kappa' = \dfrac{\kappa}{1-c/v}$ and $\tau = t-x/v$.   Notice that these are the single mode equations, apart from the replacement of κ' for κ.   Note also that, even if κ is very much less than $\gamma_\perp + \gamma_{\shortparallel}$ there is no such restriction on κ'; for values of the phase velocity v less than c, κ' can even be negative.

The boundary conditions (2), when applied to the travelling wave forms assumed in deriving equations (1)', mean that

$$E(\tau) = E(\tau+T)$$
$$P(\tau) = P(\tau+T) \qquad\qquad (2)'$$
$$S(\tau) = S(\tau+T)$$

with

$$\tau = t-x/v$$
$$T = L/v.$$

    In other words, the travelling-wave type solutions of (1) are periodic orbits of equations (1)' with basic period T=L/v (v unknown).

    In this paper we take E and P to be real, it is then more convenient to use the simple transformations given in [1] to rewrite equations (1)' in the 'standard' form normally considered in dynamical systems, first quoted by Lorenz [5]:

$$\dot{X} = \sigma(Y-X)$$
$$\dot{Y} = (r-Z)\ X - Y \qquad\qquad (3)$$
$$\dot{Z} = XY - bZ.$$

Here $\sigma = \kappa'/\gamma_\perp = \dfrac{\gamma_\perp}{1-c/v}$ and all other symbols have their usual meaning.   Differentiation is with respect to τ.

Maxwell-Bloch equations

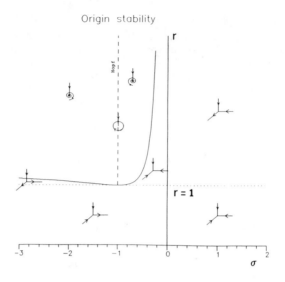

Origin stability

Figure 2. *Linear stabiilty of the origin in the Lorenz equations (schematic). There is a supercritical Hopf bifurcation at σ = -1 for all r > 1.*

Consider the superluminal case first (v>c), in which case sigma is positive. It is known [6] that when r>1 equations (3) have two satisfied) and have r as the variable control parameter. Equation (4) is quadratic in σ, the fixed points are then unstable in a region $\sigma_1 < \sigma < \sigma_2$, and it is known that these become unstable by a symmetric fixed point solutions, which are stable in the region $1 < r < r_c$,

$$r_c = \frac{\sigma(\sigma+b+3)}{\sigma-b-1} \qquad (4)$$

Several authors [6,7] have studied the Lorenz equations in r, b parameter space for a fixed value of $\sigma(\sigma=10)$. Here we fix b, κ (but not κ') and L and study the Lorenz equations with σ as an eigenvalue (i.e. σ must be such that the periodicity condition is subcritical Hopf bifurcation. It is not difficult to show that the homoclinic explosion that produces the limit cycles involved in the Hopf bifurcation occurs at two values of $\sigma(\sigma_{hc1} < \sigma_1, \sigma_{hc2} > \sigma_2)$.

Fixing r and b we can plot the period T as a function of σ for all the periodic orbits of equations (3); using the classification scheme described in the book by Sparrow [6], we show such plots for a few of the simpler orbits in Figure 1. Any such periodic orbit is an acceptable solution to the laser equations (1) provided that the period of the orbit is equal to L/v, as required by (2)'. This criterion is most usefully represented graphically by noticing that

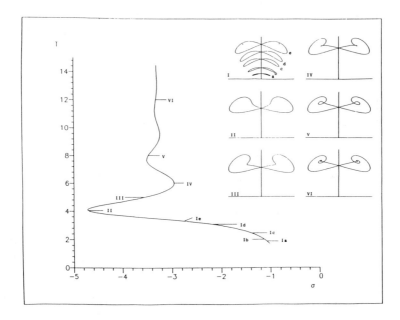

Figure 3. The symmetric principal periodic orbit for σ < 0 in the Lorenz equations. This orbit is born in a Hopf bifurcation of the origin (σ = -1) and eventually becomes the heteroclinic connection between C_1 and C_2 (t → ∞). The Z-X projection is shown for several stages in the orbit evolution, together with a plot of the period against σ. In this figure r = 16 and b = .5.

$$T_v = L/v = L/c \left\{ \frac{\sigma - \kappa/\gamma_\perp}{\sigma} \right\} \qquad (5)$$

A plot of T_v vs. σ is also shown in Figure 1. Intersections of T_v with the curves for periodic orbits give acceptable solutions, which satisfy equations (1)and (2).

As well as the positive σ solutions considered above, it is also possible to have σ<0, corresponding to subluminal travelling waves. The Lorenz equations (3) are now studied for negative σ. The stability of the origin is depicted in Figure 2. In particular, note that there is a Hopf bifurcation (supercritical) at σ = -1 for all values of r>1. Moreover, numerical studies indicate that there is a codimension one heteroclinic orbit of the Sil'nikov type [8] linking the two symmetric fixed points [9].

Our numerical investigations reveal that the periodic orbit born in the Hopf bifurcation at σ = -1 acquires extra turns round the unstable manifolds of the fixed points and eventually becomes the (symmetric) heteroclinic connection; that is, the 'principal periodic orbit' in the language of Glendinning and Sparrow [8]. This behaviour is shown in Figure 3, where the asymptotic behaviour towards the heteroclinic orbit for some specific value

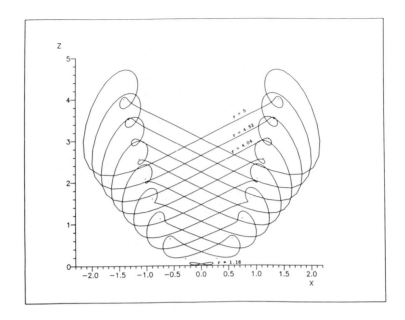

Figure 4. Periodic solutions to the laser equations for $\kappa = .5\gamma_\perp$, $L = \frac{4\pi c}{\gamma_\perp}$, $b = .5$. The subluminal travelling waves and the c.w. (fixed point) solutions are plotted for r from 1.16 to 5 in steps of .48 (Z-X projection).

of r and b is indicated. The evolution of the period against σ for this orbit is also shown in the left hand side of Figure 1. The periodicity condition (2)' must still hold; consequently we show also the plot of T_v (Eq. (5)) for $\sigma<0$. The intersection of these two curves gives again an acceptable solution which satisfies both equations (1) and (2). In addition to the symmetric principal orbit, there exists an infinite class of asymmetric and period doubled orbits winding their way up to attain homoclinic or heteroclinic status, in agreement with Glendinning and Sparrow [8].

Figure 4 shows the fixed points and the symmetric periodic solution with $\sigma<0$ for various values of the pump parameter r when all the remaining parameters (b, κ and L) are fixed, and condition (5) is satisfied.

In Figure 5, the central curve (i) is the 'lifeline' of the symmetric periodic orbit born in the Hopf bifurcation at $\sigma = -1$, which evolves towards the symmetric heteroclinic connection as discussed above. At the bifurcation point A a pair of asymmetric orbits are born. The two asymmetric orbits are mirror images of each other under the transformation $x\to -x$, $y\to -y$, $z\to z$. Each develops extra twistings around the unstable manifold of one of the symmetrically placed fixed points C_1, C_2 (Sparrow, [6]), and eventually become homoclinic connections. Figure 6 shows the z-x projection of the orbit at several stages in its development

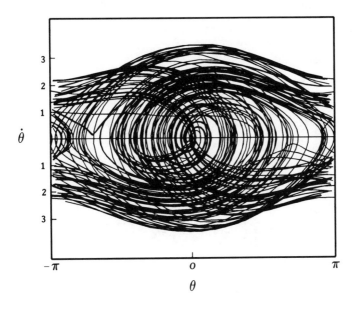

Figure 5. Bifurcation diagram for the Lorenz equations with negative σ (see text for details). Parameters are b = .5, r = 5.

(points α, β and γ).

At point B in Figure 5, a further symmetry breaking bifurcation occurs to the symmetric orbit, yielding a new pair of asymmetric orbits. The evolution of one of the members of this pair towards homoclinicity is illustrated in Figure 6 (points δ and ε). In fact, one such symmetry breaking bifurcation is observed every time the symmetric orbit completes an extra 'half twist' around the unstable manifolds of C_1, C_2. The pair of bifurcating orbits retains the same number of half twists around one fixed point and evolves towards homoclinicity at the other. This is shown in

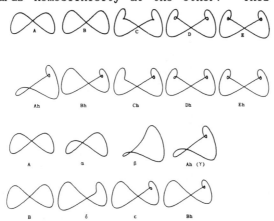

Figure 6. ZX projection of the orbits marked in Fig. 5 (see text for details).

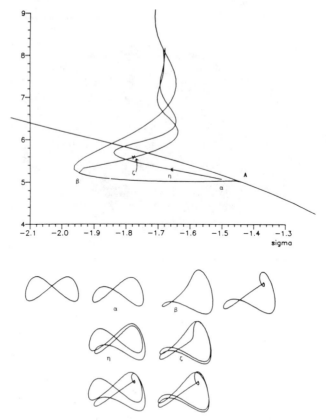

Figure 7. Period doubling of the asymmetric orbit marked A in Figure 5 (see text for details). Half the period of the bifurcating orbits is plotted against sigma together with the ZX projection at different points.

Figures 5 and 6.

The situation is still more complex; each asymmetric orbit created in this way soon bifurcates, presumably in a period doubling cascade. An example of this is shown in Figure 7, where the orbit born at point (A) period doubles at (A1), then period doubles again at (A2) and so on. Each period doubled orbit in turn winds itself round the unstable manifolds of one of the fixed points, thereby producing a whole family of asymmetric homoclinic connections, as illustrated in Figure 7.

This bifurcation pattern seems to be in agreement with the general theoretical results of Sparrow and Glendinning [8] for bifurcations of the Sil'nikov type.

Having found and classified all these types of travelling wave solutions to the equations (1) it is necessary to consider next their stability. At present, we do this by substituting the appropriate travelling wave solution into equations (1) and simply observing whether the waveform persists indefinitely. Some periodic solutions for the region $\sigma > 0$ are stable, but studies to date indicate that most are unstable. One exception is the simple pulsed solution found by Risken and Nummedal, which corresponds to a simple x-type orbit of the Lorenz equations, and which is labelled 'RN' in Figure 1. In contrast, the simplest travelling wave solutions for negative σ would apear to be stable, and have been observed to act as attractors for the system (1); integrating numerically the partial differential equations (1), (2) we have observed several instances when an initial waveform from the region $\sigma > 0$ becomes unstable, a transient regime ensues for a while before the system settles down to a different type of travelling wave which can be identified as one of the periodic orbits for the $\sigma < 0$ case. Our first evidence of the existence of some of the asymmetric and period-doubled orbits discussed above was obtained from such a study.

Acknowledgements

The authors are pleased to acknowledge many useful discussions with J.D. Gibbon, D. Wood and N. Readwin. Thanks are also due to Cathy Holmes, who first carried out the stability analysis of the origin for the negative σ case reported here. Thanks to N. Weiss, D. Broomhead, P. Glendinning and C. Sparrow. One of the authors (JBMG) is also pleased to acknowledge financial support from CONACYT (Mexico).

References

[1] H. Haken (1975). *Phys. Lett.* **53A**, 77.

[2] H. Risken and K. Nummedal (1968). *J. Appl. Phys.* **39**, 4662.

[3] P.R. Gerber and M. Buttiker (1979). *Z. Phys.* **B33**, 219; J. Zorell (1981). *Opt. Comm.* **38**, 127; P. Mandel and H. Zeglache (1983). *Opt. Comm.* **47**, 146; S.T. Hendow and M. Sargent III (1982). *Opt. Comm.* **43**, 59; L.M. Narducci, J.R. Tredicce, L.A. Lugiato, N.B. Abraham and D.K. Bandy (1986). *Phys. Rev.* **A33**, 1847.

[4] J.N. Elgin, J.D. Gibbon, J.B. Molina Garza, D. Wood (1986) in *Nonlinear phenomena and Chaos,* ed. S. Sarkar, Adam Hilger; J. Elgin, J.D. Gibbon, C. Holmes, J.B. Molina Garza and N.A. Readwin (1986). *Physica* **23D**, 19.

[5] E.N. Lorenz (1962). *J. Atmos. Sci.* **20**, 130; R. Graham (1976). *Phys. Lett.* 58A, 440 (this is the first explicit mention, to our knowledge, of the reduction to the Lorenz equation in the case of travelling waves in multimode lasers).

[6] C. Sparrow (1982). *The Lorenz Equations: bifurcations, chaos and strange attractors,* Springer Verlag, New York; K.H. Alfsen and J. Froyland (1985). *Physica Scripta* **31**, 15.

[7] A.C. Fowler, J.D. Gibbon and M.J. McGuinnes (1983). *Physica* **7D**, 126.

[8] L.P. Sil'nikov (1965). *Sov. Math. Dokl.* **6**, 163; P. Glendinning and C. Sparrow (1984). *J. Stat. Phys.* **35**, 645.

[9] The same heteroclinic connection for $\sigma < 0$ has been found independently by N. Weiss and E. Knobloch, in their studies of a model for thermosolutal convection (private communication). A similar connection exists for $\sigma > 0$ and has

been studied by Alfsen and Froyland [6] and Glendinning and Sparrow [8], but occurs at values of r probably too high to be relevant to the laser.

45   J.A. Blackburn, S. Vik, Yang Zhou-Jing, H.J.T. Smith and
      M.A.H. Nerenberg

CHAOS IN A JOSEPHSON JUNCTION MECHANICAL ANALOG

45.1 Introduction

In recent years there has been considerable interest in the
chaotic behaviour exhibited by a variety of nonlinear systems.
Chaotic domains have been predicted [1-4] in the parameter space
of a current biased Josephson junction. Direct observation of
phenomena such as period doubling sequences is virtually
impossible because of the extremely short time scale over which
Josephson oscillations occur. As an alternative, one may study
the junction dynamics by means of an electronic analog simulator
[5]. In such a circuit characteristic frequencies are effectively
rescaled to an easily accessible range.

Numerical calculations have also shown that chaos should occur
in a harmonically driven pendulum [6,7]. Again, such behaviour
has been confirmed by means of an analog simulator [8], but no
direct experimental studies have been reported. The driven
pendulum is of particular interest because the time-dependent
equations governing the quantum mechanical phase variable in a
current biased Josephson junction, and the angular coordinate of
the damped pendulum driven by an external torque, are isomorphic.
The pendulum can thus serve as a mechanical analog of the
superconducting device with characteristic frequencies reduced by
a factor of approximately 10^{12}.

We report here on experiments carried out on a Josephson
junction mechanical analog. Our apparatus was designed so that
its physical parameters (mass, moment of inertia, damping
coefficient, etc.) resulted in a system with moderate hysteresis.

45.2 Apparatus

In the RSJ model, a Josephson junction, biased by a current I, is
governed by the following equation:

$$\frac{\hbar C}{2e} \ddot{\theta} + \frac{\hbar}{2eR} \dot{\theta} + I_c \sin(\theta) = I \qquad (1)$$

where $\hbar$ is Planck's constant divided by 2π, e is the electronic
charge, C is the junction capacitance, R is the shunt resistance,

I_c is the critical Josephson supercurrent, and θ is the phase difference across the junction. If current is now measured in units of I_c and time is normalized to ω_p^{-1} where $\omega_p = [2eI_c/\hbar C]^{1/2}$ is the plasma frequency of the junction, then we obtain

$$\ddot{\theta} + (1/Q)\dot{\theta} + \sin\theta = i_0 + i_1 \sin(\Omega t) \qquad (2)$$

where the right hand side explicitly expresses the presence of both dc and ac bias currents, and $Q^2 = 2eI_c R^2 C/\hbar = \beta_c$.

As a mechanical analog of the above system, consider a pendulum consisting of a rigid disc of moment of inertia I_d. Suppose a mass m is attached to the disc at a distance r from the central axis about which rotations will occur. Let an external torque τ be applied. The equation of motion is then

$$I\ddot{\theta} + b\dot{\theta} + (mgr)\sin\theta = \tau \qquad (3)$$

where $I = I_d + mr^2$ is the total moment of inertia of the system, b is the damping coefficient, and θ is the angular coordinate of the mass m measured counter-clockwise from a vertical reference line. The undamped natural frequency $\omega_o = [mgr/I]^{1/2}$ corresponds to the Josephson plasma frequency. Let time be normalized to units of ω_o^{-1}, and measure torque in terms of mgr ($:\gamma = \tau/mgr$). Then equation (3) becomes

$$\ddot{\theta} + (1/Q)\dot{\theta} + \sin\theta = \gamma_0 + \gamma_1 \sin(\Omega t) \qquad (4)$$

where $Q^2 = I\ mgr/b^2$ and the right hand side explicitly expresses the presence of both dc and ac applied torques. Equation (4) clearly is a direct analog of the normalized Josephson equation (2). The angular velocity of the disc and mass, $\dot{\theta}$, corresponds to the junction voltage (which is proportional to $\dot{\theta}$). The analog of a junction current-voltage characteristic thus becomes a plot of applied torque, γ_o, versus $\langle\dot{\theta}\rangle$, where $\langle\rangle$ indicates a time average.

The apparatus consisted of a solid aluminium disc, 2 mm in thickness, and 12.76 cm in diameter having a moment of inertia of $i_d = 1.44 \times 10^{-4}$ kg-m^2. A small mass (m = 6.53 grams) was attached to the disc at a distance of 4.18 cm from its rotational axis. The combined moment of inertia was $I = 1.56 \times 10^{-4}$ kg-m^2 and the "plasma frequency" was $\omega_o = 4.15$ rad/sec (or 0.66 Hz).

The hysteresis parameter $[\beta_c = Q^2 = I\ mgr/b^2]$ was 17.6. The aluminium disc was fixed to the shaft of a brushless resolver. This device is, essentially, a precision rotating transformer

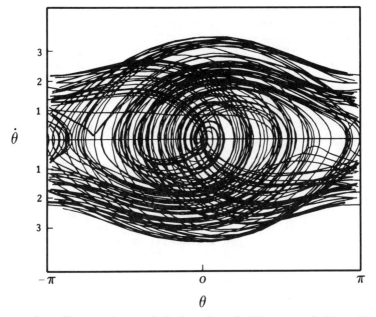

Figure 1.   Phase plane plot for Ω = 0.67, γ_1 = 0.81.   Vertical scale is in units of ac torque frequency.

having a single input winding and two pairs of outputs.   When the input is energized with a sine wave whose amplitude is a few volts and at a frequency of 5 kHz, the two output signals are modulated according to the sine and cosine of the resolver shaft angle, respectively.   Demodulation of these output voltages was performed

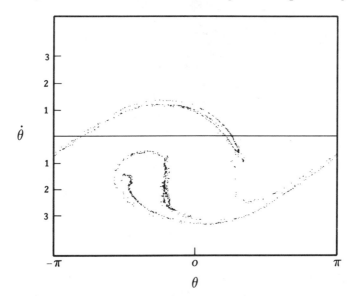

Figure 2.   Poincaré section corresponding to the motion depicted in Fig. 1;   1024 points are plotted.

by a commercial integrated circuit known as hybrid, tracking, resolver-to-digital converter. Specifications for this particular combination of resolver and 14 bit converter are as follows: resolution 1.32 arc min, system accuracy 8.78 arc min, tracking rate 42.50 rev/sec. The resolver itself has low frictional torque (3 gram-cm) and moment of inertia ($:3.5$ gram-cm^2). Velocity dependent damping was provided by eddy current induction – small permanent magnets were positioned close to the rotating aluminium disc. The damping coefficient appearing in Eq. (3) was found to be: $b = 1.54 \times 10^{-4}$ kg-m^2/sec.

External torques, either steady or alternating, were generated by a brushless, slotless linear motor whose rotor section was also fixed to the resolver shaft. The stator containing the motor drive coils was rigidly mounted to the base of the apparatus and was carefully positioned with respect to the rotor. The two halves of the motor were not physically connected to one another, thus completely eliminating motor bearing friction. It was determined that the motor torque was very nearly a linear function of drive voltage, the proportionality constant being 2.20×10^{-3}kg-m^2/sec^2-volt. An input signal consisting of dc and/or ac voltages generates corresponding steady and/or alternating torques.

Binary data from the resolver-to-digital converter was transferred to a desktop computer which performed additional

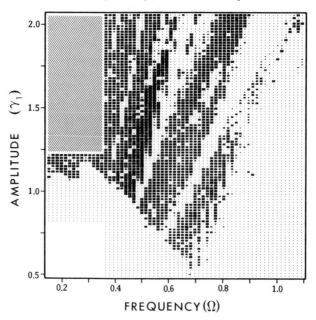

Figure 3. State diagram for the drive pendulum. Data from 5250 experimental runs are presented. Solid rectangles indicate chaotic behaviour. Vertical and horizontal labels refer to the applied ac torque. The shaded region was not sampled, but it presumed to be mainly chaotic.

processing and plotting.

45.3 **Results**

Experiments were carried out on the pendulum without a dc torque component (γ_0=0). High quality phase plane plots and Poincaré sections have been obtained. In this brief report we can only include a few examples of the observed behaviour. As expected, for sufficiently small drive amplitudes the pendulum undergoes periodic oscillations, whereas at large drive levels complex periodic motion can result. When driven with an appropriate torque, chaos is observed as illustrated in the phase plane plot (Fig. 1). Here the pendulum rotates both clockwise and counterclockwise in a completely unpredictable fashion. For this type of orbit, computed power spectra were found to be largely composed of elevated broad band noise. The associated Poincaré section is shown in Fig. 2; its dimension was determined to be 1.89.

A systematic study was undertaken of the pendulum motion as a function of two variables – the amplitude (γ_1) and frequency (Ω) of the applied ac torque. The results of this investigation are presented in an experimental state diagram (Fig. 3), which closely resembles the corresponding figure in reference [8].

References

[1]   B.A. Huberman, J.P. Crutchfield and N.H. Packard (1980). *Appl. Phys. Lett.* **37**, 750.

[2]   R.L. Kautz and R. Monaco (1985). *J. Appl. Phys.* **57**, 875.

[3]   D.C. Cronemeyer, C.C. Chi, A. Davidson and N.F. Pedersen (1985). *Phys. Rev.* **B31**, 2667.

[4]   M. Octavio (1984). *Phys. Rev.* **B29**, 1231.

[5]   Y.H. Kao, J.C. Huang and Y.S. Gou (1986). *J. Low Temp. Phys.* **63**, 287.

[6]   A.H. MacDonald and M. Plischke (1983). *Phys. Rev.* **B27**, 201.

[7]   W.C. Kerr, M.B. Williams, A.R. Bishop, K. Fesser, P.S. Lomdahl and S.E. Trullinger (1985). *Z. Phys. B. - Condensed Matter* **59**, 103.

[8]   D. D'Humieres, M.R. Beasley, B.A. Huberman and A. Libchaber (1982). *Phys. Rev.* **A26**, 3483.

SOLITON DYNAMICS IN LONG INHOMOGENEOUS JOSEPHSON JUNCTION

46.1 Introduction

Soliton dynamics in long Josephson junction (LJJ) has received great attention recently because of a number of interesting nonlinear phenomena and useful applications [1-8]. It is well established by now that the solitons in LJJ are responsible for so-called zero-field steps (ZFS) on the current-voltage characteristics (CVC) of LJJ. A number of recent works were devoted to the investigation of dynamic and static phenomena in the LJJ with non-uniform critical current density (CCD) distribution in the junction [2-6]. The soliton movement through a periodic arrangement of inhomogeneities causes the generation of plasma waves in the junction [2,3]. But up to now the calculations have not been brought up to the experimentally observable feature - CVC of the junction with the inhomogeneities (i.e. dependence of mean soliton velocity on bias current). Besides, as will be demonstrated below, the essential processes in the LJJ exist when the interaction of the soliton with inhomogeneities is not weak and, consequently, not described by the perturbation theory. A numerical simulation is needed in this case even for a qualitative description.

According to the one-dimensional model of the LJJ the phase difference $\varphi(x,t)$ is described by the perturbed sine-Gordon equation [1,2]:

$$\varphi_{xx} - \varphi_{tt} = f(x) \sin \varphi + \alpha\varphi_t + \gamma, \qquad (1)$$

where the coordinate x is normalized to the Josephson penetration depth λ_j and the time t is normalized to the inverse Josephson plasma frequency ω_o^{-1}. Here $f(x) = J_c(x)/J_{co}$ is a function which characterizes the distribution of the CCD along the junction; J_{co} is the average value of the CCD per unit length; α is the dissipation parameter; $\gamma = J/J_{co}$ corresponds to the uniformly distributed bias current J, which is supposed to be independent of the coordinate x.

The purpose of the present paper is to investigate an interaction between the soliton and regularly sited inhomogeneities in the absence of reflections from junction boundaries. The annular junction geometry which is possible to realize experimentally [7] is the best way to fit the situation. When such a junction has "frozen" the magnetic flux quantum (soliton) the following periodic condition for the phase difference occurs:

$$\varphi(L,t) = \varphi(0,t) + 2\pi. \qquad (2)$$

46.2 Results and Discussion

The CVC for the LJJ of the annular geometry with given parameters was simulated by means of numerical solution of the Eq. (1) with $f(x)$ in form:

$$f(x) = 1 + \text{sign}(f_o) \sum_{n=1}^{N} (1 - \text{th}^2(2(x - an)/f_o)) \qquad (3)$$

using periodic boundary conditions (2) and initial conditions corresponding to the travelling soliton. Here $f(x)$ is a periodic function with the period a, and N is the number of inhomogeneities. The soliton initial velocity was chosen in accordance with the relation following from the energy balance between the external current γ and dissipation α.

The modified leap-frog method [8] was used in the numerical solution routine. The steps were chosen to be $\Delta x = \Delta t = 0.05$; the accuracy was controlled by halving and doubling the steps. The accuracy of the average voltage determination was about 0.2-2.0%. The functions $\varphi(x,t_k)$ and $\varphi_t(x,t_k)$ corresponding to the stable oscillations in the previous point of the CVC were used as the initial conditions at the next point. Figure 1 shows the calculated region of the CVC for the LJJ with the following parameters: $L = 10$, $N = 5$, $\alpha = 0.01$, $a = 2$, $f_o = -0.4$. Dots show the CVC for the junction with uniform CCD distribution and with the same parameters L and α. These two characteristics are seen to be considerably different. CVC of the inhomogeneous junction displays a set of pronounced singularities at fixed voltage values.

Provided $f_o \ll 1$, CCD distribution in case of $L \to \infty$ can be represented as follows:

$$f(x) = 1 + f_o \sum_{n=-\infty}^{n=+\infty} \delta(x - an). \qquad (4)$$

The solution of Eq. (1) will be a superposition of the travelling soliton and a radiation wave having an amplitude $\varphi_1(x,t)$, which is calculated by the Green function method [3]. The wave frequency obeys the plasma dispersion law:

$$\omega_p^2 = 1 + K^2, \qquad (5a)$$

$$\omega_p = \mu\beta/q^2 \pm (\beta/q)(\mu^2\beta^2/q^2 - 1)^{1/2}, \tag{5b}$$

$$K = \pm\beta/q + (\mu\beta^2/q)(\mu^2\beta^2/q^2 - 1)^{1/2}. \tag{5c}$$

Here β is the soliton velocity. In the coordinate system associated with the soliton for $\beta > \beta_{th} = (1 + \mu^2)^{-1/2}$ plasma waves dissipate at the length

$$\ell_\alpha = \begin{cases} (q/\alpha)^{1/2}, \beta \simeq \beta_{th}, & \text{(6a)} \\ 2q/\alpha, \quad \beta \simeq 1. & \text{(6b)} \end{cases}$$

For $\beta < \beta_{th}$ the plasma waves are located near the soliton within the region $(1 - \mu^2\beta^2/q^2)^{-1/2}$. Here $\mu = 2\pi/a$, $q = (1-\beta^2)^{1/2}$.

According to the results of Ref. [3] the energy flux being transferred by the radiation approaches the maximum value at $\beta=\beta_{th}$ and becomes zero at $\beta = \beta_o = 1/u$. This energy flux determines the radiation part of the soliton energy losses. Comparison of the two CVCs shown in Fig. 1 allows us to suppose the existence of some additional mechanisms of energy dissipation in the

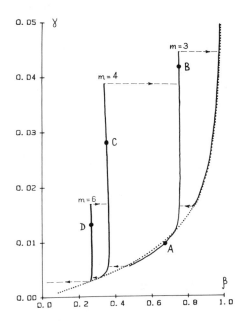

Figure 1. *Solid lines - numerically calculated region on the first ZFS for the annular LJJ with periodically modulated CCD. Arrows show the transitions between different branches. Dots show the same region for homogeneous LJJ with the same parameters L and α.*

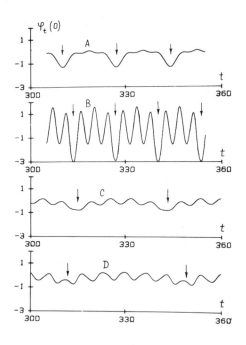

Figure 2. Voltage evolution at x = 0 for the inhomogeneous LJJ biased at the step with m = 3,4 and 6 (see Fig. 1). The curves A, B, C, D correspond to the points A, B, C, D, respectively. Arrows show the moments of soliton passing.

inhomogeneous LJJ. However, taking into account solely the direct radiation loss wouldn't allow to explain the numerically calculated CVC structure even in a qualitative way.

To clear up the origin of the above-mentioned singularities we investigated numerically the time-dependence of voltage $v(t) = \partial\varphi/\partial t$ at fixed points of the LJJ. Pictures corresponding to the points A,B,C,D in Fig. 1 are shown in Fig. 2. As seen from curves A and B the high frequency oscillation amplitude becomes larger with increasing γ along the CVC branch. Besides, for each step the oscillation period is m times smaller than the soliton circulation period (m is an integer). So the steps on the CVC appear if the following condition holds:

$$m\omega_{sol} = \omega_p, \qquad (7)$$

where $\omega_{sol} = 2\pi\beta/L$ is the frequency of the cyclic soliton motion along the junction and ω_p is determined by (5b). So the physical mechanism of the soliton energy dissipation in inhomogeneous LJJ is a resonance interaction of the soliton with the plasma waves provided the condition (7) is satisfied. The perturbation theory allows only to evaluate the positions of the steps on the CVC from (5b), (7):

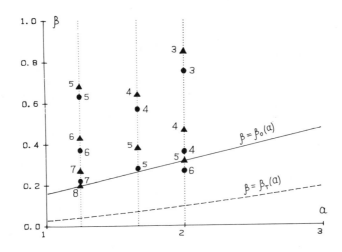

Figure 3. Positions of the steps on the CVCs as a function of the parameter a for L = 10.

$$\beta_m = (1 - 2m\mu g - (\mu^2+1))/m^2g^2)^{1/2}, \quad g = 2\pi L, \quad m < \mu g q^2. \quad (8)$$

It is interesting to note that the step with m = 5 is absent on the CVC given in Fig. 1. This may be related to the fact that the step position according to (8) turns out to be at $\beta \simeq \beta_o$.

We have also investigated the CVCs for different distances a between the inhomogeneities. The calculated positions of the steps β_m as a function of the parameter a for L = 10, f_o = -0.4, α = 0.01 are shown in Fig. 3. Black circles present the numerical results, triangles present the β_m values according to (8). Integer numbers indicate corresponding values of m. Equation (8) is seen to reproduce qualitatively the step positions β_m with an accuracy 10-20% depending on a and m. Numerical results have shown that variation of α and f_o doesn't affect considerably the step positions. Besides, the structure on the CVC is preserved even for α as large as 0.1. Note that the steps in Fig. 1 have distinct negative differential slope. We did not investigate this phenomenon in detail, but we believe it to be due to the soliton deceleration by the influence of the radiation.

The limiting case of an arbitrary arrangement of the inhomogeneities was studied in [9]. The junction with "in-line" geometry was studied experimentally in [10]. The steps observed on the first ZFS must have had the same origin as the steps investigated in the present work. This origin is the interaction of the soliton with plasma waves generated when it is reflected by the junction edges. The junction length corresponds to the inhomogeneity spacing in this case.

46.3 Conclusions

In our numerical experiment we have observed the strong interaction between a soliton and plasma waves in the annular Josephson junction of finite length. Such phenomena affect crucially the shape of the CVC of the junction, giving rise to current steps at fixed voltages. The perturbation theory approach allows the evaluation of the step positions.

Acknowledgements

We want to thank A. T. Filippov, Yu. S. Galpern, Yu. S. Kivshar, K. K. Likharev, B. A. Malomed, M. B. Mineev, V. V. Ryazanov for the helpful discussions.

References

[1]   A. Barone, F. Esposito, C. J. Magee and A. C. Scott (1971). *Nuovo Cim.* **12**, 227.

[2]   D. W. McLaughlin and A. C. Scott (1978). *Phys. Rev.* **18A**, 1652.

[3]   G. S. Mcrtchyan and V. V. Schmidt (1979). *Sol. St. Comm.* **30**, 791.

[4]   Yu. S. Kivshar and B. A. Malomed (1985). *Phys. Lett.* **111A**, 427.

[5]   M. Cirillo (1985). *J. Appl. Phys.* **58**, 3217.

[6]   A. T. Filippov and Yu. S. Galpern (1983). *Sol. St. Comm.* **48**, 665.

[7]   A. Davidson, B. Dueholm, B. Kryger and N. F. Pedersen (1985). *Phys. Rev. Lett.* **55**, 2059.

[8]   R. K. Dodd, J. C. Eilbeck, J. D. Gibbon and H. C. Morris (1983). *Solitons and Nonlinear Wave Equations,* Academic Press, New York.

[9]   M. B. Mineev and V. V. Schmidt (1981). *Sov. Phys. JETP* **54**, 155.

[10]   M. Scheuermann, C. C. Chi, N. F. Pedersen et al. (1986). *Appl. Phys. Lett.* **48**, 189.

47   *K.M. Mayer, R.P. Huebener, R. Gross, J. Parisi and J. Peinke*

AVALANCHE BREAKDOWN AND TURBULENCE IN SEMICONDUCTORS

Abstract

The complex spatial structures generated in homogeneously p-doped germanium during low-temperature avalanche breakdown have been investigated by means of a scanning electron microscope equipped with a liquid-He stage.   Two-dimensional imaging of current filaments has been performed and electron-beam-induced current oscillations have been found in the boundary regions of the filaments.   Low temperature scanning electron microscopy provides new insight into the physics of the chaotic temporal behaviour and the complex turbulent structures of electric current flow in semiconductors.

Spontaneous oscillations and chaotic behaviour in semiconductors during impurity impact ionization induced avalanche breakdown have been investigated by different groups [1-3].   Whereas so far all experiments are concentrating on the complex temporal behaviour (using standard electronic measuring techniques), we report on the first high-resolution experiments for observing the complex spatial behaviour in p-doped germanium at 4.2 K.   In a scanning electron microscope equipped with a liquid-He stage [4] we obtain two-dimensional images of current filaments that are nucleated in the homogeneous semiconductor during avalanche breakdown [5].   A scheme of the experimental set-up is presented in Fig. 1.   Our experiments were performed with polished p-doped Ge single crystals (indium, acceptor concentration 10^{14} cm^{-3}, acceptor level 11.6 meV), provided with ohmic aluminium contacts.   At the low temperatures all carriers are frozen out, and the samples remain in an electrically insulating state.   We observed avalanche breakdown at an electrical field of about 4 V/cm.

For imaging the two-dimensional spatial structures using low temperature scanning electron microscopy (LTSEM), the sample is locally perturbed by the electron beam, and the beam-induced electric current change $\Delta I(x,y)$ as the proper response signal is recorded as a function of the coordinates (x,y) of the beam focus.   The beam injects hot carriers into the semiconductor, which induce a significant current increment in the voltage-biased sample as long as the beam is directed to a non-conducting region.   In the regions occupied by the current filaments, where most of the shallow acceptors are ionized already, no significant current

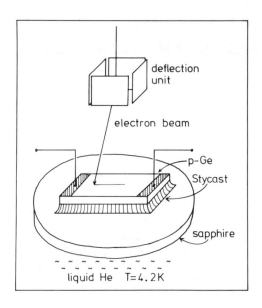

*Figure 1. Principle of the experimental set-up: the semiconductor (Ge) has been glued on a sapphire disc of 1 mm thickness using Stycast cement for good thermal contact to the liquid-He bath. The top surface of the samples was scanned by the electron beam. Typical dimensions of the Ge crystals were about 9 mm * 3 mm * 0.26 mm.*

increment is expected. Therefore, the filaments appear as dark regions in the images (Fig. 2) and the current-free parts of the specimen are corresponding to the bright regions. The experiments were performed using 26 kV beam voltage and about 100 pA beam current. This beam power was found to be small enough not to affect the filamentary structure of the sample. To improve sensitivity the electron beam was modulated and the signal was recorded using standard lock-in- or boxcar-techniques. The measured signal decay times were 10-30 µs in different samples. Therefore, the 5 kHz modulation frequency used in the experiments appears adequate.

We observed the nucleation and evolution of current filaments in correlation with the nonlinear current-voltage characteristics. Steep rises in the (voltage-controlled) characteristic were connected with the formation of a multifilamentary structure in the sample. Smallest filament diameters were found to be 30-40 µm. The growth of the filament size seems to be approximately linear with the sample current. Instabilities of the current were observed in different samples, whenever new filaments were about to be nucleated.

Electron-beam-induced current oscillations were found in the boundary regions of the filaments (Fig. 3): when the beam was focussed exactly upon the boundary region, the beam-induced current signal ΔI(t) often was modulated by current oscillations in the 10-130 kHz frequency regime, depending on the beam position

1 mm

Figure 2. *Typical spatial structure observed in a Ge-sample: the dark regions in the two-dimensional image correspond to the current filaments (the dark areas on top and bottom of the image belong to the contacts). The bright regions represent highly resistive parts of the sample. The bias voltage was 2.720 V.*

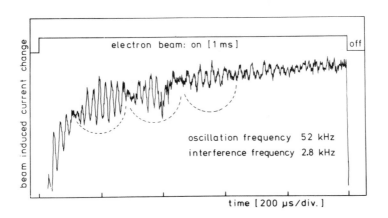

Figure 3. *Electron-beam-induced current oscillations observed when the electron beam was focussed upon the boundary region of a current filament: the upper trace shows the pulsed electron beam power. The lower trace represents the signal ΔI(t) and shows the time-resolved current changes in the sample (modulation frequency of about 52 kHz and beat frequency of about 2.8 kHz).*

along the filament boundary and the beam power. The observed beam-induced current oscillations support the "breathing" model of the current filaments, for explaining the spontaneous chaotic temporal behaviour in homogeneous semiconductors [6].

In conclusion, LTSEM provides valuable new insight into the phenomenon of avalanche breakdown and the mechanisms of the chaotic temporal resistance behaviour and the formation of complex spatial structures in semiconductors at low temperatures.

References

[1]   S. W. Teitsworth, R. M. Westervelt and E. E. Haller (1983). *Phys. Rev. Lett.* **51**, 825; S. W. Teitsworth and R. M. Westervelt (1986). *Phys. Rev. Lett.* **56**, 516.

[2]   J. Peinke,   A. Mühlbach, R. P. Huebener and J. Parisi (1985). *Phys. Lett.* **108A**, 407;   B. Röhricht, B. Wessely, J. Peinke, A. Mühlbach, J. Parisi and R. P. Huebener (1985).   *Physica* **134B**, 281.

[3]   K. Aoki and K. Yamamoto (1983).   *Phys. Lett.* **98A**, 72;   K. Aoki, O. Ikezawa, N. Mugibayashi and K. Yamamoto (1985). *Physica* 134B, 288.

[4]   H. Seifert (1982).   *Cryogenics* **22**, 657.

[5]   K. M. Mayer, R. Gross, J. Parisi, J. Peinke and R. P. Huebener, to be published.

[6]   E. Schöll (1985).   *Physica* **134B**, 271;   E. Schöll, *Phys. Rev. B,* to be published;   E. Schöll, J. Parisi, B. Röhricht, J. Peinke and R. P. Huebener, *Z. Phys. B,* to be published.

OSCILLATORY OXIDATION OF CO ON Pt

48.1 Introduction and Summary

Much work has been dedicated to the oscillatory oxidation of carbon monoxide on platinum catalysts in the last years [1]. In recent studies on well-defined Pt(100) single crystal surfaces, the mechanism was elucidated in terms of a coupling between surface reaction steps and a structural phase transition of the topmost platinum layer [2,3].

A Pt(110) surface is also known to undergo an adsorbate-induced phase transition, which is completed at a CO coverage of 0.5 [4], but unaffected by oxygen [5]. Kinetic oscillations on this surface were detected and recorded by means of work function ($\Delta\varphi$) measurements. Oscillation conditions between 440 and 590K and in a pressure range from 10^{-5} to 10^{-3} Torr were established. A variety of oscillation forms, including well-defined transitions from periodic to aperiodic oscillations were found. LEED studies suggested that the 1x2 → 1x1 structural transformation of the surface is involved.

48.2 Experimental

The experiments were carried out in a standard UHV chamber containing facilities for Auger electron spectroscopy (AES), low energy electron diffraction (LEED), and work function ($\Delta\varphi$) measurements on a cylindrical single crystal with a surface of about 0.5 cm^2. The details of operation and the cleaning procedure of the sample have been described elsewhere [6]. The conditions, i.e. sample temperature and partial pressures of CO and O_2, were stabilized by feedback regulation. Thus the experiments were performed strictly isothermally in a gradient-free flow reactor.

The pressure was continuously monitored with an ionization gauge, but more precise determination of the pressures was achieved by appropriate setting of the pressure in the gas inlet system [6], to which the pressure in the chamber was found to be proportional.

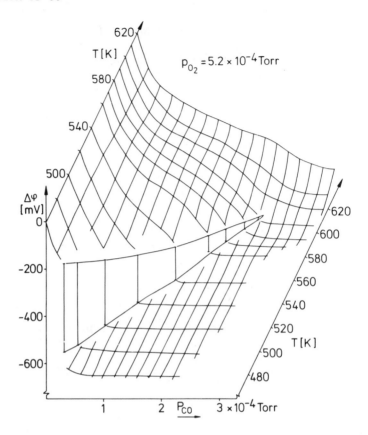

Figure 1. $\Delta\varphi$ of the Pt(110) surface as a function of p_{CO} and T at constant p_{O_2}. Vertical bars indicate the highest amplitude of oscillations at the respective temperature.

48.3 Results

Oxygen caused an increase in $\Delta\varphi$ of about 800 mV, carbon monoxide one of 150 mV. Oscillation conditions were determined by choosing fixed oxygen pressure p_{O_2} and temperature T and stepwise increasing p_{CO}. During this procedure the work function decreased smoothly until at a certain point (α) there was a sharp drop to a lower value (β) corresponding roughly to the value of a CO-covered surface and after which $\Delta\varphi$ did not change significantly any more. Between α and β, oscillations could be obtained over a narrow range in p_{CO} (about 2% at low and up to 12% at high temperature). A typical set of data (at $p_{O_2} = 5.2\times10^{-4}$ Torr) is reproduced in Fig. 1. In the same way the conditions were determined at p_{O_2}

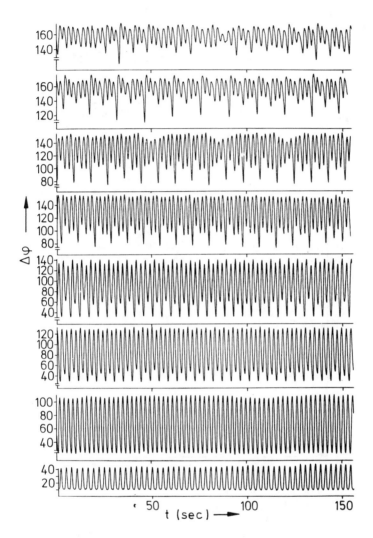

Figure 2. Change in oscillatory behavior when going from near β (bottom) to near α (top) by decreasing p_{CO}.
$T = 540K$; $p_{O_2} = 7.5x10^{-5}$ Torr; $p_{CO} = 3.84, 3.66, 3.63, 3.57,$
$3.51, 3.49, 3.42$ and $3.38x10^{-5}$ Torr (from bottom to top).

between 10^{-3} and 10^{-5} Torr.   No definite low pressure nor low temperature limit was detected, but the oscillations became so slow that further studies were dismissed for practical reasons.

At constant p_{O_2}, p_{CO} increased with increasing T in a way that an ln p_{CO} vs. $1/T$ curve came close to a straight line (Fig. 3). It should be noted that the high temperature limits of

Oxidation of CO

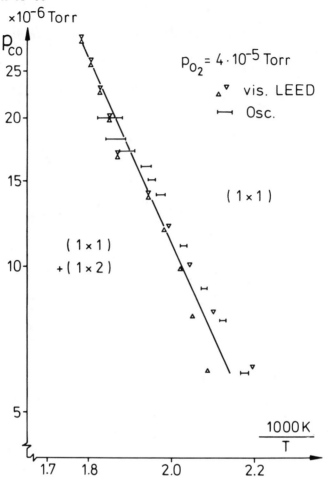

Figure 3. Completion of (1x2) → (1x1) phase transition and oscillation conditions in a ln p_{CO} vs. 1/T plot at p_{O_2} = 4x10^{-5}

Torr. $\Delta(\nabla)$ denote increasing (decreasing) temperature during visual LEED inspection.

oscillations (at various p_{O_2}), where the amplitudes vanished, coincide with the isostere of CO coverage 0.5 (in the absence of oxygen).

A large variety of oscillation forms was found, depending on the choice of conditions. At low temperature harmonic oscillations of amplitudes up to 500 mV were obtained, at intermediate temperatures square-wave forms prevailed [7]. Towards the high temperature limit rapid, small oscillations were detected. Careful reduction of p_{CO} (going from β to α) led from simple periodic to irregular oscillations via period doubling (Fig. 2). Also period tripling could occasionally be observed [7].

At fixed p_{O_2} = 4×10^{-5} Torr and various CO pressures the temperature was continuously increased and decreased, while the appearance and disappearance of the (1x2) spots in the LEED pattern was visually observed. The resulting data are shown in Fig. 3 to coincide with the oscillation conditions. The slope of the straight line yields (8 ± 0.5) kcal, which corresponds to the activation energy of the surface reaction at high oxygen coverage [8].

48.4 Discussion

The microscopic mechanism of the oscillations on Pt(110) is presumably similar to that on Pt(100) [9], since there is again evidence that the surface phase transition is involved, namely the coincidences of the high temperature limit with the θ_{CO} = 0.5 isostere and of the (dis) appearance of the 1x2 structure with the oscillation conditions.

The much narrower oscillation region on Pt(110) as compared to Pt(100) [6] could be due to the fact that the difference in sticking probability of oxygen between the reconstructed and unreconstructed surface is much smaller on Pt(110) [10].

References

[1]   L.F. Razón and R.A. Schmitz (1986). *Catal. Rev.-Sci. Eng.* **28**, 89.

[2]   M.P. Cox, G. Ertl, R. Imbihl and J. Rüstig (1983). *Surf. Sci.* **134**, L517.

[3]   M.P. Cox, G. Ertl and R. Imbihl (1985). *Phys. Rev. Lett.* **54**, 1725.

[4]   T.E. Jackman, J.A. Davies, D.P. Jackson, W.N. Unertl and P.R. Norton (1982). *Surf. Sci.* **120**, 389.

[5]   S. Ferrer and H.P. Bonzel (1982). *Surf. Sci.* **119**, 234.

[6]   M. Eiswirth, R. Schwankner and G. Ertl (1985). *Z. phys. Chemie N.F.* **144**, 59.

[7]   M. Eiswirth and G. Ertl (1986). *Surf. Sci.* **177**, 90.

[8]   J.R. Engstrom and W.H. Weinberg (1985). *Phys. Rev. Lett.* **55**, 2017, 1985.

[9]   R. Imbihl, M.P. Cox, G. Ertl, H. Müller and W. Brenig (1985). *J. Chem. Phys.* **83**, 1578.

[10]   N. Freyer, M. Kiskinova, G. Pirug and H.P. Bonzel (1986). *Surf. Sci.* **166**, 206.

49 *M. Hennenberg, A. Rouzaud, J.J. Favier and D. Camel*

UNIDIRECTIONAL SOLIDIFICATION USING THE CONCEPT OF A BOUNDARY LAYER OF SOLUTE DIFFUSION

Abstract

Coupled hydrodynamic and morphological instabilities during crystal growth are considered. The gradients of the different physical quantities (solute concentration, momentum and temperature) being orders of magnitude apart from one another, one introduces the notion of a boundary layer of solute diffusion where both the inner and the outer limits are deformable. For the marginal case of exchange of stabilities, the finite width of the deformable boundary layer results in an abrupt increase of the concentration threshold for hydrodynamic instability and a lowering of the criterion for absolute morphological stability.

49.1 Introduction

Solidification is one of the oldest exploited technological processes. Its large number of applications ranges from foundry up to computer technology. Among the fundamental problems this process has brought forward, one is that of the stability of the planar growth front during the upward directional solidification of a dilute binary alloy from its melt at an imposed velocity V in the earth gravitational field or in a microgravity environment. For the experimentalist and the theoretician alike, this subject is a challenge. Indeed its complete description takes into account the transfer of energy, mass and momentum across the liquid and the solid. Moreover, it is a typical example of a free boundary problem [1], since the liquid-solid interface position and morphology are coupled with the solute and the temperature fields.

49.2 The Deformable Solute Diffusion Boundary Layer

The problem to solve couples thus hydrodynamic and morphological instabilities since bulk convection is encountered in most practical cases. Nevertheless there is a fundamental point one cannot overlook. As is well known, the ranges of the kinematic viscosity ν, the diffusion coefficient of the solute D and the thermal diffusivity k are orders of magnitude apart from one another, so that one has the fundamental inequality

$$D \ll \nu \ll k \qquad (1)$$

As a consequence, the extents of the related inhomogeneities in the liquid phase are quite different for the solute, the momentum and the heat. This clearly suggests a description in terms of a boundary layer theory [2], applied to composition. One divides accordingly the liquid phase in two zones. The inner layer of extent δ, ahead of the solid-liquid interface keeps track of the changes in the concentration profile. Only inside this layer will thermosolutal instability be meaningful, since the outer layer is completely spatially homogeneous with respect to concentration. As has been argued in [3], for the simpler case of no hydrodynamic instability, this fundamental hypothesis should be valid as well for the unperturbed case as for the perturbed one, so that morphological instability results in the deformability of both boundaries of the inner layer. This happens in such a way as to neglect any inhomogeneous distribution of solute outside of this domain. Thus, the outer layer changes its shape so that there is no spatially dependent solute fluctuation inside it. This is a different point of view altogether from the classical one [4].[5] for which the outer boundary of the inner layer is rigid. In that last approach, the solute disturbances disappear along this undeformable border, but this reflects a situation which no longer exists and thus it induces parasite reflections.

49.3 Consequences of a Deformable Diffusion Boundary

At the interface between both concentrated phases, we suppose a Nernstian distribution of the solute, the repartition coefficient K_o being linked to local thermodynamic equilibrium. Furthermore, the interface temperature T_Σ^*, the solute concentration C_L^* and the pure liquid temperature T_M^* are linked by the Gibbs Thompson law [3]

$$(1 - \gamma K)T_M^* + m_L \, C_L^* \, |_\Sigma = T_\Sigma^* \qquad (2)$$

where K is the local curvature of the interface Σ, γ a capillary constant and m_L the slope of the liquidus in the phase diagram.

We suppose a quasi-steady temperature distribution in both phases. The density of the liquid phase varies with the temperature and the solute content as

$$\rho(z) = \rho(0) \, [1 - \beta_T(T^*(z) - T^*(0)) - \beta_C \, (C^*(z) - C^*(0))] \qquad (3)$$

where z is the distance away from the reference unperturbed surface, defined as $z = 0$ and where β_T and β_C are the thermal and solutal coefficients of expansion respectively. The boundary conditions at the interface Σ express the balance of energy, total mass, and solute. Neither accumulation of matter nor solute are supposed along Σ.

A linear stability analysis is performed as detailed in [6]. Along the perturbed surface whose equation is

$$\delta z = \varepsilon \ z_{sl} \qquad (4)$$

where z_{sl} is the normalized deformation amplitude of the solid liquid interface, any perturbed quantity $f^{*}(z)$ is now

$$f^{*}(\delta z) - f^{o^{*}}(0) = \varepsilon \ [z_{sl} \ \frac{df^{o^{*}}(z)}{d z} \ |_{z=0} + \bar{f}^{*} \ |_{z=0} \ ] \qquad (5)$$

where $\varepsilon\bar{f}$ is the first order term of the perturbation of the quantity $f^{*}(z)$ and $f^{o}(z)$ the reference unperturbed value.

According to the above discussion, the boundary condition along the perturbed outer limit of the boundary layer becomes

$$c^{*}(\Delta + \delta z) = c^{o^{*}}(\Delta) \qquad (6)$$

where

$$\Delta = \frac{\delta V}{D} \ . \qquad (7)$$

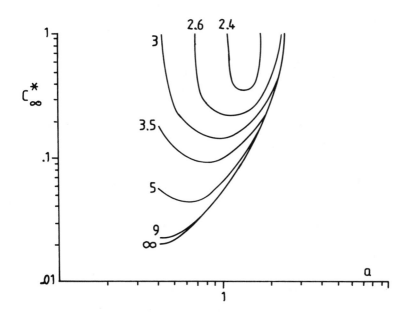

Figure 1. Non-oscillating hydrodynamic instability. Bulk concentration as function of wavenumber a for various Δ. (V = $2*10^{-3}$ cm/sec.)

49.4 Conclusions

Then one works out the compatibility condition which depends now essentially on Δ. In the limit $\Delta \to \infty$, we recover the classical results [4]. The results of the first numerical intensive calculations at finite Δ are shown in Fig. 1 and Fig. 2. For the hydrodynamic non-oscillating case, one has now a dependence of the critical wavenumber a upon Δ. The solute concentration C_∞^* in the bulk of the liquid phase necessary to trigger an unstable behaviour is much higher for Δ small than its value in the classical case (see Fig. 1). A much larger concentration can

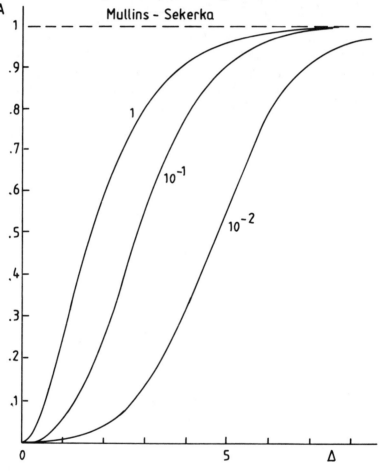

Figure 2. The absolute stability function $A = \dfrac{T_M V K_o^2}{D_L(1-K_o)C_\infty^* m_1}$ as a function of Δ for various K_o.

correspond to a quiescent liquid.

This approach induces changes on the morphological instability too. In their original approach [7], Mullins and Sekerka showed that the critical physical parameter needed to eliminate any possible surface deformation had to be at least equal to one. Now the same parameter has its value markedly affected by Δ and the Nernstian repartition coefficient K_o (see Fig. 2). The value given previously becomes the asymptotic one.

We are thus showing that a consistent treatment of the deformable boundary layer gives rise to quite unexpected results which are in better agreement with physical experiments and are free of the physical inconsistencies induced by a rigid boundary layer approach [4], [5].

References

[1]   J. Crank (1984).   *Free and Moving Boundary Problems,* Oxford Sciences Publications, Clarendon Press.

[2]   C. Bender and S.A. Orszag (1984).   *Advanced Mathematical Models for Scientists and Engineers,* MacGraw Hill Publishers, Singapore.

[3]   J.J. Favier and A. Rouzaud (1983).   *J. Cryst. Growth* **64,** 367.

[4]   S.R. Coriell and R.F. Sekerka (1981).   *PhysicO-Chemical Hydrodynamics* **2,** 281.

[5]   S.R. Coriell, D.T.J. Hurle and R.F. Sekerka (1976).   *J. Cryst. Growth* **32,** 1.

[6]   M. Hennenberg, A. Rouzaud, D. Camel and J.J. Favier, *Journal de Physique,* to appear.

[7]   W.W. Mullins and R.F. Sekerka (1963).   *J. Appl. Physics* **34,** 323.

BIFURCATIONS IN 3-DIMENSIONAL TURING SYSTEMS: PREPATTERN SIMULATION ON SUPERCOMPUTERS

Spontaneous emergence of stable spatial order in biological systems (morphogen prepatterns) may arise through bifurcations in the reaction-diffusion equations, describing biochemical control systems (Turing, 1953):

$$c_t = F(k,c) + \text{div } D \text{ grad } c \qquad (1)$$

Here, the components of vector c are the concentrations in an open, dissipative, autocatalytic biochemical network, with nonlinear rates $F(k,c)$, where vector k represents rate constants (enzyme activities). If k and D are constant in space one has a Turing system of the first kind. If one spatial prepattern induces rates and apparent diffusion constants as functions of space, a Turing system of the second kind arises. Let c_0 be the usual, homogeneous, stationary solution to Eq. (1). Inhomogeneous stationary solutions may be constructed from bifurcation theory applied to Eq. (1). With $z = c-c_0$, one seeks solutions to

$$0 = L(z) + N(z) \qquad (2)$$

where $L(z) = \text{div } D \text{ grad } z + Mz$, with $M = \partial F/\partial c$ (the Jacobian), and $N(z)$ is the remaining nonlinear part of F.
   Bifurcations may occur where

$$L_0 z = 0 \qquad (3)$$

admits nontrivial solutions $z \neq 0$. This may occur for certain values k_c or D_c, which inserted in L defines L_0. When k or D exceed these values in parameter space, an exchange of stabilities (primary bifurcation) may take place between c_0 and new, stable inhomogeneous solutions z_p, which can be constructed by small parameter expansions in ε:

Here, α is a particular rate constant among the elements of k and α_c is the value, where Eq. (3) is obtained.

Collecting terms of the same order in ε, one obtains to first order in ε

$$L_0 \phi_0 = 0 \qquad (5)$$

and thus

$$\phi_0 = d_{nml} j_n(\kappa_{nl} r/R) \sum_{m=-n}^{n} \nu_m Y_{nm}(\Omega). \qquad (6)$$

Here $j_n(\kappa_{nl} r/R)$ $Y_{nm}(\Omega)$ are the solutions to $\Delta \psi = -k^2 \psi$ with no-flux boundary conditions $\partial \psi / \partial r = 0$ at $r = R$, in a sphere. The particular value of the integers n, m, l entering Eq. (6) are limited to the particular region in parameter space, where Eq. (3) has nonzero solutions, i.e. to the particular value of α_c (nml). The ν's in Eq. (6), and thus the actual geometry of the first order solution ϕ_0, are obtained from the terms arising to second order in ε, which are:

$$L_0(\phi_1) + \tau_1 L_1(\phi_0) + N_2(\phi_0) = 0. \qquad (7)$$

Here $L_1(\phi_0)$ is a constant matrix multiplying ϕ_0.

Let $[u, v]$ denote the inner product of vectors u and v, i.e. the integration of $u^* \cdot v$ over the sphere. Consider the adjoint operator L_0^+, with solution space $\phi_0^+(m) = d_{nme}^+ j_n(\kappa_{nl} r/R) Y_{nm}$, $m \in (-n, n)$. Eq. (7) is of the form $L_0(\phi_1) = f$, and Fredholm's solvability condition is $[f, \phi_0^+(m)] = 0$, $m \in (-n, n)$. These equations and the normalization $[\phi_0, \phi_0] = 1$ yields $2n + 2$ algebraic equations in the $2n + 1$ ν_m's and τ_1.

Evaluation of the inner products by use of tables of 3-j or Clebsch-Gordan coefficients, yields algebraic bifurcation equations, which in turn yield ν's **independent** of the chemistry in Eq. (1), whereas the amplitude, measured by τ_1, is dependent on the chemistry. The result is that out of the 5 Y_{2m}'s, only one emerges (Y_{20}), the other ν's vanish. This **selection rule** is then independent of the chemical network used. A similar analysis of bifurcations from z_p to new patterns z_s (secondary bifurcation) yields a related result: the structure of the algebraic bifurcation equations is determined by the vanishing or nonvanishing of integrals over eigenfunctions to the Laplacian. Thus, selection rules are likely to be common to a broad class of chemical networks, at least for the first few bifurcations. This makes it meaningful to study such bifurcations numerically, using a model system. For further details and references, see Hunding (1982).

Numerical solution of the time evolution equation for c, Eq. (1), is achieved by discretization of the Laplacian (the method of lines), and solution of the resulting stiff system of several thousands ODE's by Gears predictor-corrector code. The corrector step was handled using sparse matrix iteration. A particularly effective code was obtained by ordering the mesh points in generalized chess board fashion throughout 3-dimensional space. This "red-black" ordering yields sparse matrix structures, which vectorize well on modern supercomputers. Indeed, the overall code performs at a sustained rate of 150-195 MFLOPS on the CRAY X-MP, and over 350 MFLOPS on the AMDAHL VP1200. The development of the stiff code made the simulation of 3-dimensional Turing structures possible. With the advent of the vectorized version, such calculations have been made economical: nonstiff codes result in execution times (VAX11/750) of several weeks, the stiff code cuts this to 10 min (Univac 1100/92) and the vectorized code cuts it to a mere 4 sec (VP1200). This development opens the possibility of treating 3-dimensional Turing structures with much higher spatial resolution. It also makes it possible to compare numerical predictions of the behaviour of real systems with experimental results.

The results of biological interest found so far are the following. A new theory of cell division (mitosis and cytokinesis) has been formulated, and it has been suggested that the mechanisms underlying these phenomena could evolve into a spatial control system for multicellular organisms (blastulas and early events in morphogenesis such as segmentation).

The prepattern theory of mitosis and cytokinesis suggests that bipolarity in cells arises as bipolar Turing structures of the

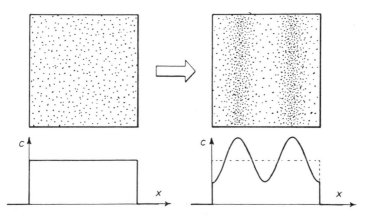

Figure 1. In autocatalytic chemical networks, fluctuations amplify exponentially, but diffusion tends to destroy them. Balance is obtained when a Turing structure has emerged, to first order given as an eigenfunction to the Laplacian: $c_t \sim + \lambda c + D\Delta c = 0$ *if* $\Delta c = -k^2 c$. *Spontaneous formation of stable inhomogeneous concentration prepattern results.*

form j_2Y_{20}.   Centrioles   are   not   needed,   as   demonstrated
experimentally: laser destruction, or micromanipulation, of
centrioles show that normal spindles develop independently of the
presence of centrioles at the poles.   Modern theories of mitosis
are consequently centriolar free models, mostly relying on
postulated   forces   arising   a.o.   through   interactions
(cross-bridging) of spindle fibres.   However, such theories
predict   forces   proportional   to   spindle   length,   whereas
experimental measurements on asymmetrically placed, trivalent
chromosomes   strongly   suggest   that   this   is   not   true   for
equator-crossing fibres.   In the prepattern theory, by contrast,
spindle forces arise through interactions between local fibre
entities and a force field, given by the prepattern.   Much better

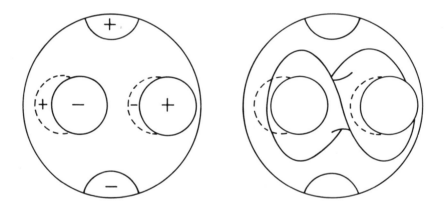

*Figure   2.   Highly   symmetrical   saddle   shaped   chromosome
distributions, photographed in Aulacantha Scolymantha, may be
Turing   structures   on   display: the   null   region   of   the
numerically recorded prepattern $j_1Y_{10} + j_2Y_{22}$ (left) is a saddle
shaped structure (right).   In general, selection rules governing
bifurcations in 3-dimensional reaction-diffusion systems have
been   recorded   numerically,   and   a   number   of   biological
experiments have been explained on this basis.*

values for observed spindle forces result.

The prepattern theory may also explain Scott's experiment: when dividing sea-urchin eggs (metaphase) were compressed, occasionally direct quadripartition is obtained. One egg cleaves directly to four cells, two with asters, and two without. This result is difficult to encompass with most theories of cytokinesis based on spindle forces. The prepattern theory, however, has been shown to yield a quadripolar prepattern when the bipolar pattern $j_2 Y_{20}$ in the sphere is compressed.

Finally, axis-tilting is explained by Turing structures: most cells contain a spatial axis known as the animal-vegetal axis. The first spindle forms perpendicular to this axis, thus resulting in cell cleavage strictly through the AV-poles, a result which has gone unexplained by mitosis/cytokinesis models based on internal spindle properties. The prepattern theory, however, yields a bipolar prepattern exactly perpendicular to the AV axis in the first and second cleavage, and spindle of orientation parallel to the AV-axis in the third cleavage, all in agreement with experiments.

Turing structures may be on display in the nucleoplasm of the radiolarian Aulacantha Scolymantha. During nuclear division, more than 1000 chromosomes are crudely segregated to the two daughter nuclei under formation. During this process, the chromosomes confine themselves to highly symmetrical regions within the nucleoplasm, apparently for no obvious reason. However, the prepattern theory suggests that these regions are connected to observed computer generated Turing structures, which may thus explain such phenomena as the upfolding of the metafase plate, confining more than 1000 chromosomes to a saddle-shaped, highly symmetrical region, photographed repeatedly in Aulacantha Scolymantha. These patterns have been found stable in both oblate and prolate spheroids.

Finally, Turing structures may govern the segmentation, observed in Drosophila embryos. Before any visible changes of cells are seen, a prepattern can be detected experimentally. Seven stripes (maxima) of transcripts of the fushi tarazi gene are observed to arise spontaneously.

Thus, the interesting possibility exists that nature has exploited Turing structures, first in crude cell division (cytokinesis) as seen in bacteria, then to orderly chromosome separation (mitosis) and finally as a spatial organizing principle in multicelluar embryos. All these processes can be simulated by existing software, as briefly described above.

Acknowledgement

Cray computer time at the Kernforschungsanlage Jülich by grant no. 11-5359 from the Danish Natural Science Foundation.

References

[1]   A.M. Turing (1952). *Phil. Trans. Roy. Soc. London Ser. B* **237**, 37.

[2]   A. Hunding (1980). *J. Chem. Phys.* **72**, 5241.

[3]   A. Hunding (1982) in *Evolution of Order and   Chaos in Physics, Chemistry and   Biology,* ed. H. Haken, Springer Series on Synergetics, 13, Springer, Heidelberg.

[4]   A. Hunding (1985).   *J. Theoret. Biol.* **114**, 571.

[5]   J. Slack (1984).   *Nature* **310**, 364.

51   *B. Röhricht, J. Parisi, J. Peinke and O.E. Rössler*

SYMMETRY-BREAKING PHASE TRANSITIONS AND BOILING-TYPE TURBULENCE OF A SIMPLE MORPHOGENETIC REACTION-DIFFUSION SYSTEM

Abstract

Representing a general dynamical model for symmetry-breaking morphogenesis, the Rashevsky-Turing theory claims for the most convenient prototype model of many different synergetic systems in nature. In the simplest case it can be realized by a two-cellular symmetrical reaction-diffusion system, consisting of cross-inhibitorily coupled, potentially oscillating two-variable subsystems (4-D flow). We report on numerical evidence of symmetry-breaking nonequilibrium phase transitions from phase-locked coherent to phase-lagged differentiated behaviour of the two subsystems. We further indicate the structural change of the evocated system flow from stable morphogenesis to boiling-type turbulence. Finally, we demonstrate experimental evidence that the spatio-temporal nonlinear behaviour during impurity-impact-ionization-induced avalanche breakdown of semiconductors can qualitatively be described by the above 4-D reaction-diffusion model.

The well-known Rashevsky-Turing (RT) theory [1,2] has originally been designed to account for a qualitative understanding of the spontaneous occurrence of differentiation in multicellular systems based on identical, individually stable cells. Assuming a symmetrically built, homogeneous reaction system, symmetry-breaking RT behaviour implies evocation of instability of the symmetrical steady state under a parametric change. Following the original Turing reaction-diffusion-transport equations [2] as the first concrete dynamical model of biological morphogenesis, several more examples have been described in literature [3-7]. Turing-type equations are of interest, therefore, for mathematical, chemical, physical, biological and biochemical systems.

As the essential behavioural characteristic of RT systems, breakdown of symmetry can in the simplest case be realized by a two-cellular symmetrical morphogenetic system consisting of cross-inhibitorily coupled, potentially oscillating two-variable subsystems (4-D flow) [5,6]. The reaction scheme of such a simple RT system is sketched in Fig. 1. The two morphogens A are self-inhibiting via B and, to a lesser extent, cross-inhibiting each other via the symmetrical coupling between the two morphogens B. The excess of self-inhibition over cross-inhibition is compensated by a path of self-activation (autocatalysis of A) which is not mediated to the other side. The effects of constant pools and reaction partners are comprised in the effective rate constants $K_1, \ldots, K_5$. Following the argumentation introduced in [5], the system of Fig. 1 can be described under the assumption of a Michaelis-Menten-type kinetics by the following set of simultaneous ordinary differential equations:

$$\dot{a} = (K_1 - K_3) \, a - K_2 \, b \, \frac{a}{a+K} \quad + K_5$$

$$\dot{b} = \quad K_3 \, a - K_4 \, b \quad\quad + D \, (b' - b)$$

$$\dot{b}' = \quad K_3 \, a' - K_4 \, b' \quad\quad + D \, (b - b')$$

$$\dot{a}' = (K_1 - K_3) \, a' - K_2 \, b' \, \frac{a'}{a'+K} \quad + K_5$$

(1)

where a, b, a', b' denote the concentration of the two morphogens A and B in compartment 1 (unprimed) and 2 (primed), respectively. D is the diffusion coefficient for the morphogen B, and K represents the phenomenological Michaelis-Menten constant. We note that the above equations presuppose isothermy, homogeneity in either compartment, and fast relaxation of intermediate products as usual.

The above two-compartment structure of the 4-D flow implies the two essential characteristics of any RT system, namely the spontaneous occurrence of evocation behaviour in the symmetrical homogeneous reaction system and the existence of a certain delay in the inner dynamics of each subsystem [5]. Based on these main phenomena, the RT reaction-diffusion model described in Eq. (1) is capable of eliciting symmetry-breaking nonequilibrium phase transitions from phase-locked coherent to phase-lagged differentiated behaviour of the two subsystems. Moreover, the evocated system flow may further bifurcate from stable morphogenesis to boiling-type turbulence. Numerical evidence on the spatio-temporal correlation of the two subsystems during the above transition from the symmetrical state of phase-locked periodic oscillations to symmetry-breaking structures of phase-lagged periodic and chaotic oscillations has been presented elsewhere [6] in detail. It is emphasized that the scenario from the fully differentiated state towards irregularly recurring differentiation ends up with a sort of screw-type bichaotic flow – two screw-type chaotic regimes separated by a symmetrical saddle-limit cycle – as described in [5]. Such kind of turbulent morphogenesis in 4-D state space admits 3-D cross-sections of generalized horseshoe type [5,6].

Since the above self-organizing cooperative processes appear to reflect a general property of real synergetic systems, let us now see whether recent experimental investigations on spatio-temporal

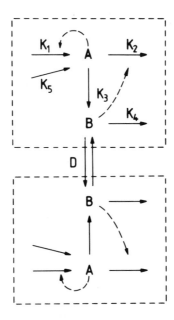

Figure 1. Reaction scheme of a simple Rashevsky-Turing system (constant pools omitted from the scheme, catalytic rate control indicated by dashed arrows).

nonlinear transport phenomena in the avalanche breakdown regime of current-carrying semiconductors [8] can be interpreted in terms of the main behavioural characteristics of the RT reaction-diffusion model. Our experimental semiconductor system consists of a single homogeneously p-doped germanium crystal, electrically driven into the post-breakdown regime due to impurity impact ionization at liquid-helium temperatures. Spatially separated and coupled sample subsystems are realized by means of an appropriate arrangement of ohmic contact probes, dividing the semiconductor crystal into different parts which by themselves show nonlinear dynamical behaviour. As discussed elsewhere [8] in detail, the current transport phenomena in semiconductors include spontaneous formation of a rich variety of both spatial and temporal dissipative structures. Experimental evidence of analogous symmetry-breaking nonequilibrium phase transitions and various kinds of boiling-type turbulence suggests a reaction scheme for the semiconductor system qualitatively similar to that of the phenomenological RT model indicated in Fig. 1. Apparently, the rather complicated nonlinear behaviour of impurity-impact-ionization-induced avalanche breakdown in semiconducting germanium can directly be projected on to the basic spatio-temporal possibilities of the present 4-D RT model [6]. Reflecting general behavioural characteristics of many different synergetic systems in nature, the RT reaction-diffusion model may acquire a rather general significance, perhaps.

Acknowledgements

The authors gratefully acknowledge financial support from the Stiftung Volkswagenwerk and stimulating discussions with R.P. Huebener, E. Schöll and H.G. Purwins.

References

[1] N. Rashevsky (1940). *Bull. Math. Biophysics* **2**, 15, 65, 109.

[2] A.M. Turing (1952). *Phil. Trans. Roy. Soc. London* **B237**, 37.

[3] I. Prigogine and G. Nicolis (1967). *J. Chem. Phys.* **46**, 3542; G. Nicolis and I. Prigogine (1977). *Self-organization in Nonequilibrium Systems*, Wiley, New York.

[4] A. Gierer and H. Meinhardt (1972). *Kybernetik* **12**, 30; H. Meinhardt and A. Gierer (1974). *J. Cell Sci.* **15**, 321.

[5] O.E. Rössler and F.F. Seelig (1972). *Z. Naturforsch* **27b**, 1444; O.E. Rössler, *Z. Naturforsch* **31a**, 1168.

[6] B. Röhricht, J. Parisi, J. Peinke and O.E. Rössler, *Z. Phys. B - Condensed Matter*, to be published.

[7] O. Sporns and F.F. Seelig (1986). *Biosystems* **19**, 83; O. Sporns, S. Roth and F.F. Seelig, *Physica D*, to be published.

[8] J. Peinke, A. Mühlbach, R.P. Huebener and J. Parisi (1985). *Phys. Lett.* **108A**, 407; J. Peinke, B. Röhricht, A. Mühlbach, J. Parisi, Ch. Nöldeke, R.P. Huebener and O.E. Rössler (1985). *Z. Naturforsch* **40a**, 562; B. Röhricht, B. Wessley, J. Peinke, A. Mühlbach, J. Parisi and R.P. Huebener (1985). *Physica* **134B**, 281; J. Parisi, J. Peinke, B. Röhricht and R.P. Huebener (1986) in *Proc. 18th Intern. Conf. on the Physics of Semiconductors*, ed. O. Engström, World Scientific, Singapore; B. Röhricht, J. Parisi, J. Peinke and R.P. Huebener, *Z. Phys. B - Condensed Matter*, to be published; K.M. Mayer, R.P. Huebener, R. Gross, J. Parisi and J. Peinke, these Proceedings.

HOW TO DISTINGUISH BETWEEN CHAOS AND AMPLIFICATION OF STATISTICAL FLUCTUATIONS IN A CHEMICAL OSCILLATOR

So far it has been difficult to distinguish between deterministic chaos and amplified statistical fluctuations in an experimental system.   As a guide for future experiments we numerically investigated the response behaviour of the periodically forced Brusselator in its chaotic and periodic regions in phase space [1-3].   The idea was to superimpose random fluctuations of increasing amplitude on the two concentration variables X and Y and to calculate the value of the largest Lyapounov exponent as a function of fluctuation amplitude.   Chaos is characterized by several features that include for example, the route to chaos which may occur via subharmonic bifurcations (period doubling) according to the sequence described by Feigenbaum or via intermittent bursts of large amplitude.   Theoretical methods to identify chaos include broad band Fourier spectra, one dimensional return maps (Poincaré sections), the largest Lyapounov exponent and the dimensionality of the trajectories.   Inevitably, there exist local fluctuations in concentration and temperature in any experimental system, particularly in a CSTR (Continuous Flow Stirred Tank Reactor) at the point of in-flow.   These fluctuations may be amplified by the overall mechanism.   If the system is located close to a bifurcation at which period doubling may occur, for example, a small fluctuation may be sufficient to move the system back and forth through the bifurcation.   The bifurcation may even be slightly shifted as a function of fluctuation size [4].   Thus an irregular sequence of simple and multiple oscillations may be observed in this case, whose appearance seems to be chaotic, although it is not to be identified with the above mentioned chaos, which is deterministic in nature.

The maximum Lyapounov exponent is known to be a particularly useful tool in the characterization of chaos.   It is positive for chaos and zero for periodic motions, respectively.   As the superimposed fluctuations increase in amplitude the maximum positive Lyapounov exponent is expected to increase further for chaos;   for periodic motions the maximum Lyapounov exponent (L.E.) is expected to become larger than zero.   This is exactly what we have found in numerical simulations of the periodically driven Brusselator.   The analogy to the experiment is provided by the stirring rate in a CSTR.   A low stirring rate is likely to produce large local concentration fluctuations of long duration at the point of in-flow of reactants into the CSTR. At high stirring rates the concentration gradients and fluctuations in the reactor are expected to be reduced in magnitude and shortened in duration. Thus it is appropriate to extrapolate the experimental L.E. to infinite stirring rates. In the limit of high stirring rates the L.E. is expected to remain positive for chaos, whereas it should approach zero for periodic motions, respectively. These predictions were studied experimentally in a chaotic region of the Belousov Zhabotinsky (BZ) reaction in a CSTR.

52.1 Numerical Integrations of the Brusselator Model and Random Fluctuations

The Brusselator mechanism may be written as

$$A \rightarrow X \qquad\qquad Y + 2X \rightarrow 3X$$

$$B + X \rightarrow D + Y \qquad\qquad X \rightarrow E$$

Here we periodically perturb X [1,2]. All rate constants are assumed to be unity. A basic uniform random number generator superimposes statistical fluctuations of a given average amplitude on the concentration variables X and Y where X' and Y' are given as

$$X' = X(1 + \alpha' f_1)$$
$$Y' = Y(1 + \alpha' f_2)$$

f_1 and f_2 are random numbers between -1 and 1 and α' is the maximum relative fluctuation amplitude in ppm. The rate law is

$$\frac{dX'}{dt} = A + (X')^2 Y' - X' - BX' + \alpha \cos \omega_p t$$

$$\frac{dY'}{dt} = BX' - (X')^2 Y'$$

where A and B are constants, α is the amplitude of the periodic perturbation and ω_p is the frequency of the periodic perturbation. This procedure actually simulates multiplicative white noise.

The calculated time series of the Brusselator was treated as quasi-experimental data. From these data an attractor is reconstructed according to [5]. A delay time T is applied successively to the oscillations to generate all the necessary higher dimensions of the phase portrait according to $X'(t_k)$, $X'(t_k + T)$, $X'(t_k + 2T)$, ... $X'(t+(n-1)T)$, where $t_k = k\Delta T$ and $k = 1, 2,$ ... ∞. The so-called attractor of the system is the n-dimensional curve describing the motion which is usually viewed in a two-dimensional projection. For periodic motion the attractor represents a closed loop whereas for chaos an infinite number of

loops is nestled together in a "strange" attractor. The largest
L.E. is defined as [6,7]

$$\lambda_{max} = \lim_{n \to \infty} \frac{1}{t_n - t_o} \sum_{k=1}^{n} \ln \frac{L'(t_k)}{L(t_{k-1})} .$$

For the calculation of λ_{max} we use the same algorithm (a
modified "fixed evolution time program" [8]) for the theoretical
model data as for the experimental data. The calculation starts
by determining the minimum distance $L(t_o)$ between two closely
lying trajectories on the attractor. Next, $L(t_o)$ is propagated
along the attractor point by point until the resulting distance
vector $L'(t_k)$ approximately reaches the size of the attractor
itself. Then $L'(t_k)$ is replaced by a shorter vector $L(t_{k+1})$ whose
orientation is most similar to the previous one. Time averaging
yields λ_{max}. Periodic or quasiperiodic motions are characterized
by a global convergence of the trajectories towards the attractor.
This is indicated by a non-positive largest L.E. Chaotic motions
display divergence of closely lying points in phase space which is
indicated by at least one positive L.E.

Calculations were done in the known chaotic region of the
Brusselator [1,2] as a function of α' (Fig. 1a) where $A = 0.4$, $B =
1.2$, for the perturbation amplitude $\alpha = 0.080$ and the perturbation
frequency $\omega_p = 0.852$ rad/time. For comparison several
calculations of the periodic behaviour were carried out in the
designated periodic regions of the phase diagram [1] close and far
from any bifurcations. For illustration we show two calculations
for fundamental entrainment close to a quasiperiodic region where
$A = 0.4$, $B = 1.2$, $\alpha = 0.040$ and $\omega_p = 0.290$ rad/time (Fig. 1b) and
far from any bifurcation at high α where $\alpha = 0.400$ and $\omega_p = 0.250$
rad/time (Fig. 1c). The integrations were performed with a 5th
order Taylor expansion in double precision. Fluctuations were
imposed on both variables after each integration step. One
oscillation period was approximated by about 2000 integration
steps. Figure 1a shows that λ_{max} is always positive for chaotic
motion and its magnitude increases with increasing fluctuation
amplitude α' as expected. Even for periodic motion close to a
quasiperiodic boundary λ_{max} becomes positive at a certain
threshold value of the superimposed fluctuation amplitude (Fig.
1b). This threshold value depends upon the distance to the
nearby bifurcation point: for large distances from a bifurcation
the threshold at which λ_{max} becomes positive is shifted to larger
values of the fluctuation size (Fig. 1c). Also the absolute
values of λ_{max} are smaller by a factor of 5-10 than for a point
closer to a bifurcation. The scatter in the λ_{max} values is
considerably less in Fig. 1c where the distance to any bifurcation

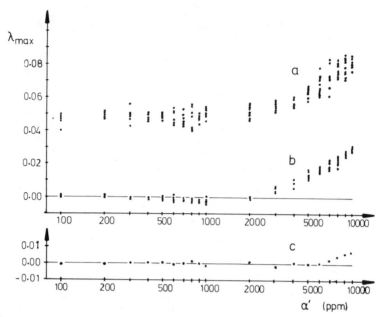

Figure 1. Maximum Lyapounov exponents as a function of fluctuation amplitude α'(ppm) for the driven Brusselator: (a) in its chaotic region where α = 0.080 and ω_p = 0.852 rad/time; (b) in a periodic region close to the quasiperiodic region far from any bifurcation, where α = 0.400 and ω_p = 0.250 rad/time. The scatter in the λ_max values which belong to one α' value is due to different seeds of the random number generator. For all calculations A = 0.4 and B = 1.2.

is large. Thus positive L.E. are observed even in periodic regions where large fluctuations are amplified in the neighbourhood of bifurcations.

52.2 Experiments

The following BZ experiments have been designed to test our theoretical prediction in a chaotic system. The CSTR (25.00 ± 0.01°C) consists of a 1 cm spectrophotometric cell of 1.82 ± 0.01 ml volume that contains an efficient spherical stirring magnet allowing stirring rates of up to 1500 rpm. An Infors precidor syringe pump employs two 50 ml syringes whose flow rate is regulated by a desk computer and a highly precise stepping motor. All flow rates were kept constant at 0.122 min^{-1} corresponding to a residence time of 8.20 min. One syringe contained a solution of 0.175 M $KBrO_3$ in 0.75 M H_2SO_4 and the other syringe 2 x 10^{-3} M Ce^{+3} in 0.60 M malonic acid. Long time experiments yielded up to 200 oscillations in 4 hours corresponding to 7200 data points.

Our single experimental observable is the absorption of Ce^{+4} at 350 nm. Data points are taken every 2 seconds for oscillation

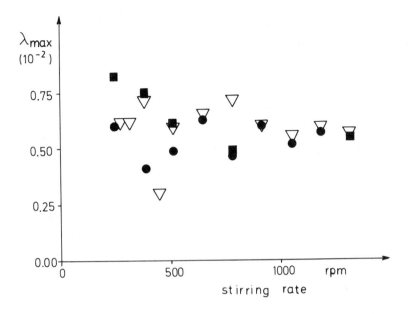

Figure 2. Experimental dependence of λ_{max} on stirring rate (in ppm) for the BZ reaction in a CSTR (see text) at a constant flow rate of $k_f = 0.122$ min^{-1}. The different symbols refer to different sets of experiments.

periods of about 70-150 seconds. Figure 2 shows the experimental λ_{max} as a function of stirring rate in a chaotic region at a relatively high flow rate of $k_f = 0.122$ min^{-1}. Extrapolation to high stirring rates is seen to retain the positive value of λ_{max} indicating the presence of deterministic chaos in the experimental system. It is noteworthy that the scatter of the λ_{max} values is found to be higher at low stirring rates. Experiments have been initiated in the periodic range. They confirm the above theoretical predictions.

52.3 Conclusions

On the basis of these results we suggest measuring the dependence of the maximum Lyapounov exponent λ_{max} on fluctuation size as an appropriate method to distinguish between chaos and amplified fluctuations. The extrapolation to infinitely small fluctuations will yield positive λ_{max} values for chaos and zero for periodic motions. Since experiments in general are beset with relatively large noise, λ_{max} will most likely be positive even for periodic motion, particularly if this motion is close to a bifurcation.

Therefore a single positive value of λ_{max} obtained from experimental data cannot be regarded as a reliable indication for deterministic chaos. To distinguish between chaos and amplified fluctuations in CSTR experiments a wide-range variation of the stirring rate as a parameter controlling the fluctuation size is desirable.

Acknowledgement

We are grateful for financial support by the Volkswagenstiftung and the Fonds der Chemischen Industrie.

References

[1]   T. Kai and K. Tomita (1979). *Prog. Theor. Phys.* **61**, 54.
[2]   B.-L. Hao and S.-Y. Zhang (1982). *J. Stat. Phys.* **28**, 769.
[3]   A. Freund, Th. Kruel and F.W. Schneider (1986). *Ber. Bunsenges. Phys. Chem.* **90**, 1079; F.W. Schneider (1985). *Ann. Rev. Phys. Chem.* **36**, 347m.
[4]   R. Lefever and J.Wm. Turner (1986). *Phys. Rev. Lett.* **56**, 1631.
[5]   F. Takens (1981) in *Lecture Notes in Mathematics*, 898, p. 366, eds. D.A. Rand and L.S. Young, Springer, Berlin, Heidelberg, New York, Tokyo.
[6]   I. Shimada and T. Nagashima (1979). *Prog. Theor. Phys.* **61**, 1605.
[7]   G. Benettin, L. Galgani, A. Giorgilli and J.-M. Strelcyn (1980). *Meccanica* **15**, 9.
[8]   A. Wolf, J.B. Swift, H.L. Swinney and J.A. Vastano (1985). *Physica* **16D**, 285.

53   *B. Turcsányi*

PARAMETER SPACE STRUCTURE OF A MODEL CHEMICAL OSCILLATOR

53.1 Introduction

The important goal of the reseach on oscillatory chemical reactions is the description of the system in terms of the chemical kinetics.   The resulting differential equations system incorporates inevitably simplifications and approximations and can be regarded as a model of the real system.   Such modelling is generally considered as successful, if for some set of parameters (incorporating concentration and rate constant values) the solution of the D.E.S. matches closely the experimentally observed time variation of the components.

Obviously, further information could be gained from the comparison of the global behaviour of the model and the real system.   It is, however, in most of the cases a very complicated task.   Nevertheless, in the last years many results were published on the bifurcation characteristics of CSTR systems, i.e. reactions realized in continuously flowing, stirred tank reactors.   Under such circumstances, with constant inflow of reactants, far-from-equilibrium states can be maintained and the determination of transitions between various regimes in function of outer parameters (mean residence time τ, nominal reactant concentrations $[R_i]_0$) becomes relatively easy.   For the two-dimensional subspace some generalizations were made too [1]. The results on the connection of bistable and oscillatory regions in the parameter plane (generally referred to as the "cross-shaped phase diagram") was proved to be very helpful for the experimental research.

The aim of this paper is to give a full characterization of the parameter space of an oscillatory model system.   The model discussed here was developed [2] for the description of the dioxygen/sulfide/Methylene Blue (MB) oscillatory reaction, reported by Burger and Field [3].   Because of the relatively simple structure (two variables, three parameters) it was hoped that a relatively easy analysis can be made, giving illustrative results.

53.2 The Model

The net chemical change corresponds to the oxidation of sulfide ions by dioxygen producing sulfur and hydrogen peroxide.   The free enthalpy change associated with this process is the driving force of the system. The rate of the direct reaction is, however, negligibly small in relation to the rate of the mediated process:

$$HS^- + MB^+ \rightarrow HMB + S \qquad (1)$$

$$O_2 + HMB \rightarrow MB^+ + HOO^- \qquad (2)$$

where MB^+ and HMB stand for methylene blue and leuco methylene blue, respectively.

It was assumed that both of these reactions are autocatalytic, with the following rate laws:

$$\text{rate (1)} = k_{red}[HS^-][MB^+]^{0.5}[HMB]^{0.5} \qquad (3)$$

$$\text{rate (2)} = k_{ox}[O_2][MB^+][HMB] \qquad (4)$$

For a CSTR system, if the sulfide concentration can be taken as constant, the dynamics of the model is given by the following differential equations:

$$w_x \equiv \frac{1}{k_o}\dot{x} = 1 - x - c_1\sqrt{x(1-x)} + c_2 x(1-x)y \qquad (5)$$

$$w_y \equiv \frac{1}{k_o}\dot{y} = 1 - y - c_3 x(1-x)y \qquad (6)$$

where $x \equiv [MB^+]/[MB^+]_0$ and $y \equiv [O_2]/[O_2]_0$ are relative concentrations (the index 0 refers to the absence of any chemical change), k_0 is the flow rate (the reciprocal of the mean residence time) and the $c_1 - c_3$ parameters are defined by the relations

$$c_1 = k_{red}[HS^-]_0/k_0$$
$$c_2 = k_{ox}[MB^+]_0[O_2]_0/k_0 \qquad (7)$$
$$c_3 = k_{ox}[MB^+]_0^2/k_0.$$

The phase portrait of the system, for the case of a single oscillatory state, is given by Fig. 1.

53.3 General Discussion

For a given point of the (c_1, c_2, c_3) parameter space the behaviour of the system is determined by the stability of the corresponding stationary points. The $(x;y) = f(c_1, c_2, c_3)$ stationary points are the solutions of the $w_x = 0; w_y = 0$ algebraic equation system. The corresponding fifth-grade equation with multiple roots, whose stability must be determined separately, can

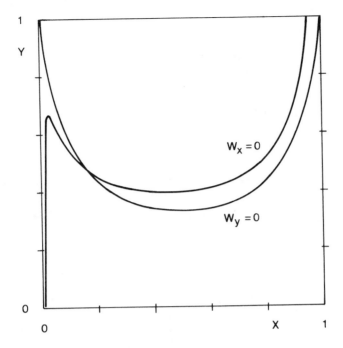

Figure 1.   Phase portrait for c_1 = 12, c_2 = 55, c_3 = 8.   The single stationary point is unstable (case limit cycle oscillations).

be solved only numerically and there is little hope to obtain information on the global structure by this direct way.

Taking into account, however, that the variables and parameters are interchangeable, we can seek the solutions in form $(y; c_1)$ = $f(x_1, c_2, c_3)$. Then, y is uniquely determined from w_y = 0 by (8)

$$y = \frac{1}{1 + c_3 x(1-x)} \qquad (8)$$

and c_1 from w_x = 0 by (9)

$$c_1 = \sqrt{\frac{1-x}{x}} + c_2 y \sqrt{x(1-x)} \qquad (9)$$

where (and in the forthcoming discussion) y is merely a shorthand notation for the right side of (8).

Taking c_3 as constant, we can determine the stability domains of the stationary point in the (x, c_2) plane.   The Jacobian of the system for the stationary point becomes:

$$J = \begin{matrix} (c_2(1-2x)y - 1/x)/2 & c_2x(1-x) \\ -\dfrac{(1-2x)(1-y)}{x(1-x)} & -1/y \end{matrix} \qquad (10)$$

For det = |J| < 0 we have a saddle instability. The condition of this can be given by the relation

$$c_2 > \frac{1}{x(1-2x)y(2y-1)} . \qquad (11)$$

The region of stability is determined by the relations det > 0 and tr = J_{11} + J_{22} < 0. The second relation is fulfilled for any positive c_2 if x > = 1/2 and, in case of 0 < x < 1/2 if

$$c_2 < \frac{2x + y}{x(1-2x)y^2} . \qquad (12)$$

In case of equality (11) and (12) are analytical expressions for

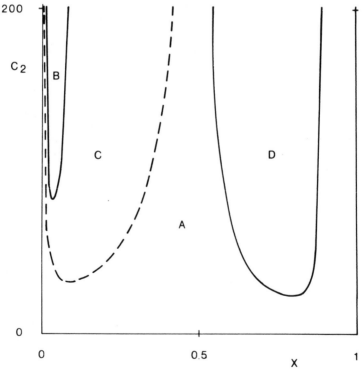

Figure 2. Stability regions in the (x, c_2) plane for c_3 = 10. A: stability, B: saddle instability 1, C: unstable focus or node, D: saddle instability 2. Full lines correspond to det = 0, dashed line to tr = 0.

the borderlines of stability domains.

It was found, that for $c_3 < 1$ only saddle instability can exist. for greater values of c_3 both of the relations det > 0, tr > 0 can be fulfilled, corresponding to instable focus or node. This domain surrounds the first saddle instability domain.

For $c_3 > 4$ a second saddle instability domain appears at x > 1/2 values. This is always directly connected to the domain of stability. Figure 2 shows the structure in the (x, c_2) plane for this case.

From this, the structure in the (c_1, c_2) plane (Fig. 3) can be obtained by the transformation (9). Under this transformation, the straight lines x = constant remain straight lines, the (x, c_2) sheet will be, however, partly folded. The cusp-shaped domains of sheet multiplicity correspond to the areas of stationary point multiplicity. The borderlines of these are the transformed det = 0 lines.

The line tr = 0 is transformed into a loop, the closed part of

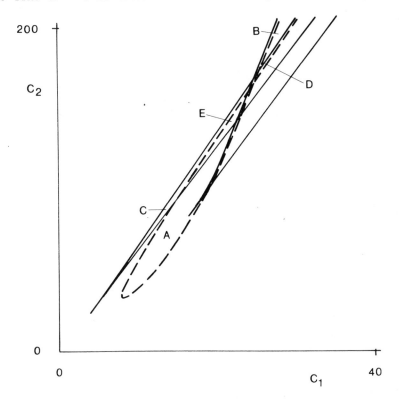

Figure 3. Section of the parameter space at c_3 = 10. A: oscillations, B; bistability 1, C: bistability 2, D: tristability, E: bistability with one state oscillating. Lines as on Fig. 2.

which is the oscillatory region. The point of self-crossing connects the oscillatory and bistable regions. The latter does not coincide with the corresponding region of multiple stationary points, i.e. of the first saddle instability. The domain of the second saddle instability is, however, entirely multistable.

The structure becomes more complex because some regions overlap. The overlap of the two bistability area determines a domain of tristability. The joint region of oscillations and the second bistability region correspond to a bistability with one oscillatory state. The existence of the latter was verified by integration of the D.E.S. (5,6) under appropriate conditions.

53.4 Summary

For the dynamic system studied here, direct computation of the instability regions was possible in the mixed parameter-variable space. Transformation into the parameter space gave well understandable results on the remarkably complex structure. The method seems to be generally applicable for dynamic systems which are linear in parameters.

References

[1]   J. Boissonade and P. De Kepper (1980). *J. Phys. Chem.* **84**, 501.

[2]   B. Turcsányi, posters presented at: (a) Gordon Research Conference, July 1985, Plymouth, N.H.; (b) Bunsentagung, May 1986, Heidelberg.

[3]   M. Burger and R.J. Field (1984). *Nature* **307**, 720.

CHAOS AND BIFURCATIONS IN DYNAMICAL SYSTEMS OF NERVE MEMBRANES

54.1 Introduction

Nonlinear responses of a neural oscillator to periodic force are studied experimentally with squid giant axons and numerically with the Hodgkin-Huxley equations [1]. Squid giant axons in a state of self-sustained oscillation [2] are periodically driven by sinusoidal stimulation under space-clamp condition. The space-clamp condition makes it possible to compare the experimental results with numerical solutions of the Hodgkin-Huxley ordinary differential equations. The amplitude A and the frequency F of the sinusoidal stimulation are changed as bifurcation parameters.

54.2 Sinusoidally Forced Oscillations in Nerve Membranes

The sinusoidally forced oscillations in the dynamical systems of both squid giant axons and the Hodgkin-Huxley axons can be qualitatively classified into (1) synchronized oscillations, (2) quasi-periodic oscillations, and (3) chaotic oscillations by examining the Poincaré sections, the return maps, the power spectra and the Lyapunov exponent [3-6]. Figure 1 shows an example of a chaotic oscillation in squid giant axons. Figures 1(a) and (b) correspond to the waveform and the stroboscopically displayed Poincaré sections of the chaotic oscillation, respectively. The Poincaré sections in Fig. 1(b) elucidate transforming processes of stretching, folding and compressing peculiar to chaotic attractors.

54.3 Routes to Chaos in Nerve Membranes

Various synchronized oscillations are produced with changing the amplitude A and the frequency F of the force. The region of each synchronization forms an Arnold tongue in the parameter space A x F. When A is sufficiently large or small, almost all the responses are synchronized oscillations or quasi-periodic oscillations, respectively. In the intermediate range of A, chaotic oscillations can be observed with adjusting A and F. The routes from synchronized oscillations to chaotic oscillations are successive period-doubling bifurcations or intermittency [4,6]. Moreover, there exists the third route with collapse of a

a

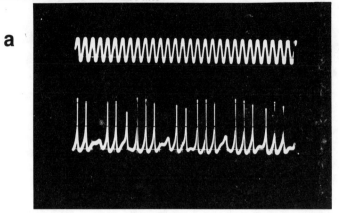

b

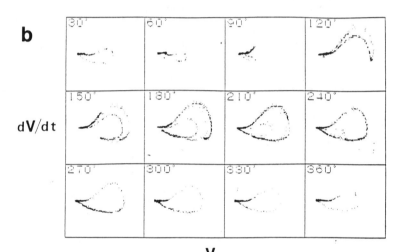

dV/dt

V

Figure 1. A chaotic oscillation experimentally observed in squid giant axons (A = 2 μA and F =288 Hz). (a) Waveforms (above: stimulation, below: response of the membrane potential). (b) Poincaré sections on the V x dV/dt plane. Each number (°) shows the phase of the sinusoidal force at which the corresponding section is plotted.

2-dimensional torus from quasi-periodic oscillations to chaotic oscillations (Aihara et al., in preparation). Figure 2 demonstrates a 2-dimensional torus just before the collapse.

54.4 Discussion

While we have reported the nonlinear responses of the sinusoidally forced neural oscillator, another periodic force such as a train of pulses also produces not only periodic but nonperiodic responses in both self-oscillating and resting membranes. Therefore, it seems that the nerve membranes generate abundant

dV/dt

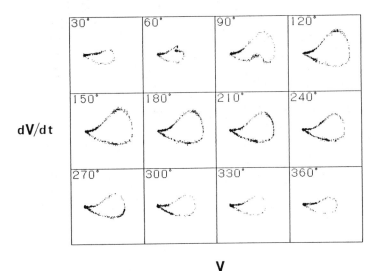

V

Figure 2. Poincaré sections of a quasi-periodic oscillation just before collapse of a 2-dimensional torus in squid giant axons (A = 1 μA and F = 250 Hz).

time series of graded action potentials (see Fig. 1(a)) in many situations. Since information in neural networks is believed to be carried by trains of action potentials, it is important to describe characteristics of the time series by simpler models phenomenologically. From this point of view, responses of a self-sustained biological oscillator stimulated by a train of pulses have been successfully analysed with circle maps of the phase from S^1 to S^1 (e.g. [7]) on the condition that the stable limit cycle representing the spontaneous oscillation has orbital stiffness and fast contraction along the radial direction after perturbation of each pulse stimulation. On the other hand, Nagumo and Sato [8] examined responses of a resting neuron to a train of pulses with a difference equation based on the Caianiello's neuronic equation [9] and revealed that the model has response characteristics of complete devil's staircases. The following difference equation (1), which is our modified version of the Nagumo-Sato model, can describe not only devil's staircases [8] but also alternating periodic-chaotic sequences [10].

$$x(t+1) = f(A(t) - \sum_{r=0}^{t} b^{-r} g(x(t-r)) - \theta) \qquad (1)$$

where $x(t) (\subset [0,1])$ is the output of the neuron, or a graded action potential of the neuron at the discrete time t; $A(t)$ is the stimulation at the time t; θ is the threshold; f is a function describing the relationship between the internal state $y(t) (=A(t)$

$-\sum\limits_{r=0}^{t} b^{-r} g(x(t-r)) - \theta)$ and the output $x(t+1)$; g is a function describing the relationship between the output and the magnitude of the refractoriness to the following stimulation; $b(>1)$ is the parameter representing the temporal decay of the refractory memory.

By introducing the internal state $y(t)$ defined above, equation (1) is transformed as follows:

$$y(t+1) = b^{-1}y(t) + a(t) - g(f(y(t))) \qquad (2)$$

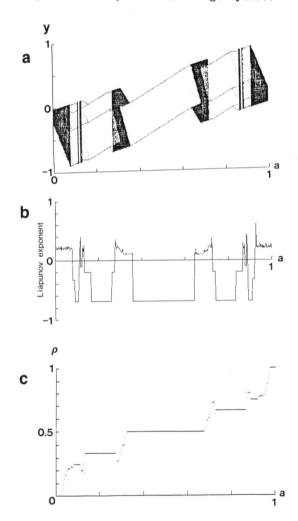

Figure 3. *Response characteristics of equations (2) and (3) where b = 2.0, ε = 0.05 and y(0) = 0.1. (a) The phase diagram. (b) The Lyapunov exponent. (c) The average firing rate where h(z) = 1(z≥0.5) and = 0(z<0.5).*

$$x(t+1) = f(y(t)) \tag{3}$$

where

$$a(t) = A(t+1)-b^{-1}A(t)-\theta(1-b^{-1}). \tag{4}$$

When $f(z) = 1$ for $z \geq 0$, and $f(z) = 0$ for $z < 0$ and $g(z) = z$, equations (2) and (3) are reduced to the Nagumo-Sato model. In general, f is an increasing function reflecting graded responses of the nerve membranes. Figure 3 demonstrates a response of equations (2) and (3) where $f(z) = 1$ for $z \geq \varepsilon$, $f(z) = z/(2\varepsilon)+1/2$ for $-\varepsilon < z < \varepsilon$, and $f(z) = 0$ for $z \leq -\varepsilon$ and $g(z) = z$. Figures 3(a) and (b) show the phase diagram and the Lyapunov exponent, respectively. Figure 3(c) is the characteristic of the average firing rate ρ defined below

$$\rho = \lim_{N \to \infty} \sum_{t=1}^{N} h(x(t))/N \tag{5}$$

where h is a function describing threshold and shaping effects of the axons on propagating action potentials. An alternating periodic-chaotic sequence as shown in Fig. 3 is also observed experimentally in squid giant axons (Matsumoto et al., in preparation). The detailed dynamics of equations (2) and (3) will be reported elsewhere (Aihara et al., in preparation).

Acknowledgements

The authors would like to thank A.V. Holden and L. Glass for their valuable comments at the MIDIT 1986 Workshop. They also wish to thank T. Takabe for his help in the numerical calculation of Fig. 3.

References

[1]   A.L. Hodgkin and A.F. Huxley (1952). *J. Physiol.* **117**, 500.

[2]   K. Aihara and G. Matsumoto (1982). *J. Theor. Biol.* **95**, 697.

[3]   G. Matsumoto, K. Aihara, M. Ichikawa and A. Tasaki (1984). *J. Theor. Neurobiol.* **3**, 1.

[4]   K. Aihara, G. Matsumoto and Y. Ikegaya (1984). *J. Theor. Biol.* **109**, 249.

[5]   K. Aihara, T. Numajiri, G. Matsumoto and M. Kotani (1986). *Phys. Lett.* **A116**, 313.

[6]   K. Aihara and G. Matsumoto (1986) in *Chaos,* ed. A.V. Holden, Manchester University Press, Manchester.

[7]   M.R. Guevara, L. Glass, M.C. Mackey and A. Shrier (1983). *IEEE.SMC-13,* 790.

[8]   J. Nagumo and S. Sato (1972). *Kybernetik* **10**, 155.

[9]   E.R. Caianiello (1961). *J. Theor. Biol.* **2**, 204.

[10]   K. Aihara, G. Matsumoto and M. Ichikawa (1985). *Phys. Lett.* **A111**, 251.

ASPECTS OF QUANTUM MECHANICAL THERMALIZATION IN THE ALPHA-HELIX

In order for Davydov solitons actually to transport energy in biological systems, it is a necessity that the soliton is stable enough. We must therefore question, if the soliton is stable in the following cases:

(a) Spatial fluctuations. Spatial fluctuations will appear since biological systems are not really built up regularly.

(b) Quantal fluctuations. Ignoring spatial fluctuations the system should be translational invariant. But then quantum theory tells us that in general a zero point motion of the soliton should occur, and a localized state should have only a limited lifetime. We will treat this problem in a further paper.

(c) Thermal fluctuations. This is the problem we want to deal with now.

In two recent papers [1,2] the existence of Davydov solitons at 300 K is questioned. Results given by other authors [3,4,5,6] confirm the existence of solitons at this temperature. Therefore, we believe that it is necessary to clarify some aspects of thermalization.

Of course, the alpha-helix in biological systems is not isolated, it is in thermal contact with other material, the so-called heat bath. For thermal equilibrium, we need only to know some of the parameters of the heat bath: for example the temperature. If the system is also classical, we can simulate the heat bath by random forces and friction terms acting directly on the particles.

If we excite the system to get a time-dependent reaction we need more information about the heat bath. The relaxation time T_R, that is the time needed for the heat bath to get into equilibrium, seems to be the most important quantity. It has to be compared with T_S, the time necessary for the soliton to run through the alpha-helix. In general, we can distinguish two limiting cases:

(a) $T_R \gg T_S$. The heat bath is very slow, it cannot follow the system. We have no dynamical backflow of the heat bath on the system.

(b) $T_R \ll T_S$. This is the adiabatic case. The action of the system is very slow if it is seen from the heat bath. Thus the heat bath will always remain in thermal equilibrium. It will, however, act back on the system, but only in an adiabatic way, resulting in time-independent potentials acting additionally on the particles of the system.

For the following, we use the hamiltonian (1) in 56.1. We want to start the soliton at a certain time, say t=0. We assume, therefore, that the system for t<0 is in thermal equilibrium. Since $\varepsilon \gg kT$, (ε is the excitation energy of the amide-I-oscillators, T is the temperature), only the phonon part of the system is then thermally excited. The amide-I-oscillators remain in the ground state and we have no interaction between the subsystems. So with probability ρ_n

$$\rho_n \sim e^{-\dfrac{E_n}{kT}}; \quad \sum_n \rho_n = 1$$

we find the system in the following state

$$|\xi_o \rangle = |o\rangle |E_n\rangle.$$

Here, $|E_n\rangle$ is the phonon state with energy E_n, $|o\rangle$ is the ground state for the amide-I-oscillators.

We now excite the system at t=0. In the nonadiabatic regime (a) we could just switch off the heat bath without any difference for the system. Therefore, we can describe the time development of the system in the following way. Again with the probability ρ_n, we will find the state

$$|\xi_n(t)\rangle = e^{\dfrac{1}{i\hbar} \sum_k (P_k q_k(t) - \hat{Q}_k P_k(t))} e^{\sum_k \{a_k(t)b_k^+ - a_k^*(t)b_k\}} |o\rangle |E_n\rangle$$

For T=0 this gives the old Davydov ansatz, because we then have

$$\rho_o = 1, \quad \rho_n = 0 \text{ for } n \neq 0.$$

$|\xi(t)\rangle$ should obey the normal time dependent Schrödinger equation, and no coupling to the heat bath will occur. So the time dependent variational principle gives the equations of motion for the parameters q(t), p(t), a(t). In this paper we use the hamiltonian of [7] with the same definitions. We obtain then

$$i\hbar \frac{\partial}{\partial t} a_n = [\chi(q_n - q_{n-1}) + \varepsilon]a_n - J(a_{n+1} + a_{n-1})$$

$$\ddot{q}_n = \frac{1}{m} \{K(q_{n+1} - 2q_n + q_{n-1}) + \chi(|a_{n+1}|^2 - |a_n|^2)\}$$

which are exactly the same equations as for a ground state theory. So for this limiting case we see in principle the soliton unchanged.

For calculating the adiabatic case (b) we need to know more about the heat bath. H_B should be the hamiltonian of the heat bath, H_I $(Q_1, Q_2, ...)$ the interacting hamiltonian. For simplicity we assume that the heat bath is coupled only to some "coordinates" Q of the system.

Since we are describing the adiabatic case we must now use a treatment very similar to the Born-Oppenheimer approximation in solid state physics. So we first have to calculate the free energy $F_B(Q_1, Q_2, ..., T)$ of the heat bath dealing with the coordinates Q as macroscopic variables. Then $F_B(Q_1, Q_2, ..., T)$ has to be used as an effective potential for the system

$$H_S^{eff} = H^S + F(Q_1, Q_2, ..., T).$$

H_S^{eff} now has to be used in a similar way, as for case (a). Although the exact form of F_B is not know, it could even have a localizing effect on the system, resulting in Davydov solitons.

To study the influence of the heat bath in the general case $T_R \approx T_S$, we need even more information, for example, if we model the bath with harmonic oscillators, we need the frequency distribution $f(\omega)$ of the bath. Even in this case it is known, from the problem of quantum tunnelling, that a localization can be the effect of the heat bath.

In the second part of this paper we want to comment on calculations where the heat bath is simulated by random forces together with friction terms acting on the coordinates of the lattice, as it is done in [1,2]. In both simulations the Davydov soliton became unstable with a very short lifetime at room temperature. Moreover, Lawrence et al. [2] reported that the lifetime is even shorter if the coupling parameter χ is increased, which is indeed a very fancy result.

As we saw before, such a procedure cannot give the right answer for dynamical processes in the system since the dynamics of the heat bath is not inhibited. Moreover, we now want to show that not even the right thermal equilibrium is reached by such a procedure.

We show this by a simple example. To simulate our two subsystems, one classical (the phonon part), the other quantal (the oscillators), let us consider the exact integrable model of two interacting harmonic oscillators:

$$H = \frac{\hat{P}_1^2}{2} + \frac{1}{2} \omega_1^2 \hat{Q}_1^2 - \lambda \hat{Q}_1 \hat{Q}_2 + \frac{\hat{P}_2^2}{2} + \frac{1}{2} \omega_2^2 \hat{Q}_2^2.$$

Since we assume $\hbar\omega_1 \gg kT$ and $\hbar\omega_2 \ll kT$, this is a quantal and a quantal system coupled together. Of course, we know the exact canonical density operator for every temperature, since the system is exactly integrable.

Thermalization in alpha helix

To use the classical thermalization scheme, we use now Q_2, P_2 as classical variables. Then the effective hamiltonian H_1^{eff} for the subsystem 1 is:

$$H_1^{eff} = \frac{\hat{P}_1^2}{2} + \frac{1}{2}\omega_1^2\hat{Q}_1^2 - \lambda Q_2(t)\hat{Q}_1.$$

The solution of the time-dependent Schrödinger equation with H_{eff} as hamiltonian is well known [8]. If we start, for example, at $t=0$ with the groundstate($|o>$, $H_1^{eff}(t)$ produces the quasiclassical states

$$|q_1(t), p_1(t)> = e^{\frac{1}{i\hbar}(q_1(t)\hat{P}_1 - p_1(t)\hat{Q}_1)}|o> \equiv U(t)|o>.$$

The equations of motion for the coefficient q_1 and the classical quantities Q_2 are

$$\ddot{q}_1 + \omega_1^2 q_1 = \lambda Q_2(t)$$

$$\ddot{Q}_2 + \omega_2^2 Q_2 = \lambda q_1(t) - \gamma\dot{Q}_2 + K(t).$$

But these are equations to thermalize a pure classical system with coordinates Q_2 and q_1. This system will therefore go into a classical thermal equilibrium, resulting in classical distributions for the parameters q_1 and p_1. Indeed, the whole system 1 behaves more or less classically. To show this, let us consider any operator A of the system 1, which can always be expressed as function of $\hat{Q}_1$, $\hat{P}_1$.

$$\hat{A} = A(\hat{Q}_1, \hat{P}_1).$$

The time-dependent expectation value is easily calculated

$$<\hat{A}> = <o|U^{-1}A(\hat{Q}_1, \hat{P}_1)U|o> = <o|A(\hat{Q}_1 + q_1(t), \hat{P}_1 + p_1(t))|o>.$$

Time averaging then gives the "thermal" expectation value $A_{c\ell}^{eff}$. For many quantities this is almost equal to the pure classical thermal expectation value $A_{c\ell}$. For example, let us now assume a small coupling constant λ. System 1 and 2 are then independent, but system 2 works as a heat bath for system 1. The "thermal" expectation value of H_1

622

$$H_{1c\ell}^{eff} = \frac{1}{2} \hbar\omega_1 + kT$$

should then be just the internal energy U(T) of the quantum oscillator 1 in this classical thermalization scheme. (T is the temperature.) But the exact U(T) is

$$U(T) = \frac{1}{2} \hbar\omega_1 + \frac{\hbar\omega_1}{\hbar\omega_1/kT} .$$

As a result of our simple example we see: the classical thermalization procedure greatly overestimates the thermal excitation energy in the quantum oscillator and therefore gives wrong answers for the equilibrium state.

If we use the classical thermalization scheme to the alpha-helix we see a new artificial effect. The interaction hamiltonian

$$H_{int} = \chi \sum_n (q_n - q_{n-1}) b_n^+ b_n$$

only allows energy transport to the amide-I-oscillators if we already have excitation energy in these oscillators $\langle b_n^+ b_n \rangle \neq 0$.

If we start again with the groundstate, random force terms can never excite the oscillators, not even in a quasi-classial way described in our example.

So, before we start the soliton, this procedure gives the right equilibrium state, but only since the energy transportation between the two subsystems is blocked. If we would open any additional channel for a possible energy transfer we would have energy flux to excite the oscillators to the same "classical" temperature as the lattice.

If we now excite the first oscillator to start the soliton, we just open such a channel. The system wants to form the soliton and at the same time it wants to bring "thermal" energy into the oscillators in the same way as described in our example. Since these two effects happen at the same lattice site, the soliton is destroyed

| t<0 no soliton excited | Phonons T=300 K | no heat transfer | amide osc. T=0 K |

| t~0 try to create soliton | Phonons T=300 K | heat transfer $\longrightarrow$ | $\|a_n\|^2 \neq 0$ T~0 |

Of course the heat transfer is increased if we increase the coupling parameter χ. The soliton seems to be more unstable in this case as it was reported in [2]. In our opinion this is therefore an artificial effect. On the contrary we expect that the soliton is stabilized if we increase the coupling parameter or use a more quantum soliton. To answer the question about the thermal stability of the Davydov soliton, we need more calculations with realistic assumptions for the heat bath.

Acknowledgements

The author would like to thank Dipl.-Phys. M. Opper and Dipl.-Phys. R.D. Henkel for stimulating discussions. He also wants to thank Prof. A. Scott and Prof. P. Christiansen for the possibility of discussing some aspects of this paper at MIDIT.

References

[1]   P.S. Lomdahl and W.C. Kerr (1985). *Phys. Rev. Lett.* **55(11)**, 1235-1238.

[2]   A.F. Lawrence, J.C. McDaniel, D.B. Chang, B.M. Pierce and R.R. Birge (1986). *Phys. Rev.* **A33(2)**, 1188-1199.

[3]   A.S. Davydov (1980). *Sov. Phys. JETP* **51**, 397.

[4]   J.H. Jensen (1986). *Thesis Technical University of Denmark.*

[5]   A.C. Scott (1986). *On Davydov Solitons at 310 K,* to appear in the *Festschrift* for H. Fröhlich on his 80th birthday, Springer-Verlag.

[6]   L. Cruzeiro, J. Halding, P.L. Christiansen, O. Skovgaard and A.C. Scott, *Temperature Effects on Davydov soliton,* preprint.

[7]   H. Bolterauer, R.D. Henkel and M. Opper, *Resonant and Quasiclassical Excitations of Solitons in the Alpha-Helix,* this volume.

[8]   R.J. Glauber (1963). *Phys. Rev. Lett.* **10**, 84.

RESONANT AND QUASICLASSICAL EXCITATIONS OF SOLITONS IN THE ALPHA-HELIX

56.1 Basic Theory

Considering one of the three spines of the alpha-helix we start with the following hamiltonian [1]

$$H = H_{lat} + H_{osc} + H_{int} \qquad (1)$$

where

$$H_{lat} = \sum_k \frac{\hat{P}_k^2}{2M} + V(\hat{Q}_k - \hat{Q}_{k-1})$$

$$H_{osc} = \sum_k b_k^\dagger (\varepsilon b_k - J\{b_{k-1} + b_{k+1}\}) \qquad (2)$$

and

$$H_{int} = \chi \sum_k b_k^\dagger b_k (\hat{Q}_k - \hat{Q}_{k-1}). \qquad (3)$$

Here H_{lat} describes the basic lattice structure composed of hydrogen bonds between rigid peptide groups. The potential V of the hydrogen bonds is in general anharmonic [2]. We use in this paper a harmonic potential V with $\sqrt{K/m}$ as frequency. H_{osc} describes coupled amide-I-bond vibrations and H_{int} accounts for the nonlinear interaction between the two subsystems.

H commutes with N

$$N = \sum_k b_k^\dagger b_k \qquad (4)$$

the number of amide-I-bond quanta. Resonant energy transfer results in states $|\xi\rangle$ which are eigenstates of N. The standard Davydov-theory treats only one quantum state, a fact which has been overlooked by almost all authors. The biological energy is about two times larger than the energy associated with the Davydov ansatz. Therefore, it is necessary to study two or more quanta excitations. However, since the explicit excitation mechanism is not known, we may also discuss the possibility of a quasiclassical mechanism. This would lead to a total coherent state where the number of quanta is not conserved. We will show that in both cases soliton solutions exist.

56.2 Time-dependent Variational Method

Since the lattice is almost classical, a product ansatz is appropriate for the states $|\xi\rangle$

$$|\xi(t)\rangle = |\phi(t)\rangle \, |\psi(t)\rangle \tag{5}$$

where $|\phi(t)\rangle$ is the lattice state and $|\psi(t)\rangle$ the state of the oscillators. The time-dependent variational principle

$$\delta \int_{t_1}^{t_2} \langle\xi| i\hbar\partial_t - \hat{H}|\xi\rangle dt = 0 \tag{6}$$

is a systematic procedure to optimize the ansatz (5). Furthermore, we use the quasiclassical state

$$|\phi\rangle = e^{\frac{1}{i\hbar} \sum_n (q_n(t)\hat{P}_n - P_n(t)\hat{Q}_n)} |o\rangle \tag{7}$$

for the lattice. Application of the time-dependent variational principle results in

$$\dot{q}_n = \frac{P_n}{m}$$

$$\ddot{q}_n = \frac{1}{m} \{K(q_{n+1} - 2q_n + q_{n-1}) - \chi\langle\psi| b_n^{\dagger} b_n - b_{n+1}^{\dagger} b_{n+1} |\psi\rangle\}$$

$$i\hbar|\dot{\psi}\rangle = h(t)|\psi\rangle$$

$$h(t) = \sum_n \{[\varepsilon + \chi(q_n(t) - q_{n-1}(t))] b_n^{\dagger} b_n - J b_n^{\dagger}(b_{n+1} + b_{n-1})\}$$

$$= \sum_{nn'} \alpha_{nn'} \, b_n^{\dagger} b_{n'} \tag{8}$$

In the following, we discuss two possible forms of $|\psi(t)\rangle$, the general multiquantum state and the total coherent state.

56.3 The General Multiquantum State

In this case, $|\psi\rangle$ has the form

$$|\psi\rangle = \sum_{m_1 \ldots m_N} f(m_1 \ldots m_N) b^{\dagger}_{m_1} \ldots b^{\dagger}_{m_N} |o\rangle. \tag{9}$$

We observe that $h(t)$ is a one-particle operator. Therefore, the ansatz

$$f(m_1 \ldots m_N, t) = \sum_p a_{1,m_1}(t) \ldots a_{N,m_N}(t) \tag{10}$$

gives an *exact* solution, with

$$i\hbar \, \dot{a}_{\ell,m} = \sum_{m'} \alpha_{mm'} \, a_{\ell,m'}. \tag{11}$$

Specializing on the two quantum state

$$f(m_1 m_2) = a_{1,m_1} a_{2,m_2} + a_{1,m_2} a_{2,m_1} \tag{12}$$

we have

$$i\hbar \frac{\partial}{\partial t} a_{\ell,n} = [\varepsilon + \chi(q_n - q_{n-1})] a_{\ell,n} - J(a_{\ell,n+1} + a_{\ell,n-1}); \quad \ell = 1,2$$

$$\ddot{q} = \frac{1}{m} K(q_{n+1} - 2q_n + q_{n-1}) -$$

$$- 4\chi[|a_{1,n}|^2 \langle 2|2\rangle + |a_{2,n}|^2 \langle 1|1\rangle + a^*_{1,n} a_{2,n} \langle 2|1\rangle + a_{1,n} a^*_{2,n} \langle 1|2\rangle]$$

$$+ 4\chi[|a_{1,n+1}|^2 \langle 2|2\rangle + |a_{2,n+1}|^2 \langle 1|1\rangle + a^*_{1,n+1} a_{2,n+1} \langle 2|1\rangle$$

$$+ a_{1,n+1} a^*_{2,n+1} \langle 1|2\rangle]$$

$$\langle 2|1\rangle = \sum_n a^*_{2,n} a_{1,n} \tag{13}$$

These two equations have to be solved with the condition:

$$\langle \psi|\psi\rangle = 1 = 4\langle 1|1\rangle \langle 2|2\rangle + 4|\langle 1|2\rangle|^2 \tag{14}$$

It is easy to show that the matrix A

$$A = \begin{bmatrix} \langle 2|2\rangle & \langle 2|1\rangle \\ \langle 1|2\rangle & \langle 1|1\rangle \end{bmatrix} \tag{15}$$

is conserved in time.

Let us first consider the multiquantum state with total overlap. Here we have

$$a_{2,n} = a_{1,n} = a_n. \tag{16}$$

The normalized state is thus

$$|\psi\rangle = \frac{1}{2\sqrt{2}} \sum_{\substack{m_1 \\ m_2}} a_{m_1} a_{m_2} b^{\dagger}_{m_1} b^{\dagger}_{m_2} |o\rangle; \quad \sum_n |a_n|^2 = 2 \qquad (17)$$

with the equations of motion

$$\ddot{q}_n = \frac{1}{m} \{K(q_{n+1} - 2q_n + q_{n-1}) + \chi(|a_{n+1}|^2 - |a_n|^2)\}$$

$$i\hbar \frac{\partial}{\partial t} a_n = [\chi(q_n - q_{n-1}) + \varepsilon] a_n - J(a_{n+1} + a_{n-1}) \qquad (18)$$

These are the same equations as for the Davydov one quantum state. Only the norm is changed to

$$\sum_n |a_n|^2 = 2. \qquad (19)$$

Previous numerical calculations, starting from Davydov's one quantum state used such a normalization assuming that it should be an approximation to two quantum states. We have shown here that it is an exact result in the total overlap case.

To treat the general case we apply, for simplicity, a continuum approximation

$$e^{(\frac{i}{\hbar}(-2J)t)} a_{\ell,n}(t) \to \phi_{\ell}(x,t). \qquad (20)$$

Neglecting the term $\ddot{q}_n$, which is possible for slowly moving solitons, we obtain

$$i\hbar \frac{\partial}{\partial t} \phi_{\ell}(x,t) = -J \frac{d^2}{dx^2} \phi_{\ell} - \frac{\frac{\lambda}{4m\chi^2}}{k} \sum_{\ell'\ell''} A_{\ell'\ell''} \phi^*_{\ell'} \phi_{\ell''} \phi_{\ell} \qquad (21)$$

Since A is conserved, the transformation matrix $\tilde{U}$

$$\tilde{U}^{\dagger} A \tilde{U} = 1 \qquad (22)$$

can be calculated from the initial state. Defining $\tilde{\phi}$

$$\tilde{\phi}_{\ell}(x,t) = \sum_{\ell'} \tilde{U}^{\dagger}_{\ell\ell'} \phi_{\ell'}(x,t) \qquad (23)$$

equation (21) reduces to the generalized nonlinear Schrödinger equation

$$i\hbar \frac{\partial}{\partial t} \tilde{\phi}_\ell = -J \frac{d}{d_x{}^2} \tilde{\phi}_\ell - \lambda \sum_{\ell'} |\phi_{\ell'}|^2 \phi_\ell. \qquad (24)$$

This equation is exactly integrable, as shown by Manakov [3] and yields a variety of soliton solutions.

56.4 The Total Coherent State

To describe a quasiclassical excitation we use the total coherent state vector

$$|\psi\rangle = e^{\sum_k \{a_k(t)b_k^\dagger - a_k^*(t)b_k\}} |\psi_o\rangle \qquad (25)$$

This type of state vector should be the result of an interaction with a classical source. This was shown by Glauber [4] in the case of a radiation field. The time-dependent variational principle then again yields the previously obtained equations of motion (18). Note that the parameters now have a quite different meaning than in section 56.3. Especially the norm N

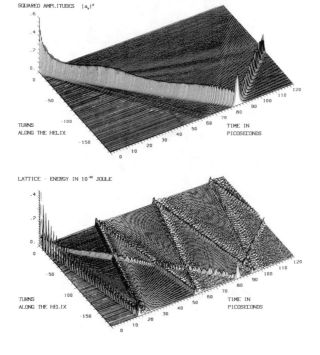

Figure 1. (a) *Formation of the soliton seen in the amide-I-oscillators.* (b) *The same, seen in the lattice.*

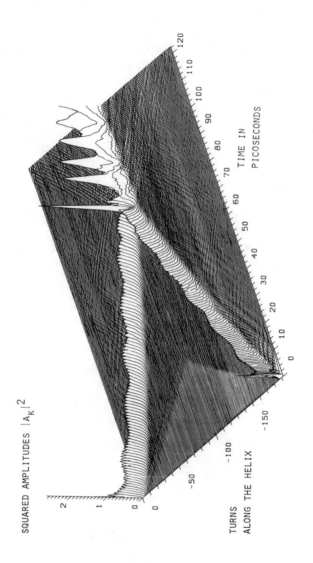

Figure 2.   Fusion of two solitons.

$$N = \sum_{n} |a_n|^2 \qquad (26)$$

now has no restriction at all.

56.5 Numerical Simulations

Solving equation (18) numerically, we obtained the soliton solutions suggested by the continuum analysis. Restricting the norm to integer values, the numerical results are applicable to both types of state vectors [5], the two quantum state with total overlap, and the total coherent state.

Figure 1 shows the formation of the soliton seen from the oscillator system (1a) and from the lattice system (1b). Note the radiation in the lattice which moves quickly away without disturbing the soliton.

Figure 2 can be interpreted in two different ways. If we use the forward time direction it shows the fusion of two solitons running together, resulting in a soliton, which is in rest but with internal vibrations. If we reverse the time direction it shows how a soliton can become unstable if it is internally excited with a specific frequency. At the present time we don't know if this is a resonant effect or not.

References

[1]   A. S. Davydov and N. I. Kislukha (1973).   *Phys. Stat. Sol.*
       B59,   465.
[2]   H. Bolterauer and R. D. Henkel (1986). *Physics Scripta* **T13**,
       314.
[3]   S. V. Manakov (1973).   *Sov. Phys. JET* **38**, 248.
[4]   R. J. Glauber (1963).   *Phys. Rev. Lett.* **10**, 84.
[5]   R. D. Henkel (1986).   *Diploma Thesis*.

57 *S. De Filippo, M. Fusco Girard and M. Salerno*

NUMERICAL ANALYSIS OF THE ORDER-CHAOS-ORDER TRANSITION FOR THE N=3 DISCRETE SELF-TRAPPING EQUATION

Abstract

Some new numerical results on the order-chaos-order transition for the n=3 Discrete Self-Trapping Equation by means of Lyapunov Characteristic Exponents analysis are presented. An order window corresponding to a small range of nonlinearity parameter is exhibited at low energy values.

In a recent paper by J.C. Eilbeck, P.S. Lomdahl and A.C. Scott, the Discrete Self-Trapping Equation

$$i\dot{A} = +\Gamma D(|A|^2)A + \epsilon M A \tag{1}$$

was introduced as a model to describe the dynamics of small polyatomic chains such as water, ammonia, etc., as well as large polymers such as acetanilide [1]. In Eq. (1), A is a complex n-component vector, $D(|A|^2)$ denotes the diagonal matrix diag $(|A_1|^2, |A_2|^2, \ldots, |A_n|^2)$, M is a real symmetric matrix, and Γ and ϵ are real constants.

This equation has an hamiltonian structure, with two independent (for n>1) conserved quantities: the energy

$$H = \frac{1}{4} \Gamma \sum_{j=1}^{n} |A_j|^4 + \frac{1}{2} \epsilon \sum_{j,k=1}^{n} m_{jk} A_j^* A_k \tag{2}$$

and the conserved norm

$$N = \sum_{j=1}^{n} |A_j|^2, \tag{3}$$

this obviously leading to integrability for n=2. As shown in Ref. [1], in general, both for Γ/ϵ going to 0 and to ∞, the system exhibits regular behaviour, corresponding, respectively, to the linear and the uncoupled limits; for n>2, power spectrum analysis shows that as Γ/ϵ increases from zero, regularity regions shrink, as expected by the KAM theorem, and chaotic ones grow; a further increase of Γ/ϵ beyond a certain value restores regular behaviour.

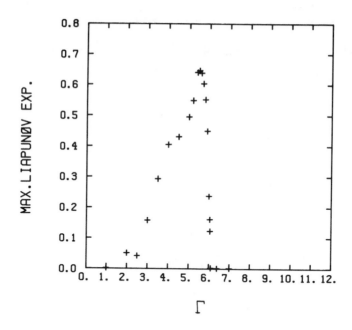

Figure 1. Finite time approximation of the first LCE vs. Γ for n=3 DST Eq. (1) with initial conditions A_1=1, A_2 = .001 i, A_3 = 0, and an evolution time t = 10000.

In particular, for n=3 and for the initial conditions A_1 = 1, A_2 = .001 i, A_3 = 0, a sharp transition occurs for Γ/ϵ > 6. This is also found in Ref. [2], by computing the correlation exponent for the same initial conditions. This refers to the choice for the matrix M: m_{jk} = $1-\delta_{jk}$.

In a previous paper [3] these results were confirmed by the Lyapunov characteristic exponents (LCE) method [4,5].

Data in Fig. 1 reproduce numerical results from Ref. [3], showing the behaviour of the maximal LCE versus the nonlinearity parameter Γ. Here too a clear-cut order-chaos-order transition is observed. In particular the sharp chaos-order transition was exhibited at the same value of Γ as in Refs. [1,2], by using the same initial conditions.

The present numerical analysis refers to a more detailed investigation of this transition for several energy values. In particular the presence of a small plateau in the diagram of the maximum LCE versus Γ is shown to be associated with the presence of an order window at lower energies; here the behaviour of the maximum LCE versus Γ seems to denote a richer structure than the one suggested in the previous papers.

The results reported below were obtained according to the following scheme. Equation (1), with the same choice of the M matrix, m_{ij} = $1-\delta_{ij}$, and its variational equation for δA are numerically integrated in double precision arithmetic by a fourth-order Runge Kutta algorithm on a DEC Vax 8200. The maximum LCE is then estimated by the ratio $\ln||\delta A||/t$ when t ($\sim$ 10000) is large enough to make it approximately constant.

Computations for varying initial conditions at fixed energy values near the transition value of Γ seem to exhibit a substantial independence on the particular initial conditions (apart from small regions whose presence cannot be excluded without further analysis).

As to the energy and norm dependence, data in Figs. 1, 2 and 3, respectively, refer to $E_1 = .25$ Γ, $N_1 = 1.000001$, $E_2 = .08 + .1024$ Γ, $N_2 = .65$, and $E_3 = .07 + .06$ Γ, $N_3 = .5$.

A first evident feature is the increasing of the transition value Γ for lowering energies. This is obviously not surprising since at lower energies the nonlinear term plays a less relevant role, the system being less excited with respect to the equilibrium. On the other hand the changing behaviour of the maximum LCE on the left transition point is worthy of further analysis. In fact the presence of a plateau around $\Gamma\sim4$, which is plainly seen in Fig. 1, becomes more evident at lower energies as clearly shown by Fig. 2. At still lower energies the plateau develops to an auxiliary maximum and a new minimum appears, which suggests the appearance of an order window. This conjecture is confirmed by data in Fig. 3, where a more detailed numerical analysis was performed around the value $\Gamma\sim9$. The width of the window is of the order of $\Gamma\sim.07$ as appears from the following

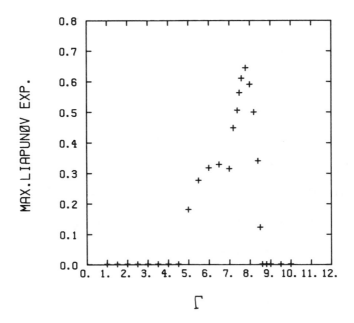

Figure 2. The same as Fig. 1, except for initial conditions
$A_1 = .8$, $A_2 = .1$, $A_3 = 0$.

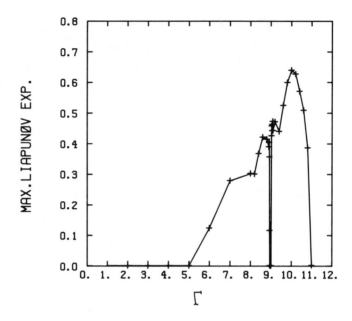

Figure 3. The same as Fig. 1, except for initial conditions $A_1=.7$, $A_2=.1$, $A_3=0$.

table:

Γ	Max LCE	Γ	Max LCE
8.93	.356	8.98	.00104
8.94	.116	8.99	.00107
8.95	.00107	9.00	.00108
8.96	.00107	9.01	.00105
8.97	.00109	9.02	.445

In Fig. 3, the linear interpolation curve is shown just in order to make data reading more transparent, this not excluding the existence of richer structures (such as other narrow windows around different values of Γ) after a finer resolution. This possibility is presently under investigation both by LCE analysis and alternative methods. Finally it is worth observing that the maximum of the curves in Figs. 1,2,3 has the same value (~.64) independently of initial conditions, due to their being close to satisfying the scaling condition: $H \to s^2 H$, $N \to s^2 N$, $\Gamma \to s^{-2}\Gamma$.

References

[1]   J.C. Eilbeck, P.S. Lomdahl, A.C. Scott (1985). *Physica* **16D**, 318, 338.
[2]   J.H. Jensen, P.L. Christiansen, J.N. Elgin, J.D. Gibbon and O. Skovgaard (1985). *Phys. Lett.* **110A**, 429, 431.
[3]   S. De Filippo, M. Fusco Girard and M. Salerno (1986).

Lyapunov Exponents for the n=3 Discrete Self-Trapping Equation, preprint, University of Salerno.

[4]   G. Benettin, L. Galgani, A. Giorgilli, J.M. Strelcyn (1980). *Meccanica* **15**, 9, March; ibid., 30, and references therein.

[5]   J.P. Eckmann and D. Ruelle (1985). *Rev. Mod. Phys.* 57, 617, 656, 1985, and references therein.

COHERENCE, CHAOS, AND CONSCIOUSNESS: NONLINEAR DYNAMICS AT "MACRO" AND "MICRO" EXTREMES OF BRAIN HIERARCHY

58.1 Introduction

The brain/mind appears to be organized as layers of parallel processing systems. Dynamic activities within a hierarchy ranging from structural biochemistry (i.e. intracellular cytoskeleton), membrane electrodynamics, neuron and synaptic behaviour, collective effects of neural networks, and cortical potentials recorded as electroencephalography (EEG) are related to cognitive functions [1]. Nonlinear dynamic activities at both extremes of the brain's organizational hierarchy may have elements of coherence and chaos relevant to consciousness. In this paper we describe, at the "macro" level, human EEG plotted in phase space with dimensional analysis in both awake and anaesthetized states. Our findings suggest that human EEG goes from relative chaos in the awake state to relative order in an anaesthetized state. At the "micro" level, we examine a cellular automaton model of information processing within cytoskeletal polymers (microtubules) which comprise a dynamic network within all living cells including neurons [2-4]. This model is based on coherent nanosecond oscillations of microtubule subunits [5,6] and demonstrates dynamic patterns which generate self-trapping "soliton" gliders, organizing patterns capable of computing and chaotic behaviour, and mechanisms for cytoplasmic organization, information representation, and biomolecular cognition.

58.2 Phase Space EEG

Electroencephalography (EEG) is an important derivative of brain function, and can be a useful intraoperative monitor of brain integrity and anaesthetic depth during surgical procedures. Anaesthetic depth is a convenient paradigm to observe correlates of consciousness. "Awake" EEG consists of polymodal frequencies ranging from several Hz to about 50 Hz. Intraoperative EEG is commonly displayed and analysed in a "power spectrum" frequency distribution mode where, in general, power at higher frequency decreases as anaesthetic depth increases [7]. Accordingly, univariate descriptors such as spectral edge (the frequency below which 95% of the EEG power occurs) are purported to be clinical indices of anaesthetic depth. However, these univariate descriptors are clinically useful only with unimodal frequency distributions (virtually nonexistent in EEG), and change inconsistently among various anaesthetics. Thus EEG monitoring of anaesthetic depth may be aided by geometric and qualitative techniques which have evolved from mathematical theories on deterministic chaos. The dynamic behaviour of the brain/mind may be modelled as a complex, nonlinear system [8].

Characterization of a dynamic system on the order and chaos spectrum involves considering an "n" dimensional phase space into which the variable set [x(t), x(t+T)] can be mapped. In the case of EEG, x is the voltage amplitude, t is time, and T is a fixed time interval (phase lag). An instanteous state of such a system becomes a point in the phase space, whereas a sequence of such states followed in time defines a curve: the phase space trajectory. As time grows, a system whose dynamics are reducible to a set of deterministic laws reaches a permanent state indicated by the convergence of families of phase space trajectories towards an attractor, a subset of the phase space.

Nonlinear dynamic theory and phase space analysis of EEG can attempt to answer the following questions:

1. Do phase space trajectories demonstrate easily recognizable differences among EEG states: (can an EEG attractor be identified for a given state of consciousness)?

2. If an attractor exists, what is its dimensionality "d" and does "d" correlate with anaesthetic depth and level of consciousness?

EEG data were collected from twenty patients undergoing surgery

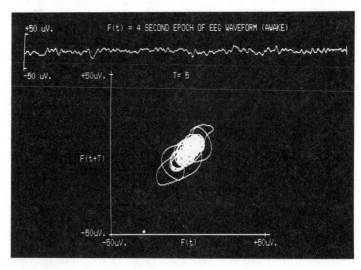

Figure 1. A four second epoch of patient's awake (sedated) EEG. Raw EEG is shown at top, and "simultaneous" phase space display [digitization frequency 300 Hz, phase lag (T) = 5] is at bottom. The phase space display is fairly dense and relatively "chaotic".

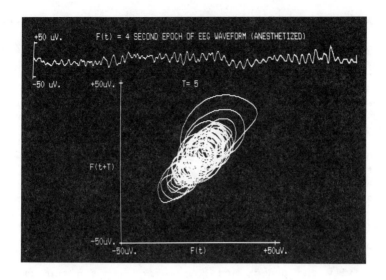

Figure 2. A four second epoch of patient's anaesthetized EEG. Raw EEG is shown at top, and "simultaneous" phase space display [digitization frequency 300 Hz, phase lag (T) = 5]. Compared to Figure 1, the phase space display is less dense, eccentric, has larger area, and is contoured suggesting a geometric pattern or attractor. Dimensional analysis shows it is more "ordered" than Figure 1.

and general anaesthesia.

After preoperative sedation (morphine 10 mg, diazepam 5 mg) EEG electrodes were placed; the C3 to P3 electrode pair was monitored and recrded on FM tape using a low frequency cutoff at 1 Hz, and high frequency cutoff at 35 Hz. Data were recorded during sedation and preparation for induction of anaesthesia (AWAKE), and following anaesthetic induction, intubation, and maintenance with isoflurane and fentanyl (ANAESTHETIZED). The data were digitized at 300 Hz for phase space reconstruction and dimensional analysis according to the method of Grassberger and Procaccia [9].

"Raw" EEG and corresponding phase space trajectories for the AWAKE and ANAESTHETIZED states are shown in Figures 1 and 2. The "raw" EEG patterns are typical and demonstrate reduced frequency distribution as anaesthetic depth increases. The AWAKE state phase space trajectory (Fig. 1) shows a relatively tight, dense pattern mainly centred around the central part of the graph. The ANAESTHETIZED phase space trajectory (Fig. 2) shows a more "unravelled" pattern with geometry suggesting an attractor. The patterns appear to be visually robust in that marked differences among the two states are apparent at a glance. Dimensionality for the patient presented as 2.15 and 2.07 for the awake and anaesthetized states, respectively. Awake and anaesthetized state dimensionality values for the twenty patients were significantly different (p < 0.05). Thus dimensionality of EEG was observed to decrease slightly as awake patients became anaesthetized.

58.3 Automaton Model of Dynamic Organization in Microtubules

Interiors of **all** living cells are organized by dynamic networks of parallel arrayed protein lattices including microtubules (MT), filaments, and microtrabecular lattice. MT are involved in a wide range of biological functions (ciliary movement, mitosis, axoplasmic flow, synaptic plasticity) involving information processing functions. Cellular automata are dynamical systems that can generate and process patterns and information by way of rules based on lattice neighbour interactions [2]. Lattice structure and apparent information processing capabilities have suggested an MT-automaton behaviour based on coherent dipole oscillations within MT subunits [3,4].

Various models of dynamic activities in biomolecules predict functional organization and regulation. Among these, Frohlich has theorized that nonlinear dipole excitations among subunits of biomolecular arrays result in coherent oscillations (10^{-9} to 10^{-11} sec fluctuations), lower frequency metastable states, long-range cooperativity and order, and coupling of protein conformational states to dipole oscillations within protein hydrophobic regions [5,6]. Individual subunit conformation within a lattice array would thus depend on various factors (GTP, Ca^{++}, primary structure, lattice neighbour electrostatic interactions) influencing oscillation phase. Based on MT geometry we have calculated MT-lattice neighbour electrostatic influences on dipole oscillations as rules for an MT-automaton computer simulation.

MT subunit states at each "clock tick", or generation were determined by neighbour states at the previous generation

$$\text{state} = \alpha, \text{ if } \sum_{i=1}^{n} f(y) > 0, \text{ state} = \beta, \text{ if } \sum_{i=1}^{n} f(fy) < 0,$$

where n=7 as the number of neighbours, and f'(y) the force from the "ith" neighbour in the y direction

$$f(y) \propto \frac{\sin \theta}{r^2}, \text{ } \sin \theta = \frac{y}{r}; f(y) \propto \frac{y}{r^3}.$$

Thus states may be determined by the summation of subunit neighbour dipole forces in proportion to y/r^3, and relative values calculated from known MT structure (Fig. 3).

Figure 3 shows a microtubule (MT) automaton model showing coherent **order** evolving towards **chaos**. Automaton patterns were obtained from computer simulation based on known structure of MT lattice, conformational/dipole coupling among neighbours, and coherent nanosecond MT subunit excitations. In the lower left of the figure, a local neighbourhood of MT subunits is shown with neighbour distances from which automaton "rules" are derived from calculated electrostatic forces. x=5 nm, y=4 nm, r=6.4 nm. Dynamic computer simulations demonstrate stable coherent patterns, oscillators, structures which dissipate to chaos, and kinks which travel at 8 nanometers per nanosecond generation (self-trapped

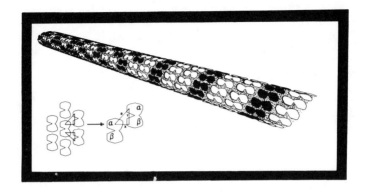

Figure 3. Microtubule (MT) automaton model showing coherent order evolving into chaos. Automaton patterns are obtained from computer simulation based on MT structural lattice, conformational/dipole coupling among neighbours, and coherent nanosecond excitations of subunits. The patterns evolve from right to left at a maximal velocity of 8 nanometres per nanosecond. Lower left: hexagonal local neighbourhood of MT subunits showing neighbour distances from which automaton "rules" are derived from calculated electrostatic forces. x=5 nm, y=4 nm, r=6.4 nm. MT and other cytoskeletal filaments form complex parallel networks within all living cells.

"solitons"?). In automaton nomenclature these travelling kink patterns are known as gliders which can be used for information transfer.

MT and other cytoskeletal filaments form complex parallel networks within all living cells, including brain neurons. Stable and moving patterns of MT-subunit conformational phase could orchestrate intracellular activities, function as specific sites of ion or protein binding and transport, and represent dynamic information in cytoskeletal substrates of memory and consciousness.

References

[1]   A.C. Scott (1977). *Neurophysics,* Wiley, New York, 184-226.

[2]   *Cellular Automata, Proc. of an Interdisciplinary Workshop, Los Alamos,* eds. D. Farmer, T. Toffoli and S. Wolfram, North Holland, Amsterdam, 1984.

[3]   S.A. Smith, R.C. Watt and S.R. Hameroff (1984). *Physica* **10D,** 168.

[4]   S.R. Hameroff (1987). *Ultimate Computing - Biomolecular Consciousnessand Nanotechnology,* Elsevier-North Holland, Amsterdam.

[5]   H. Frohlich (1970). *Nature (London)* **228.**

[6]   H. Frohlich (1975). *Proc. Natl. Acad. Sci. U.S.A.* **72,** 4211.

[7]   I. Pichlmayr, U. Lips and H. Kunkel (1984). *The electroencephalogram in anaesthesia - Fundamentals, Practical Applicatio;ns, Examples,* Springer-Verlag, Berlin.

[8]   A. Babloyantz, J.M. Salazar and C. Nicolis (1985).   *Phys. Lett.* **111A**, 152.

[9]   P. Grassberger and I. Procaccia (1983).   *Physica* **9D**, 189.

59   *K.M.C. Da Silva*

FERROELECTRIC PROPERTIES OF BILAYER LIPID MEMBRANES

59.1 **INTRODUCTION**

The response to alternating electric fields of heavy liquid dielectrics has traditionally been described by an empirical impedance relationship due to Cole and Cole [1]:

$$Z(j\omega) = R_{\infty} + \frac{R_p}{1 + (j\omega\tau)^{[1-(\alpha/90)]}} \tag{1}$$

in which R_p and R_{∞} are, respectively, the dielectric resistance between the measuring electrodes and the series resistance of the measuring circuitry, $\tau = R_p.C$, C is the dielectric capacitance, α the dielectric angle, $\omega = 2\pi f$ and f is the frequency of the electric stimulating signal.   Equation (1) is represented geometrically by the diagram of Fig. 2(b) and has been interpreted in terms of a population of dielectric relaxation rates [2].   In a formally different approach we start from the concept of dielectric hysteresis to deduce for the impendance locus a description which provides a better insight on the ferroelectric properties of these materials.   Preliminary results on the influence of temperature on the electrical parameters of these material structures are presented.

59.2 **THEORETICAL CONSIDERATIONS**

Two methods are used to correlate the Cole & Cole diagram with a hysteretic phenomenon:

59.2.1 *Conformal mapping*

The circular part of the transfer impedance locus may be defined by the following complex variable transform:

$$Z(s) = \frac{R_p}{1 + \tau.e^{-j\alpha}.s} \quad (s=j\omega) \tag{2}$$

which maps the positive imaginary axis of the s-plane on the

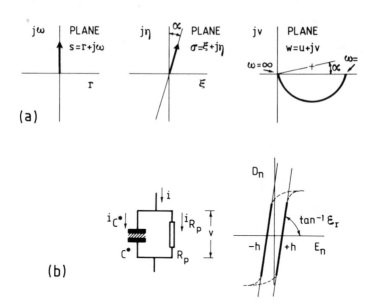

Figure 1. (a) Mapping of the positive imaginary axis of the s-plane on the sector of the w-plane centred at $u_0 = R_p/2$, $v_0 = (R_p/2) . \tan \alpha$ and radius $(R_p/2) . \sec \alpha$; $\sigma = e^{j\alpha} . s$ and $\omega = \frac{1}{1+\tau . \sigma}$. (b) Circuit diagram and hysteretic loop of a leaky capacitor: voltage (v), currents (i), relative permittivity (ε_r) and loop width (2h).

circular sector of the w-plane as illustrated in Fig. 1(a). In this function, $\tau = R_p . C$ is a time constant and the factor $e^{-j\alpha}$ represents a constant frequency-independent time delay. This transfer function also fits the experimental data, as shown in Fig. 2(b).

59.2.2 *Describing function method*

The transfer impedance of the network shown in Fig. 1(b) is deduced from the following relationships:

$$i = i_{R_p} + i_C^*; \quad i_{R_p} = \frac{v}{R_p}; \quad i_C^* = \frac{S}{4\pi} \cdot \frac{\partial D_n}{\partial t}; \quad E_n = \frac{v}{\delta}; \quad v = V . e^{j\omega t} \quad (3)$$

where E_n and D_n are the normal components of the field and of the electric displacement, respectively, S is the capacitor area, δ the dielectric thickness and v, the voltage across the capacitor, is represented in symbolic form. If, for simplicity's sake, the hysteretic cycle is taken as a simple backlash, the corresponding describing function will be of the form [3]:

$$N(V) = \frac{D_n}{E_n} = \sqrt{A_1^2 + B_1^2} \cdot e^{-j \tan^{-1}(B_1/A_1)} \tag{4}$$

where A_1 and B_1 are parameters which depend only on ε_r, h and V.

Finally, making $C^* = \frac{\varepsilon_r \cdot S}{4\pi\delta} \cdot \sqrt{A_1^2 + B_1^2}$, $\tau = R_p \cdot C^*$ and $\alpha = \tan^{-1}(B_1/A_1)$, the transfer impedance becomes:

$$Z(j\omega) = \frac{R_p}{1 + j\omega\tau \cdot e^{-j\alpha}} \tag{5}$$

59.3 EXPERIMENTAL RESULTS

Preliminary studies of the influence of the temperature on the electrical parameters of the frog (*Rana Ridibunda*) skin were done

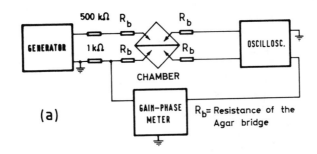

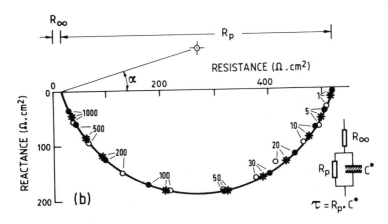

Figure 2. (a) Experimental set-up; (b) Impedance of the frog skin: Nyquist diagram and corresponding equivalent circuit. Conditions and parameters: $T=10.5^{\circ}C$; $R_{\infty}=10$ Ω; $R_p=502$ Ω; $C^*=4.7$ μF ($\bullet$); $C^*=5.0\mu F$ ($*$); $\alpha=16.2^{\circ}$. Numbers represent frequency in Hz and symbols represent experimental data (o), computed from formulae (1) ($\bullet$) and (2) ($*$).

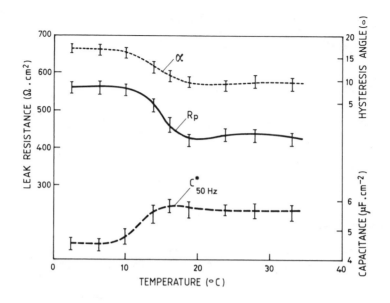

Figure 3. Influence of the temperature on the electrical properties of the frog skin.

with the set-up shown in Fig. 2(a) which consists of a thermostatic Üssing-type chamber [4], a wideband sinusoidal generator, an oscilloscope and a gain-phase meter. The ion transport properties of the skin are determined by those of the serosal membrane of a single cell layer (*stratum granulosum*) [5] and constitute a reasonable model of a bilayer lipid membrane. The circular sector of the impedance locus is fitted to the data by the least square error method to determine the values of R_p, R_∞ and α; C^* is determined from Eq. (1) and (2) for f=50 Hz. The experimental and calculated points are shown in Fig. 2(b).

Figure 3 shows a phase transition between 14°C and 16°C which affects the values of R_p, C^* and α as well as the activity of the Na^+-pump, reflecting a loosening of the phospholipid packing density and a change in the bilayer thickness [6,7].

59.4 DISCUSSION

Both equations (1) and (2) share the same values of α and R_p and both map the axis $s=j\omega$ on the circular sector of Fig. 2(b). Indeed they only differ in that $(\omega\tau)$ is raised to the power $[1-(\alpha/90)]$ in the second of these equations leading to a slightly different distribution of the corresponding computed points over the circular sector. However, neither formula fits adequately the data for the simple reason that the value of C is not constant with the frequency as it is assumed in the computation of the points. There are two reasons for this: (a) the impedance of the skin drops with the increase in the frequency and, being stimulated by a current source, the voltage across it drops also;

(b) the hysteretic characteristic has a sigmoidal shape and small amplitude cycles are less steep than larger ones [8] and consequently the equivalent permittivity is larger for lower frequencies.

REFERENCES

[1]   K. S. Cole and R. H. Cole, *J. Chem. Phys.* 9, 341, 1941.
[2]   W. Kauzmann, *Rev. Mod. Phys.* 14, 12, 1942.
[3]   J. Gibson, *Nonlinear Automatic Control,* McGraw Hill, New York, 1963.
[4]   C. H. Jesus and M. W. Smith, *J. Physiol.* 243, 211, 1974.
[5]   H. H. Üssing and U. Lassen, *Ann. Rev. Physiol.* 36, 17, 1974.
[6]   J. F. Nagle, *Proc. Natl. Acad. Sci. U.S.A.* 70(I), 3443, 1973.
[7]   V. Luzatti and A. Tardieu, *Ann. Rev. Phys. Chem.* 25, 79, 1974.
[8]   A. R. von Hippel, *Dielectric Materials and Applications,* The M. I. T. Press, Cambridge, Massachusetts, 1966.

SELF-TRAPPING OF INFRARED ENERGY ABSORBED IN ACETANILIDE

Acetanilide (ACN) is an organic solid made of paired chains of molecules with nearly planar hydrogen-bonded amide groups, whose bond lengths and angles are very close to those observed in most polypeptides, in particular in α-helical regions of proteins. ACN has an anomalous low temperature IR absorption band, which has been attributed to a self-trapped state of molecular vibrational energy in the amide group (Davydov-like soliton) [1]. In order to get an insight into this phenomenon, a classical model for a segment of an ACN chain has been set up with the purpose of analysing the role of the nonlinear hydrogen bond in the molecular energy exchange. In each molecule, only those degrees of freedom of the amide group which are involved in the H-bond have been taken into account: the stretching (u) of the C=O bond, the stretching (ρ) and the bending (θ) of the N-H bond (Fig. 1); the rest of the molecule is kept fixed. The dynamics of the model has been investigated by integrating the equations of motion using a standard molecular dynamics technique.

The model includes three interactions. The main force acting on the atoms is harmonic and directed towards the equilibrium position: to a first approximation, it represents the restoring action of the medium. The second force is due to the hydrogen bond, which is the main intermolecular interaction and directly affects the degrees of freedom considered here. The third force derives from an intramolecular interaction between the charge excesses of the oxygen and of the hydrogen in the amidic group: the movement of one of these two atoms about its equilibrium position generates a change in the charge distribution along the electronic structure of the group, which in turn exerts a force on the other atom. The equilibrium value of the last two interactions must be subtracted, because it is implicitly accounted for in assigning the equilibrium positions.

The characteristic frequencies for the three degrees of freedom are taken as follows: $\tilde{\nu}_1$ = 1665 cm^{-1} (C=O stretching); $\tilde{\nu}_2$ = 3350 cm^{-1} (N-H stretching); $\tilde{\nu}_3$ = 1600 cm^{-1} (N-H bending).

The harmonic forces acting on the coordinates of the i-th molecule have then the expressions:

$$F_{h,u}^{(i)} = -M_1\omega_1^2(u_1-u_0), \quad F_{h,\rho}^{(i)} = -M_2\omega_2^2(\rho_1-\rho_0),$$

$$F_{h,\theta}^{(i)} = -M_2\omega_3^2\rho_0^2\theta_i.$$

here M_1 is the mass of the oxygen, M_2 is the mass of the hydrogen and $\omega_\ell = 2\pi c\tilde{\nu}_\ell$. The force acting on θ is a generalized force in a lagrangian formulation.

The hydrogen bond is represented in our model by a Coulomb interaction between the negative charge excess on the oxygen of a molecule and the positive charge excess on the hydrogen of the neighbouring one. In order to obtain explicit expressions for this force, the distance between the oxygen in the i-th molecule and the hydrogen in the (i+1)-th molecule is written as a function of u_i, ρ_{i+1}, and θ_{i+1} (see Fig. 1). A power expansion up to 2nd order in the displacements from equilibrium positions yields the following expressions, in which the equilibrium values have been subtracted:

$$F_{hb,u}^{(i)} = \frac{3\cos^2\alpha - 1}{r_0^3}|qq'|(u_i-u_0) + \frac{2\cos\alpha}{r_0^3}|qq'|(\rho_{i+1}-\rho_0)$$

$$+ \frac{\rho_0\sin\alpha}{r_0^3}|qq'|\theta_{i+1} + \frac{15\cos^2\alpha - 9}{2r_0^4}\cos\alpha|qq'|(u_i-u_0)^2$$

$$+ \frac{3\cos\alpha}{r_0^4}|qq'|(\rho_{i+1}-\rho_0)^2$$

$$- \frac{3\rho_0^2 + 2\rho_0 r_0}{2r_0^4}\cos\alpha|qq'|\theta_{i+1}^2$$

$$+ \frac{9\cos^2\alpha - 3}{r_0^4}|qq'|(u_i-u_0)(\rho_{i+1}-\rho_0)$$

$$+ (\frac{1}{r_0^3} + \frac{3\rho_0}{r_0^4})\sin\alpha|qq'|\theta_{i+1}(\rho_{i+1}-\rho_0)$$

$$+ \frac{3\rho_0\sin\alpha}{r_0^4}|qq'|(u_i-u_0)\theta_{i+1}$$

$$F_{Hb,\rho}^{(i)} = \frac{2}{r_0^3} |qq'|(\rho_i - \rho_0) + \frac{2\cos\alpha}{r_0^3} |qq'|(u_{i-1} - u_0)$$

$$+ \frac{3}{r_0^4} |qq'|(\rho_i - \rho_0)^2 + \frac{9\cos^2\alpha - 3}{2r_0^4} |qq'|(u_{i-1} - u_0)^2$$

$$- \frac{3\rho_0^2 + 4\rho_0 r_0 + r_0^2}{2r_0^4} |qq'|\theta_i^2$$

$$+ \frac{6\cos\alpha}{r_0^4} |qq'|(\rho_i - \rho_0)(u_{i-1} - u_0)$$

$$+ (\frac{1}{r_0^3} + \frac{3\rho_0}{r_0^4}) \sin\alpha |qq'|(u_{i-1} - u_0)\theta_i$$

$$F_{Hb,\theta}^{(i)} = -\frac{\rho_0^2 + \rho_0 r_0}{r_0^3} |qq'|\theta_i + \frac{\rho_0 \sin\alpha}{r_0^3} |qq'|(u_{i-1} - u_0)$$

$$+ \frac{3\rho_0 \sin 2\alpha}{2r_0^4} |qq'|(u_{i-1} - u_0)^2$$

$$+ (\frac{1}{r_0^3} + \frac{3\rho_0}{r_0^4}) \text{in}\alpha |qq'|(\rho_i - \rho_0)(u_{i-1} - u_0)$$

$$- \frac{3\rho_0^2 + 4\rho_0 r_0 + r_0^2}{r_0^4} |qq'|(\rho_i - \rho_0)\theta_i$$

$$- \frac{3\rho_0^2 + 2\rho_0 r_0}{r_0^4} \cos\alpha |qq'|(u_{i-1} - u_0)\theta_i.$$

Here q is the excess charge on oxygen ($-1.5417 \, 10^{-10}$ esu), and q' the excess charge on hydrogen ($4.8990 \, 10^{-11}$ esu); $\pi - \alpha$ is the angle between C=0 and the H-bond at equilibrium; r_0 is the equilibrium length of the H-bond.

The intramolecular force has been simulated by means of a

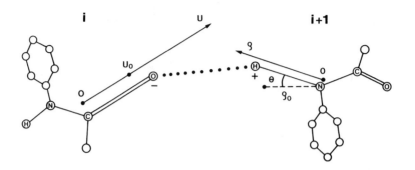

Figure 1.   Two molecules in an ACN chain, with the set of coordinates. u_0 is the equilibrium distance between C and O (1.2187 Å); ρ_0 is the equilibrium distance between N and H (1.0800 Å); $\theta=0$ at equilibrium.

charge-dipole interaction.   In order to obtain a manageable expression for it, the geometry of the peptide bond has been "straightened";   the O, C, N and H atoms have been placed on a straight line, at mutual distances equal to those in the equilibrium configuration of the crystal.   A dipole $\vec{p}$ is located at the centre of the C-N bond;   the dipole-charge forces acting on the two atoms are:

$$F_{dc,u} = - \frac{2p|q|}{(u+d)^3} \qquad\qquad F_{dc,\rho} = - \frac{2p|q'|}{(\rho+d)^3} .$$

The sign of the forces refers to the axes along which the displacements u and ρ are measured;   d is half the length of the C-N bond.   The dipole $\vec{p}$ is, in its turn, generated by the charge excesses of the two atoms, and its component along the C-N bond is

$$p = a[|q|(\frac{1}{(u+d)^2} - \frac{1}{(u_0+d)^2}) + |q'|(\frac{1}{(\rho+d)^2} - \frac{1}{(\rho_0+d)^2})]$$

where a is the polarizability of the C-N bond (a = 2 Å^3).   The equilibrium value of the induced dipole has been subtracted because it has already been taken into account in the balance of all forces which determine the equilibrium position of the atoms.

These expressions are developed in powers of $u-u_0$ and $\rho - \rho_0$ up to terms of the second order.   Inserting the expansion of p into the expansion of the forces and keeping again terms up to the second order yields:

$$F_{dc,u}^{(i)} = \frac{4aq^2}{(u_o+d)^6}(u_i-u_o) + \frac{4a|qq'|}{(u_o+d)^3(\rho_o+d)^3}(\rho_i-\rho_o)$$

$$- \frac{18aq^2}{(u_o+d)^7}(u_i-u_o)^2 - \frac{6a|qq'|}{(u_o+d)^3(\rho_o+d)^4}(\rho_i-\rho_o)^2$$

$$- \frac{12a|qq'|}{(u_o+d)^4(\rho_o+d)^3}(u_i-u_o)(\rho_i-\rho_o),$$

$$F_{dc,\rho}^{(i)} = \frac{4aq'^2}{(\rho_o+d)^6}(\rho_i-\rho_o) + \frac{4a|qq'|}{(u_o+d)^3(\rho_o+d)^3}(u_i-u_o)$$

$$- \frac{18aq'^2}{(\rho_o+d)^7}(\rho_i-\rho_o)^2 - \frac{6a|qq'|}{(u_o+d)^4(\rho_o+d)^3}(u_i-u_o)^2$$

$$- \frac{12a|qq'|}{(u_o+d)^3(\rho_o+d)^4}(u_i-u_o)(\rho_i-\rho_o).$$

The results reported here refer to several molecular dynamics runs performed on a segment of five molecules. Each run was 1.6×10^5 time steps long, corresponding to a physical time of 8 ps. If the energy was initially given only to few degrees of freedom, for example, to u_2, ρ_3 and θ_3, it did not spread, remaining confined in that region of the chain. The degrees of freedom initially excited exchanged energy in a fairly regular, periodic way; the dynamics was quite ordered, showing steady oscillations of all degrees of freedom over very long times, even at high temperature.

The ρ_i and θ_i coordinates of a N-H bond exchange a large amount of energy between them; the energy initially given to the ρ vibration is passed to the θ coordinate and then periodically regained, with a characteristic period of about 0.2 ps. The transfer through the H-bond of the energy initially given to the u coordinate is of smaller amplitude, but still periodical; this period, which is reflected as a modulation in the ρ-θ exchange, is about 0.4 ps. In general, energy exchanges are lower and slower at lower temperatures.

In a series of runs energy was initially assigned only to ρ_3 and θ_3; at various temperatures the fraction of the total energy initially given to the stretching ρ_3 was varied. In Fig. 2 the fraction of total energy periodically exchanged between stretching and bending is plotted as a function of the fraction of total energy initially present in the ρ_3 coordinate. In all these runs the energy did not pass in any significant amount to the u_2 or u_3 coordinates, which were initially at rest. This is a general feature of these runs: despite the presence of couplings, the dynamics of the initially excited degrees of freedom is not very much affected by the presence of other potentially available degrees of freedom, if these are initially at rest.

The effect of the nonlinearity is clearly shown in Fig. 2 by the temperature dependence of the curves. In a set of purely linear oscillators, changing the temperature would only rescale the energies; so all points in this plot, where fractional energies are reported, should lie on the same curve. But this is not the case. Also note the two extreme cases: when the whole energy is initially in the bending, this energy can be exchanged with the stretching apparently at all temperatures; but when the whole energy is initially in the stretching, exchange with the bending takes place only at quite a high temperature. This asymmetrical behaviour too is a manifestation of the nonlinearities, as

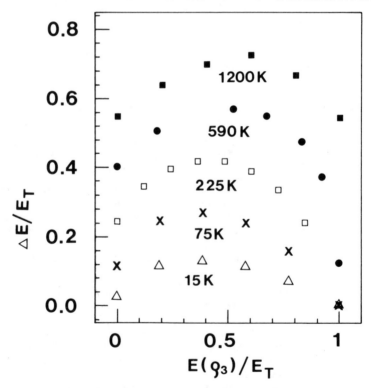

Figure 2. ΔE *is the energy periodically exchanged between stretching* (ρ_3) *and bending* (θ_3) *of a N-H bond.* $E(\rho_3)$ *is the energy initially given to* ρ_3; E_T *is the total energy in the excited region, corresponding to the different temperatures reported.*

explored by increasing the temperature. There is a sort of threshold for the exchange of energy, when the energy is concentrated in the stretching, and apart from a slight temperature effect, it is evident that the exchange is most efficient when the two degrees of freedom have about the same energy.

Runs were also performed in which a fraction of the total energy was initially given to the u_2 vibration, while ρ_3 and θ_3 equally shared the remaining quantity. Figure 3 shows the amount of energy periodically exchanged by u_2 as a function of the fraction of total energy initially given to it. No energy exchange takes place when the u coordinate on one side, or the ρ and θ coordinates on the other are initially at rest; energy can be efficiently transferred through a hydrogen bond only if all the degrees of freedom are similarly excited. Thus, at all temperatures there is a sort of exchange window, which favours energy transfer between molecules at similar temperatures, and

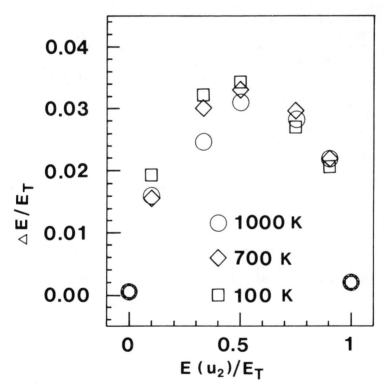

Figure 3. ΔE is the energy periodically exchanged between a C = O bond (u_2) and a N-H bond (ρ_3, θ_3); $E(u_2)$ is the energy initially given to u_2; ρ_3 and θ_3 equally shared the remaining energy at the beginning of the run. E_T is the total energy in the excited region, corresponding to the different temperatures reported.

lowers energy transfer between "cold" and "hot" molecules.

Thinking of an ACN chain, attention is drawn to the phenomenon of energy trapping, or soliton excitation in connection with the absorption of IR radiation [1,2]. Indeed, the characteristic features of the energy exchange through the hydrogen bond might contribute substantially to this phenomenon. This can be seen in the following way, considering the results of Figures 2 and 3: absorption of an IR photon of 1650 cm^{-1} amounts to an energy increase corresponding to $\Delta T = hc\ \tilde{\nu}/k_B = 2380$ K; suppose that the absorption process leads to a sudden transition in the energy of one of the vibrational degrees of freedom. If the system is initially at a temperature of 20 K, as a result of the photon absorption this degree of freedom will increase its energy to 2400 K, while the other two will stay at 20 K. The intermolecular region entailing the three degrees of freedom has a total energy corresponding to 2440 K; therefore, it can be thought of as being at an average local temperature of about 815 K. The initial energy being concentrated mainly on one degree of freedom, the horizontal coordinate in Fig. 3 has a value near 0 or 1; as a consequence, the transfer of energy through the hydrogen bond between the oxygen and the hydrogen in the excited region almost vanishes. Moreover, Fig. 2 shows that if the excited degree of freedom is ρ or θ, also the energy exchange between the two is reduced. Altogether, quantum-like absorption of radiative energy modifies the dynamics of the excited region in such a way that the dispersion of this energy is minimized. The nonlinearity of the H-bond generates thus a self-trapping mechanism.

Acknowledgement

The financial support by the European Economic Community through contract STI-059-J-C (CD) is acknowledged.

References

[1]   G. Careri, U. Buontempo, F. Galluzzi, A.C. Scott, E. Gratton and E. Shyamsunder (1984). *Phys. Rev.* **B30**, 4689.
[2]   A.C. Scott (1985). *Phil. Trans. R. Soc. Lond.* **A315**, 423.

Acetanilide, 140, 633, 651-658
Activation networks, 200-203
Ahlfors-Bers integration, 47
Alloys, microstructure, 245
Alpha-helix, 541, 619-624,
    625-631
Anderson localization, 140,
    339, 344
Anharmonicity, 139, 162
Artificial intelligence, 199-
    206
Asymmetric patterns, 289
Attractor, 429-430, 560, 640
Autism, 117-118
Autocatalysis, 351-352, 371-
    381, 385, 459-460
Avalanche breakdown, 575-
    578, 599
Axisymmetrization, 465-466

Basin boundaries, 485-490
Belousov-Zhabotinskii reaction,
    234, 288, 457-463, 602-
    604
Benjamin-Feir instability,
    306
Benjamin-Ono equation, 227
Benzene, 151-157, 144
Bethe ansatz, 61, 66-67, 72
Bifurcation, 25-26, 29, 108-
    111, 213, 301-308, 347,
    473-474, 477, 491-495,
    503, 553, 559-560, 591-
    592, 598, 607, 613
Bilayer lipid membrane, 645-
    649
Biochemical control systems,
    591
Biological clocks, 318
Bistable system, 123-137
Blow-up, 541-546
Born-Oppenheimer approximation,
    621
Bose-Fermi equivalance, 61
Boundary layer, 585
Breather, 24-27, 34, 38, 59,
    68, 72, 106, 161-168
Brusselator, 601-604
Burgers' equation, 306

Caianello's neuronic equation,
    615
Cantor set, 485, 487
Capillary effects, 245
Carbon monoxide, 579-583
Cardiac tissue, 195, 461
Carnot engine, 531
Casimir operators, 523-527
Catalytic feedback cycles,
    324-329
Catastrophe theory, 179
Cell division, 593-597
Cellular automaton, 639,
    642-643
Cerebral cortex, 113, 118,
    195, 461
Chaos
    in artificial intelligence
        systems, 206
    in chemical oscillators, 601
    in CSTR system,, 124
    in dissipative oscillator,
        485-490
    in DST equation, 141, 147,
        157, 633-637
    in economics, 404-406
    in excitable media, 185
    in Josephson junctions, 108,
        207, 427-439, 563-568
    in near-integrable systems,
        23-39
    in nerve membranes, 185, 613
    in semiconductors, 575-578
Chaotic
    attractors, 23-25, 29
    behaviour, 315, 330
    dynamics, 60, 237, 293-297,
        347, 372, 380
    oscillators, 473-479, 485
Charged hard spheres, 415
Chemical oscillator, 601,
    607-612
Chemical rate equations, 523
Chemical waves, 457
Classical diffusion, 517-521
Coarsening process, 246-249
Coding, 291-293
Coherence, 93, 113-115,
    120-121, 639

Collective coordinate methods, 25, 441-456
Condensed matter systems, 535
Consciousness, 639
Complex systems, 3, 23, 441
Contour dynamics, 468-470
Continuous Flow Stirred Tank Reactor, 123-126, 353-363, 384, 601-606, 607-612
CRAY-2, 41-43
Crystal growth, 585
Cyclic AMP, 460
Cyclic symmetry, 376, 385
Cytokinesis, 593-597

Darwinian selection, 10
Davydov soliton, 139-140, 541, 619-624, 625
deBroglie wavelength, 161
Debye-Hückel law, 415
Dendritic microstructures, 245-255
Devil's staircase, 233, 485, 489, 615
Diffusion in random media, 333-345
Discrete Self-Trappig equation, 139-159, 633-637
Dislocation theory, 77, 85
Dispersive systems, 541
Dissipation rate quotient, 483
Dissipative
  dynamics, 301-314
  oscillator, 485-490
  systems, 207, 301-314, 481
DNA, 287, 298, 316, 349, 368
Dynamic scaling, 333
Dynamical systems, 24, 34, 60, 123, 205, 245, 4645, 613

Easy-plane ferromagnetic chain, 535-540
Ecological systems, 523
Ecology, wave propagation in, 233
Economic long wave, 389-414
Economics, theoretical, 384
Ehrenfest theorem, 497
Eigenvalue
  simple zero, 302-309
  double zero, 309-313
Electroencephalography, 639

Entrainment, 465
Euler equations, 465, 468, 482
Evolution, 3-20, 315-330, 347-349
  rate, 16
Evolutionary
  adaptation, 370-371, 381
  selection process, 361
Excitable media, 185-197
Excitability, 458-460
Experimental results
  avalanche breakdown, 575-578, 599
  Belousov-Zhabotinskii, 604-606
  bilayer lipid membrane, 647-649
  coarsening studies, 249-252
  continuous flow stirred tank, 123-126
  dendritic microstructures, 249-251
  economic long wave, 407-408
  Josephson junction, 98
  circulation junction oscillation, 102
  mechanical analog, 563-568
  microsignal effect, 111
  noise in, 430-438
  reaction diffusion system, 234
  oscillation oxidation of CO on Pt, 579-583
  optical fibre soliton, 257, 263-265, 280
  RNA, 317, 249-252
  solidification studies, 249-252
  soliton interactive forces, 274-279
  soliton laser, 265-272
  squid axon, 613-618
  turbulence in semiconductors, 575-578
Extended Curie principle, 288

Feigenbaum bifurcations, 347, 381, 485, 489
Feynman propagator, 63
Filamentation, 465-466
Fisher's selection equation, 374-375, 385
Fiske steps, 105
Fluid dynamics, 465
Fluxon dynamics, 101

Fokker-Planck equation, 85,
181-183, 220, 333-345,
455,517
Functional-differential
systems, 511-515
Functional integral methods,
61, 71, 74

Gene network, 315-331
Genetic coding, 316
Gibbs' principle, 172
Gibbs-Thomson equation,245,
247-248, 255
Ginzburg Landau equation, 133,
301, 304, 333
Global stability, 207, 217
Globular protein, 140
Gradient intensification, 465

Harmonic oscillator, 500
Hodgkin-Huxley equation,
188-190, 613
Holomorphic surgery, 46-48
Homoclinic orbits, 34-39,
473-479, 503, 557-559
Hopf bifurcations, 189-190,
301, 302 305-309, 347,
376, 511-515, 556-558
Hypercycles, 317-324, 379-380
Hysteresis, 172, 126-133, 309,
563, 645-649

Information, 287, 291, 296-298
biological, 297
compression, 297
Information-rich structures,
287-299, 315-317, 327
Integrability, analytic,
227-232
Intebrale dynamical system,
59, 227, 442, 523
Inverse scattering, 60, 279
Irreversibility, 289-291, 297,
529
Iterated maps, 491

Jaynes' principle, 172, 174-175
Jordan-Wigner transformation,
62
Josephson junction, 93-112,
209, 213, 216-217, 569-575
mechanical analog, 563-568
noise in, 428-430
Josephson transmission line, 98

KAM theory, 60, 633
Kink, 25, 59, 68, 77-83,
442-444
quantum dynamics of, 453-454
Kink-antikink pair, 37, 83-87,
442, 448-453
Kinetic equations, 366
Klein-Gordon equation, 59, 62,
302, 310,442-444, 448
Kondratiev cycle, 389
Korteweg-deVries equation, 227
Kuramoto-Sivashinsky equation,
305, 306

Landau Ginzburg equation, 133,
301, 304, 333
Landau Lifschitz model, 62
Langmuir solitons, 541
Langevin equation, 333, 445,
517-518
Laser
colour centre, 547
single mode, 176, 553
soliton, 257-258, 264-272,
547-552
Limit cycles, 206, 290, 326,
376, 379, 389, 405, 556
Local mode analysis, 151
Long wave, 389-391, 395-399
Lorenz equation, 553, 560

Mandelbrot set, 45-49
Maximum information entropy
principle, 177-81, 182
Maxwell Block equations, 547,
553-563
McCumber solution, 94
Melnikov theory, 503-510
Metropolis method, 416
Microstructure of solidified
material, 246
Microtubule, 639, 642-643
Minimization problem, 481
Mitosis, 593-597
Model complementarity, 465-471
Molecular dynamics, 157
Mushy zone in two phase system,
245
Mutation, 297, 347-388

Nagumo-Sato model, 615
Navier Stokes equation, 481-484
Near-integrable system, 23, 39
Neo-Darwinism, 3-6

Neural network, 113-122, 186, 615, 639
Neural tissue, 461
Neurons, 170, 639
Neutron scattering, 62
Nerve membranes, 185, 613
Nonlinear relaxation time, 127-129
Nonlinear Schrödinger equation, 27-28, 34-36, 60-63, 140, 141, 151, 227, 269-260, 302, 307, 497-451, 541-546, 549
Nonlinear spectral methods, 25-27, 34-34, 38
Nonlinear Sturm-Liouville problem, 499-500
Nonlinear wave equation, 24
Normal mode analysis, 151
Numerical diagnostics, 465-470

Optical
  communications, 274
  computing, 275
  fibres, 257, 549-550
Optimization, 3, 361, 384-385
Order-chaos-order transition, 633-637
Ornstein Zernike equation, 422
Oscillatory oxidation of CO, 579-583
Ostwald ripening, 246-248

Pacemaker neurons, 120
Painleve test, 227-232
Parallel processing, 639
Pattern selection, 23
Peierls hill, 78
Pendulum equations, 23-39, 209, 211, 427-428, 503-510, 563, 564
Period doubling, 102, 107-108, 214, 381, 560-561, 563, 601, 613
Perturbation theory, 274
Phase locking in nonlinear oscillators, 105
Phi-four equation, 442, 448, 451
Phonons, 59, 72, 446, 449
Polaron, 139, 443, 454
Polynucleotide synthesis, 364
Prebiotic evolution, 315-324
Protein synthesis, 317

Population
  dynamics, 3, 7-9, 12, 18, 511
  genetics 347,373-376, 385

Quantum inverse method, 61, 67, 70
Quantum localization, 333

Raman gain, 272
Random graph theory 315, 319-324, 330
Random media, 333-345
Rashevsky-Turing theory, 597-601
Reaction-diffusion system, 223-243, 288, 301-305, 309-310, 591, 597-601
Reaction-diffusion-convection system, 240-241
Relaxation kinetics, 123-137
Renormalization-group techniques, 333
Replication kinetics, 347-388
Resonant ring cavity, 553
RNA, 315-330
  replication, 347-353
Rössler's system, 477-478, 494

Selection rule, 592
Self
  focusing, 306
  ordering, 397-401, 408
  organization, 169, 315, 481-484, 598
  trapping, 651-658
Semiconductors 233, 575-578
Schlögl model, 373-375, 385
Schrödinger equation, 306, 333-345, 443, 517-521, 620, 622
Shannon information formula, 173, 291, 296
Shilnikov's theorem, 473-479
SH-kinetics, 233, 235
Sil'nikov type bifurcation, 557, 560
Sine-Gordon equation, 24, 27-28, 34, 37, 39, 59-63, 73, 74, 77-91, 93-97, 98-100, 102-104, 161-168, 227, 442, 447, 448, 535-536, 569
Singularity theory, 301
Slaving principle, 169, 174-175

Slime mould, 460
Smale's horseshoe, 473, 487,
    505, 507
Solidification, 245, 249, 254,
    585-590
Solid-liquid interface,
    245-248, 251
Solitary waves, 190-192, 541,
Solitons, 227
  in biological molecules,
    61, 619-624, 625-631
  collective coordinate
    methods, 25, 451-456
  in condensed matter systems,
    535-537
  Davydov, 139-140, 541,
    619-624, 625
  in DST equation, 144-145
  dynamics of, 451-456
  in Josephson junction
    systems, 93, 103, 569-574
  Langmuir, 541
  laser 257-258, 264-272,
    547-552
  in microtubules, 639
  noise, in presence of, 454
  nonlinear spectra transform
    method, 25, 38-39
  in non-integrable systems,
    451-456
  in optical fibres, 257-286,
    547, 549
  in polymers, 454
  quantum, 68, 70
  self frequency shift, 260,
    272-274, 277
  soliton-soliton collisions,
    283
  in telecommunications, 257,
    258, 278-284
Specific heat, 536
Sectral transform, 60
Spin lattices, 368
Spin model, 535
Spiral galaxies, 462
Spiral waves, 457-463
Statistical mechanics, 59,
    199, 422
Stiff code, 593
Strange attractor, 27, 206,
    381, 386, 473-474
String theory, 61
Supercomputers, 41-43, 591-597
Superconductivity, 93

Surgery algorithms, 468
Symmetry breakig morphogenesis,
    597
Synergetics, 169-184, 597
System Dynamics National Model,
    390

Thermal explosion problem, 123
Thermalization, 619-624
Thermodynamic properties of
    electrolyte solutions,
    415-426
Thermodynamics, finite-time,
    529-534
Toda lattice, 227
Total collapse, 543
Transfer Internal Method, 64,
    69, 73
Travelling waves, 459
Turbulence, 305, 466, 575, 597
Turing system, 245-597
Two phase system, 245-255
Two stroke oscillator, 49-56

Vectorization, 42, 593
Visualization, 465, 469-470
Vortices, 465-471 484

Wave propagation, 233

Zero field step, 96-97, 105,
    569